HARVEST OF FISH AND WILDLIFE

New Paradigms for Sustainable Management

Edited by
Kevin L. Pope and Larkin A. Powell

CRC Press
Taylor & Francis Group
Boca Raton London New York

CRC Press is an imprint of the
Taylor & Francis Group, an informa business

First edition published 2021
by CRC Press
6000 Broken Sound Parkway NW, Suite 300, Boca Raton, FL 33487-2742

and by CRC Press
2 Park Square, Milton Park, Abingdon, Oxon, OX14 4RN

CRC Press is an imprint of Taylor & Francis Group, LLC

Library of Congress Cataloging-in-Publication Data
Names: Pope, Kevin L., 1969- editor. | Powell, Larkin A., editor.
Title: Harvest of fish and wildlife : new paradigms for sustainable management / Kevin L. Pope, Larkin A. Powell.
Description: First edition. | Boca Raton, FL: CRC Press, 2021. | Includes bibliographical references and index. | Summary: "Harvest of Fish and Wildlife: New Paradigms for Sustainable Management unites experts in wildlife and fishery sciences for an interdisciplinary overview of harvest management. This book presents unique insights for embracing the complete social-ecological system to ensure a sustainable future. It educates users on evolutionary and population dynamics; social and political influences; hunter and angler behavior; decision processes; impacts of regulations; and stakeholder involvement"-- Provided by publisher.
Identifiers: LCCN 2020053905 (print) | LCCN 2020053906 (ebook) | ISBN 9781032002002 (paperback) | ISBN 9780367436667 (hardback) | ISBN 9781003009054 (ebook)
Subjects: LCSH: Wildlife management. | Sustainable fisheries. | Fishing. | Hunting. | Adaptive harvest management.
Classification: LCC SK355 .H28 2021 (print) | LCC SK355 (ebook) | DDC 639.9--dc23
LC record available at https://lccn.loc.gov/2020053905
LC ebook record available at https://lccn.loc.gov/2020053906

ISBN: 978-0-367-43666-7 (hbk)
ISBN: 978-1-032-00200-2 (pbk)
ISBN: 978-1-003-00905-4 (ebk)

Typeset in Times
by MPS Limited, Dehradun

Cover photo: Green-winged teal at Seedskadee National Wildlife Refuge in Wyoming, USA.
Photo by Tom Koerner, U.S. Fish and Wildlife Service.

To Susan and Kelly, important people in our lives who also happen to be good friends with each other.
And those are delightful things.

Table of Contents

Preface

An august group of wildlife scientists gathered at the Caesar Kleberg Wildlife Research Institute in Kingsville, Texas in October of 1983. The proceedings from their third international symposium were published as *Game Harvest Management* (Beasom and Roberson 1985), and that volume has surprisingly remained the only dedicated book for the wildlife profession on harvest management. Graeme Caughley (Caughley 1985) used his introductory statements in *Game Harvest Management* to suggest that the social and political processes used to set harvest regulations for wildlife were responsible for more conservative approaches to harvest management than the maximum sustainable yield approach used in fisheries biology. Yet, no fisheries biologist was present to defend their field of study. A budding scientist, David Anderson (Anderson 1985) elaborated on a then-new theoretical approach to harvest mortality that evaluated competing predictions for population change under compensatory and additive harvest mortality assumptions. And a team of scientists led by Daniel Decker (Decker et al. 1985) demonstrated how systematically collected public input—longitudinal surveys, in this case, of landowner attitudes toward a dangerous species—could be used to guide wildlife management, representing a growing field of research in human dimensions of wildlife management that was gaining a foothold in the 1980s.

Game Harvest Management, evaluated with hindsight, is a time capsule of ideas—the compilation of thoughts and strategies that can be traced to the design of a modern paradigm for harvest management, which blossomed in the 1990s (Conroy 2021 [Chapter 1]). Given the advances in the past four decades, we believed a reexamination of approaches and theories surrounding harvest management was overdue.

We proposed this book to provide new insights into a traditional area of emphasis for fisheries and wildlife management. No manual has guided the process, but for more than a century, natural-resource managers in North America have sought to set effective and efficient regulations for harvest of game animals that meet population-level goals for fish and wildlife. Indeed, harvest management is omnipresent as a decision mandate for state, provincial, and federal agencies, and is closely tied to funding of state and provincial conservation agencies. Until recently, efforts to manage harvested species were primarily focused on predicting population-level effects of regulations using models of population growth.

We are now in a new era of harvest management. Population biologists have new modeling tools that can be applied to harvest questions. Evolutionary biologists have measured effects of harvest that go beyond simple changes in population size, and we can evaluate the potential for selective mortality through harvest to affect populations and species. Social scientists have begun to look reflectively at behaviors of anglers and hunters, especially as anglers and hunters respond to changing densities of fish and game. And, tenets of decision science have proven useful as improved frameworks to select regulations for harvested species in social and political climates that are often hostile toward consumptive uses of fish and wildlife. In sum, harvest management has broadened beyond its traditional roots to embrace information provided by genetics and advanced population-dynamics modeling as well as insights obtained through consideration of human dimensions.

There are multiple settings throughout the world for harvesting fish and wildlife, such as indigenous subsistence harvest and the European Landowner-centric Model. We intentionally and implicitly asked authors to frame harvest management from a North American perspective, as a way to focus chapters. Though far from perfect, the North American Model of Wildlife Conservation is the cornerstone of modern wildlife conservation throughout the United States of America and most of North America (Figure 0.1).

FIGURE 0.1 Anne and Tobin Armstrong hunt in a field in Texas in December 1967. Photo by Toni Frissell, from the collections of the Library of Congress (LC-F9-04-6712-034).

Despite the widespread need for fish and wildlife managers to justify their harvest recommendations, there has been no prior guide for the process of setting harvest regulations. We designed this volume to provide a summary of current, cutting-edge research and management approaches to decisions related to harvest regulations of recreational fishing and hunting. It highlights advances in population biology and modeling and the applications of evolutionary biology, social science, and decision science on the field of harvest management. Importantly, it also brings the combined knowledge of fisheries and wildlife scientists together for an interdisciplinary overview of harvest management.

Structure

The reader will find the chapters to be divided among three sections (five sub-sections). The book begins with *Section I, Setting Regulations*, an exploration of the biological, social, and political processes used to set harvest regulations and how hunters and anglers respond to regulations. Six teams of authors describe *Harvest Management Paradigms*—the context provided by the public trust doctrine for harvest management in the United States of America as well as critical components of human dimensions in modern management of harvests for fish and wildlife. Then, five author groups provide an in-depth description and assessment of the *Harvest Management Decision Process*—a synthesis of decision frameworks applied to harvest management with special attention to stakeholder integration and metrics to use to compare alternative regulation strategies for harvest management.

Section II, Harvest Outcomes is a critical assessment of outcomes of harvest on species and populations. Three author teams tackle evaluations of *Evolutionary and Population Dynamics for Harvested Species*—providing guidance to managers regarding situations in which evolutionary effects of harvest may manifest and delving into a modern, quantitative description of the "doomed surplus." Five groups of authors then provide impressive investigations into the *Efficacies of Harvest Regulations*—providing answers to our basic questions of whether our regulations for harvest result in the desired effect on the population.

The book ends with *Section III, Management Alternatives*, descriptions of management alternatives typically considered by fish and wildlife managers with justifications and potential impacts of variations in regulations. We engaged five teams of authors with extensive background with on-the-ground management of harvest to provide applied descriptions of current *Harvest Regulation Paradigms for Wildlife and Fisheries*. We also asked these groups of authors to push our profession forward with roadmaps for the future of harvest management.

We designed the sections and subsections to be useful for students seeking understanding of applications of evolutionary theory, social science, decision science, and advanced population biology to harvest management. We also believe the structure and approach to content will be useful for managers needing to make and justify harvest recommendations to supervisors, agency commissioners, and the public.

Each of the three sections begins with a commissioned drawing from a talented artist, Jennifer Clausen. We hope the Jennifer's work exemplifies the role of creativity in the decision-making process for harvest management, and her images provide an artist's vision of new paradigms for sustainable management of our natural resources.

Themes

An amazing group of experts in our field responded to our invitation with contributions of original research results and reviews of published information. We followed the model of the editors of *Game Harvest Management* by bringing together many of our author teams in a symposium at a joint meeting of The Wildlife Society and the American Fisheries Society in Reno, Nevada during September 2019; yet, this volume is much more than proceedings of a symposium. As expected, the interaction between fisheries and wildlife scientists created some impressive synergies that you should see on display throughout the book.

Less than three months after the symposium, the initial versions of the invited chapters began to arrive. Although we had been purposeful in our invitations, the book took shape with an Odum-like "whole greater than the sum of the parts." Indeed, as a giant puzzle assembling itself, we began to witness something akin to the organic process of ecological self-organization. Challenged to point toward the future paradigm for harvest management, author teams gestured with like minds toward the importance of human dimensions, decision making, embracing uncertainty and complexity, and new models to support management. Thirty-six years after the publication of *Game Harvest Management*, the framework of structured decision making has become integral in our field. Indeed, our colleague and chapter author Mike Runge (Runge 2021 [Chapter 7]) noted that the set of chapters in our digital folders supported each step in the decision process (Figure 0.2).

The PrOACT (problem, objective, alternatives, consequences, tradeoffs) model for decision making (Gregory et al. 2012, McGowan 2013) begins with defining the problem. Similarly, we begin and end this volume with a focus on framing the history and context of the decision—a broad focus at the beginning (Conroy 2021 [Chapter 1], Hiller et al. 2021b [Chapter 2]) and a taxa-specific focus at the end (Dahlgren et al. 2021 [Chapter 21], Diefenbach et al. 2021 [Chapter 22], Gangl 2021 [Chapter 24], Hiller et al. 2021a [Chapter 23], Vrtiska 2021 [Chapter 20]). Continuing the iterative loop, our authors explore the definition of objectives (Fuller et al. 2021 [Chapter 8], Paukert et al. 2021 [Chapter 18], Robinson et al. 2021 [Chapter 9]), principles for development of alternatives (Melstrom 2021 [Chapter 5], Morina et al. 2021 [Chapter 19]), analyses designed to evaluate consequences on harvested animals as well as the human harvesters (Arnold 2021 [Chapter 14], Feiner et al., 2021 [Chapter 16], Festa-Bianchet and Arlinghaus 2021 [Chapter 12], Gruntorad and Chizinski 2021 [Chapter 4], Graham et al. 2021 [Chapter 6], Harms and Dinsmore 2021 [Chapter 15], Keamingk et al. 2021 [Chapter 3], Sylvia et al. 2021 [Chapter 17], Wszola and Fontaine 2021 [Chapter 13]), and models designed to identify a preferred alternative (Cummings and Bernier 2021 [Chapter 10], Tyre and Tenhumberg 2021 [Chapter 11]), all within an overall decision framework (Runge 2021 [Chapter 7]; Figure 0.2).

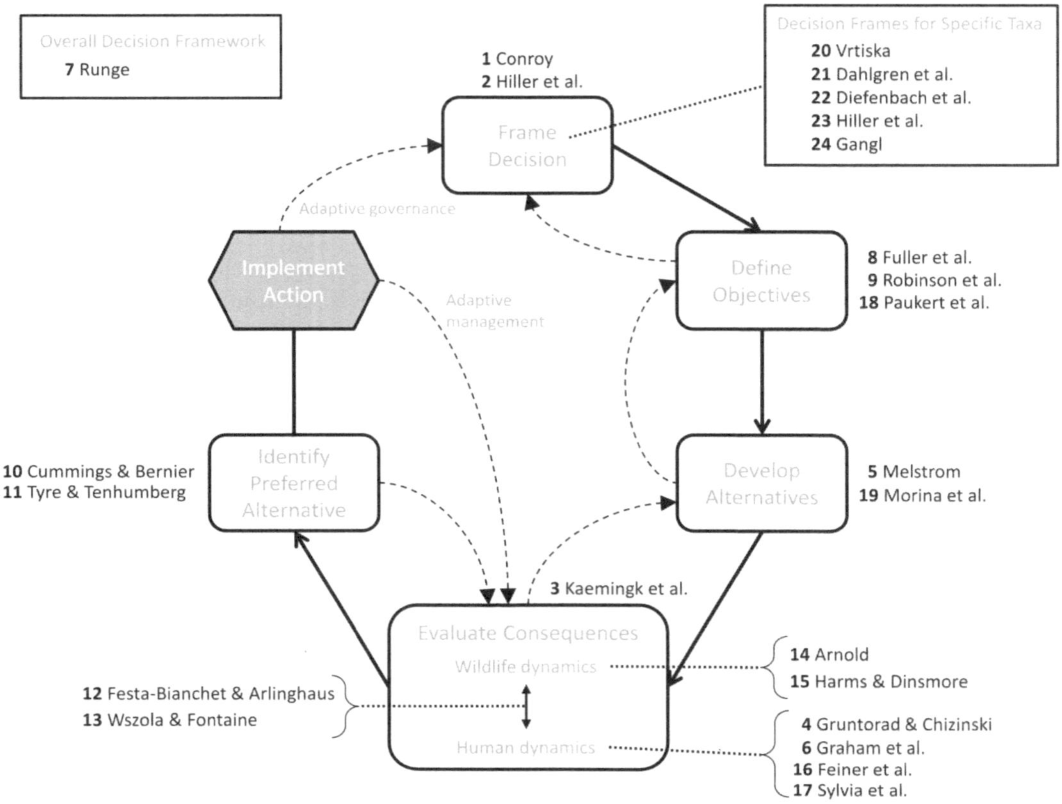

FIGURE 0.2 An overview of where each chapter in this book fits within the PrOACT (problem, objective, alternatives, consequences, tradeoffs) model for decision making (image adapted from Runge 2021 [Chapter 7]).

As you begin your journey through this volume, we leave you with a poetic summary of the concepts we believe you will find in the pages ahead (Figure 0.3). Indeed, if you are here because you thought harvest management was simply about wildlife, there are some lessons ahead.

Harvest Management

We once considered the potential loss of game animals
Real was the threat of overharvest
We now consider the prospective loss of game pursuers
Real is the threat of apathy

Doing the job well requires balance and skill
Tradeoffs absolute
Knowledge is required in complicated fields
Sustainability firm

How much, What segments, and By whom
Questions to answer
Drake or hen, Spring or fall, Tag or license
Decisions to make

Regulating those exuberant to pursue game
Tools abound
Inspiring those keen to manage fish and wildlife
Creativity key

Integrating across social and ecological disciplines
Opportunity blooms
Embracing complexity of social-ecological systems
Managing people

FIGURE 0.3 An angler waits in the early morning. Photo by Ed Dunens.

Acknowledgments

Dedication is the hallmark of a scientific author. Authors completed their tasks in timely fashion, even though much of the work on these chapters was completed during the complex work-from-home, isolated dynamics of the COVID-19 pandemic of 2020. Most authors were also attempting to find ways to continue field and laboratory research or to restructure their university teaching efforts. Thus, the editors are especially grateful for their contributions.

Peer review is the heart of scientific publication. The editors engaged an impressive and willing set of reviewers to provide the scientific rigor and peer review desired in this volume, and the product is better for their efforts: M. Allen (University of Florida), T. Arnold (University of Minnesota), G. Balkcom (Georgia Department of Natural Resources), B. Chaffin (University of Montana), D. Clapp (Michigan Department of Natural Resources), W. Clark (Iowa State University), N. Cole (U.S. Geological Survey), T. Cooper (U.S. Fish and Wildlife Service), J. Cummings (University of Massachusetts—Dartmouth), D. Dahlgren (Utah State University), D. Diefenbach (U.S. Geological Survey—Pennsylvania Cooperative Fish and Wildlife Research Unit), S. Demarais (Mississippi State University), P. Devers (U.S. Fish and Wildlife Service), M. Eichholz (Southern Illinois University Carbondale), T. Harms (Iowa Department of Natural Resources), C. Hasler (University of Winnipeg), J. Heffelfinger (Arizona Game and Fish Department), D. Humberg (Retired, Duck Unlimited), L. Hunt (Ontario Ministry of Natural Resources and Forestry), S. Gangl (North Dakota Game and Fish Department), C. Goble (South Dakota Game, Fish and Parks), K. Malone (University of Missouri), C. McGowan (U.S. Geological Survey), E. Metcalf (University of Montana), C. Moore (U.S. Geological Survey—Georgia Cooperative Fish and Wildlife Research Unit), V. Moore (Idaho Department of Fish and Game), J. Murdoch (University of Vermont), C. Reeling (Purdue University), R. Regan (Association of Fish and Wildlife Agencies), M. Runge (U.S. Geological Survey), G. Scholten (Iowa Department of Natural Resources), M. Schroeder (Washington Department of Fish and Wildlife), S. Schroeder (University of Minnesota), A. Sylvia (Iowa State University), D. Smith (U.S. Geological Survey), N. Stewart (Minnesota Department of Natural Resources), B. Tenhumberg (University of Nebraska-Lincoln), V. Travnichek (Missouri Department of Conservation), K. Vercauteren (U.S. Department of Agriculture), J. Vokoun (University of Connecticut), M. Weber (Iowa State University), B. White (Association of Fish and Wildlife Agencies), and L. Wszola (University of Nebraska-Lincoln).

The editors are grateful to R. Brehm (CRC Press), J. Tanner (CRC Press), and R. Radhika (Taylor & Francis) for their patience and guidance through the publication process. The editors appreciate the creative efforts of J. Clausen, who drew images at start of each section. The Nebraska Cooperative Fish and Wildlife Research Unit is jointly supported by a cooperative agreement among the U.S. Geological Survey, the Nebraska Game and Parks Commission, the University of Nebraska, the U.S. Fish and Wildlife Service, and the Wildlife Management Institute. L. Powell was supported by Hatch Act funds through the University of Nebraska Agriculture Research Division, Lincoln, Nebraska. Any use of trade, firm, or product names is for descriptive purposes only and does not imply endorsement by the U.S. Government.

Editors

Kevin L. Pope earned his PhD in biological sciences at South Dakota State University and is leader of the U.S. Geological Survey—Nebraska Cooperative Fish and Wildlife Research Unit housed at the University of Nebraska-Lincoln, where he also serves as professor for the School of Natural Resources. Dr. Pope's research program concentrates on understanding complex social–ecological systems, especially inland recreational fisheries. He has written more than 115 journal articles. Dr. Pope received the Award of Excellence from the Fisheries Management Section of the American Fisheries Society for inspirational leadership in the fishery profession.

Larkin A. Powell earned his PhD in ecology at the University of Georgia and is a professor in the School of Natural Resources at the University of Nebraska-Lincoln, where he also serves as Associate Dean for the College of Agricultural Sciences and Natural Resources. Dr. Powell's research program focuses on landscape dynamics, animal demography and movements, and decisions made by private landowners. He has written more than 120 journal articles and authored, coauthored, or contributed to seven books. Dr. Powell received the Excellence in Wildlife Education Award from The Wildlife Society.

Contributors

Roger D. Applegate
Tennessee Wildlife Resources Agency
Nashville, Tennessee

Robert Arlinghaus
Department of Biology and Ecology of Fishes
Leibniz-Institute of Freshwater Ecology and Inland Fisheries
and
Division of Integrative Fisheries Management
Faculty of Life Sciences
Humboldt-Universität zu
Berlin, Germany

Todd W. Arnold
Department of Fisheries, Wildlife, and Conservation Biology
University of Minnesota
Saint Paul, Minnesota

Chris Bernier
Vermont Fish and Wildlife Department
Springfield, Massachusetts

Erik J. Blomberg
Department of Wildlife, Fisheries, and Conservation Biology
University of Maine
Orono, Maine

Christopher J. Chizinski
School of Natural Resources
University of Nebraska-Lincoln
Lincoln, Nebraska

Michael J. Conroy
Warnell School of Forestry and Natural Resources
University of Georgia
Athens, Georgia

Jonathan W. Cummings
Vermont Cooperative Fish and Wildlife Research Unit
Rubenstein School of Environment and Natural Resources
University of Vermont
Burlington, Vermont

David K. Dahlgren
Wildland Resources Department
Utah State University
Logan, Utah

Duane R. Diefenbach
U.S. Geological Survey
Pennsylvania Cooperative Fish and Wildlife Research Unit
Department of Ecosystem Science and Management
Pennsylvania State University
University Park, Pennsylvania

Stephen J. Dinsmore
Department of Natural Resource Ecology and Management
Iowa State University
Ames, Iowa

R. Dwayne Elmore
Department of Natural Resource Ecology and Management
Oklahoma State University
Stillwater, Oklahoma

Zachary S. Feiner
Wisconsin Department of Natural Resources
Office of Applied Science
Madison, Wisconsin

Marco Festa-Bianchet
Département de biologie
Université de Sherbrooke
Québec, Canada

Joseph J. Fontaine
Nebraska Cooperative Fish and Wildlife Research Unit
School of Natural Resources
University of Nebraska-Lincoln
Lincoln, Nebraska

Drew N. Fowler
Wisconsin Department of Natural Resources
Office of Applied Sciences
Madison, Wisconsin

Angela K. Fuller
U.S. Geological Survey
New York Cooperative Fish
and Wildlife Research Unit
Cornell University
Ithaca, New York

R. Scott Gangl
North Dakota Game and Fish Department
Bismarck, North Dakota

Katherine A. Graham
School of Natural Resources
University of Nebraska-Lincoln
Lincoln, Nebraska

Matthew P. Gruntorad
School of Natural Resources
University of Nebraska-Lincoln
Lincoln, Nebraska

Christian A. Hagen
Department of Fisheries and Wildlife
Oregon State University
Corvallis, Oregon

Christopher P. Hansen
W.A. Franke College of Forestry and
Conservation
University of Montana
Missoula, Montana

Lonnie P. Hansen
Resource Science Division
Missouri Department of Conservation
Columbia, Missouri

Corbin D. Hilling
Department of Fish and Wildlife Conservation
Virginia Polytechnic Institute and State University
Blacksburg, Virginia

Tyler M. Harms
Wildlife Research Section
Iowa Department of Natural Resources
Boone, Iowa

Tim L. Hiller
Wildlife Ecology Institute
Helena, Montana

Keith L. Hurley
Nebraska Game and Parks Commission
Lincoln, Nebraska

Michael L. Jones
Department of Fisheries and Wildlife
Quantitative Fisheries Center
Michigan State University
East Lansing, Michigan

Valerie K. Jones
College of Journalism and Mass Communications
University of Nebraska-Lincoln
Lincoln, Nebraska

Mark A. Kaemingk
Department of Biology
University of North Dakota
Grand Forks, North Dakota

W. Matthew Knox
Virginia Department of Wildlife Resources
Richmond, Virginia

Alexander W. Latzka
Wisconsin Department of Natural Resources
Bureau of Fisheries Management
Madison, Wisconsin

Brandon Maahs
Department of Natural Resource Ecology
and Management
Iowa State University
Ames, Iowa

Richard T. Melstrom
Institute of Environmental Sustainability
Loyola University
Chicago, Illinois

Joshua J. Millspaugh
W.A. Franke College of Forestry and
Conservation
University of Montana
Missoula, Montana

Daniel L. Morina
W.A. Franke College of Forestry
and Conservation
University of Montana
Missoula, Montana

Michael S. O'Brien
Fur Institute of Canada
Ottawa, Ontario

Craig P. Paukert
U.S. Geological Survey
Missouri Cooperative Fish and Wildlife Research Unit
School of Natural Resources
University of Missouri
Columbia, Missouri

Kelly A. Perkins
New York Cooperative Fish and Wildlife Research Unit
Department of Natural Resources
Cornell University
Ithaca, New York

Nathaniel B. Price
School of Natural Resources
University of Nebraska-Lincoln
Lincoln, Nebraska

Kelly F. Robinson
Department of Fisheries and Wildlife
Quantitative Fisheries Center
Michigan State University
East Lansing, Michigan

Christopher S. Rosenberry
Pennsylvania Game Commission
Harrisburg, Pennsylvania

Michael C. Runge
U.S. Geological Survey
Patuxent Wildlife Research Center
Laurel, Maryland

William F. Siemer
Cornell University Center for Conservation Social Sciences
Cornell University
Ithaca, New York

Joshua C. Stiller
New York State Department of Environmental Conservation
Albany, New York

Andrea Sylvia
Department of Natural Resource Ecology and Management
Iowa State University
Ames, Iowa

Brigitte Tenhumberg
School of Biological Sciences
University of Nebraska-Lincoln
Lincoln, Nebraska

Andrew J. Tyre
School of Natural Resources
University of Nebraska-Lincoln
Lincoln, Nebraska

Mark P. Vrtiska
Nebraska Game and Parks Commission
Lincoln, Nebraska

Elisabeth B. Webb
U.S. Geological Survey
Missouri Cooperative Fish and Wildlife Research Unit
School of Natural Resources
University of Missouri
Columbia, Missouri

Michael J. Weber
Department of Natural Resource Ecology and Management
Iowa State University
Ames, Iowa

Lyndsie S. Wszola
School of Biological Sciences and School of Natural Resources
University of Nebraska-Lincoln
Lincoln, Nebraska

Max H. Wolter
Wisconsin Department of Natural Resources
Bureau of Fisheries Management
Hayward, Wisconsin

Section I

Setting Regulations

Artist's Impression: Society has a diverse array of people who all use the same natural resources, but the way they use those resources varies. Wildlife and fisheries managers need to not only take into account the wants and needs of society when creating management plans, but also the limitations of the ecosystem to strike a balance.—Jennifer Clausen

Section IA

Harvest Management Paradigms

A duck hunter surveys her decoy spread in a wetland in North Dakota, United States of America. Photo by Tina Shaw, U.S. Fisheries and Wildlife Service.

1 Some Perspectives on the Development of a Paradigm for Modern Harvest Management

Michael J. Conroy

CONTENTS

INTRODUCTION

Harvest management, the management of offtake of a renewable fish or wildlife resource, has long been a staple tool for fishery and wildlife managers. Given its importance and even dominance as a means of regulating fish and wildlife in a sustainable manner, it is important to consider how the discipline has arrived at its current *state of the art*. I selectively review the development of the conceptual and mathematical framework that is the basis of modern harvest management. During much of my career, I was involved with migratory bird management, largely focused on the management of waterfowl harvest. The specifics of harvest management differ among taxa, yet there are important, overarching concepts that are shared across taxa of exploited wildlife and fish, and with renewable resources in general.

I identify several overlapping phases of development of a harvest management paradigm: (1) a *protophase*, during which harvest management, if considered, was essentially *ad hoc*, intuitive, and largely free of feedback from data; (2) an *early, developmental phase*, during which the basic underpinnings of sustainability developed; (3) a *middle phase*, involving explicit hypotheses and mechanistic models, coupled with a stronger empirical bases; and (4) a *modern phase* with more formal treatment of population dynamics, non-equilibrium conditions, stochasticity, and other forms of uncertainty. I end with some thoughts on an *emerging phase*, which must deal with the realities of complex population structure, multiple stocks, multiple resource objectives, changing social values, and profound system change and extreme uncertainty as envisaged under, for example, climate change.

PROTOPHASE: BEGINNINGS

Ideas of resource limitation and sustainability have existed for centuries in North America, even millennia in indigenous cultures (Nakashima et al. 2000), but have been less prevalent until recent history after European settlement. From the standpoint of European colonizers of North America, game resources were seemingly without limit, with robust or lightly exploited stocks. Nevertheless, as early as the first half of the nineteenth century there dawned a recognition that rapidly increasing human populations and technology were on a collision course with natural resources. James Fenimore Cooper was a visionary for the concept of sustainability, as voiced by fictional characters such as Judge Temple and Natty Bumppo (Cooper 1823, 1827). By the late nineteenth and early twentieth centuries, several cautionary examples of overexploitation were on exhibit in North America, including decimation of American bison (*Bison bison*; Gates and Aune 2008), extinction of the passenger pigeon (*Ectopistes migratorius*; Halliday 1980), and collapse of the northwestern Atlantic cod (*Gadus morhua*) fishery (Hutchings 1996). However, during this early period there emerged gems of the stronger conceptual and theoretical foundations of sustainable harvest to follow.

EARLY PHASE: CONCEPTUAL AND THEORETICAL FOUNDATIONS

Errington (1946) proposed the idea of a surplus of animals that would die from predation or starvation if not harvested. For game birds, the reduction of the fall population by harvest to a "normal" breeding density would result in population equilibrium. Ideas such as doomed surplus were intuitively derived but had a degree of empirical support (e.g., Errington 1945), and implicitly invoked both notions of carrying capacity and density dependence. These conceptual frameworks led to later, more formal ideas about density regulation and harvest compensation, as well individual variability in populations, with some individuals more likely to succumb to natural threats than others (Arnold 2021 [Chapter 14]). A more formal understanding of the relationship between harvest offtake and population dynamics began to emerge in the twentieth century, with the application of growth models that incorporated these concepts. These models have the common feature of being based on differential equations that describe somatic growth (e.g., length of individual organisms), population growth (number of organisms), or combination (e.g., total volume or mass of a forest), and are generically described by the von Bertalanffy family of growth models (e.g., Beverton and Holt 1957: 32). Here I focus on population growth and the logistic special case, introduced by Verhulst (1845):

$$\frac{dN_t}{dt} = r_{max} N_t \left[1 - \frac{N_t}{K} \right], \tag{1.1}$$

where r_{max} and K are, respectively, the maximum per-capita growth potential and an upper limit to abundance, with K often interpreted as carrying capacity. This expression may be cast in discrete time and re-arranged to provide the equation for population growth (Fig. 1.1a):

$$N_{t+1} = N_t + r_{max}N_t\left[1 - \frac{N_t}{K}\right]. \tag{1.2}$$

The role of harvest in this dynamic can be seen by removing an annual harvest (yield), H_t

$$N_{t+1} = N_t + r_{max}N_t\left[1 - \frac{N_t}{K}\right] - H_t.$$

Setting $N_{t+1} = N_t$ produces a stable, unchanging population, resulting in

$$H_t = r_{max}N_t\left[1 - \frac{N_t}{K}\right]. \tag{1.3}$$

By definition, all values of annual harvest that satisfy (1.3) are *sustainable*, that is, they will lead to indefinite maintenance of the population, given the model assumptions and specified parameter values. A plot of H versus N produces a yield curve (Fig. 1.1b), whose maximum with respect to N occurs when $N_t = N_{msy} = K/2$. Substituting $K/2$ for N_t into (1.3) provides the *maximum sustainable* (or *sustained*) *yield* (MSY) of $H_{msy} = r_{max}K/4$. The yield curve is unimodal, and hence yield will be below maximum at any N above or below N_{msy}, with maximum sustainable yield achieved at the unique abundance N_{msy} (Fig. 1.1b). Finally, maximum sustainable yield will be maintained by fixing the per-capita yield (harvest rate) at $h_t = \frac{H_t}{N_t}$ at $r_{max}/2$ when the population is at $K/2$, with higher rates balancing growth below this value, and vice versa (Fig. 1.1c). This relationship leads to a simple rule for dynamic harvest management:

- For $N_t < K/2$, harvest at a per-capita rate h_t below the balancing line (Fig. 1.1c), thereby allowing the population to grow to $K/2$.

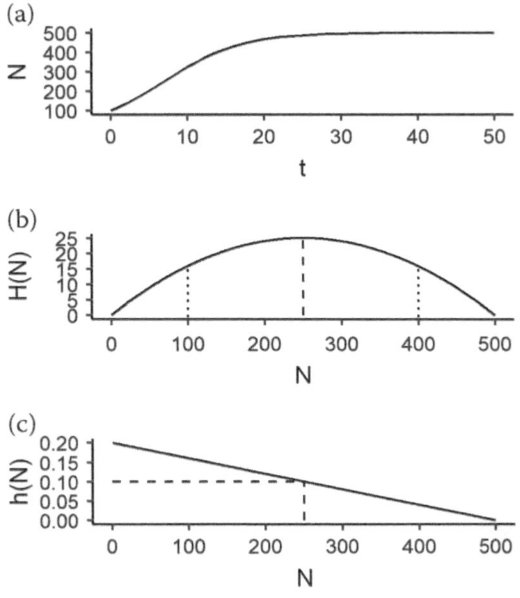

FIGURE 1.1 Maximum sustainable yield. (a) Logistic growth curve for $r_{max} = 0.2$, $K = 500$, initial population size $N_0 = 100$. (b) Yield curve for logistic model with $r_{max} = 0.2$, $K = 500$. Provides maximum yield of $H = r_{max}K/4 = 25$ occurring at $N_{msy} = K/2$ (dashed line). Lower values of yield provided by N values below or above $K/2$; e.g., $N = 100$ or 400 provide $H = 16$ (dotted lines). (c) Sustainable per-capita yield. Maximum yield maintained by fixing per-capita harvest rates at $h = r_{max}/2$ when the population is at $K/2$, $h = 0.10$, $N_{msy} = 250$ in the example.

- For $N_t > K/2$, harvest at a per-capita rate h_t above the balancing line (Fig. 1.1c), thereby causing the population to decline to $K/2$.
- At $N_t = K/2$, fix per-capita harvest at rates $h_t = r_{max}/2$, thereby maintaining the population at $K/2$ and providing maximum sustainable yield.

The logistic model is more commonly used in wildlife management, whereas fisheries managers traditionally have used the analogous Ricker (1954) and Beverton-Holt (1957) models. The latter focus on relationships between the stock size and recruitment, and harvest (catch) in relation to effort (time, costs, or other units) in seeking to identify a maximum equilibrium yield at which a fishery can be maintained without depletion, also termed maximum sustainable yield, or sometimes *optimal sustainable yield*, particularly when economic goals and constraints are explicitly incorporated (Clark 1990).

The logistic model and closely related fisheries models, and the resulting derivations of maximum sustainable yield, admittedly are crude representations of population dynamics. These models ignore age and spatial structure, as well as other forms of population heterogeneity, collapsing prediction down to a few parameters (for the logistic, r_{max} and K) for which it is assumed we have known values. The models are deterministic, so demographic, environmental, and other forms of stochasticity are ignored. Maximum sustainable yield inherits these characteristics because of the assumption that population growth follows the logistic or similar model.

On the positive side, the maximum sustainable yield paradigm formalized a mathematical theory underlying population behavior with respect to exploitation, and instilled several important ideas. First, the logistic and related models can mimic population behavior over the long term, especially for long-lived, so-called K-selected organisms existing in reasonably static environments, and are supported by classic examples such as the George Reserve white-tailed deer (*Odocoileus virginianus*) herd (McCullough 1980). Second, by proposing an equilibrium solution to harvest, an indefinitely long-time frame is assured—the essence of sustainability. Given that the underlying form of the growth model and parameter values are known, sustainability is thus "guaranteed." The approach also connects well to notions of optimality, and economic models of resource exploitation (e.g., Clark 1990).

On the negative, as noted above, maximum sustainable yield invokes strong assumptions about the functional form and parameter values of the growth model. Furthermore, the mechanisms giving rise to model behavior (specific forms of density dependence) are largely implied and not empirically tested. Finally, maximum sustainable yield can be problematic for populations that may be far from equilibria (often, unknown), or exist in highly stochastic environments (e.g., prairie-breeding ducks). Also, by focusing on annual density compensation, the model ignores seasonal compensation, likely important for species with high reproductive potential (Boyce et al. 1999). Forward thinking fisheries and wildlife ecologists have longed recognized these and other serious deficiencies in maximum sustainable yield: "*Principle: You cannot determine the potential yield from a fish stock without overexploiting it. .....The hardest thing to do in fisheries management is reduce fishing pressure*" (Hilborn and Walters 1992). Larkin (1977) tendered an epitaph for maximum sustainable yield:

"Here lies the concept, MSY.
It advocated yields too high
And didn't spell out how to slice the pie
We bury it with the best of wishes
Especially on behalf of fishes.
We don't know yet what will take its place,
But hope it's as good for the human race."

In spite of these deficiencies, and the simplifications noted above, the logistic and maximum sustainable yield models continue to form guiding paradigms in the management of wildlife and fish and other renewable resources. Despite their obvious limitations, these classical harvest models can be useful caricatures for describing the potential behavior of populations in response to exploitation and set the stage for more realistic (and hopefully more useful) approaches. Unfortunately, in my experience, wildlife and fisheries education and professional training often presents maximum sustainable yield simplistically and uncritically. My recommendation to educators (particularly of graduate students) is to, instead, treat this approach as a first approximation, subject to refinements as elaborated below.

MIDDLE PHASE: TOWARD A HYPOTHETICAL DEDUCTIVE BASIS FOR HARVEST MANAGEMENT

This phase was notable for two overlapping, but ultimately co-dependent developments. The first was the implementation of several large-scale, long-term population monitoring programs. The second was the use of ecological theory and models to formulate and test hypotheses about population-level factors, including the influence of exploitation on dynamics. The latter is impossible (or at least, untestable) without the former; the former is (arguably) irrelevant without the latter. In hindsight, it is clear that these two developments were essential to the emergence of a modern paradigm for harvest management. They were also critical elements in the *pre-adaptation* of management to recent advances, such as adaptive harvest management (Nichols 2000, Johnson 2006).

Development of Important Monitoring Programs for Informing Harvest Management

There are many examples of monitoring programs that emerged during the twentieth century, mostly following World War II, including programs developed by state agencies, non-governmental organizations, and volunteer groups. Here, I focus on programs that involve collection and analysis of data over long-time horizons and broad geographic regions, and are or were at least in part directed at informing the decisions about regulation of offtake. In my view, the single best examples of such efforts are the programs directed at management of North American migratory waterfowl, and those directed at the regulation of Great Lakes and oceanic fisheries. Both sets of programs are continental or hemispheric and thus international in scope, and typically involve the formal cooperation of national, state, and provincial governments, non-governmental organizations, and sporting or industry advocates. My work lends itself to a deeper knowledge of programs directed at North American migratory waterfowl management (e.g., Conroy and Eberhardt 1983, Conroy and Blandin 1984, Nichols et al. 1984a, Krementz et al. 1987, 1988. Conroy et al. 2002), and so will cover these in some depth.

North American Migratory Waterfowl

Beginning in the early twentieth century, and increasingly so following World War II, aircraft were used by federal and state agencies to conduct surveys of waterfowl populations on key wintering and breeding areas in the United States of America, Canada, and later, Mexico. Led by pioneers such as Frederick Lincoln and Walt Crissey (Hawkins et al. 1984), these surveys were not based on statistical sampling designs, lacked correction for incomplete and heterogeneous detection, and thus did not provide valid abundance estimates (Anderson and Henny 1972, Pospahala et al. 1974, Williams et al. 2002). Breeding ground surveys, initially focused on mid-continent prairie and parkland habitats, were initiated c. 1947 and operational by 1955. Improvements in these surveys by the 1960s included a statistical design and correction for detection, providing valid abundance estimates by species, counts of ephemeral wetlands critical to predicting reproductive success, and number of broods, a leading indicator of fall-flight forecasts, a key element of regulation setting (Anderson and Henny 1972, Pospahala et al. 1974). Large-scale banding programs (of limited utility prior to c. 1950 due to inconsistent record keeping) and new developments in statistical modeling provided trustworthy estimates of annual survival by sex and age, and of harvest rates for key species such as the mallard (*Anas platyrhynchos*; Anderson and Burnham 1976, Brownie et al. 1985), and the geographic distribution and derivation of the harvest (Munro and Kimball 1982). Band-recovery data provide estimates of recovery rates (f; Brownie et al. 1985) and, after adjustment for band reporting, of harvest rates (Henny and

Burnham 1976, Conroy and Blandin 1984). Mail questionnaire surveys of migratory-bird hunters provided information on hunter activity and numbers of ducks and geese harvested, with a subsample of hunters mailing wings of ducks or tails of geese, used to assess the species, sex, and age composition of the harvest (Benson 1968, 1970, Martin and Carney 1977, Environment and Climate Change Canada 2018). These data are corrected for relative vulnerability using band recovery data, providing estimates of the fall population age ratio (young-to-adult), and thus inference on annual recruitment rates. More recently, the Hunter Information Program (Ver Steeg and Elden 2002), an obligatory survey of all migratory-bird hunters, conducted at license points of sale, has provided additional information on hunter activity and success.

Collectively, these data provide a vast and long-term data set on abundance, distribution, habitats, vital rates, and rates of harvest for key species of migratory birds in North America. This seemingly placed managers in a very favorable position to regulate exploitation of this resource using information feedback. These data also have been extensively mined in search of patterns among these rates, for instance in search of evidence supporting or refuting hypotheses of harvest compensation and density dependence (e.g., Geis and Taber 1963, Geis and Crissey 1969, Geis 1972, Anderson and Burnham 1976, Conroy and Eberhardt 1983, Nichols and Hines 1983, Nichols et al. 1984a, Krementz et al. 1987, 1988). I will discuss below some of the reasons why these efforts have been relative failures, and where this leaves us today.

Oceanic and Great Lakes Fisheries

Parallel developments provided long-term databases for management of oceanic and other fisheries, notably in the Great Lakes. The National Oceanic and Atmospheric Administration conducts regular population assessment of oceanic fisheries stocks, typically based on abundance, catch, and "biology" data (National Oceanic and Atmospheric Administration 2019). Catch data include log books, observers on vessels, dockside monitoring, samples of recreational anglers, and tag recoveries. Abundance data may include tagging recoveries or recaptures, hydroacoustics, and visual sampling. Information on fish movements, demography, and other aspects of biology informative to managers are gleaned from tagging and sampling of individual fish to measure size, age, and growth rates. Stock assessment models are periodically fit to the data to estimate abundance, age structure, and offtake from catch. Finally, model simulations provide assessments of changes induced by harvest pressure and to provide feedback into management-decision making (Maunder 2012, National Oceanic and Atmospheric Administration 2019, Runge 2021 [Chapter 7]). Similar types of data and modeling approaches have been informative for management of Great Lakes fisheries, with added complications of multi-jurisdictional (and binational) regulatory bodies, and the inclusion of a significant recreational fishery (Taylor and Ferreri 1999, Great Lakes Fisheries Commission 2019).

Development and Testing of Hypotheses about Harvest Effects

Overlapping with the development of important large-scale databases for exploited fish and wildlife resources was development of hypotheses about the interplay between exploitation and population growth, formalized as testable predictions using quantitative models. By the 1970s, this led to a rigorous development of a hypothesis of compensatory mortality, contrasted to a hypothesis that exploitation was additive to other sources of mortality (Anderson and Burnham 1976). Similar formulations of these fundamental relationships emerged in models of density-dependent mortality and recruitment in exploited fish populations (Beverton and Holt 1957), foreshadowed by catch equations and models of competing risks by Baranov (1918). I will cover these ideas in a bit of formal depth because of the importance of this issue to an understanding of dynamics under exploitation, focusing on migratory waterfowl, and leaning heavily on Anderson and Burnham (1976) and Williams et al. (2002).

Additive and Compensatory Mortality

Informally, harvest compensation depends on the assumption that there is a baseline level of natural mortality, m_O, that (on average) occurs in unexploited populations, take for example m_O = 0.4. For migratory waterfowl, the period following harvest (winter, in North America) is thought to exhibit the greatest potential for natural mortality. If mortality from harvest were simply additive to this baseline, then each increment of exploitation mortality would increase total mortality correspondingly. Instead, compensation dictates that within limits the sum of natural and exploitation mortality is constant, so that annual survival remains unchanged. Both cases are covered by a general relationship between annual natural and harvest mortality (Anderson and Burnham 1976, Burnham and Anderson 1984, Williams et al. 2002: 227, equation 11.9):

$$m(t) = m_0 + \beta h(t) \tag{1.4}$$

for $0 \le h(t) \le m_0$, where $m(t)$ is annual natural (not related to exploitation) mortality, m_0 is natural mortality in the absence of harvest, $h(t)$ is annual harvest mortality (including crippling loss), and β is a coefficient representing compensation, with $\beta = -1$ providing complete compensation and β approaching $-\ m_0$ complete additivity of harvest and non-harvest sources, with all rates finite rather than instantaneous (see Anderson and Burnham 1976: 4 for a discussion of finite versus instantaneous rates). Annual survival is obtained by

$$S(t) = 1 - h(t) - m(t). \tag{1.5}$$

Under compensatory mortality, natural mortality decreases from a maximum of m_0 to 0 at a threshold value of $h(t) = m_0$, at which point all compensatory potential has been exhausted. Under additive mortality, natural mortality decreases linearly to 0 with increasing $h(t)$, solely as a consequence of the fact that given a mortality from hunting, an individual is no longer available for either natural or harvest mortality (Fig. 1.2). Likewise, under complete compensation total survival remains constant $S(t) = 1 - m_0$ for values of harvest up to a threshold of $h(t) = m_0$, after which survival decreases linearly to 0. Intermediate degrees of compensation are specified by β between the compensatory and additive endpoints, resulting in a partially compensatory relationship (e.g., $\beta = -0.75$, Fig. 1.2). Importantly, compensation is not to be confused with *substitution mortality*

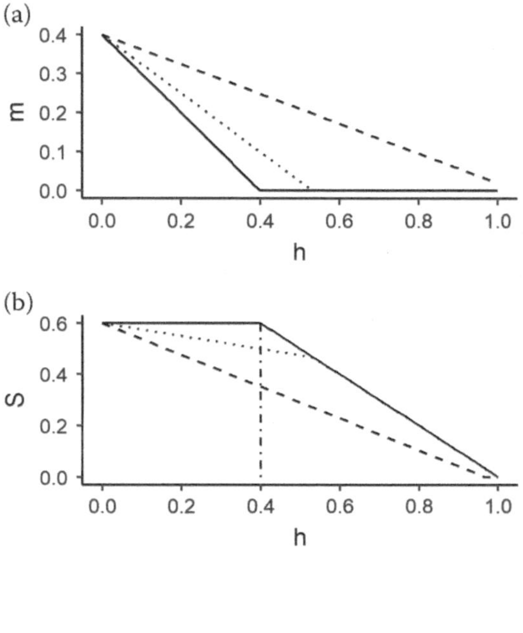

FIGURE 1.2 Alternative forms of harvest compensation. (a) Relationship between per-capita harvest rate (*h*) and natural mortality (*m*). (b) Relationship between per-capita harvest rate (*h*) and total survival (*S*). Dashed lines indicate additive mortality, solid lines indicate complete compensation to threshold of *h* = 0.4, and dotted lines indicate partial compensation.

or competing risk. Quite obviously, if a duck is killed by a hunter, it cannot subsequently die of starvation, or vice versa. The phenomenon of competing risk clearly requires only timing, whereas compensation requires density dependence, the reason that non-harvest mortality declines under both compensatory and additive mortality (Fig. 1.2a).

The above relationships represent *phenomenological* relationships, meaning that they describe patterns that occur in nature (Williams et al. 2002: 228, Cooch et al. 2014). However, they do not capture the underlying mechanisms thought to induce the observed patterns, and thus are not completely satisfactory. Alternatively, a density-dependent mechanism can be used to motivate compensation (but see Lebreton 2005). To do so, I constructed a logit-linear model of non-harvest mortality as a function of density post-harvest. Johnson et al. (1993) and Williams et al. 2002: 228) present similar formulations, but in terms of survival rather than mortality:

$$m(t) = \frac{1}{1 + e^{-[b0+b1N(t)]}}, \tag{1.6}$$

where $N(t)$ is population size following harvest given a pre-harvest population of N_0, that is $N(t) = N_0[1 - h(t)]$. With total survival defined by (1.5), we can produce graphs of the compensatory relationships, with compensation now directly based on density dependence. For illustration, I specified values of $b_0 = -4.88$, $b_1 = 0.045$ and $N_0 = 100$ (Fig. 1.3). The resulting natural mortality and survival curves resemble the earlier phenomenological models (Fig. 1.2), but lack the sharp (and arguably unrealistic) thresholds, producing compensation intermediate between the Anderson and Burnham (1976) compensatory and additive mortality model and additive mortality models (Fig. 1.3b, c). I will return to this below as relevant to the statistical testing of hypotheses of compensation.

Empirical Tests for Harvest Effects

Early efforts to test for evidence of compensation versus additivity in waterfowl populations preceded the formal development of a mathematical theory of compensation by Anderson and

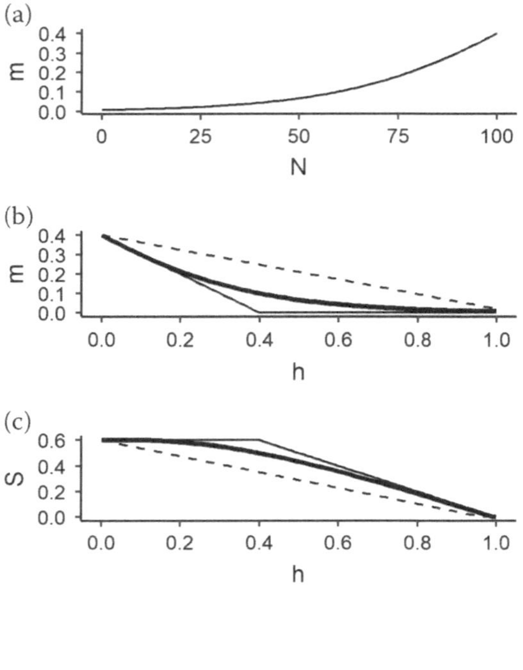

FIGURE 1.3 Density mechanism for compensation. (a) Natural mortality as a function of population size. (b) Relationship between per-capita harvest rate (h) and natural mortality (m). (c) Relationship between per-capita harvest rate (h) and total survival (S). (b, c) Dashed lines indicate additive mortality, heavy solid lines indicate compensation modeled by density-dependent mechanism, and solid lines indicate compensation modeled per Anderson and Burnham (1976).

Burnham (1976) and others. Geis and Taber (1963), Geis (1972), and others used band recovery data to purportedly support the additive mortality hypothesis in mallard using regressions between estimates of annual survival (*S*) and band recovery (*f*) rates, with significantly negative coefficients taken as evidence of additive mortality. Anderson and Burnham (1976) showed that these tests were flawed for several reasons, most fatally because the negative association between the estimates of *S* and *f* could be explained as an artifact of the estimation process. Burnham and Anderson (1984) later proposed an approach for statistical modeling, known as utrastructural modeling, that directly incorporates the relationship between *h* and *S* under the competing hypotheses (but see criticisms, e.g., Otis and White 2004, Sedinger et al. 2010, Servanty et al. 2010).

Anderson and Burnham (1976) analyzed data from mallard banded in North America prior to hunting seasons during 1961–1971, and after correcting for the statistical issues they identified in previous analyses, found little evidence of additive mortality. This seminal work and subsequent analyses (e.g., Burnham and Anderson 1984) have frequently been taken as definitive proof of compensation, and dismissive of additive mortality, although Anderson and Burnham were careful to qualify their conclusions.

Nichols et al. (1984*a*) revisited the issue and clarified several points. First, as I noted earlier, this type of analysis focuses on the *phenomenon* of *S* versus *h* (Fig. 1.2), but glosses over the actual *mechanism* of compensation: the relationship between *m* and *N* (Fig. 1.3). Logically, if compensatory mortality is true, then there must exist a density-dependent mortality mechanism. This duality also raised the important issue of statistical burden of proof. For instance, if the test is based on regression of *S* and *h* (or comparison of *S* between high or low harvest years), then failure to detect a significant slope or difference supports the compensatory mortality hypothesis only to the extent that it fails to reject it as the statistical null hypothesis. By contrast, when testing is based on *m* versus *N*, additive mortality is now associated with the statistical null hypothesis, to be retained unless sufficient evidence is marshalled in favor of the prediction of compensatory mortality of negative density dependence. These difficulties are to some extent avoided by approaching the problem from an estimation rather than hypothesis testing viewpoint, in which the focus is on estimating the strength of compensation and corresponding measures of uncertainty (e.g., Burnham and Anderson 1984, Williams et al. 2002, Conroy et al. 2002).

Additionally, most of the previous analyses and conclusions were based on mallard, and are not necessarily portable to other species of ducks with different life-history strategies (e.g., Patterson 1979). Nichols et al. (1984a) reviewed evidence for compensation for several duck species including American black duck (*Anas rupribes*) and ring-necked duck (*Aythya collaris*), with mixed results including some support for additivity. Further analyses, again mainly of mallard data, removed concerns about circular statistical reasoning identified earlier, but were unable to unequivocally demonstrate either compensation or additivity (Nichols and Hines 1983). Similarly, analyses for American black duck (Krementz et al. 1987, 1988) were generally unsatisfactory in providing a definite answer as to the contribution of harvest to total mortality, and especially, evidence for a density-dependent mechanism. Consequently, the role of harvest versus other factors, notably competition from and genetic introgression by mallard, remained in dispute (Ankney et al. 1987, 1989, Conroy et al. 1989).

Overall, retrospective analyses such as these have proved unsatisfactory for discriminating between alternative hypotheses about harvest compensation. In part, this is because of confounding effects inherent in time series of data. For example, over the same span of years that harvest rates were generally declining because of restrictive regulations, waterfowl populations were also declining, as were habitat, making causal inferences from resulting correlations problematic (Sedinger and Rexstad 1994, Conn and Kendall 2004, Sedinger and Herzog 2012). Further efforts to model variation in harvest and survival rate, including Bayesian approaches, were partially successful at distinguishing causal *signal* from statistical *noise* (Otis and White 2002, 2004, Conroy et al. 2005, Sedinger et al. 2010, Servanty et al. 2010). In the early 1980s the U.S. Fish and Wildlife Service promulgated a program of stabilized regulations, deliberating confining regulation to a narrow range, in an effort to understand natural variation in populations (Patterson and Sparrowe 1987). Although much was learned from these efforts, it became clear that they were unhelpful in determining the degree of compensation in populations, precisely because the resulting range in harvest rates was too narrow to allow strong testing. Harvest policy became increasingly confrontational in the 1980s and 1990s, and included political intervention contrary to Flyway Council recommendations for the 1994–1995 waterfowl season. These failures, in turn, lent support to burgeoning efforts to actively learn while managing the essence of adaptive harvest management (Nichols 2000).

Finally, I have focused on the mortality side of the population dynamic equation, especially mortality following hunting season (winter), traditionally considered the limiting time of year for waterfowl. Obviously, density regulation can also occur through reproduction rates, for example, competition for habitats during the breeding season. For waterfowl, the connection of reproductive compensation to harvest is apparently less direct than that of winter mortality, and may be more important for longer-lived, so-called *K-selected* species. Nonetheless, density-dependent reproduction is very relevant to understanding the response of populations to harvest, and is considered in a more holistic, life-cycle framework, below.

Model-based Approaches for Managing Migratory Waterfowl Harvest

Mathematical description of the compensatory relationships between harvest and natural mortality accounted for annual variation in survival of waterfowl, but did not account for other aspects of dynamics, notably reproduction. By the 1980s, several approaches had been proposed for modeling the annual life cycle of mallard and other waterfowl. These typically started with an initial spring (or sometimes fall) population, to which was applied specified survival and recruitment (young produced per adult, or per adult female) rates and assumed or estimated compensatory or other functional (e.g., habitat) relationships (e.g., Anderson 1975, Brown et al. 1976, Johnson et al. 1988). By folding in the relationship between harvest and natural mortality and the assumed degree of compensation, the models could be used to predict population growth under hypothetical harvest scenarios, now in context of other factors. Further, by retaining elements of mechanistic thinking, these approaches would allow forecasts under alternative assumptions about harvest compensation and other functional relationships, a key "pre-adaptation" to adaptive harvest management (Nichols 2000).

An important feature of this modeling approach was the close connection of model states and parameters to quantities that were observable in annual surveys and other monitoring efforts, leading to the ability to parameterize models and make testable predictions, particularly for mallard and other species for which data were sufficient. I will focus on the seminal work by Anderson (1975) on mallard because of its importance to future developments. The subsequent models of mallard, American black duck, and other species have the common features of representing dynamics in terms that (1) captured key aspects of the annual life cycle; (2) included state variables and parameters that were readily observable from existing monitoring data; and (3) had close linkage to management controls, particularly harvest regulations. In the case of mid-continent mallard, the two obvious and observable state variables were the annual numbers of breeding adults (N_t) and the number of ponds present (P_t), both derived from annual aerial surveys (Pospahala et al. 1974). These state variables provide the basic, annual inputs to model the population, with the population evolving through the annual cycle influenced by density, habitat conditions (indexed by pond numbers), and natural and harvest mortality (Fig. 1.4). Data archives were used to estimate key model parameters and relationships, and model predictions formulated under key alternative hypotheses, including compensatory relationships. A similar approach was taken in modeling the life cycle of American black duck, with mallard numbers replacing ponds as a habitat driver (Conroy et al. 2002).

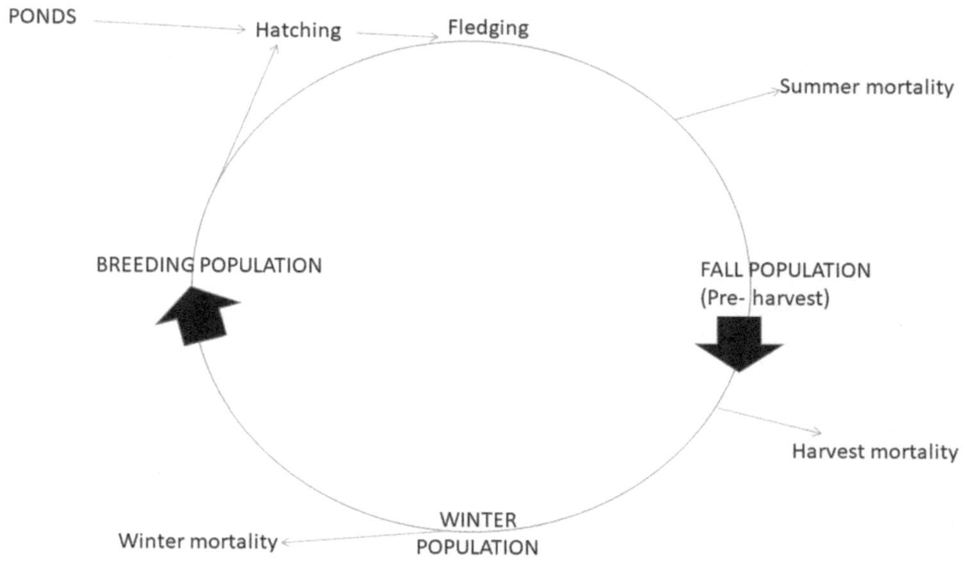

FIGURE 1.4 Life-cycle model for mallard (modified after Johnson et al. 1988).

The survival and reproduction submodels for mallard and American black duck allow expression of several combinations of alternative hypotheses relevant to each species' management. For American black duck, the respective roles of harvest and competition from mallard can be expressed by coefficients quantifying the degree of compensation (additive or compensatory) and the effects of mallard density on American black duck reproduction (present or not), resulting in four combinations of these factors. Historical population, harvest, banding, and habitat surveys could then be used to estimate model parameters and attempt to discriminate among hypotheses. Conroy et al. (2002) fit a series of predictive models to historical data, and found empirical support for mallard, habitat, and density-dependent effects on reproduction rate, but little evidence for density-dependent survival. These results were thus incapable of rejecting either mallard or harvest as being influential on American black duck dynamics. A similar construct for mid-continent mallard (Williams et al. 1996) formed combinations of additive or compensatory mortality and strongly or weakly density-dependent reproduction, and was similarly incapable of unequivocal inference about any one factor.

In summary, retrospective analyses such as those conducted for mallard and American black duck, however well underpinned by testable hypotheses, were generally incapable of providing firm answers about the role of harvest in dynamics. By the 1990s, it was clear that there were no unequivocal answers about the role of harvest and non-harvest mortality in total mortality (compensatory or additive) or the importance of harvest versus habitat and other factors as limiting to populations. Clearly, there was a need to move away from *proof* (*disproof*) of hypotheses, to how to conduct harvest management given uncertainty about the role of harvest.

MODERN PHASE: DYNAMIC AND ADAPTIVE DECISION MAKING

A Formal Optimization Framework for Harvest

Key elements of a modern approach for the management of migratory waterfowl harvest have existed for decades. Initially, emphasis was on approaches largely based on the maximum sustainable yield concept and derivations thereof. For example, Brown et al. (1976) developed a fairly sophisticated life-cycle model for mallard, and via a series of mathematical derivations proposed algorithms for deriving optimal rates of exploitation. In common with maximum sustainable yield theory, this approach treated mallard populations as deterministic, and solutions were based on equilibrium rather than transient conditions. By this time, Anderson (1975) had elucidated an alternative and more realistic perspective, and proposed a strategy that would, among other features, (1) achieve maximum sustainable yield, but as a long-term objective, (2) realistically incorporate environmental and other sources of uncertainty, (3) provide for frequent (e.g., annual) feedback of information about populations and environmental states, (4) allow harvest decisions to be based on currently observed population and environmental states, yet still be optimal for achieving a long-term objective, and (5) deal with political or socio-economic constraints limiting the range of exploitation to be entertained.

Anderson (1975) showed that harvest management for waterfowl could be encompassed by two mathematical properties: the Markov Property, and the Principle of Optimality. By the former, the dynamics of the system follow the premise that events at time $t+1$ are stochastically dependent only on events at time t. This property, a feature of all the preceding model formulations, is critical to rendering dynamic decision making tractable. The latter, conceptually stated as: "*An optimal policy has the property that whatever the initial state and decision are, the remaining decisions must constitute an optimal policy with regard to the state resulting from the first decision*" (Bellman 1957: 83). Parsing this a bit more crudely: "*if it ain't optimal in the future, it won't be now.*" This sensible but seemingly useless statement (how can we possibly know the future effects of our current decisions?) is rendered useful by working *backwards* from a distant future decision known to be optimal to the current one under contention. The Markov principle allows us to do this in steps or "stages," wherein each stage has influence specifically on the next, but also in the aggregate on all the remaining. The *dynamic programming* solution to the optimal harvest problem is obtained in a stepwise fashion using backwards iteration by:

1. Selecting the decision that provides the maximum value for current plus future harvest value at some arbitrarily distant future time. This ordinarily results in selection of the maximum feasible or permissible harvest rate at the last stage because there is no future value to be obtained from the resource,
2. Taking one step back in time and selecting the decision that maximizes current plus future return, taking advantage of the fact that optimum decision and value from the vantage point of the first step has already been determined,
3. Continuing this process, at each step incorporating the optimum decisions and values from the previous vantage point.

At each stage, the optimal decision and its value depend both on time and on the system state. For many (but not all) problems, repetition of this backward induction process will eventually result in a *stationary policy*—one that still depends on the system state but no longer depends on time. At this point, the policy maker has available a rule set that provides the optimal decision for achieving sustainable harvest at any given current state, and fully anticipates the influence of current decisions on future system states (ducks, ponds), and decision opportunities (achievable harvest). This formulation is apparently restrictive, in that it assumes that system dynamics are deterministic. However, the dynamic programming algorithm is readily generalized to stochastic situations expressing the value function as an expectation, averaged over stochastic distributions representing environmental and other sources of uncertainty. This more generalized, stochastic formulation is referred to as *stochastic dynamic programming* (Lubow 1995, Williams et al. 2002). As with dynamic programming, stochastic dynamic programming strives to find stationary decision strategies, so for example, harvest rates that depend only on the current system state, but are nevertheless optimal with respect to harvest over an arbitrarily long-time frame, now taking into account stochastic effects.

Casting harvest decision in this stochastic, dynamic framework was a profound change in the way of thinking about harvest decisions. The approach shifted away from the simplistic maximum sustainable yield paradigm, with its emphasis on equilibrium conditions, while avoiding a reactive (e.g., "regulations chasing ducks") syndrome, and allowed for decision making that was optimal over a long-time horizon in the face of a stochastic environment, given basic information about population dynamics and knowledge of the current system state. The stochastic dynamic programming framework also relaxes the restriction of ordinary dynamic programming that optimal policies assume that the decision maker always carries out optimal actions in the future. Such a situation, also known as *partial control*, occurs because managers generally do not directly manipulate harvest rates or numbers; rather, quotas, bag limits, season lengths, and other regulations are promulgated that are thought to influence harvest rates, but in an inexact, that is, probabilistic manner. Partial control is straightforward under stochastic dynamic programming, via the interposing of probability distributions between the decision (harvest regulations) and the intended outcome (harvest rates; Williams et al. 1996, 2002). Somewhat more problematic is the fact that we generally do not have an exact knowledge of the current system state, but rather must infer that state via data collection and statistical inference. *Partial observability* is complex and formally handled by a generalization of Markov decision processes known as partially observable Markov decision processes (Chadès et al. 2008). Many applications, in practice (e.g., the current adaptive harvest management program for waterfowl), ignore partial observability, and treat the system state as directly observable.

Recognizing and Reducing Structural Uncertainty

The above framework assumes that we are operating from an agreed upon understanding of the influences of harvest on population dynamics. A key impediment to advancing harvest management has been the issue of *structural uncertainty* about the specific mechanisms and relationships underlying population dynamics. Frequently, political disagreements about resource values masquerade as structural uncertainty. As noted above, efforts to discriminate among alternative harvest hypotheses using historical data, models, and statistics have been only partially successful. Experimental approaches (e.g., Anderson et al. 1987) to resolve key questions have failed to materialize for logistic, political, and other reasons. An alternative approach, known as *adaptive resource management* (Walters 1986, Williams et al. 2002, Conroy and Peterson 2013), incorporates two closely connected concepts.

First, as with stochastic (e.g., demographic, environmental, or control) uncertainty, structural uncertainty decreases the value that would have been obtained from a resource decision in the absence of uncertainty (Williams et al. 2002, Conroy and Peterson 2013). However, structural uncertainty is fundamentally different than most of the above forms of uncertainty, being a type of reducible or *epistemic* uncertainty, whereas for example environmental variation is a form of irreducible or *aleatoric* uncertainty (Conroy and Peterson 2013, Tyre and Tenhumberg 2021 [Chapter 11]). Explicitly including system (structural) uncertainty in the harvest optimization framework is obtained by positing a (usually discrete) list of alternative models, each incorporating different structural assumptions, and collectively thought to span the range of structural uncertainty critical to management. For the mallard example, structural uncertainty currently is characterized by four models involving combinations of additive versus compensatory mortality, and strongly versus weakly density-dependent recruitment (U.S. Fish and Wildlife Service 2019a). The alternative models are each assigned probability weights representing the current support among models, with weights updated periodically by monitoring feedback using Bayes' Theorem, forming a parallel *information state*.

Second, optimal decision making proceeds as before under stochastic dynamic programming, but now with the value function averaged over both the distributions representing environmental and other sources of irreducible (aleatoric) uncertainty and the distribution representing structural uncertainty. We already know that the evolution of the system state depends, in part, on the decisions about harvest that are made through time. It turns out that the information state *also* is influenced by these decisions, by means of Bayes' Theorem, and is thereby changed once a new system state is observed following implementation of a decision. Optimization under this framework is conducted either by *stochastic dynamic programming* or *adaptive stochastic dynamic programming* (Williams et al. 2002), depending on how the evolution of structural uncertainty in treated. Under *active adaptive management* the influence of current decisions on future information states is explicitly expressed via the active stochastic dynamic programming algorithm, which models state and information transitions through time. Thereby, current decision making accounts for the potential to reduce future structural uncertainty, thus enhancing resource value. Under *passive adaptive management* the information state is not modeled dynamically. Rather, it is treated as fixed at the current model weights, with the value functions averaged over these weights within the architecture of the ordinary stochastic dynamic programming. Thus, under passive adaptive management, current decisions do not take into account potential reductions in structural uncertainty. In theory, the active form of adaptive stochastic dynamic programming provides greater resource value (i.e., is globally optimal) compared to passive; in practice, analyses have suggested that differences between the two are small (Johnson et al. 2002). Under either form, the current information weights are updated, using monitoring feedback and Bayes' Theorem. The current adaptive harvest management program of the U.S. Fish and Wildlife Service is based on passive adaptive management (Johnson et al. 2002, U.S. Fish and Wildlife Service 2019a). Finally, experimental approaches or deliberate system probing (Walters and Holling 1990) can be undertaken in an effort to learn more rapidly, but are theoretically suboptimal compared to either the active or passive forms of adaptive management, given an objective of maximizing long-term harvest (Williams et al. 2002, Conroy and Peterson 2013, but see Johnson et al. 2002).

Development of Adaptive Harvest Management for Migratory Waterfowl

All of the essential technical ingredients to enable adaptive harvest management, including monitoring data, alternative models, optimization, and updating algorithms, were in place well before 1990, yet as of that date and later harvest regulations for waterfowl in the United States of America still were being promulgated on a largely reactive, rather than long-term prospective basis—let alone one that could easily accommodate uncertainty (and disagreement) about key processes, particularly the role of harvest in regulating populations. In addition, the regulation setting environment was increasingly subject to political pressures, culminating in direct intervention in the 1994–1995 season, in which political considerations trumped technical recommendations (Nichols 2000). The resulting frustration among managers helped to stimulate the implementation of an objective and data-driven adaptive harvest management process for mid-continent mallard (Nichols 2000, Johnson 2006). Inspired by the technical and collaborative success of mallard adaptive harvest management, efforts were untaken to implement adaptive harvest management for other species of waterfowl, with varying degrees of success (e.g., Conroy et al. 2002, Runge et al 2009, U.S. Fish and Wildlife Service 2019a). Implementation of adaptive harvest management for American black duck has been complicated by the binational nature of regulation and the large proportion of the harvest occurring in Canada, and profoundly different views among stakeholders about the relative importance of harvest as a driving factor. Nonetheless, adaptive harvest management provided a framework for dealing with these issues in a coherent way, by allowing separation of stakeholder values from scientific uncertainty about mechanisms (Conroy and Peterson 2013).

EMERGING PHASE: UNRESOLVED ISSUES

Complexity: How Much Is Needed?

The availability of detailed, long-term data sets, sophisticated statistical analyses, simulation models, and optimization tools creates the temptation to build ever more complex and realistic models, including details such age, size, spatial structure, or even individual attributes, and to make matching, finely resolved harvest decisions. Managers may be advised, however, to take a step (or several steps) back and do some reality checking. Of course, collecting, storing, and analyzing data have real costs, and acquiring data at arbitrarily fine scales may or may not have utility. Related to this, detailed and realistic models would require dramatic increases in sampling intensity for empirical support. Finally, these efforts beg the question of whether the benefits in improved decision making outweigh costs of surveys or implementation of management at fine scales (Johnson 2006, Conroy and Peterson 2013, Cummings and Bernier 2021 [Chapter 10]).

Other Values

The above theory and applications of a sustainable and scientifically supported harvest paradigm are generally directed at maximizing harvest (or economic gain) over some time horizon. However, alternative social values may dictate lower or higher harvest rates and correspondingly different ideal population levels than maximum sustainable yield or its dynamic counterparts. Examples include reduction of wildlife numbers in urban areas, avoidance of vehicle collisions, considerations of age- and sex-specific structure, and maintenance of lower- and higher-quality (e.g., trophy) populations (Gruntorad and Chizinski 2021 [Chapter 4], Hiller et al. 2021a,b [Chapters 2, 23], Kaemingk et al. 2021 [Chapter 3], Morina et al. 2021 [Chapter 19], Paukert et al. 2021 [Chapter 18], Robinson et al. 2021 [Chapter 9]). Indeed, the size of the harvest may become disconnected from hunter and angler satisfaction, which often depends more on the perceived quality of the experience including cost, access, and seclusion from competing hunters and anglers (Gruntorad and Chizinski 2021 [Chapter 4]). Beyond these competing or ancillary values, recreational harvest faces an uncertain future, with changing societal values resulting in declines in hunting and angling traditions (Hiller et al. 2021b [Chapter 2]).

System Change and Extreme Uncertainty

The theory and mathematics supporting maximum sustainable yield depend on assumptions of stable environments with limited if any stochastic variation, and relative certainty about system dynamics. Optimization under stochastic dynamic programming removes these constraints, and the further construct of adaptive stochastic dynamic programming allows for optimization in the face of uncertainty about system dynamics and the response of populations to harvest and other factors. Nevertheless, successful application of these approaches requires that the system behaves in a *stationary* manner, essentially varying about a long-term average. Temporal shifts in system states, for example under global climate change, create a moving target, and must somehow be anticipated and dealt with to derive harvest policies that remain optimal in the long term, particularly because ensuing extreme uncertainty may very well exceed anything in recent human experience (Johnson 2006, Conroy et al. 2011).

SUMMARY

Modern harvest management has its roots in intuitive ideas about resource limitation and population self-regulation. These were later formalized in simple models of population dynamics and a focus on achieving and maintaining equilibrium conditions. Although unrealistic, these models and associated ideas of maximum sustainable yield were useful as a jumping-off place for more realistic and mechanistic models, which in turn led to the first major advance of expressing harvest impacts in terms of testable, alternative hypotheses. Now monitoring data could be used to challenge model predictions, setting the stage for information feedback into harvest decisions. The second major advance was fully incorporating the dynamic and stochastic behavior of populations into an optimization framework, together with the explicit recognition and reduction of uncertainty under adaptive resource or harvest management.

The evolution of modern harvest management is by no means complete, with significant remaining challenges, including dealing with spatial and other system complexities, multiple resource objectives, and changing societal values (Hiller et al. 2021b [Chapter 2]). Properly accounting for and including human valuation of natural resources is a challenge faced by the science of human dimensions (Decker et al. 2001), is addressed several places in this book (Gruntorad and Chizinski 2021 [Chapter 4], Fuller et al. 2021 [Chapter 8], Hiller et al. 2021b

[Chapter 2], Kaemingk et al. 2021 [Chapter 3], Melstrom 2021 [Chapter 5]), and is explicitly addressed in the most recent North American Waterfowl Management Plan (North American Waterfowl Management Plan 2018). Perhaps most daunting is dealing with rapidly changing systems, as predicted (and partially realized) under climate change, and the challenges of decision making under novel conditions and extreme uncertainty. Finally, although I have used harvest management for waterfowl as the illustrating example, the above theory and methods are generalizable to other species of fish and wildlife, or for that matter, any dynamic resource decision-making problem.

ACKNOWLEDGMENTS

This chapter was improved by the input of the editors and two reviewers, T. Arnold and C. Moore; the author is grateful.

2 The Social and Political Context of Harvest Management

Tim L. Hiller, Roger D. Applegate, and Larkin A. Powell

CONTENTS

INTRODUCTION

Harvest refers to the consumptive use (e.g., food, fur) of fish and wildlife resources for avocation or commercial purposes. The goals of harvest management have evolved in scope and complexity since the early years of wildlife management. Early goals of management were designed to prevent unsustainable harvest (or overharvest) once regulated harvest became established during the 1930s (Leopold 1933; Conroy 2021 [Chapter 1]). Contemporary goals of harvest management decisions vary and may involve complexities and uncertainties. For example, most state fish and wildlife agencies may be directed first to conserve wildlife, and then to provide opportunities for sustainable harvest through fishing, hunting, and trapping for the appropriate fish and wildlife resources. Harvest regulations may be set based on the best available information with regard to population size and realized harvest rates, but monitoring through time may contribute to a formal or informal decision process to reduce uncertainties and increase the probability of long-term sustainable use of resources (Conroy 2021 [Chapter 1], Runge 2021 [Chapter 7]). Alternatively, a goal may be to use harvest strategies to reduce population abundance (especially at a local scale) of either invasive or native species that are causing undesired impacts on other species, unacceptable economic and other losses (e.g., exceeding social carrying capacity), or environmental damage (e.g., exceeding ecological carrying capacity).

Harvest of fish and wildlife also provides an opportunity for collection and analysis of data from consumptive users to inform decision making (Cummings and Bernier 2021 [Chapter 10]). For example, the collection of data from hunters in the United States of America dates back to at least 1952, when the U.S. Fish and Wildlife Service Waterfowl Harvest Survey tallied the response of hunters to questions about their success, and the Cooperative Parts Collection Survey followed in 1961 (Bolen 2000); such efforts may represent the oldest forms of citizen science in North America. Collection of data on individual sex, age, and location of harvest for a species may be used to estimate population abundance and trends, estimate reproduction and survival, and assess effects of harvest (e.g., Carpenter 2000). Despite the benefits of acquiring such relatively low-cost, high-yield data for management and conservation purposes, there remain situations in which our profession can better use scientific data for making management decisions (Artelle et al. 2018, but see Mawdsley et al. 2018).

Harvest management decisions may often be subject to a complex set of ecological, social, political, and economic considerations, resulting in a balancing act. Ironically, the ecological considerations (e.g., species' population characteristics, predator–prey relationships, ecological carrying capacity) are likely the most obvious to wildlife professionals, as this has been the focus of traditional coursework at universities, and what many future wildlife professionals had assumed to be driving such decision making. However, given the public-trust responsibilities of state (and for some species, federal) government for management of fish and wildlife, harvest management necessitates a moderate level of social and political activity (Fig. 2.1). Indeed, harvest regulations are policies, and a *good policy decision* is consistent with knowledge and uncertainties, alternative decisions, and desires of the decision makers or their constituencies (Michaels and Tyre 2012). Therefore, social and political considerations may at times take precedence during decision making, including at the extreme to constrain (or eliminate) harvest of a particular species irrespective of ecological considerations and evidence that supports otherwise. Here, we explore some dynamics that are currently pushing harvest management decisions into higher levels of social and political activity and engagement, which we believe is a critical dynamic to be considered by managers.

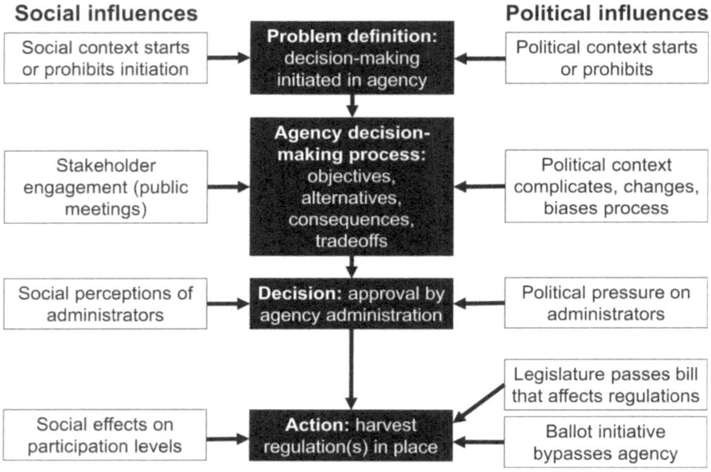

FIGURE 2.1 Decision-making process following PrOACT (a process of defining problem, determining objectives, and exploring alternative solutions as well as their consequences and tradeoffs) model for harvest regulations for a state agency (dark boxes). White boxes indicate potential social (left) or political (right) influences on the decision-making process.

Our goals for this chapter are to (1) inform future and current wildlife professionals of the complex and technical decision-making environment, (2) provide examples of how social and political considerations have affected management decisions, and (3) provide pragmatic guidance on how to navigate situations where these considerations must be integrated into decision making. Given the broad array of such topics within fish and wildlife management, we must necessarily limit our discussion to a broad overview, but our hope is that readers are able to apply this knowledge to meet their specific needs.

THE ECONOMICS OF CONSUMPTIVE USE

Economics plays a critical role in the social and political context of harvest management. For example, revenue from license sales results in substantial and critical support for state fish and wildlife management agencies (Fig. 2.2). Many state agencies may have little or no dedicated funding (e.g., general fund) through their state legislature that can be used for managing harvested species, and much of the state-level revenue is derived from license sales. Funds received from the sale of hunting and fishing licenses are required, by law, to be used only for agency management programs. Non-federal funds, such as revenue from license sales, are required to leverage substantial sources of federal funding for the states through the U.S. Fish and Wildlife Service's Wildlife and Sport Fish Restoration Program, which collects excise taxes on firearms, ammunition, archery equipment, fishing tackle, and boat fuel. Coupled with revenue from license sales, these funds provide substantial support to manage fish and wildlife resources, mainly for state agencies. The allocation of federal Wildlife and Sport Fish Restoration Program funds to a state agency is based on a formula that includes number of paid licenses sold and numbers of hunters and anglers. These federal programs have match requirements, typically for every US$3 in federal funds requested, $1 in non-federal funds must be used as match. Annual fluctuations in both the available state and federal amounts results in a challenging scenario for annual planning, regardless of the funding program or source. Some sources of state and federal funds (or budgets) must be annually approved by legislative bodies, and consequently may at times be constrained by political processes. States may lack sufficient non-federal match in some years to fully utilize the available federal funds.

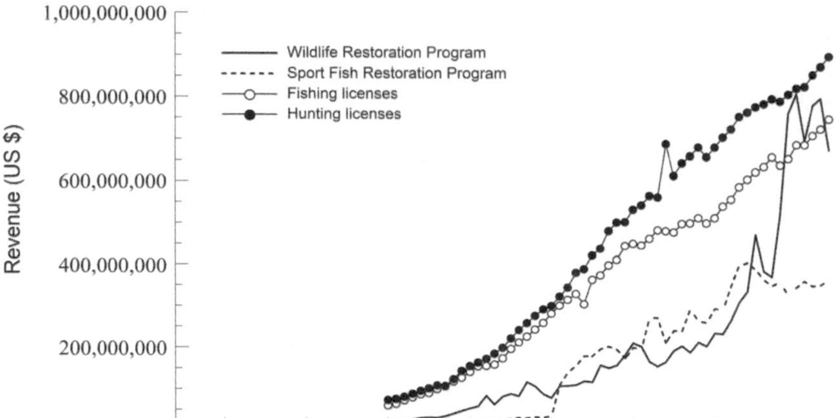

FIGURE 2.2 Annual revenue from two major federal fish and wildlife programs and number of hunting and fishing licenses sold during 1939–2019 in the United States of America (U.S. Fish and Wildlife Service 2019*b*).

Traditionally, consumptive-user groups have realized funding advantages because hunters and anglers paid for most efforts for fish and wildlife management and conservation. However, diversified funding for wildlife management with additional focus on non-game species has broadened stakeholder appeal beyond consumptive-user groups. One example is the availability of congressionally appropriated funds through State Wildlife Grants (U.S. Fish and Wildlife Service 2019*c*), which focus funds to prevent species from becoming threatened or endangered (Lauber et al. 2009). Under this program, states prepare a comprehensive State Wildlife Action Plan (Association of Fish and Wildlife Agencies 2019*b*), which guides how grant funds will be used to achieve conservation goals. Each State Wildlife Action Plan includes lists of, and information about, species that state agencies have prioritized based on conservation concerns, and these lists may include both game and non-game species. For example, the Tennessee Wildlife Resources Agency identifies several currently harvested species in its State Wildlife Action Plan (Tennessee State Wildlife Action Plan Team 2015), such as the northern bobwhite (*Colinus virginianus*), ruffed grouse (*Bonasa umbellus*), eastern spotted skunk (*Spilogale putorius*), Appalachian cottontail (*Sylvilagus obscurus*), and red squirrel (*Tamiasciurus hudsonicus*). State Wildlife Action Plans typically also include management and conservation of fish, invertebrates, and other animal taxa. The current consideration by the 116th Congress of the Recovering America's Wildlife Act (H. R. 3742, 2019) would, if passed, provide over $1.4 billion annually to states, territories, and tribes to further strengthen strategies described in State Wildlife Action Plans. To receive State Wildlife Grant funds, states must also provide match that is derived from either license fees or from other non-federal sources. Additionally, federally recognized tribal governments may apply for grants through the Tribal Wildlife Grants Program to help support wildlife conservation and management efforts (U.S. Fish and Wildlife Service 2020*c*). Some agencies have realized increased funding from state general revenue to support marketing, law-enforcement public-safety assistance, and enforcement of boating, snowmobile, and all-terrain vehicle regulations.

Additionally, the Migratory Bird Joint Ventures provide an opportunity for state, federal, and non-governmental partners to develop programs to be funded under a variety of grant programs, including the North American Wetlands Conservation Act, Neotropical Bird Conservation Act, and the Urban Conservation Treaty for Migratory Birds (U.S. Fish and Wildlife Service 2015*b*, 2018*d*). These funding sources can include management for harvested migratory birds as well as the many species that are not harvested. The paradigm for funding wildlife management and conservation is now shifting toward more support from all Americans to help meet the increasingly critical needs that consumptive users traditionally continue to support. Such a trend is also supported through innovative funding sources that result in substantial revenue for state fish and wildlife agencies. For example, the Conservation Sales Tax in Missouri was established via ballot initiative in 1977 as a 0.125% tax and has provided a minimum of $100 million/year (Thorne 2016).

THE REGULATORY PROCESS

STATE

Public Trust Doctrine in the United States of America has established that fish and wildlife are the property of states. As such, a history of court decisions and common law directs state governments to manage wildlife in the public trust of the citizens. State wildlife agencies are typically given this authority (Powell 2020). We use the term commission to refer to the various bodies (e.g., advisory committee, board, commission) that U.S. states have for guiding policy of fish and wildlife agencies, which is a key aspect of state fish and wildlife agency administration. Regardless of terminology, the function of such groups is essentially the same: to provide oversight to agency administrators and channel public input into policy and regulatory decisions (Towell 1979).

Commissions arose largely from the need to make management decisions timelier than would be the case if state legislatures had to be involved. As noted by Simmons and Dvorin (1977: 436), legislation does not necessarily result in direct translation of legal and constitutional values to agency administration that represents public values. Thus, commissions are important in providing a bridge between the public and the fish and wildlife management agency for policy matters. Finally, commissions are intended to remove party politics from fish and wildlife management (Towell 1979); however, commissioners are sometimes appointed by state governors or legislatures, and may be confirmed by legislatures, and many individuals have perspectives and opinions that contribute to their decision making, including individuals who do not have an interest in harvest of fish and wildlife.

The original concept of commissions was formalized as a model *game law* that was published by the International Association of Game, Fish and Conservation Commissioners in 1934 (Trefethen 1964). In this model, commissions would hire a professionally trained director who would essentially serve as the corporate executive officer for the agency (Batcheller et al. 2018). The agency director, variously termed as commissioner, secretary, and other titles, would then lead the technical and support personnel of the agency.

Although commissions have varying authority depending on the state, most provide a channel for public input and oversight of harvest management decisions. State agency personnel develop harvest management recommendations (often with alternatives and a set of expected outcomes) and commissions confirm or adopt recommendations, request alternatives based on receiving comments from the public, or simply reject both the recommendation and any alternatives. Public comments are solicited formally or informally by commissions, and often presented to commissions during public hearings.

Stakeholder groups (e.g., hunting and trapping organizations, conservation organizations, animal-rights organizations) commonly provide input during proposed regulatory changes. For example, hunting organizations may request that the agency consider increasing (or decreasing) hunting opportunities in a specific management unit due to real or perceived population increases (or declines) for a particular harvested species. Such a proposal may be initiated formally or informally, but support from these organizations, whether or not initiated by them, is often an important consideration during the rule-making and decision-making processes. Alternatively, our experience has been that animal-rights groups typically challenge many proposed changes through use of testimony (and submitting public comments) at commission meetings. Obviously, the underlying goal of animal-rights groups is at the opposite end of the continuum from those of consumptive-user groups, and depending on the specific decision under consideration, can result in a very controversial and occasionally confrontational situation. Animal-rights groups may also attempt to bypass the commission-agency rule-making process and utilize a legislative approach (e.g., ballot initiatives), as described below.

Agencies in many states solicit public comments on fish and wildlife harvest regulations, and use that input, along with biological and social science data, to develop harvest regulation proposals. State attorneys may be consulted to ensure language is consistent with intent, and that proposed regulations are not in conflict with any state statutes. Those proposals are then presented to the commission for preliminary vetting. Commissioners and the public provide comments that are considered in potentially revising the regulation. In addition, the law enforcement component of the agency may provide input to address potential issues associated with enforcing a particular regulation. Agency staff makes appropriate changes to the regulations and formally presents the revised regulation at a subsequent commission meeting. Typically, members of the general public may provide written or oral testimony to the commission during the meeting, the commission may ask questions of the agency staff, then discuss or possibly modify the regulation and vote to accept or reject the regulation, or table it if additional input or information is requested before making a decision. If approved by the commission, the chair of the commission signs the proclamation and that version then becomes the legally binding regulation. Unless there are extenuating circumstances (e.g., disease outbreaks) that require emergency rule making, the regulatory process usually occurs about five to six months prior to when the next harvest season occurs. Some states may also require some level of legislative review and approval at some point in the process before regulations can be implemented.

Tribal

Tribes have a unique relationship with U.S. government. Lands that were reserved by treaty by the federal government, or in a few cases state government, are under management authority of specific tribes. Therefore, the tribes make regulatory decisions for harvest fish and wildlife independent of state or federal regulations, as long as this harvest occurs on reserved lands (Freyfogle et al. 2019:160). Tribes work with the U. S. Fish and Wildlife Service to protect species that are listed as threatened and endangered under the Endangered Species Act (U. S. Fish and Wildlife Service 2020*d*) and migratory birds protected under the Migratory Bird Treaty Act (U. S. Fish and Wildlife Service 2004). Tribes have also been granted use of non-reserved lands for harvesting fish and wildlife, and are generally exempt from state and federal regulations. An example of this type of harvest regulation is the use of traditional fishing areas. Outside of reserved lands or designated non-reserved areas, tribal members are subject to all state and federal harvest regulations (Freyfogle et al. 2019:164–166).

Non-tribal members, defined by each tribe, cannot hunt or fish on tribal reserves or at other designated non-reserved areas unless specifically permitted to do so by the tribe (Freyfogle et al. 2019: 164–166). For example, the Penobscot Indian Nation in Maine offers several permits each year to non-tribal members to harvest moose on tribal lands. The tribe defines a bidding fee, minimum bid amount, and a permit fee. Successful bidders must obtain the permit ($3000 for 2020) and hire a guide who is a tribal member (Penobscot Nation 2020). As with states, harvest regulations on tribal lands can be complex and may vary between jurisdictions. In Idaho, for example, anglers may purchase either a tribal or state fishing permit for steelhead trout (*Oncorhynchus mykiss*) for the sections of the Clearwater River that cross tribal land (Idaho Department of Fish and Game 2003).

Pursuant to their responsibilities to manage fish and wildlife on their lands, tribal agency biologists conduct surveys and otherwise collect data on fish and wildlife populations and may also hire or collaborate with non-tribal biologists. Data are then used to assist in setting seasons, harvest limits, and other regulations. Such seasons are then enacted by the Chief or designee. Tribes also hire officers to enforce regulations.

Federal

Although the Public Trust Doctrine of the United States of America provides for states to manage fish and wildlife in the public trust of their citizens, there are three situations in which the federal government is given authority for management (Powell 2020). All three situations require coordination beyond the state level: (1) birds or fish that migrate across state lines or use off-shore, marine waters; (2) situations involving commerce or law enforcement of illegal activity that occurs when animals or animal products are transported across state lines; and (3) imperiled species in need of federal protection through the Endangered Species Act. Federal management of harvest could apply in each of these contexts, but we will discuss the former two.

The U.S. Fish and Wildlife Service in the Department of the Interior uses a flyway system (i.e., Pacific, Central, Mississippi, and Atlantic Flyways) to manage harvest of migratory birds based on migration patterns. A Flyway Council is populated by a representative from each agency within the states located in the particular flyway as well as scientists with the U.S. Fish and Wildlife Service. Since 1986, the Councils have operated under the framework of the North American Waterfowl Management Plan, which has provided guidance for harvest decisions in Canada, the United States of America, and Mexico (Williams et al. 1999). In addition, in 1995, the U.S. Fish and Wildlife Service began to use a formal program of adaptive management to manage harvest of mallard

(*Anas platyrhynchos*) in North America (Nichols et al. 2007). Councils now also receive guidance from the federal coordinators, and Councils set general frameworks for state-specific regulations. Managers within state agencies can then make decisions that are more restrictive, but not more liberal, than the Council frameworks for a given year. Thus, states may select harvest limits lesser than or equal to the maximum allowed, or states may adjust timing of seasons within the frameworks set by the Council.

The National Marine Fisheries Service of the National Oceanic and Atmospheric Administration in the Department of Commerce has responsibility for management of marine fisheries within the U.S. exclusive economic zone, which extends from three to 200 nautical miles from the coastline of the United States of America. State agencies are typically responsible for management of marine fish species from their coastline out to three miles. Management in the exclusive economic zone is divided spatially into eight regional Fishery Management Councils (i.e., North Pacific, Pacific, Western Pacific, Gulf of Mexico, Caribbean, South Atlantic, Mid-Atlantic, and New England). Councils include representatives from the commercial and recreational fishing sectors, as well as university scientists, environmental consultants, and state and federal agency biologists (Abbott-Jamieson and Clay 2010).

THE SOCIAL SIDE OF HARVEST MANAGEMENT

Much of the decision process for harvest management has potential to be impacted by social factors (Fig. 2.1). As noted above, it is common for previous license holders to be consulted in public meetings or harvest surveys with regard to alternatives for future harvest regulations (Johnson and Martinez 1995, Miller and Graefe 2001), although stakeholders attending public meetings may be more experienced and successful than those who do not attend (Johnson et al. 1993). However, the impact of social dynamics goes well beyond the public meeting.

Game Species Designation

The decision process for harvest management must address two questions: (1) should this species be designated a "game species" (we define traditionally as a species to be legally hunted, trapped, or angled), and (2) if so, what regulations will influence how, when, where, and how many individuals may be taken. With regard to the first decision for game species definition, ecological principles may be influential. For example, greater prairie chicken (*Tympanuchus cupido*) has largely been extirpated from Iowa, but is a relatively common species in the rangelands of Nebraska; the species is a game species in Nebraska, but not in Iowa. In contrast, harvest of sandhill crane (*Grus canadensis*) is allowed in all states of the Central Flyway with the exception of Nebraska. The reason for the lack of a season in Nebraska is not ecological, but social. In the spring, over 500,000 sandhill cranes stop along the Platte River during an impressive migratory event, and public opinion has kept the state agency from designating the species as a game species in Nebraska. Similar public opinion, fostered by livestock mortality, has led many western state agencies to initiate hunting and trapping seasons for gray wolf (*Canis lupus*) as soon as the political hurdle of protection under the Endangered Species Act was removed (Bruskotter 2013).

Social traditions also influence the range of regulations for a harvested species. Social customs have been critical to establishment of goals for fisheries management for Native communities in Hawaii (Freidlander et al. 2002). The use of dogs to aid with hunting deer is supported by those from regions in the southeastern United States of America, where the practice has a strong history (Cook et al. 2015). Women who are mothers and residents of Minnesota can fish without a license on the opening day of walleye season, which often corresponds with the weekend of Mother's Day. Tradition is responsible for a start time of 1000 hours in midmorning during the pheasant season in South Dakota, whereas nearby states start at sunrise. Officials point to social reasons for this start time—restaurant owners want hunters to eat breakfast before the hunt or farmers wish to finish chores before hunters arrive at their farms.

Participation Levels

A key uncertainty found in regulatory decisions for harvest management is partial controllability of harvest rate, which is the loose connection between a harvest regulation (e.g., liberal or restrictive harvest limits) and the resulting proportion of the population harvested (Williams et al. 1996). Managers control the regulations, but the same regulation set may result in a range of harvest rates over a period of years. The dynamic may be due to weather and nonlinear numerical responses of hunters, anglers, and trappers to population sizes of fish or wildlife as examples. Similarly, a decrease in the number of active consumptive users and changes in the values of society are causing managers to incorporate new approaches to management of the "social habitat" for hunting, angling, and trapping (Larson et al. 2014).

Urban and Rural Population Trends

The distribution of the human population in the United States of America continues to shift toward urban areas. Since 1910, the population of the United States of America has become more concentrated during every decade, with the exception of the 1970s (Otterstrom 2001). In 1960, approximately 70% of the population was urban, and in 2017, it increased to 82% (Ritchie and Roser 2020). The 1920 census marked the first time that over half of the people living in the United States of America were urban residents. Regionally, the Northeast has had a majority of urban residents well before the twentieth century, whereas the Great Plains, Rocky Mountains, and Southeast were majority rural through the middle of the twentieth century. However, the latter three regions are currently showing the strongest trends of urban amplification in the United States of America (Otterstrom 2003).

Urbanization can have a direct impact on angling, hunting, and trapping through three mechanisms: (1) removal of previously occupied fish or wildlife habitat due to development associated with urban sprawl (Theobald et al. 1997); (2) geographically longer distances for anglers and hunters to travel to participate in their respective activities (Poudyal et al. 2008); and (3) indirect effects resulting from changes in the composition of stakeholder experiences and values (Zinn et al. 2002). In response, fish and wildlife agencies have worked in recent decades to provide incentives to landowners for open-access lands for angling, hunting, and trapping, and priority may be given to providing opportunities on lands near urban areas (Wszola et al. 2020*a*). In addition, management for angling in urban lakes has become an effective tool for engagement with a unique urban stakeholder group. However, the indirect effects of urbanization may be the most important dynamic currently affecting harvest management. Residents from rural areas versus urban centers tend to have different value orientations toward wildlife (Manfredo et al. 2003, Manfredo et al. 2018, Gruntorad and Chizinski 2021 [Chapter 4]) and social issues (Kron 2012). Thus, increasing rates of urbanization in previously rural-dominated states have the potential to change the political landscape with regard to stakeholder opinions about harvest management decisions.

Trends in Value Orientations and Levels of Participation

Human-dimensions scientists often use a combination of two value orientations to describe the degree to which people prioritize utilitarian (at the extreme: human well-being prioritized over wildlife and a belief that wildlife should be used and managed for human benefit) and mutualism (at the extreme: considering wildlife as part of an extended family and a belief that people and wildlife should live side by side without fear; Manfredo and Zinn 2008, Teel and Manfredo 2010, Gamborg and Jensen 2016). A body of research suggests that younger age groups and urban residents are more likely to be less utilitarian than older age groups and rural residents (e.g., Manfredo and Zinn 2008, Manfredo et al. 2018). This may in turn suggest that decisions for harvest management in the future will take place in a continually evolving landscape of value orientations of stakeholders that differ from traditional, consumptive, utilitarian orientations. Changes in levels of education and income may also affect value orientations of stakeholders; for example, Manfredo et al. (2003) reported that study participants with value orientations labeled as neutral or protectionist had generally higher levels of education, but slightly lower incomes than other groups.

Value orientations also have been demonstrated to vary spatially within the United States of America. For example, in a survey of western states, >45% of residents of Alaska, Idaho, Montana, North Dakota, Oklahoma, South Dakota, and Utah were classified as utilitarian, whereas >35% of residents from California, Hawaii, and Washington were mutualists. The latter set of states had <32% utilitarian residents, whereas the former set had <20% mutualists (Teel and Manfredo 2010).

Participation in angling and hunting varies within the United States of America as well. Nationwide, 14% of people in the United States of America currently participate in angling. Regional levels of participation range from a low of 8%–11% in the Pacific coastal states, New England, and Middle Atlantic states to a high of 20% in the East South-Central region (Alabama, Kentucky, Mississippi, and Tennessee). Eleven percent of residents of cities with human populations >1 million participate in angling, compared with 26% of residents in or near towns of <50,000 people. Participation in hunting is lower (4% of the U.S. population) than participation in angling, and regional trends are similar between activities (Gruntorad and Chizinski 2021 [Chapter 4]). Participation levels range from a low of 2%–3% on both west and east coasts to a high of 8% in the West North Central region (northern Great Plains states and Iowa, Minnesota, and Missouri) and the East South Central region. Hunting participation by residents in cities with populations >1 million is 2%, whereas for residents of small towns or rural areas, 17% engage in hunting (U.S. Fish and Wildlife Service and U.S. Census Bureau 2016). Per capita license sales in Ohio were lower in urban areas than in rural areas (Karns et al. 2015). Hunters that purchase a license may not hunt as much as managers might believe; for example, only 15%–30% of duck hunters in the Central Flyway of the United States of America harvested ducks on five or more days during a hunting season (Haugen et al. 2015).

As value orientations become less utilitarian in nature, participation in angling and hunting may decrease and public perceptions of wildlife-related issues may change (Teel et al. 2005). Half of the respondents in a recent survey of residents of the United States of America related that hunting would be an unpleasant experience (Wilkins and Miller 2018). In Europe, strong social reactions to the morality of angling, an activity often considered less controversial than hunting or trapping in the United States of America, have resulted in regulations to limit recreational angling (Arlinghaus et al. 2012). Managers working in this new paradigm of value orientations of stakeholders will need to continually assess and appreciate the reasoning behind conflicting viewpoints. Simultaneously, agencies will need to be engaged in higher levels of political support and lobbying than in the past, while working to implement practices that reconcile with contemporary views on animal welfare (Arlinghaus et al. 2012). Examples include recent efforts to train anglers to engage in rapid kill rather than letting a fish die slowly by hypoxia, providing for education on internationally acceptable trapping devices and methods (Vantassel et al. 2010, White et al. 2021), and banning the use of lead ammunition during hunting (Pain et al. 2019).

POLITICS AT PLAY

By definition, political influences are part of the reality in government decision making. State fish and wildlife agencies are no exception, and in fact, harvest management decisions that involve controversy are particularly susceptible. Decisions within an agency are often meant to address a problem that has been defined within the agency, by a stakeholder group, or perhaps through another agency or entity (Fig. 2.1). However, the previously described state-level decision-making process can be substantially affected, interrupted, or nullified through political intervention, which has been described as erosion of state management authority of fish and wildlife (e.g., Batcheller et al. 2000). Geist and Organ (2004: 49) more explicitly listed, "*Privatization, game ranching, unsustainable land use practices, and animal rights [as] examples of assaults on the [Public Trust Doctrine].*"

For states that have a ballot initiative (or referenda) legislative process in place, their use has affected harvest management of fish and wildlife by bypassing state agencies (Fig. 2.1). The ballot initiative process was adopted by many states during the late nineteenth and early twentieth centuries in response to concerns about state-level government corruption, and their use has increased over the decades, including for management of natural resources (Manfredo et al. 1997, Minnis 1998, Williamson 1998). Basically, an individual or a group of citizens drafts a policy concept, and if they collect the required number of signatures via petition to support their concept, it is placed on a general ballot for the voting public. If the majority of voters support the policy concept, it becomes adopted and implemented. Specifically related to harvest management of fish and wildlife, these initiatives are often directed at the prohibition of harvest-related activities such as fishing, hunting, and trapping, whether specific or broad in impact. The end result has often been special-interest groups in the minority but with substantial financial resources successfully influencing the often-uninformed general public and placing unnecessary restrictions that would not otherwise be considered or implemented on consumptive use (Williamson 1998). Ironically, one of the intents of the direct democracy approach of ballot initiatives was the prevention of legislative control by special-interest groups (Minnis 1998), who now seem quite adept and organized when pursuing this legislative option.

For harvest management, special-interest groups (e.g., animal-rights groups) have been strongly focused on trapping furbearers and on hunting large carnivores (e.g., use of hounds or bait), with periodic success in prohibition despite support and regulation by state agencies. For example, trapping has been prohibited or severely restricted in several states (e.g., Arizona [DeVos et al. 1998], Colorado [Manfredo et al. 1997], Massachusetts [Deblinger et al. 1998]) as the result of the ballot initiative process. In 1994, voters in Oregon approved the prohibition of hunting black bear (*Ursus americanus*) and mountain lion (*Puma concolor*) with the aid of dogs, and the hunting of black bear with the use of bait through a ballot initiative (Measure 18), despite these methods being the most effective and selective (Oregon Department of Fish and Wildlife 2017). These trends continue to the present day.

State agency wildlife professionals may experience frustration by the ballot initiative process in relation to harvest management for several reasons. In most if not all states, the state fish and wildlife agency cannot intervene in a ballot initiative process, and are limited only to providing requested information (Whittaker and Torres 1998). Additionally, when special-interest groups draft the language for their ballot initiative, they rarely involve agency staff and focus instead on acquiring specific data that might support their policy concept. Finally, agencies attempt to make the most informed and defensible decisions possible based on the best available science. Ballot initiatives from special-interest groups attempt to change that decision model.

The use of ballot initiatives for wildlife management has been described as problematic, where uninformed voters may be directing policy on complex and technical topics for which they are not trained and ultimately would not be representative of the majority (Williamson 1998). The majority of voters may simply be unwilling or unable to expend the necessary time to fully appreciate the consequences of their decision, especially if they are not invested in the topic (e.g., Manfredo et al. 1997). This leaves limited opportunity to provide brief yet accurate information to the general public, especially with the limited financial resources of consumptive-user groups, as state agencies typically cannot intervene at that point. However, it seems that wildlife professionals from other state agencies are not restricted from assisting if their state agency is supportive.

It is not uncommon for state legislative bodies to initiate (and occasionally pass) legislative bills, including to address potentially contentious harvest management decisions that bypass the state agency normally responsible for such decisions. For example, members within the state legislature in Oregon routinely draft bills designed to overturn Measure 18, described above, but to date have been unsuccessful.

Lastly, the use of science to make decisions seems to be increasingly challenged by a segment of the general public, and this may also be affecting fish and wildlife management. People are increasingly developing attitudes and beliefs based on information presented via social media, of which some information is clearly misleading or erroneous. The prevalence and rate of spread of such information is unprecedented and will undoubtedly need to be considered by wildlife professionals during decision making, particularly when politics are at play.

Case Study 1: Grizzly Bear—To Be Hunted or Not?

Social and political influences over management of large carnivores in North America have been occurring since before the birth of modern wildlife management (Musiani et al. 2004). Early on, these influences proved detrimental by extirpating or greatly reducing populations of large carnivores, especially in the contiguous United States of America. Although conservation efforts have been proving effective for population recovery for decades, a high level of controversy is still associated with management of large carnivores, including social, political, economic, and ecological considerations. Such considerations stem from balancing the intrinsic values of large carnivores with real or perceived issues associated with human safety, livestock depredation, property damage, competition with hunters for harvest of ungulate populations, and others (e.g., Fascione et al. 2004).

The grizzly bear (*Ursus arctos horribilis*; a subspecies of brown bear) was listed as threatened in the coterminous United States of America in 1975 under the U.S. Endangered Species Act (U.S. Fish and Wildlife Service 2018*b*). This species has been among the more controversial topics associated with management and conservation of large carnivores. Before European presence, grizzly bear inhabited 16 states, but due primarily to predator control efforts, their populations were limited to four states and less than 2% of their former range by the 1970s (Servheen 1999). Of the six recovery areas identified in the Grizzly Bear Recovery Plan, the U.S. Fish and Wildlife Service delisted the grizzly bear in the Greater Yellowstone Ecosystem in 2007, were challenged in court and lost, later delisted this species in 2017, and despite meeting recovery goals, the delisting was vacated by the U.S. District Court of Montana (U.S. Fish and Wildlife Service 2018*b*). They were in the process of evaluating whether delisting in the Northern Continental Divide is warranted, but current litigation following the most recent delisting decision for Greater Yellowstone Ecosystem is currently delaying that decision. Undoubtedly, any future delisting in any of the recovery areas will be met with litigation.

Litigious approaches to delisting often include challenges to non-compliance and to ambiguities in the Endangered Species Act (Miller 2007), the latter of which may ignore having met pre-established recovery goals and other ecological aspects of the species under consideration. Delisting results in management authority returning to the states and tribes, a situation that causes concern for some, including the potential implementation of harvest management programs, although carefully implemented programs may prove beneficial (Mincher 2002).

In Canada, management of grizzly bear (and brown bear) is under the jurisdiction of each province or territory, which includes the designated western population within Alberta, British Columbia, Manitoba, Saskatchewan, the Northwest Territories, Nunavut, and Yukon (Committee on the Status of Endangered Wildlife in Canada 2019). Where hunted, each jurisdiction has complex harvest regulations, and monitoring harvest levels is critical to ensure sustainability (McLellan and Banci 1999). In British Columbia, grizzly bear inhabit about 80% of the province, which is about 90% of its historic range in that province (Ministry of Water, Land, and Air Protection 2002). Although the estimated 15,000 grizzly bears in British Columbia is less than the historical population, the province does contain about one-quarter of the North American population of this species; of the 56 units where grizzly bear is extant, nine units are classified as threatened (Environmental Reporting BC 2012). Despite the determination that human-caused habitat loss, not hunting, was the greatest threat to grizzly bear populations in British Columbia, the overall management framework, including the lack of a mandated management plan, and responsibilities between two co-management ministries, were deemed inadequate and not transparent (Office of the Auditor General of British Columbia 2017). The end result was the province-wide prohibition of hunting brown bear starting in November 2017.

Interestingly, McLellan and Banci (1999: 49) stated, "*Whether society continues to support the hunting of brown bears due to ethical issues must be addressed from a neutral viewpoint.*" Harvest of brown bear in British Columbia seems to have been prohibited based largely on social and political considerations that essentially constrained any decisions related to ecologically based sustainable harvest. When such substantial prohibitions occur outside of the state or provincial agency process that is typically used, it is often exceedingly difficult to reverse those decisions.

Case Study 2: Bobcat and Trapping: Erosion of State Agency Responsibilities?

Bobcat (*Lynx rufus*) populations in the United States of America have putatively increased in both abundance and distribution in recent decades (Roberts and Crimmins 2010), resulting in increased post-recovery harvest opportunities (and population monitoring via harvest data collection) in many states. These state-led programs to manage harvest of bobcat have been described as very advanced (Nowell and Jackson 1996). Coupled with the implementation of programs in Canada (Agreement on International Humane Trapping Standards; Fur Institute of Canada 2019) and the United States of America (Best Management Practices for Trapping; Association of Fish and Wildlife Agencies 2019*a*, White et al. 2021) starting over two decades ago to test traps based on international humane testing standards, the majority of arguments against trapping bobcat would seem to have been pragmatically settled. However, we use this charismatic species as an example of how regulatory processes have in many instances deviated away from scientific evidence due to strong social and political influences that affected harvest management, particularly via trapping.

First, we provide information at the global and U.S. scale that shaped state regulations based on international policies. The Convention on International Trade in Endangered Species of Wild Fauna and Flora (CITES) listed bobcat under Appendix II (e.g., abundant species with similar appearance to endangered species) during 1977. Consequently, state agencies substantially increased collection of harvest data for bobcat to document population status for an evaluation of a non-detriment finding to help ensure global trade persisted; the U.S Fish and Wildlife Service managed exportation permits and carried out responsibilities on behalf of the Secretary of the Interior. The United States of America challenged the CITES listing, as neither state agencies nor recognized experts on this species had been consulted prior to this listing decision (Mech 1978). However, all efforts to remove bobcat as an Appendix II species failed, with decades of claims that the listing decision was based on political bias rather than biological information, and that misuse of the treaty had occurred (Association of Fish and Wildlife Agencies 2014). Further, animal-rights organizations have consistently challenged the CITES process in the United States of America (with the underlying goal to eliminate harvest) through litigation, including the standard used for export approval, and that the CITES program is a major federal action that requires provisions under the National Environmental Policy Act; these challenges were directly addressed through congressional amendment of the Endangered Species Act, an Environmental Assessment, and other policy approaches (see Anderson and Lovallo 2003, Hansen 2007).

Several examples exist at the state scale of harvest management, but we selected a specific state with a history of strong social and political influences on trapping, including bobcat. During 1998, Proposition 4 (a ballot initiative) passed and resulted in, among other things, the prohibition of essentially all types of traps, except box or cage traps, for trapping furbearers. About 15 years later, the original goal of the Bobcat Protection Act of 2013 (Section 4155, Fish and Game Code) was to prohibit all trapping of bobcat, but it was amended to prohibit trapping of bobcat only in certain areas near Joshua Tree National Park. The act also directed the California Fish and Game Commission to initiate rule making that would prohibit trapping bobcat in areas adjacent to the boundaries of state and national parks, national monuments, and wildlife refuges; trapping bobcat was already prohibited within the boundaries of these areas. However, social and political influences later again overrode science through the decision by the California Fish and Game Commission to prohibit all recreational and commercial trapping of bobcat in that state despite the support from the Governor and members of the California legislature for conducting a bobcat population survey prior to implementing any regulatory changes (California Fish and Game Commission 2015). In 2019, the subsequent Governor signed the Wildlife Protection Act of 2019, which prohibited all recreational trapping and the sale of fur through other forms of harvest in California (California Legislative Information 2019).

Case Study 3: Paddlefish Management in the Ohio River Watershed

Politics at the international level have added complexity to management of harvest of American paddlefish (*Polyodon spathula*) in river systems of the central United States of America. In 1991, the dissolution of the Soviet Union resulted in the decentralization of management of beluga sturgeon (*Huso huso*) in Eastern Europe and Northern Asia. The functional end to enforcement of harvest regulations led sequentially to a population decline in sturgeon, CITES protection for sturgeon, international restrictions on trade of processed roe (sturgeon eggs, or caviar), and a worldwide shortage of caviar from beluga sturgeon (Association of Fish and Wildlife Agencies 2015*a*). Given that roe of American paddlefish from the Mississippi River watershed had competed with sturgeon roe in quality and price since the late 1800s, it was not surprising that the demand from the international market led to legal harvest and illegal overexploitation of paddlefish by commercial operations in the United States of America (Rider et al. 2019).

Paddlefish is a migratory, planktivorous, long-lived species. Even before the recent pressure from the international caviar market, American paddlefish stocks had declined due to blocked migration routes, habitat loss, overharvest, pollution, and modifications to river flow regimes (Carlson and Bonislawsky 1981, Jennings and Zigler 2009). The species is not protected at the federal level, although several states list it as threatened or endangered at the state level. Although 18 states have allowed commercial harvest of paddlefish, by 2018 only eight states had commercial regulations: Alabama, Arkansas, Illinois, Indiana, Kentucky, Mississippi, Missouri, and Tennessee (Rider et al. 2019).

A voluntary, multi-state management group, the Mississippi Interstate Cooperative Resource Association (MICRA) has coordinated sampling and complied a common set of data, which led to an understanding that individual paddlefish may range between the Missouri, Mississippi, and Ohio Basins of the Mississippi River watershed (Pracheil et al. 2012). State regulations for commercial and recreational harvest of paddlefish differ dramatically, and paddlefish have high potential to be exposed to the effects of harvest regulations of more than one state during one year. Throughout the Mississippi River watershed, several combinations of neighboring states have contrasting regulations for harvest of individuals swimming between opposing banks of the same river. For example, Ohio lists paddlefish as a threatened species, whereas on the opposite side of the Ohio River, Kentucky has both recreational and commercial seasons (Pracheil et al. 2012). Although some states, such as Mississippi and Louisiana, coordinate protective regulations on shared rivers (Rider et al. 2019), the water laws and harvest regulations on shared rivers, such as for Indiana and Kentucky, have provided loopholes and other opportunities for poachers to continue operations (Evans 2019).

Harvest management of paddlefish is complicated by the juxtaposition of an active, lucrative commercial market for paddlefish with the interest in the unique opportunity for recreational anglers (Neely et al. 2015). Whereas commercial operations use nets to catch paddlefish, recreational anglers use snagging techniques with large hooks that are cast into areas with concentrations of spawning fish, typically below dams. The danger of jeopardizing the economic livelihoods of commercial operators places firm social and political pressure on state agencies during regulatory decision processes. Describing a series of emotional public meetings during annual harvest decisions in Tennessee in the period of high prices for paddlefish roe, Bettoli et al. (2007: 390) wrote, "*The need to compromise with the fishing industry meant that not all of the measures proposed to protect the fisheries from overfishing were enacted...*"

Despite the social and political influences on harvest management decisions for paddlefish, state-level managers have found ways to put more restrictive regulations in place. In Tennessee, biologists' recommendations were often not approved by members of the Tennessee Wildlife Resource Commission because of political influence. The political pressure was caused because the proposals of biologists and managers were hotly disputed by the fishing community. Over time, managers learned that open public meetings were not conducive to problem solving when the opposition to additional regulations was strong and organized, which is true for any such controversial topic in our field. Officials began to use structured, facilitated meetings with commercial permittees, which resulted in more constructive conversations and incremental agreements to restrict regulations (Bettoli et al. 2007). In this scenario, commissioners were more likely to approve the recommendations. Mississippi and Alabama recently undertook a series of proactive approaches designed to protect the future of recreational angling for paddlefish and allow for some commercial harvest, including training requirements for harvesters with state personnel. Using meetings with similar structure to those held in Tennessee, Alabama and Mississippi have improved relationships with the commercial fishing industry, which has allowed for an increased level of stakeholder support for constraints on commercial fishing areas and seasons, restrictions on harvest, increased licensing fees, and a common harvest-reporting method (Rider et al. 2019). These examples show the benefits of incorporating stakeholder views, despite the potential for conflict, into the decision-making process (Bettoli et al. 2007). Progress toward better harvest regulations was achieved when managers embraced their public-trust responsibility (Rider et al. 2019) and socio-political dynamics with their stakeholders in constructive ways.

The future of paddlefish management would seem to be building toward the need for a *swimways* approach, similar to coordinated management of migratory waterfowl by the federal level (Pracheil et al. 2012). At present, the current status of management in states across the range of the American paddlefish exemplifies the influence of social and political influence on the process of managing harvest of an economically valuable species of conservation concern.

CONCLUSIONS

Harvest management is a process of making decisions related to a unique social-ecological system. We have described how processes and decision-making bodies vary with regard to the political context of the harvested species and the level of state, tribal, or federal authority for the decision. Our discussion and case examples provide evidence of the potential for social and political influences to have critical roles in the decision-making process as well as the eventual status of the consumptive resource, as would be expected in a coupled system (Bieg et al. 2017).

In the new paradigm for harvest management (Conroy 2021 [Chapter 1]), it is critical to distinguish uncertainties induced by ecological and social dynamics so that the decision process can fully address them (Tyre and Michaels 2011, Tyre and Tenhumberg 2021 [Chapter 11]). If we define a good policy decision as being logically consistent with (1) what is known, (2) what management alternatives are possible, and (3) what the decision makers want for themselves and their constituencies (Michaels and Tyre 2012), it is clear that defendable decisions in harvest management will never be fully informed by ecology. Many harvest management situations have low potential for social or political volatility during the harvest management decision, and thus may focus more on ecological information. However, we have also described socio-demographic changes and shifts in values regarding consumptive use that would predict enhanced potential for socio-political influence (e.g., ballot initiatives) on harvest decisions in the future. Managers must be aware of the process and work to develop strategies that bring information from biologists and social scientists to bear on harvest decisions.

ACKNOWLEDGMENTS

The authors thank R. Regan and V. Moore for constructive reviews that improved this chapter. The authors are grateful for B. Pracheil's review of the paddlefish case study, M. Lawing's review of the bobcat case study, and H. Cooley's review of the grizzly bear case study. T. Hiller was supported by Wildlife Ecology Institute for this project. L. Powell was supported by Hatch Act funds through the University of Nebraska Agriculture Research Division, Lincoln, Nebraska.

3 Linking Social and Ecological Components to Spatial and Temporal Components of Harvest

Mark A. Kaemingk, Mark P. Vrtiska, Keith L. Hurley, and Larkin A. Powell

CONTENTS

INTRODUCTION

Populations of fish and wildlife fluctuate through space and time. Managers of recreational harvest have developed increasingly effective ways to assess population fluctuations to attempt to meet ecological objectives related to either increasing, sustaining, or reducing fish and wildlife populations (hereafter referred to as wildlife; Johnson and Martinez 1995; Williams and Johnson 1995; Carpenter 2000; Conroy 2021 [Chapter 1]). It follows then that establishing objectives and evaluating management strategies (e.g., bag and size limits, seasons, special regulations) could be placed entirely in the context of ecological theory (e.g., compensatory and additive mortality, density dependence; Allen et al. 1998, Link 2002).

In the past three decades, harvest managers have identified a key uncertainty for harvest management: the ability to control harvest rate (Williams et al. 1996, Johnson et al. 1997). Although a given set of regulations may have potential to increase, decrease, or maintain the previous harvest rate, the response of hunters or anglers can vary according to many factors.

Managers may attempt to adjust regulations in time (e.g., season dates) or space (e.g., zones, area-specific regulations) to reduce uncertainties because sportsperson populations (i.e., anglers, hunters, trappers) also fluctuate through space and time. The application of specificity of temporal or spatial resolution for harvest regulations varies by species in accordance with meeting ecological objectives (Fig. 3.1). However, the location of zones and timing of season lengths can be controversial among participants (Heberlein 2004), leading to energetic responses at public meetings. Harvest managers rarely have evaluations of the spatial or temporal behavior of sportspersons or the effectiveness of spatial or temporal regulations.

Another gap in our current management paradigm is the lack of explicit objectives to address the social component of these systems (Degnbol et al. 2006, Symes and Phillipson 2009). A majority of higher education coursework and professional certification (American Fisheries Society, The Wildlife Society) is devoted to mastering population biology and ecological principles (Gigliotti and Decker 1992, McMullin et al. 2016). This training has perhaps unintentionally narrowed the focus on wildlife populations, leading to an expectation that well-educated ecologists can effectively manage complex social–ecological systems (Jacobson and McDuff 1998). Wildlife management is a relatively new profession and the development and discussions of including social components into wildlife management is even more recent. Wildlife professionals have not generally been adequately equipped to consider or obtain information regarding social components. Indeed, as Johnson et al. (2015) noted with respect to waterfowl management, "... *Not surprisingly, waterfowl biologists seem more comfortable assessing potential biological impacts than crafting management objectives and regulatory alternatives that reflect social values. The general lack of understanding about what satisfies and motivates waterfowl hunters makes the task all the more difficult.*"

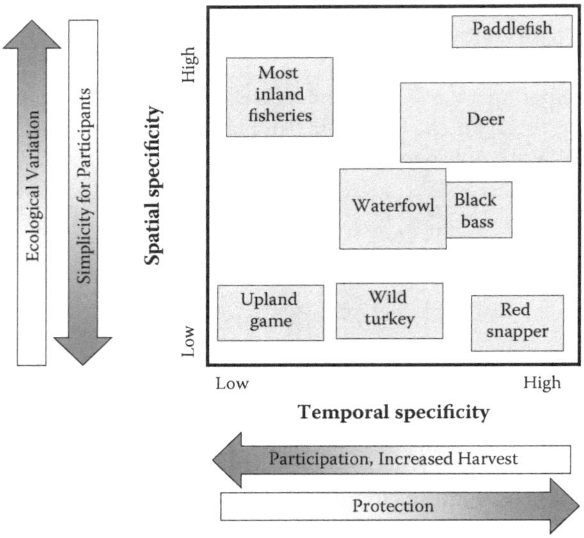

FIGURE 3.1 Specificity of temporal and spatial resolution for harvest management and examples of different species (e.g., paddlefish *Polyodon spathula*) or systems (e.g., inland fisheries) in attempt to meet ecological objectives. Temporal specificity may be increased with shorter seasons, which may serve to decrease harvest levels or participation by sportspersons to protect a population or species. Spatial specificity may be increased with regional or local zones (e.g., lake-specific regulations for inland fisheries, waterfowl harvest zones), which may respond to spatial variation in population levels but complicate regulations for participants.

A recent decline in participation among sportspersons has caught the attention of many state wildlife agencies because of the potential consequences it has for wildlife conservation and management (Enck et al. 2000, Vrtiska et al. 2013). This decline has led managers to ask questions concerning public motivations for resource use. Do new anglers find lake-specific angling regulations too confusing (Page and Radomski 2006)? Does the lack of spatial control of upland game harvests lead to empty fields that discourage new hunters (Gruntorad et al. 2020)? Does the increasingly urban-based group of hunters approach the hunting season differently because of reduced access to private lands (Wszola et al. 2020*b*)? As these questions illustrate, harvest must be viewed as managing a linked social–ecological system (Levin et al. 2013) through space and time. This is especially important as we enter a new era of changing public participation, transparency, perception, attitudes, urbanization, as well as declining and shifting wildlife populations, and resource scarcity (Gruntorad and Chizinski 2021 [Chapter 4], Hiller et al. 2021*b* [Chapter 2]).

The current approach to harvest management has not fully and explicitly recognized and integrated aspects of the social dimension, which has artificially simplified the system and consequently reduced the complexity of management options and subsequent decisions (Enck et al. 2006). For example, managers commonly implement regulations that treat sportspersons as a homogenous group that will respond in a predictable manner (Radomski et al. 2001, Johnson et al. 2015). Yet, managers would also agree that sportspersons are a diverse set of users (e.g., Schroeder et al. 2006). Managers frequently discuss the importance of including sportspersons in the context of management but rarely incorporate social components into the decision-making process. There is a tendency to implement simple sportsperson regulations across spatial and temporal scales with the expectation that these regulations will be successful. The irony here is that this assumption would rarely be made about an ecological predator–prey relationship. Effective recreational harvest management will involve designing spatial and temporal plans that are capable of accounting and managing for non-linearity, emergent properties, self-organization features, and feedback loops between sportspersons and wildlife populations (Hunt et al. 2013, Ward et al. 2016, Arlinghaus et al. 2017*a*). We believe that challenging this ecological harvest paradigm by doing a better job of understanding and formally incorporating the social component into how managers view and manage recreational harvest will benefit both these managed populations and their consumptive users (Radomski et al. 2001, Enck et al. 2006, Hunt et al. 2013).

The goal of this chapter is to highlight novel spatial and temporal opportunities that will improve harvest management by thrusting the social component on an equal platform with the ecological component, resulting in a coupled social–ecological harvest management approach. We focus on how familiar ecological techniques and concepts (e.g., population estimation and models, distribution, behavior) could be applied to understand spatial and temporal dynamics of sportspersons. We believe a new social–ecological paradigm is necessary for harvest management, which arguably needs to be better equipped for a turbulent and uncertain future with respect to spatial and temporal processes. Spatial and temporal aspects of the social component provide a rich source and new frontier of unique opportunities for managers that, when combined with current ecological knowledge, should lead to more successful harvest management.

COMPARISONS BETWEEN AQUATIC AND TERRESTRIAL HARVEST MANAGEMENT

The goals of aquatic and terrestrial harvest management are often the same but the spatial and temporal approach may deviate for aquatic and terrestrial organisms. Much of our current approach to harvest management relies heavily on our knowledge of a particular species and its population dynamics (Barrows et al. 2005). We have long enjoyed and appreciated understanding how and why wildlife populations fluctuate (Darwin 1859). This dedication and focus has certainly expanded our understanding of ecological relationships and enhanced our ability to manage diffuse and dynamic wildlife populations. For example, knowing home ranges and habitat use for ring-necked pheasants (*Phasianus colchicus*) has led to opportunities to improve population numbers by identifying critical habitats during certain times of the year (Whiteside and Guthery 1983). Similarly, improved age and somatic growth estimates have greatly improved our ability to manage sportfish populations, especially tracking these patterns through time (Maceina et al. 2007).

From a spatial perspective, the most obvious reason for adopting different approaches for harvest management of wildlife relates to the distribution and mobility of organisms across different spatial units (e.g., waterbodies, patches). Aquatic organisms are typically confined to more discrete spatial units (e.g., waterbodies), which promotes spatially explicit harvest management (Johnson and Martinez 1995). In contrast, terrestrial organisms can typically move freely both within and across different spatial units (e.g., habitat patches), leading to a more spatially diffuse treatment of harvest management (Nichols et al. 2007). Such spatial approaches to harvest management have certainly considered these ecological differences and have likely guided the current ecological harvest paradigm.

From a temporal perspective, harvest seasons commonly apply to managing terrestrial organisms and are less common for managing aquatic organisms (although there are several exceptions). Hunters, compared to anglers, usually have a smaller window to capitalize on harvest opportunities. Depending on the organism, seasons afford the ability to provide or restrict opportunities for hunters. Harvest seasons may occur when wildlife populations are most abundant or concentrated (e.g., migration events) to maximize harvest opportunity (Clausen et al. 2017), or they may be closed during times when wildlife populations are particularly vulnerable or would be negatively impacted (e.g., during reproductive events; Healy and Powell 1999; Gwinn and Allen 2010). Again, harvest approaches currently focus on temporal aspects of ecological components within these complex social–ecological systems.

Another important and perhaps obvious distinction worth noting is that hunting involves the pursuit and ultimate harvest of an organism, whereas this is not always the case for those participating in fishing. Anglers can decide whether or not to harvest a fish or release it (Cooke and Schramm 2007). Therefore, fisheries managers have the option of developing and sustaining a *catch-and-release* only fishery. Harvest in aquatic systems can be restricted, while still allowing for continued participation. However, such a requirement may limit participation from anglers seeking harvest or cause conflicts between fishing groups (Arlinghaus 2007). Fisheries management has more flexibility because harvest is not a prerequisite for angler participation. In contrast, the catch-and-release option is not available for managers that seek to manage most terrestrial populations. Terrestrial harvest management options are limited in this respect because harvest is an anticipated prerequisite for hunter participation. Balancing spatial and temporal opportunities for sportspersons has therefore likely led to different harvest management approaches and attitudes between terrestrial and aquatic managers. Thus, one must appreciate and consider the spatial and temporal interactions between heterogeneous sportsperson and wildlife populations when designing and implementing harvest management strategies.

SOCIAL COMPONENT OF HARVEST

Managers, from an ecological standpoint, could consider sportspersons as apex predators that assist with controlling prey populations (Stedman et al. 2004). We acknowledge that this viewpoint reduces the complexity of human behavior but attempt to use this illustration to connect ecological principles (that are familiar to most managers) to a more comprehensive social–ecological harvest management approach. In the current ecological harvest paradigm, this viewpoint is helpful because it considers sportspersons as an integral and important part of the system. Though the social component has been acknowledged, it falls short compared to our treatment and understanding of the ecological components of the system. For example, we probably know more about the spatial and temporal dynamics of white-tailed deer (*Odocoileus virginianus*) populations than we do the spatial and temporal dynamics of white-tailed deer hunters (Stedman et al. 2004, Gruntorad and Chizinski 2021 [Chapter 4]). We have learned invaluable information by examining predator–prey dynamics; see the extensive and well-known studies of snowshoe hare (*Lepus americanus*) and Canada lynx (*Lynx canadensis*) populations (e.g., Stenseth et al. 1997), or bluegill (*Lepomis macrochirus*) and largemouth bass (*Micropterus salmoides*) populations (Hoyle and Keast 1987, Stenseth et al. 1997, Trebitz et al. 1997). What could we learn by taking this same ecological approach to understanding harvest by sportspersons? Indeed, this will involve a paradigm shift for harvest management and a greater understanding of human ecology (Berkes et al. 2000).

As an example, someone seeking to harvest chum salmon (*Oncorhynchus keta*) will likely behave differently than someone else seeking to harvest a Rocky Mountain elk (*Cervus elaphus nelson*). Vast differences exist within the sportsperson (*Homo sapiens*) population and one could further argue that several subpopulations exist simply based on the species sought (Fedler and Ditton 1994, Pope et al. 2016). Their success may vary greatly depending on their age, site selection, timing, and resources such as money and time. Furthermore, the same individual may seek to harvest chum salmon and Rocky Mountain elk within a given year. There would be immense utility in understanding human behavior by using a recreational budget, which requires spatial and temporal allocation of time and money to pursue prey (e.g., license costs for chum salmon and Rocky Mountain elk; Clawson and Knetsch 2013).

Perhaps one of the many major shortcomings of viewing sportspersons as apex predators, even if we view them as separate subpopulations, is that not all sportspersons desire to harvest (Fedler and Ditton 1994, Hunt et al. 2002) and this can vary through space and time for an individual (Schroeder et al. 2013). Even sportspersons who seek to harvest vary widely in their specializations, motivations, and preferences (Floyd and Gramann 1997, Miller and Graefe 2000). Several studies have demonstrated this variation among sportsperson populations (e.g., Schroeder et al. 2006), and it is not our purpose to review them here. However, management plans could greatly benefit by outlining specific management objectives that directly link spatial and temporal dynamics of sportspersons to wildlife. For example, the social landscape is changing with more people moving and residing in urban areas than ever before (Alig et al. 2004). Harvest attitudes and behaviors of urban and rural residents differ (Hendee 1969, Stedman and Heberlein 2001, Arlinghaus and Mehner 2004, Arlinghaus et al. 2008), but have managers formally incorporated this information into harvest management and should they in the face of a shifting social landscape?

SPATIAL AND TEMPORAL SOCIAL CONSIDERATIONS OF HARVEST

The lack of social objectives relative to ecological objectives for harvest management suggests that managers have largely treated sportspersons as a homogenous population (Degnbol et al. 2006, Symes and Phillipson 2009, Pascoe et al. 2014). The current ecological harvest paradigm by default then considers sportspersons to have equal temporal and spatial desires, opportunities, and abilities to harvest. For example, certain harvest regulations (e.g., daily and possession limits) may be implemented based on ecological considerations (e.g., population growth, high natural mortality, high emigration), with the expectation that sportspersons will uniformly respond to these harvest regulations. Managers may also assume unintentionally that the attitudes and willingness to harvest do not vary across the landscape (e.g., urban vs. rural) even though several studies have shown otherwise (Arlinghaus et al. 2008, Lyach and Čech 2019, Kaemingk et al. 2020).

Do all sportspersons desire to harvest, and how strongly does harvest relate to their preferences, motivations, and satisfaction? Evidence suggests that harvest may not be a primary motivation or central to satisfaction (Hazel et al. 1990, Gigliotti 2000, Arlinghaus 2006*a*, Hinrichs 2019, Gruntorad et al. 2020, but see Bradshaw et al. 2019). Yet, many harvest regulations are uniformly applied (e.g., statewide) or specifically focused on spatial units that may not be socially effective. Simple regulations often appeal to sportspersons by putatively maximizing opportunities and these simple regulations become even more attractive from a law enforcement perspective. However, a spatial and temporal mismatch between social and ecological components could result in unintended consequences. For example, length limits were unsuccessful for managing largemouth bass populations in the U.S. state of Mississippi because of voluntary catch-and-release behavior by anglers (Miranda et al. 2017). Managers expected that anglers would harvest largemouth bass and ultimately improve the size structure of the population by lessening the effects of density dependence; this was ultimately not the case. Furthermore, anglers across the U.S. state of Nebraska do not exhibit similar harvest patterns (Kaemingk et al. 2020), with urban anglers exhibiting a lower propensity to harvest fish compared to rural anglers (see also Arlinghaus et al. 2008). There is also great variation in the spatial and temporal harvest of ducks across the Central Flyway (Haugen et al. 2015). Successfully implementing the same set of harvest regulations at large spatial and temporal scales could be futile if there is high variation in harvest propensity among the sportsperson population, ultimately leading to unequal harvest across waterbodies and landscape patches. These patterns highlight the utility of taking into account harvest propensity and location of sportspersons relative to wildlife populations on the landscape.

There has also been a general assumption that sportspersons have equal opportunities (via mobility) to capitalize on local or distant resources that optimize their utility. Angler travel distances can vary through time, space, and across species-seeking groups (Camp et al. 2018, Wilson et al. 2020). For example, anglers targeting common snook (*Centropomus undecimalis)* stayed within close proximity of the fishery (< 30 km); however, anglers targeting red snapper (*Lutjanus campechanus)* traveled more than 200 km to reach the fishery (Camp et al. 2018). Waterfowl hunters in the Central Flyway of the United States of America desired hunting experiences that were within one hour of their residence (Slagle and Dietsch 2018*c*) and harvested ducks in fewer than two counties (Haugen et al. 2015). Likewise, urban ring-necked pheasant hunters in Nebraska traveled shorter distances and encountered fewer ring-necked pheasants than rural ring-necked pheasant hunters (Wszola et al. 2020*b*). Some regions or waterbodies have the ability to draw a wider distribution of sportspersons from the landscape (Martin et al. 2015), creating invariant social–ecological catchments (Kaemingk et al. 2021). This suggests that sportspersons indeed select sites that optimize their utility, which considers travel costs (time and money; Beardmore et al. 2011b, 2013) or other factors (e.g., interference from other sportspersons; Slagle and Dietsch 2018*c*) and further underscores the need to examine spatial and temporal relationships between sportsperson and wildlife populations.

A CASE STUDY: SPATIAL AND TEMPORAL HARVEST PROTECTION FOR UPLAND BIRDS?

Harvest dynamics of upland birds in the Midwest and Great Plains of the United States of America may serve to illustrate the potential harvest management consequences of overlooking the social and ecological landscape. Hunters of upland birds on public and private lands in the United States of America are regulated by season lengths and daily bag limits (Dahlgren et al.

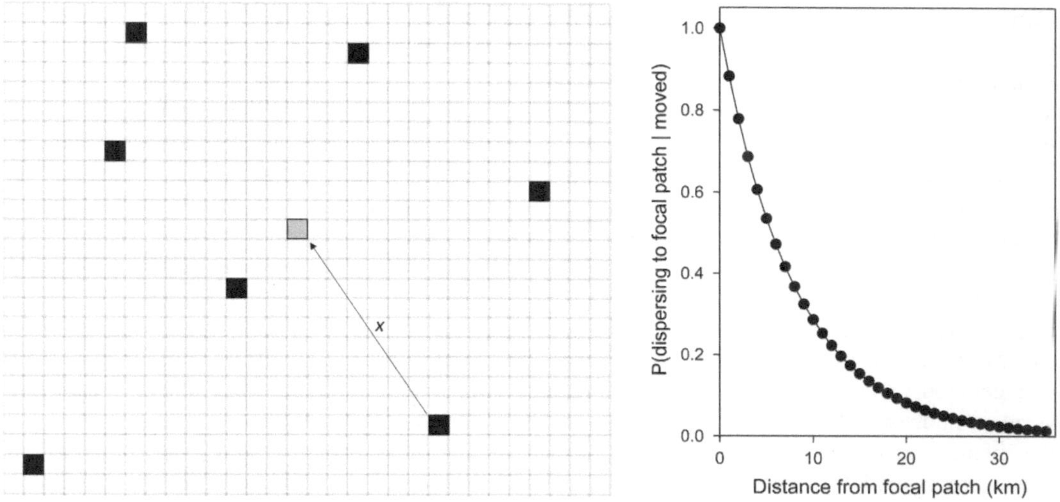

FIGURE 3.2 Features of dispersal simulation model in which (left) a focal patch (light gray) receives dispersing birds from other grassland patches (solid black) in landscape comprised 1-km^2 patches. The probability of dispersing distance x is shown (right).

2021 [Chapter 21]. Unlike those who manage fisheries, waterfowl, or big game, wildlife managers for upland game rarely design spatially explicit regulations or quotas for individual units of land (Fig. 3.1). Therefore, harvest is treated as a geo-political or metapopulation-level phenomenon, although we know harvest effects manifest in dynamics at the local level (Williams et al. 2004, Zimmer-Shaffer et al. 2014). To date, the philosophy for harvest management of upland birds has assumed that any local overharvest would be compensated with dispersal of birds from nearby patches of habitat and does not account for variation in the social landscape or context (e.g., proximity to an urban center).

McCullough (1996) warned of the potential for harvest conducted within a metapopulation as in our pheasant case study (Fig. 3.2) to increase the likelihood of local extinction and decrease the probability of dispersal between patches. Hunting pressure on local patches of land has increased as grasslands become rarer on the landscape. Increased hunting pressure has been further exacerbated with incentive payments by state wildlife agencies to willing private landowners to obtain *open fields* access rights of some grasslands for public use (Wszola et al. 2020*a*). Harvest pressure is expected to be high near concentrations of hunters, such as urban areas (Ling and Milner-Gulland 2008). For example, a 32-ha (80-ac) tract of land enrolled in an open fields program close to major metropolitan areas in eastern Nebraska could have up to 16 hunters per day for a peak period of 60 days during the upland bird harvest season compared to eight hunters per day during a peak of only 15 days in western Nebraska (Wszola et al. 2020*a*).

Coincidentally, available breeding habitat for upland birds has been converted to agriculture (Hiller et al. 2009, Johnston 2014), and fragmentation of grasslands has increased to the point that dispersal of animals could be affected. Landscape-level dynamics are important for upland birds and hunters. The effects of landscape fragmentation are fourfold: (1) loss of demographic components (leading to higher natural mortality) that existed on now non-existent patches of breeding or wintering habitat; (2) loss of sources of immigrants; (3) impediments to immigration through conversion of habitat between habitat patches (Wilcox and Murphy 1985); and (4) concentrated populations of upland birds that facilitate greater hunter–bird encounters and harvest mortality

(Connelly et al. 2003). For example, declining populations of northern bobwhites (*Colinus virginianus*) are found in landscapes that differ significantly from landscapes that support increasing populations, and urbanization characterizes landscapes of extinct populations of bobwhites (Veech 2006). Similarly, ring-necked pheasants are limited at the local scale by landscape-level habitat features such as trees (Jorgensen et al. 2014), and greater prairie chickens (*Tympanuchus cupido*) selected landscapes with prairie habitat (Winder et al. 2015). Hefley et al. (2013) predicted that northern bobwhite in eastern Nebraska had crossed an extinction threshold, which they attributed to habitat loss. Furthermore, population declines of greater sage-grouse (*Centrocercus urophasianus*) were exacerbated in fragmented habitats near metropolitan centers (Connelly et al. 2003).

Synergistic extinction drivers have potential to cause cascading effects for population dynamics, and fragmentation has led to overharvest across the globe (Brook et al. 2008). Certainly, harvest managers must continue to assess assumptions regarding population dynamics as the landscape and social components of the harvest system change. Should upland game biologists manage pheasants like inland fisheries (Fig. 3.1)? That is, are there scenarios in which upland game birds would benefit from spatially explicit harvest regulations? In this case study, we use simulations to assess the need for sociological and ecological dynamics to be integrated into regulations that are designed to prevent local overharvest.

Local Patch Extermination

How quickly can local groups of upland game birds be legally exterminated from a local patch that is hunted by multiple parties? Variation in hunter pressure can be the result of sociological factors. For example, pressure has potential to vary along a distance gradient from major cities, as harvest pressure can be heavy at patches near population centers that are easier for large numbers of users to access (Powell et al. 1998, Wszola et al. 2020*a*). To assess the potential sociological impacts of variation in hunter pressure on local abundance following hunting season for pheasants, we simulated 20, 40, 60, 80, and 100 hunter parties in a local patch that had been seeded with five males and five females. We loosely based our simulations to mimic low (20 hunters) to high (100 hunters) harvest pressure for ring-necked pheasants in a single 1-km^2 patch of grassland in eastern Nebraska. Each hunting party entered the patch independently in our simulations. During the hunt, each male had a probability of being harvested of 3%–5%, which we chose to represent potential scenarios of small and large hunter groups or the absence or presence of a dog with the hunter group. Thus, the probability of survival during a group's hunt, S, was 0.97 or 0.95. A random value for a variable, x, was selected from a uniform distribution between 0 and 1.0 for each male, i, in simulation j. If $x > S$, the male was harvested, and harvest probability was density independent. Likewise, for females, the hunters were given error rates (mistakenly and illegally killing) for female birds of either 1% or 2%, resulting in the probability of survival for females during a hunting party's hunt of 0.99 or 0.98. With visits from only 100 hunter groups, our simulations showed large reductions in males and females (Fig. 3.3). At rates of 16 hunters per day on open fields lands in eastern Nebraska near population centers of Lincoln and Omaha (Wszola et al. 2020*a*), a patch could exceed 100 hunters in only six days. Our simulations suggest high potential for local extinction of both males and females in areas with high hunter pressure. In contrast, in patches with low hunter pressure, perhaps those farthest from urban centers, the potential for local extinction is much lower (Fig. 3.3).

Dispersal Dynamics

At what level of fragmentation does dispersal connectivity become too low to repopulate a depopulated patch? We used components of the previous sociological exercise to simulate ecological dynamics on ring-necked pheasants. We extended the model to the landscape level to

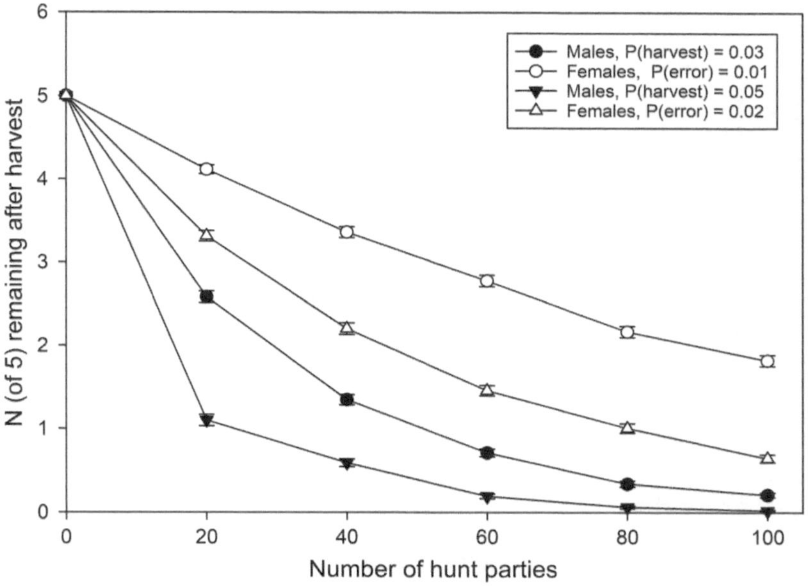

FIGURE 3.3 Mean (95% confidence intervals) number of male (solid black) and female (hollow) upland birds remaining in a habitat patch initially seeded with five birds during harvest simulations after visits by a range of hunter parties. Males are subject to two levels of legal harvest probabilities for each individual during each hunter party visit, and females are subject to removal by illegal harvest because of errors of identification.

simulate dynamics of dispersal to a single patch of breeding habitat from which an upland bird had been exterminated by local overharvest. The depopulated patch was in the center of a square landscape (50 km by 50 km) with 2500 patches, each one km^2 in area. Each 1-km^2 patch represented an area slightly smaller than a half-section of land in eastern Nebraska (130 ha, or 320 acres), and the area roughly represents the space needed to hold five to eight female pheasants. The availability and correlated fragmentation level of grassland in the surrounding landscape varied in each simulation as grass was assigned to 1500 (60% grassland, low levels of fragmentation and short distances between available grassland patches), 1000, 500, 250, 125, 50, 25, 15, or 5 (0.2% grassland, high levels of fragmentation and long distances between available grassland patches) 1-km^2 random patches. To represent variation in local abundance of pheasants throughout the landscape, the surrounding grassland patches varied with regard to the number of female pheasants in the patch, with random assignments from a uniform distribution from one to eight females per patch. During simulations, each female in surrounding patches had a density-dependent probability to disperse and leave the patch; the probability to disperse was 0 if 0–3 females were in the grassland patch, 0.10 if there were four females, 0.25 if there were five or six females and 0.5 if there were seven or eight females. Upon leaving a patch, the available patches to each bird were scored, based on the probability of the dispersing bird arriving at a given grassland patch as a function of the distance (x km). The dispersal function was a negative exponential $P(dispersal \mid distance\ of\ x\ km) = e^{-\frac{x}{8}}$, which resulted in probabilities ranging from P = 0.9 at 1 km to P = ~0 at a distance of 25 km. If a target patch was filled (eight females present), the dispersing bird was allowed to make a different choice of patch. Each level of habitat availability was simulated 1000 times, and the model output was the mean number of birds that dispersed to the central, depopulated patch from the surrounding patches.

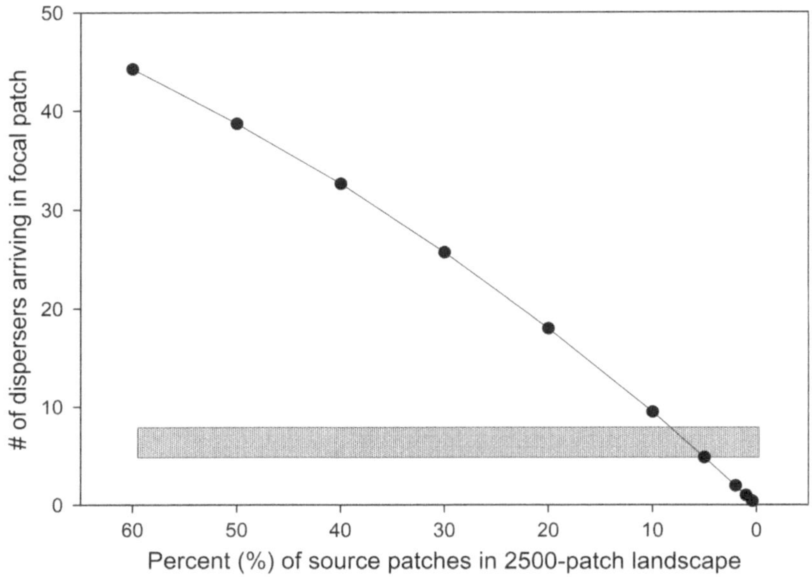

FIGURE 3.4 Mean number (n = 1000 simulations) of dispersing upland birds arriving at a depopulated patch of grassland in a simulated landscape of 2500 patches across a gradient of fragmentation. Source patches are grassland patches randomly located. Gray bar shows the goal range of five to eight dispersers needed to replace the five to eight females originally found in the depopulated patch.

As expected, the number of dispersing females arriving at the focal patch declined as fragmentation increased (Fig. 3.4). When the amount of grass patches on the landscape dropped below 10%, we observed a loss of function as dispersal from surrounding patches could no longer replace the five to eight birds lost in the focal patch.

Currently, Clay County, in south central Nebraska, has approximately 8% of its landscape in grasslands, and grasslands are extremely fragmented (Dappen et al. 2007). Our simulations suggest that the current landscape composition in many agricultural regions may not allow the dispersal of upland birds, such as pheasants, to repopulate habitat patches, should they become depleted.

Our simulation model is a simplistic view of landscape composition and upland bird dispersal, although we consider the predictions to be conservative. For example, we have only considered the effects of harvest on our hypothetical population of upland birds. The potential for local extinction may be enhanced through density-dependent responses to isolated habitat patches by other predators or density-independent factors such as extreme weather events (e.g., ice storm). We also assumed that all other source patches had at least one female remaining, which may not be the case when harvest pressure increases at the patch level at high levels of fragmentation.

The simulations in our case study suggest the potential for sociological and ecological dynamics to have effects on local populations of an upland game bird. Thus, harvest management of species such as upland birds in fragmented habitats must consider the social and ecological dynamics of landscape change to prevent overharvest (Dahlgren et al. 2021 [Chapter 21]). Establishing social goals and objectives, in addition to wildlife population objectives, would benefit the management of these dynamic systems. Our case study dealt with spatial distribution of harvest pressure, but temporal variation in harvest caused by hunter behavior and other sociological dynamics may be similarly critical for meeting harvest objectives in other situations (Araya and Dubovsky 2008, Madsen et al. 2016).

Certainly, there are logistical hurdles to managing pheasants with spatially explicit regulations like fish, including appropriate signage and law enforcement presence at individual patches of upland habitat. More critically, variation in hunter pressure through space and time can be difficult to document. However, with evidence of fragmented habitat and high hunter pressure, state agencies could designate a hunting zone or a portion of the season with lower bag limits and shorten hunting seasons to minimize the potential for local overharvest (Dahlgren et al. 2021 [Chapter 21], Vrtiska 2021 [Chapter 20]). Additionally, lower bag limits or more restrictive regulations also may manage hunter expectations, which would help alleviate hunter dissatisfaction (Brunke and Hunt 2008, Bradshaw et al. 2019). Such zones would be in keeping with spatially explicit harvest regulations used in the management of migratory waterfowl. The role of refuges (areas of habitat in which hunting is not allowed) within the landscape can be complicated (Dale 1951, McCullough 1996), but our simulations suggest a high value for available dispersers to recolonize patches that have been extirpated. Thus, some landowner incentives could be targeted to maintain refuges to provide for breeding bird sources within the landscape. Open fields lands now make up a significant portion of publicly accessible lands in some regions of the country, so agencies should concomitantly monitor upland birds and hunter behavior on open fields lands to inform management. Harvest management of these systems could also consider strategically targeting landowners and properties that provide the most potential for high-quality habitat, increased habitat connectivity, and hunter access. These considerations would leverage the social–ecological landscape to improve harvest management.

NOVEL SPATIOTEMPORAL OPPORTUNITIES FOR SOCIAL–ECOLOGICAL HARVEST MANAGEMENT

Effective harvest management is challenging and arguably will become even more challenging in the future via increased losses of quality habitat, climate change, water scarcity, shifting social norms, and a redistribution of human populations (Manfredo et al. 2003, Tilman et al. 2017). Additionally, a continued loss of sportspersons leads to commensurate loss of revenue to support conservation and management (Vrtiska et al. 2013), political support (Enck et al. 2000), and ability to control wildlife populations with sportspersons. We believe managers can address these challenges by taking advantage of viewing harvest management in the context of a social–ecological system (Ostrom 2009, McGinnis and Ostrom 2014). In particular, we assert that spatial and temporal aspects of the social component are largely unexplored and when combined with ecological information, holds the most promise for achieving success in harvest management. We propose gathering spatiotemporal information and developing tools to understand sportsperson population dynamics in relation to the ecology of these systems (Fig. 3.5). We highlight three specific topics and recognize that this is just part of a more comprehensive list that is required for effective harvest management.

Understanding How Sportsperson Populations Fluctuate through Space and Time

Estimating and tracking variation in sportsperson populations along with wildlife populations will provide critical insight that is often lacking. Traditional techniques used to study wildlife populations have great potential to help us understand sportsperson dynamics (e.g., recruitment, growth, and mortality). The sportsperson population could be estimated at a variety of spatial (residence, patch) and temporal (seasonal, annual) scales. For example, Schorr et al. (2014) developed a temporal symmetry model to estimate hunter population dynamics using license sales to forecast annual changes at the state level for Montana. Pope et al. (2017) conducted a mark-recapture study to estimate the population size of anglers during the open-water fishing

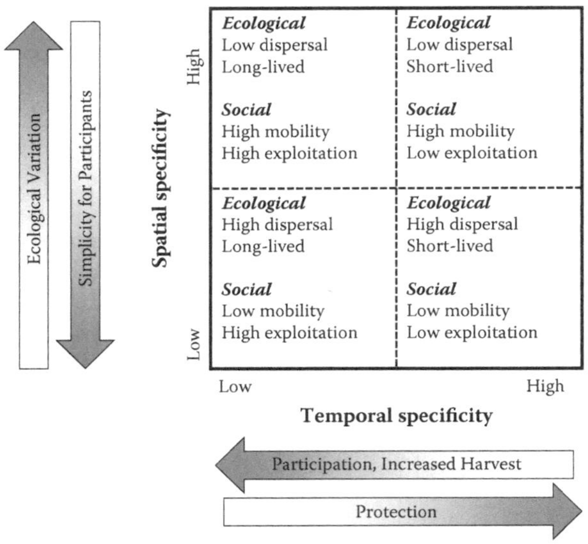

FIGURE 3.5 Harvest management should collectively consider social (mobility, harvest propensity) and ecological (dispersal, lifespan) attributes of the system to determine the appropriate spatial and temporal specificity of regulations. Linking the population dynamics and behavior of sportspersons and wildlife populations will enable creative opportunities to manage harvest as a coupled social–ecological system. Qualifiers (e.g., low, high) of each attribute will likely be context specific and most useful for contrasting social and ecological components of the system to identify the appropriate set of harvest regulations.

season at the waterbody level. Existing data sets, such as sportsperson license databases and creel and hunter surveys, provide opportunities to track spatial and temporal fluctuations in the sportsperson population (Fig. 3.6); additional sampling techniques and information may be required to understand cross-scale spatial and temporal socio-ecological interactions. Recent technology affords the ability to track angler behavior and site choice with mobile apps (Papenfuss et al. 2015). Some agencies (e.g., Mississippi Wildlife, Fisheries, and Parks) require sportspersons to check in, using mobile application, to hunt certain sections of land—possibly facilitating further understanding of the link between a hunter's residence and site choice.

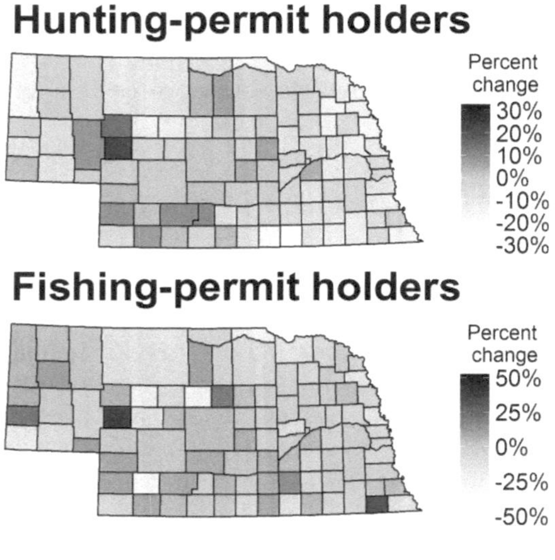

FIGURE 3.6 Percent change (regression of permit holders against year) in hunting- (top) and fishing-permit (bottom) holders in Nebraska from 2010 to 2019 (note different ranges in percent change).

Technologies and information, such as this, will enable managers to understand changes in both sportsperson and wildlife populations.

Predicting Cross-scale (e.g., Patch, Regional, Annual, Decadal) Outcomes Caused by Behavioral Variation between Sportspersons and Wildlife

Once we understand the population dynamics of sportspersons, we then need to capture information about how sportspersons uniquely interact with wildlife populations. Detailing behavioral tendencies such as harvest propensity and ability, site selection, travel distance, and group size will be a critical next step. Sportspersons vary widely in their attitudes, motivations, and preferences, but how does this translate to onsite sportsperson–wildlife interactions across different social–ecological settings? Characterizing and tracking these different sportsperson typologies (or predator types) through space and time will yield creative opportunities for harvest management. For example, there may be a connection between sportsperson typology and residence. After all, birds of a feather flock together, and such a dynamic may hold true for the sportsperson population and their propensity to harvest certain species (Fig. 3.7). Anglers residing in certain Nebraska zip codes had a higher propensity of harvesting fish (Kaemingk et al. 2020). In our upland bird case study, creative solutions to harvest management might involve establishing sportsperson management zones, in addition to wildlife management zones. Sportspersons residing in certain zones or areas of the state could follow unique regulations such as spatially and temporally explicit bag limits or harvest seasons, based on their harvest propensity or other important behavioral characteristics. This management approach may overcome halos of wildlife depletion near urban centers (Wilson et al. 2020). Mapping social–ecological catchments or the spatial draw of sportspersons on the landscape to a certain resource could provide further insight to sportsperson–wildlife interactions (Martin et al. 2015, Kaemingk et al. 2021). Complex harvest regulations run the risk of becoming socially unacceptable; however, this may be required for ensuring long-term sustainability and retention of key services exchanged between sportsperson populations and wildlife populations. A resilience management approach could assist with outlining the consequences of these complex harvest regulations (Anthony et al. 2015, Sterk et al. 2017). New or innovative regulations that promote

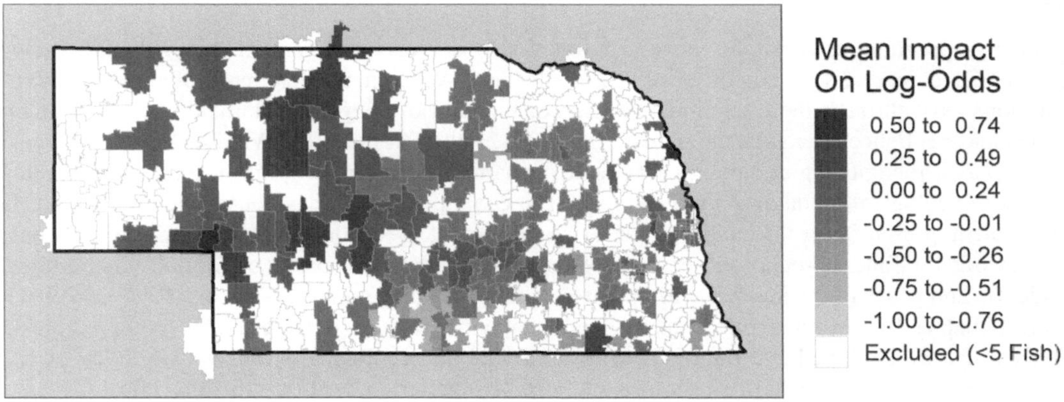

FIGURE 3.7 The impact of an angler's residence (i.e., zip code) on the angler harvest-release decision for walleye (*Sander vitreus*) in Nebraska; positive log odds indicate a higher likelihood of harvest, whereas negative log odds indicate a lower likelihood of harvest.

sportsperson participation also need consideration, with the appropriate level of evaluation of their effectiveness.

Managing for Resilience

Managers need to explicitly state or determine social and ecological objectives for harvest management, which will require moving from a monothematic ecological approach to a complex social–ecological systems approach (Carlson et al. 2020). Establishing effective harvest regulations must recognize and account for nonlinearity, emergent properties, self-organization, and feedback loops between sportspersons and wildlife populations. For example, local-level processes may scale up to form large-scale patterns (Kaemingk et al. 2018) and by applying statewide (or regional) harvest regulations it may have unintended local-scale consequences (see case study above). The success of harvest regulations will depend on dynamic feedbacks between sportsperson–wildlife interactions that operate at multiple spatial and temporal scales. Sustaining these complex relationships will necessitate developing strategies that view resilience as a desired system characteristic of harvest management. Resilience is a characteristic of complex social–ecological systems that allows the maintenance of key services (e.g., food, recreation, sustainable wildlife populations) under a range of conditions (Folke et al. 2004, Walker et al. 2004). Therefore, implementing resilience thinking for management could have tremendous value for harvest management (Camp et al. 2020). Resilience management could also assist with outlining successful strategies for addressing the burgeoning R3 efforts (sportsperson recruitment, retention, and reactivation; Byrne and Dunfee 2018) and the potential consequences of a dynamic sportsperson population in relation to dynamic wildlife populations. Establishing sustainable harvest regulations that retain critical social and ecological properties will depend on a thorough understanding of cross-scale and dynamic interactions between a heterogeneous sportsperson population and wildlife population (Yeiser et al. 2018). The upland bird case study illustrated how ignoring the sociological dynamics and number of hunters that visit a patch could lead to a decline and loss in wildlife populations, a salient property of this social–ecological system.

MOVING FORWARD

Exciting opportunities await for novel approaches to harvest management that will ultimately confront and meet future socio-ecological challenges. We contend that most of these creative solutions exist through the integration and implementation of coupled social and ecological strategies and resilience management (Folke et al. 2004, Fulton et al. 2011, Ward et al. 2016, Camp et al. 2020), recognizing complex, cross-scale, dynamic sportsperson–wildlife interactions (analogous to ecological predator–prey dynamics). The ecological toolbox, when applied to the social component, presents a new frontier for harvest management. Decision analysis affords an excellent framework for linking social and ecological components that is capable of formally addressing tradeoffs and predicting outcomes within these complex social–ecological systems (Robinson et al. 2021 [Chapter 9]). Moving into a new social–ecological harvest paradigm will require professionals that are equipped and trained in both the social and ecological sciences (McMullin et al. 2016). These new cohorts of professionals will also be required to develop skills that allow them to communicate and work effectively with other professionals from a diverse set of disciplines (Smith et al. 2019). Universities, professional societies, and employers provide great platforms to prepare a workforce capable of revolutionizing our approach to harvest management. Embracing this new frontier to harvest management will have noticeable and substantial impacts on sportsperson and wildlife populations.

ACKNOWLEDGMENTS

The authors thank G. Balkom and D. Clapp for providing excellent reviews of this chapter that ultimately resulted in a much better contribution to this book. C. Bender and C. Chizinski provided data and figures to illustrate main points of the chapter.

4 Hunter and Angler Behavior in Harvest Management

Matthew P. Gruntorad and Christopher J. Chizinski

CONTENTS

HUNTER AND ANGLER BEHAVIOR AND HARVEST MANAGEMENT

The connection between human behavior and harvest management is multifaceted and dynamic (Graham et al. 2021 [Chapter 6], Hiller et al. 2021*b* [Chapter 2], Kaemingk et al. 2021 [Chapter 3], Runge 2021 [Chapter 7]). The relationship between harvest management and hunter social dynamics can be illustrated by grouse (*Tympanuchus* spp.) hunting in Nebraska. The number of grouse harvested in the state has steadily declined since a peak during 1980s, despite there being relatively abundant grouse populations (Fig. 4.1; J. Lusk, Nebraska Game and Parks Commission, personal communications). A key question to managers is "*Why, despite being a relatively abundant game species, has harvest been low for so many years?*" Although there are likely many socio-ecological reasons influencing this decline, an important factor has been a change in hunter preferences from grouse to alternative game. The abundance of deer (*Odocoileus* spp.) and wild turkey (*Meleagris gallopavo*) has risen statewide since the mid-1970s, as has the popularity of pursuing deer and turkey in Nebraska. In 2016, deer and wild turkey hunting were the two most preferred games in Nebraska among small-, upland-, and big-game hunters with grouse falling to near the bottom (Grams 2018). Thus, the decline in grouse harvested in Nebraska over the past 30 years is less tied to the abundance grouse itself, but to other factors including an overall shift in the preferences and attitudes of Nebraska hunters.

The decline in Nebraska grouse hunters is not a unique situation; similar trends are occurring among hunters and anglers throughout the United States of America (Winkler and Warnke 2013, U.S. Fish and Wildlife Service and U.S. Census Bureau 2016). Numerous factors lay behind the decrease in hunting and fishing participation, including changes among and within individuals (e.g., self-identity), familial and mentor relationships (e.g., decrease in hunting socialization), local landscapes (e.g., increase in social distance to hunting land opportunities), and broad-scale shifts in demographics and society (e.g., urbanization, population composition of racial and ethnic minorities; Larson et al. 2013 and references therein). The decades-long decline in licensed hunters and anglers in the United States of America has resulted in ecological (Enck et al. 2000, Heffelfinger et al. 2013, Winkler and Warnke 2013), economic (Mehmood et al. 2003, Vrtiska et al. 2013,

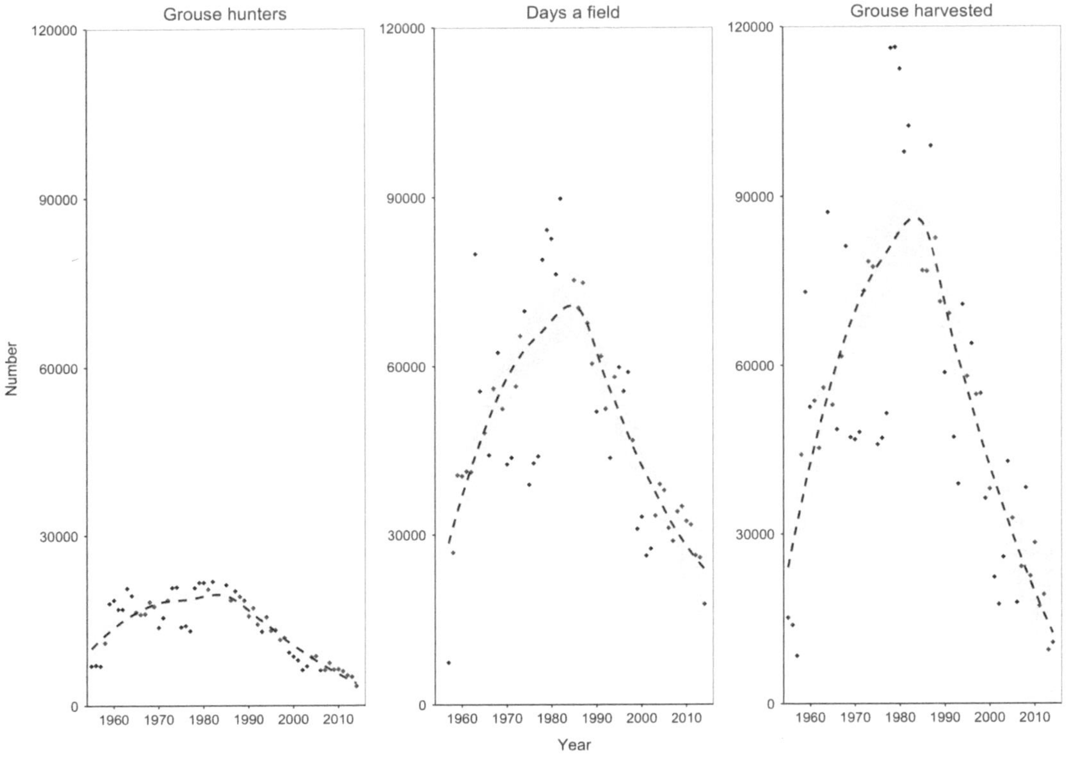

FIGURE 4.1 The number of grouse hunters, days afield pursuing grouse, and the number of grouse harvested from 1955 to 2015 in Nebraska (data courtesy of J. Lusk, Nebraska Game and Parks Commission). Points represent yearly estimates, and the line and error ribbon are curve-linear trends and 95% confidence intervals obtained from a loess regression of the estimates versus years.

Arnett and Southwick 2015), and social consequences (Peterson et al. 2010, Larson et al. 2014, Serenari and Peterson 2018). To help offset the consequences associated with the decline in hunters and anglers, state wildlife and fisheries agencies and nongovernmental organizations have expended considerable time, money, and effort to increase the hunter and angler population through recruitment, retention, and reactivation (often referred to as *R3*) programs (Council to Advance Hunting and the Shooting Sports 2016). To effectively manage hunter and angler populations, a foundational understanding of who participates and why people participate in hunting and fishing is required to increase the relevancy of wildlife-based recreation and its connection to wildlife management to more diverse stakeholders.

Effective and proactive harvest management is faced with critical challenges that stem from issues related to *human dimensions*. First, as described above, the economic and personnel resources devoted to conservation and harvest management are expected to continue to wane into the future, putting greater pressure on fish and wildlife agencies to do more with fewer resources (Mehmood et al. 2003, Winkler and Warnke 2013). Few institutions that rely on customers, whether private or public, can survive without finding new customers, and the process of finding new customers begins by learning about current customers (Miller 2015). Second, traditional harvest management has largely focused on the ecological component of the social ecological systems, while deemphasizing factors affecting humans, the crucial "predator" in the system (Kaemingk et al. 2021 [Chapter 3]). However, it is well documented that people who participate in hunting and fishing are dynamic (i.e., patterns changing over space and time; Kaemingk et al.

2020) and not homogenous (Arlinghaus et al. 2008, Beardmore et al. 2015). Such characteristics suggest that there are differences in attitudes, preferences, motivations (i.e., factors that influence the participation in an activity), and constraints (i.e., a circumstances or obstacles that inhibit or prohibit participation) that vary within and among wildlife-based recreators that ultimately influencing what, where, and how much game is harvested. More effective harvest management must equally balance the management of both the ecological and social component of these social–ecological systems. Further, hunter and angler behavior and the associated feedbacks have the potential to confound predicted effects of regulations on harvested populations (Conroy 2021 [Chapter 1]), necessitating an even greater understanding of the role of human behavior on harvest management.

Human dimensions of fish and wildlife refers to the multidisciplinary approach to scientifically describe, predict, understand, and affect human thought and behavior toward fisheries and wildlife resources (Manfredo et al. 1996*b*). The field of human dimensions has developed considerably since the 1960s, with the scope of study broadening in recent decades (Brown 2008). As such, human dimensions should be fully incorporated as an essential component in the study of harvest management. Our goals for this chapter are to (1) draw connections between human dimensions and harvest management; (2) provide a background of important concepts that help explain the connections between the social and ecological components; and (3) provide examples of how human dimensions can be included in harvest management decisions.

HUNTER AND ANGLER MOTIVATIONS

Motivations are the multitude of goals that drive interest in recreational activities prior to participation (Decker et al. 1980, Watkins et al. 2018). Understanding motivations is key to determine why people engage in leisure behavior (recreational activities like hunting and fishing) in the manner they do (Driver and Knopf 1977). Conceptually, recreation can be thought of as a psychophysiological experience that produces self-rewards during non-obligated free time (Manfredo et al. 1996*a*). For example, fishing may provide a participant temporary escape from day-to-day stress and responsibilities, and therefore provides an important reason to participate in the activity (Driver and Knopf 1976, Wellman 1979, Manfredo 1984).

Many motivations for hunting and fishing extend beyond catching or harvesting game and include the social, psychological, emotional, and physical benefits from participating in the activity (Hrubes et al. 2001). Understanding motivations allow state and federal wildlife and fisheries agencies (hereafter agencies) to minimize conflict between user groups and assess the demand for outdoor recreation (Vaske 2008). Knowledge of motivations can also aid agencies in predicting levels of support for management decisions and the development of specific programs (Schroeder et al. 2006, Ward et al. 2008, Watkins et al. 2018). Further, by recognizing the diverse reasons why hunters and anglers participate in wildlife-based recreation, agencies can tailor opportunities to distribute the varying needs and wants of these groups across a landscape with finite opportunities (Watkins et al. 2018) and more effectively manage where, what, and how much game is harvested. For example, agencies can use motivations to establish new hunting and fishing access to meet the wants of the hunters and anglers or adjust current recruitment, retention, and reactivation efforts to promote participation in additional hunting and fishing activities by appealing to the reasons that individuals participate in an activity. However, few agencies currently monitor and assess motivations of hunters and anglers (or other relevant stakeholder groups) to proactively manage opportunities in the landscape.

Management decisions rest on a critical understanding of motivations of specific leisure activities or groups (Ebeling-Schuld and Darimont 2017, Schroeder et al. 2018, 2019*a*). We find several themes that describe why individuals participate in different hunting and fishing activities: spending time with companions, being outdoors, and tradition are among the most frequently cited (Hayslette et al. 2001, Schroeder et al. 2006, Beardmore et al. 2011*b*, Gigliotti and Metcalf 2016).

Motivations vary little among activity types such as big-game hunting, small- and upland-game hunting, waterfowl hunting, or angling. Similarities of motivation among state of residence in the United States of America suggest a level of universality of certain motivation factors (Hinrichs et al. 2021). However, harvest management would benefit from greater exploration of the sources of heterogeneity among hunters and anglers through deeper research on a greater diversity of stakeholders.

Most assessments of the reasons why people hunt and fish have largely concentrated on traditional hunting and fishing stakeholders (i.e., older, rural, white males; U.S. Fish and Wildlife Service and U.S. Census Bureau 2018). Although this demographic still composes the majority of the hunter and angler population in the United States of America, this segment of the population is shrinking (Larson et al. 2013, Winkler and Warnke 2013, U.S. Fish and Wildlife Service and U.S. Census Bureau 2018). Proactive fish and wildlife management (Graham et al. 2021 [Chapter 6]) must broaden the tent of who is considered a stakeholder in conservation and focus attention on building greater relevancy and inclusion in conservation among diverse stakeholders. More than 30 years ago, Snepenger and Ditton (1985) observed a noticeable rise in participation rates of woman in recreational fishing and suggested that there was a strong need to assess women's engagement in recreational fishing activities. Despite this call, there continues to be a relatively weak understanding of why women engage in fishing and if their motivation differs from males (Gaynor et al. 2016). The participation rate of women in hunting is also increasing (Keogh George 2016). Effective engagement strategies used to appeal to women would be expected to vary from strategies that are effective with men (Schroeder et al. 2006, Metcalf et al. 2015, Rodriguez et al. 2016). Similarly, the role, inclusion, and recognition of black and Indigenous people, as well as people of color in general, have largely been ignored in wildlife conservation (Peterson and Nelson 2017), while this is an ever-growing segment of the U.S. population (Ortman and Guarneri 2009, Colby and Ortman 2015). A new paradigm of harvest management must focus on better understanding the social factors influencing the support for wildlife-based recreation that also reflects the nation's shifting demographic structure (Mehmood et al. 2003, Larson et al. 2014, Schorr et al. 2014).

HUNTER AND ANGLER SEGMENTATION AND TYPOLOGIES

Hunters and anglers are broadly heterogeneous in their value orientations, attitudes, preferences, and behaviors, and thus, there is utility for fisheries and wildlife managers to identify and partition individuals into likewise groups. The partitioning process known as *segmentation* is used to develop *typologies* (description of a group that has characteristics or traits in common) of customers, and the strategy is common in fields of communication and marketing. Segments can be identified through geographic, demographic, psychographic, and behavioral characteristics (Miller 2015). Often segments are identified by minimizing within-group variance and maximizing across-groups variance. Once segments are identified, opportunities, marketing messages, advertising, and promotions can be tailored to the preferences of each segment.

Harvest managers typically establish objectives for population levels of game species, but also attempt to maximize recreational opportunities for anglers and hunters (Driver et al. 1984, Runge 2021 [Chapter 7], Vrtiska 2021 [Chapter 20]). Inherent in this process is understanding how to balance limited agency time and money to produce the greatest benefit (Cole and Ward 1994). Segmentation allows agencies to evaluate alternative opportunity scenarios with information that defines how different groups of hunters or anglers view the resources and the recreational opportunities (Driver 1985, Metcalf et al. 2015). When management agencies provide a variety of choices to meet varying segments' preferences, they can elevate satisfaction for each group, rather than providing a single choice and greater discontent among those users whose needs are not being met. For example, in an examination of waterfowl hunters in Minnesota, Schroeder et al. (2006) suggested that managers could provide more varied waterfowl-hunting experiences through individualized management of public hunting areas and new licensing options. Specifically, managers were encouraged to enhance the experience of social waterfowl-hunting enthusiasts and longtime waterfowl-hunting participants through management that facilitates social interaction (e.g., areas that provide a high level of access, group camp sites), and individualist waterfowl-hunting enthusiasts' experience could be enhanced through opportunities for solitary waterfowl hunting (e.g., areas that limit hunter numbers). Thus, segmentation allows managers to better understand the attitudes and preferences of the hunter and angler population, which should provide managers with enhanced ability to predict how groups will be affected by management decisions (Fisher 1997).

The ability to predict how hunters or anglers may react to changes in regulations is an important tool in harvest management. In particular, the management of overly abundant (e.g., peri-urban ungulates or geese; Ward et al. 2008, Paukert et al. 2021 [Chapter 18], respectively) or nuisance species (e.g., invasive white perch *Morone americana*; Chizinski et al. 2010*a*) requires the harvest of a large number of game to keep population abundances at or below ecological or social carrying capacity. Assessments and segmentation of hunters and anglers can provide agencies with the opportunities to appeal to different groups of the population and meet harvest objectives. For example, policies that encourage harvest such as reduced license prices, liberal bag limits, and extended seasons could be designed to incentivize hunters to shoot antlerless deer (females and fawns) voluntarily (Andersen et al. 2014, Gruntorad and Chizinski 2020). Alternatively, regulations could be imposed that require hunters to harvest an antlerless deer before harvesting a buck (Brown et al. 2000), which have been successful in reducing the deer population, but were generally disliked by hunters (Van Deelen et al. 2010). Understanding human behavior is more critical than ever to tailor opportunities and incentives in order to set policy with a clear view of the hunters' and anglers' interests (Fisher 1997).

HUNTER AND ANGLER EXPECTATIONS AND SATISFACTION

Prior to the 1970s, agencies were focused primarily on habitat improvement and programs designed to maintain or increase game and fish populations (Conroy 2021 [Chapter 1]). Hunting and fishing were widely viewed as an activity for securing food, and satisfaction was estimated based on the number of game bagged. As more hunters and anglers participated, success rates decreased, and it became apparent to agencies that more explicit management would be required to produce the variety and quality of experiences desired (Hendee 1974). The framework for assessing satisfaction shifted from measuring the amount of game harvested to the number of hunting or angling days afield (Crissey 1971). However, approaches to maximize the days afield inevitably led to crowded hunting conditions that were both unsafe and unpleasant, particularly on opening day of hunting seasons. During the 1970s, agencies started to focus not only on harvest-related factors influencing satisfaction, but also incorporated multiple measures of satisfaction that tied to motivations for wildlife-based recreation (Hendee 1974, Hautaluoma and Brown 1978, Driver 1985).

Hunter and angler satisfaction research is grounded in theory originating in the business and marketing fields. Hunters and anglers are paying customers to agencies (Graham et al. 2021 [Chapter 6], Melstrom 2021 [Chapter 5]). Similar to customers of a private industry, future purchases of licenses are connected to how satisfied the customer is with the experience. Conceptually, *satisfaction* is thought to comprise an *emotional state* (with elements of joy and content) as well as a *cognitive component* (evaluation of quality over time; Oh and Parks 1996, Lee et al. 2004, Ladhari 2009). Theoretical frameworks have been developed to connect emotion and cognitions with expectations and satisfaction (Oh and Parks 1996, Burns and Graefe 2006). The *Expectancy Valence Theory* (Lawler 1973) defines satisfaction as the agreement between the expectations and outcomes of a recreational experience (Buchanan 1983, Hammitt et al. 1990, Gigliotti 2000, Miller and Graefe 2001, Brunke and Hunt 2007). This theory has been used as the theoretical framework to understand multiple recreational experiences (Manfredo et al. 1983,

Holland and Ditton 1992, Manfredo and Larson 1993, Hunt et al. 2005). For example, Holland and Ditton (1992) explored the expectations and satisfaction that included a sense of freedom, excitement, catching a fish, relaxation, enjoying the natural setting, and thinking about past fishing experiences that were important to anglers in Texas. They observed that catching fish was only important to satisfaction in a minority of respondents, highlighting the importance of non-harvest factors on the satisfaction of the fishing experience.

Examining the multiple sources of satisfaction that individuals gain from hunting and fishing indicates that harvest is an important component to a satisfying experience (Gigliotti 2000, Bhandari et al. 2006, Siemer and Decker 2015, Ebeling-Schuld and Darimont 2017, Pang 2017), although not always the most important influence on satisfaction (Hammitt et al. 1990, Hazel et al. 1990, Woods et al. 1996, Everett 2014). Rather, spending time outdoors (Mehmood et al. 2003, Reis 2009), helping control wildlife populations, observing game, experiencing camaraderie with family and friends, or improving outdoor skills can have more influence on overall satisfaction than the number of animals harvested (Decker and Connelly 1990, Duda et al. 1996, Bhandari et al. 2006, Siemer et al. 2012). The satisfaction of any outcome from a recreational experience is influenced by the perceptions and expectations of the participant. For example, residency influenced overall satisfaction of the Nebraska white-tailed deer (*Odocoileus virginianus)* season. Non-resident hunters in Nebraska were more satisfied with their season than Nebraska residents (Fig. 4.2). These results suggest that non-residents may have had different motivations and expectations of the experience than residents; for example, non-residents may have been more likely to be visiting family and friends during their hunting trip. It is also possible that they had opportunities to hunt different types of land depending on the social context of the visit. Access policies of private landowners can have an important effect on hunting-related behavior and expectations of the experience (Floyd and Gramann 1997, Diefenbach et al. 2005). Therefore, it is important to monitor and understand what drives satisfaction with the recreational experience, especially the role that private landowners may have in that experience (Gruntorad and Chizinski 2020).

Any company, public or private, should measure multiple sources of satisfaction and estimate their influence on overall satisfaction of their customers. *Importance-grid analysis* (IGA) and *penalty-reward-contrast analysis* (PRCA) have been used in marketing research to identify the factors influencing customer satisfaction (Matzler and Sauerwein 2002, Albayrak and Caber 2013).

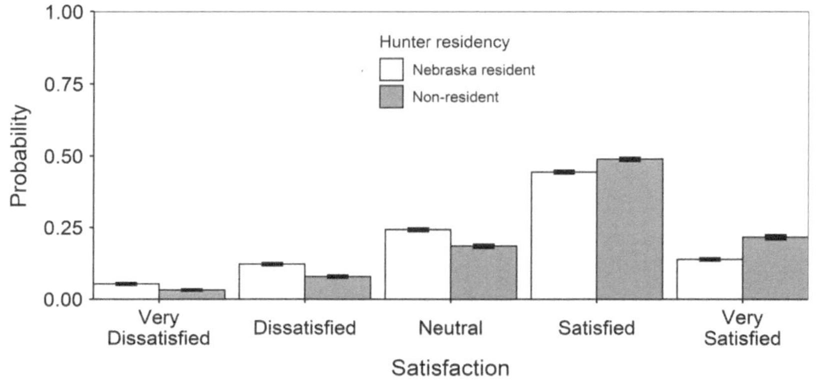

FIGURE 4.2 The probability of Nebraska residents and non-residents being satisfied with the white-tailed deer hunting experience during the 2015 Nebraska firearm season. Error bars represent standard errors. Data obtained from Jeffery J. Lusk and the Nebraska Game and Parks Commission's 2015 Nebraska deer hunter satisfaction survey.

Importance-grid analysis clarifies discrepancies between explicit and implicit importance ratings for attributes associated with an activity such as hunting or fishing. *Explicit importance* serves as a metric to explain the rational assessment of how hunters or anglers state how an attribute (e.g., seeing game) to affect satisfaction. In contrast, *implicit importance* reveals the effect of the attribute on overall satisfaction based on analyses of experiences had while hunting or fishing (Smith and Deppa 2009). For example, the majority of duck hunters state (i.e., explicit importance) that seeing ducks is an important factor influencing the satisfaction of their duck-hunting experience (Gruntorad et al. 2020). These same duck hunters also state that harvesting ducks is only of medium importance, but statistical analysis reveals (i.e., implicit importance) that harvesting ducks is actually one of the most important factors influencing overall satisfaction.

Penalty-reward-contrast analysis incorporates the *three-factor theory*, which is based on the concept that basic, performance, and excitement factors are important to customer satisfaction (Matzler and Sauerwein 2002, Deng et al. 2008, Schofield and Reeves 2015). *Basic factors* include those attributes that do not lead to satisfaction if fulfilled, but do increase dissatisfaction when not met. In the context of hunting activities, basic factors can be viewed as the minimum requirements necessary to justify continued participation. *Performance factors* will induce dissatisfaction if expectations are not met, but will lead to satisfaction if expectations are fulfilled or exceeded. The amount of expectation fulfillment among all performance factors affects overall satisfaction and subsequently, future participation. *Excitement factors* will increase satisfaction with a positive experience but will not lead to dissatisfaction when not experienced. Hunters are not likely to cease hunting activities if expectations for excitement factors, such as harvesting a trophy buck for example, go unfulfilled but they have a strong effect on overall satisfaction of the experience.

Mission statements or strategic goals for federal, state, and tribal fish and wildlife management agencies in North America often include a variation on "*stewardship of outdoor recreation resources for the people*." Therefore, managers must seek to understand satisfaction of their stakeholders with regard to potential management decisions (e.g., providing more public hunting access or lengthening the hunting or fishing season). However, the more heterogenous the stakeholders, the range of game species, or diversity in habitat, the more challenging it becomes to achieve hunter and angler satisfaction (Fuller et al. 2021 [Chapter 8]). Despite the challenges, the public has expectations of being heard and understood as a part of the public trust process (Hiller et al. 2021*b* [Chapter 2]).

Agencies may change components of hunting or fishing systems to improve satisfaction, although cited examples of resulting action are rare. When hunters and anglers are dissatisfied with regulations, license costs, or access to public areas, agencies can change regulations, adjust license costs, open more public land, or modify management regimes to accommodate hunter preferences. However, the most dissatisfying aspects of an experience are sometimes aspects that agencies have no control, such as weather conditions, migratory patterns (although these can be predicted to a degree), and failure to harvest (Gruntorad et al. 2020). Although customer satisfaction is rarely affected by a single aspect of the recreational experience (Lim et al. 2018, Rezaei et al. 2018), there are instances when participation may cease due to a single factor, including license-fee increases, requirement of non-toxic ammunition, or change in season length (Barro and Manfredo 1996, Wright et al. 2001, Mehmood et al. 2003). Further, evidence suggests that there are connections between satisfaction and customer loyalty (Anderson and Sullivan 1993, Boulding et al. 1993, Yi 2012) and willingness to pay more for opportunities (Lee et al. 2004, Ladhari 2009), which has a direct effect on an agency's ability to conserve and manage fisheries and wildlife resources. Therefore, it is critical that agencies understand the effects of fulfilling or failing to fulfill hunter expectations.

HUNTER AND ANGLER PARTICIPATION

Constraints are factors that limit or prohibit participation and enjoyment in leisure activities (Jackson 1991). Identifying these factors is an important step in developing management approaches that can be implemented to help individuals negotiate constraints and ultimately increase participation in wildlife-based recreation. Crawford et al. (1991) outlined a model of intrapersonal, interpersonal, and structural constraints. *Intrapersonal constraints* depict an individual's psychological state and guide leisure preferences and includes factors like family attitudes and perception of skills. *Interpersonal constraints* depict factors that are focused on other individuals and include familial obligations or the lack of a social hunting circle. Lastly, *structural constraints* depict resource factors and encompass factors such as finances, climate, and access opportunities. Constraints to hunting activities have been identified to include factors such as overcrowding, lack of accessible land, equipment and travel costs, work and family obligations, and skills (Enck et al. 1993, Schroeder et al. 2006, Metcalf et al. 2015, Gruntorad and Chizinski 2020).

Participation in leisure activities is not due to the absence of constraints, but more the ability of the participant to negotiate the constraints (Hubbard and Mannell 2001). Negotiation of constraints includes both cognitive and behavioral approaches that require an active investment of time and effort by potential participants (Jackson et al. 1993, Jackson 2000, Wright et al. 2001). Though constraints must be negotiated by the individual, agencies and nongovernmental organizations cannot solely depend on individuals to negotiate these constraints on their own. Agencies and nongovernmental organizations can facilitate the negotiation process through targeted programming and management efforts (Hinrichs 2019). Examples of these programs or management efforts may include providing sponsored mentor programs and family events (Responsive Management and National Shooting Sports Foundation 2017) or opening up more private land for public access through *open fields* programs (Wszola et al. 2020*a*). Understanding constraints to hunting and fishing can provide valuable insight for the creation and revisions of regulations and programs to encourage participation.

Socialization and social-support structure (or the lack thereof) are important factors influencing an individual's ability to navigate constraints (Larson et al. 2014) and participate in hunting and fishing activities (Serenari and Peterson 2018). Generational or cohort demographic effects (e.g., rigidity and flexibility in behavior, shared cohort life events) influence the likelihood of participation in hunting and fishing activities (Winkler and Warnke 2013, Schorr et al. 2014). For example, male baby boomers have been one of the largest demographic groups engaged in hunting and fishing activities (Duda et al. 1996, Dizard 1999, Presser and Taylor 2011), which has been attributed to their coming of age during the post-World War II period when people had relatively more free time and wildlife populations had begun to rebound (Winkler and Warnke 2013). Society has also historically viewed hunting and fishing as a masculine activity (McGuigan 2017), although women are participating in the activity at an increasing rate (Keogh George 2016, Rodriguez et al. 2016). Male and female hunters must negotiate many of the same constraints (Metcalf et al. 2015, Hinrichs et al. 2020), but early socialization to hunting is often different between males and females. These patterns, of course, encourage greater participation by males (particularly in rural environments; Stedman and Heberlein 2001). Participation rates for hunting and fishing are greatest in rural communities, with participation rates falling rapidly in urban populations (Vaske and Manfredo 2012). This greater participation by rural residents is linked to a heighted socio-cultural value placed on hunting and fishing in rural areas, stronger social support in those communities (Stedman and Decker 1996, Graham 2019), and more access opportunities for outdoor recreation (Hendee 1969, Marks 1991). The linkage between social support structures, urbanization, and cultures also influences the ability to navigate constraints by black and Indigenous people, as well as people of color in general. Several hypotheses have been posited to explain cultural differences in leisure. The *Ethnicity Hypothesis* suggests that the variation we observe in leisure behavior is because of unique cultural processes (e.g., norms, social organization, and value systems), rather than poverty or socio-economic marginalization (Krymkowski et al. 2014, Serenari and Peterson 2018). The *Cultural Fit Hypothesis* suggests that the variation we observe in leisure behavior is due to the degree of congruency between an individual's cultural values, beliefs, and expectations and the expected result, behavior, and cognition about hunting and fishing (Floyd et al. 1994, Serenari and Peterson 2018). For example, fishing, as compared to hunting, appeals to the socio-cultural values of a wider range of racial and ethnic groups, and thus exhibits greater participation rates (U.S. Fish and Wildlife Service and U.S. Census Bureau 2018). Alternatively, fewer racial and ethnic groups have a socio-cultural connection to hunting; therefore, hunting has lower representation by black and Indigenous people, as well as people of color in general. Building relevancy of hunting and fishing to a greater range of ethnic groups will require a change in the focus in marketing and outreach to concepts that appeal and build the social habitat to a more diverse stakeholder base (Bissell et al. 1998).

We observe considerable variation in the activity patterns for hunting and fishing. Some hunters and anglers may focus all their reactional efforts on a single type of hunting (e.g., upland-game hunting, bow hunting) or fishing (e.g., flyfishing, trophy bass fishing). However, other sportspersons may participate in a variety of hunting and fishing activities and identify equally with big-game hunting, upland-game hunting, waterfowl hunting, turkey hunting, and all types of fishing. Those hunters and anglers who participate in a greater number of activities are greater contributors from a conservation perspective, as they purchase multiple permits (known as *cross buying*) and spend more money on equipment and travel. Further, cross-buying participants are less likely to lapse in their participation (Hinrichs et al. 2020) and have increased customer loyalty (Kamakura et al. 2003). However, there is some debate as to whether customers who are more loyal are more likely to engage in cross buying, or if cross buying results in increased loyalty (Reinartz et al. 2008).

Although the potential for broad strategies to encourage adoption of other wildlife-based recreational activities exists, many agencies and nongovernmental organizations have been encouraging anglers to try hunting and hunters to engage in angling for decades (Council to Advance Hunting and the Shooting Sports 2016). The participant already engaging in a wildlife-based activity is seen as a "low-hanging fruit" that can be encouraged to participate in complementary activities. In reality, few (<10%) wildlife-based recreators actually transition between the two activities (Hinrichs et al. 2020). *Substitutability* in leisure activities has been defined as the interchangeability of recreation activities in satisfying participants' motives, needs, wishes, and desires (Hendee and Burdge 1974). Theory suggests that only when an alternative activity is perceived as satisfying the needs and providing outcomes equivalent to the original activity can it be considered substitutable (Iso-Ahola 1980). An activity should not be considered an alternative if the activity does not provide the same benefits as the original (Shelby and Vaske 1991). Understanding why individuals choose to prefer and engage in one form of wildlife-based recreation over another will be essential for increasing the stakeholder base (both traditional users and non-traditional users) for wildlife conservation and management.

HUNTER AND ANGLER MOVEMENTS

Agencies are keen to understand what influences site choice by hunters and anglers. The site that people choose to hunt or fish can be explained by a host of factors including why the sportsperson is participating in the activity (Beardmore et al. 2011*b*, 2013), target species abundances and locations (Papworth et al. 2009), social circles (Muth 2000, Hayslette et al. 2001), frequency of participation (Manfredo et al. 2004, Kerr and Abell 2016), perceived location of game (Harmon et al. 2018, Messinger et al. 2019, Wszola et al. 2019), and proximity to urban areas (Post and Parkinson 2012). In addition, disposable time and income are two of the most influential determinants of willingness to travel to distant sites (Heberlein and Kuentzel 2002, Nicolaisen et al. 2012, Metcalf et al. 2015, Shrestha and Burns 2016). Further, the social context of the trip is also important in determining site choice. Hunters and anglers motivated to participate by social aspects may be more likely to travel longer distances and take longer trips because the longer trips provide more opportunities to engage in fellowship (Stedman et al. 2008). Conversely, hunters and anglers with young children may take shorter trips that are close to home due to family responsibilities and financial constraints such as childcare. However, these relationships influencing site choice are complex and may be difficult for agencies to parse (Wszola et al. 2020*a*).

One important distinction that is often made in site choice is between public and private lands, which is tied closely to perceived expectations of the experience (Stedman et al. 2008). Lack of access to private land is often reported as the most limiting constraint to participating in hunting or fishing activities, with private land usually more preferred to public land (Wright et al. 2001, Miller and Vaske 2003, Recce 2008, Knoche and Lupi 2012, Larson et al. 2014, Metcalf et al. 2017). Hunts on public land are generally perceived to be lower quality, less satisfying, and encompass poorer habitat quality than private land hunts, or in some cases there may be no publicly accessible land to hunt (Wallace et al. 1989, Diefenbach et al. 2005, Stedman et al. 2008). Conversely, private land is perceived as higher quality because it provides ready access, is often conveniently located, and has higher quality game and lower hunter density (Siemer and Brown 1993). Fear of losing access is one of the greatest concerns of hunters (Eliason 2021) and lack of access has led to of decreased hunter participation (Miller et al. 2002). This fear is not unwarranted as access to private lands is declining (Lauber and Brown 2000, Jagnow et al. 2006), with a decrease in social connections to rural lands that hunters and anglers historically used to gain access to private land (Schulz et al. 2003, Stedman et al. 2008, Robison and Ridenour 2012). Providing more *quality* access and better sportsperson management may be important to alleviate issues surrounding land access, especially in those landscapes dominated by private lands.

Once a site is located, hunters and anglers combine multiple sources of real and perceived information while moving through a location (Decker et al. 1980, Larson et al. 2014). Hunters and anglers make decisions based on game abundance, starting location, and information gained from previous hunting and fishing experiences (Brøseth and Pedersen 2000, Kaltenborn and Andersen 2009, Lande et al. 2010, Lone et al. 2014). Another important component affecting hunter and anglers' movement with a site is the physical exertion required to move through a site, with people often seeking to minimize that exertion (Merrill and Graefe 1998, Lee et al. 2007, Ólafsdóttir and Runnström 2013). Hunter and angler movement and the subsequent degree of effort within locales at a site affect the opportunities for other hunters and anglers, but it also affects the fitness of game (Biro and Stamps 2008, Bunnefeld et al. 2009, 2011*b*, Asmyhr et al. 2012, Madden and Whiteside 2014, Madden et al. 2018, Wszola et al. 2019) and wildlife behavior such as space use, foraging, or mating (McGrath et al. 2018, Messinger et al. 2019, Wszola et al. 2019), often with fitness consequences (Wirsing et al. 2008, Madin et al. 2011, Lone et al. 2015, Festa-Bianchet and Arlinghaus 2021 [Chapter 12], Wszola and Fontaine 2021 [Chapter 13]). Therefore, wildlife agencies should be continually working to mitigate intended and unintended disturbance effects of hunting and fishing on game populations (Allendorf and Hard 2009, Proffitt et al. 2009, Turner et al. 2016, Leclerc et al. 2017).

To assess hunter and angler movement and pressure on fish and wildlife within a location, we must consider where hunters and anglers begin their activity and how hunters and anglers navigate through space for the duration of the trip (Lima and Bednekoff 1999, Lima 2002). For those hunters and anglers who use publicly available lands and waters, starting-location decisions are contingent on the distribution of public area infrastructure. Roads (and boat ramps and parking lots) are frequently utilized as starting locations because using roads reduces time and physical exertion requirements to access a site (Stedman et al. 2004). Sportspersons are subject to laws and norms when entering and moving about locations, such as postings indicating whether or not hunting and fishing are permitted (Sigmon 2004). Individuals make greater use of access points that provide cues of public-land accessibility and avoid access points with posting prohibiting harvest (Wszola et al. 2019). Locations where target species are more abundant also tend to be more utilized than spaces that require greater exertion to access (Harmon et al. 2018, Wszola et al. 2019). Strategic placement of posting for access and prohibitions, creating habitat that is highly physically demanding to access, and temporarily closing specific access points are all methods in which wildlife agencies use to mitigate adverse effects of hunting and fishing pressure on fish and wildlife populations. Hunters themselves are faced with negotiating tradeoffs among interference from other hunters, game abundance, effort required, and time to access, which in turn indirectly controls hunter density. Understanding the social and ecological factors at a site will allow agencies to better manage and produce higher quality sites for sportspersons.

ILLEGAL AND CONSERVATION BEHAVIOR

As with any activity, hunting and fishing activities are engaged in by people of varying ethical profiles. Most of the research focused on the topic of illegal hunter and angler behavior is limited in scope to local or statewide areas with few national statistics to document the issue (Elliott 1991, Crow et al. 2013). However, the capacity of hunters and anglers to occasionally partake in illegal behavior derives from three basic categories: (1) ignorance, (2) frustration, and (3) willful defiance (Grosslein 1980, Muth and Bowe 1998, Eliason 2020). Further, the *levels of offence* are characterized at the *micro-level* (such as subsistence poaching and individual acts of cruelty), *meso-level* (such as domestic trade in resident-vulnerable species and organized illegal hunts) and *macro-level* (notably import and export of endangered species for international trade; Wellsmith 2011, Kahler et al. 2013).

One of the greatest contributors to fish and game law violations stems from a belief that these laws are not "real laws." Long before immigrants settled in North America, game laws created access inequities as wildlife preserves were only legally accessible by royalty and the wealthy. Early immigrants to North America naturally resisted the idea of game laws and gun control, having recently been liberated from heavily populated monarchies (Reisner 1992, Eliason 2012). A genuine belief that game and fish laws are oppressive to hunters and anglers is likely held by some hunters and anglers today (Elliott 1991). Hunters and anglers may also view wealth as a justification for breaking game and fish laws. Some hunters and anglers pay large sums for the opportunity to hunt or fish in exclusive clubs or be guided by a professional. This attitude can be characterized by the belief that they are entitled to a bag limit or trophy because of their investment (Elliott 1991, Simon 2009). The fact that hunting and fishing frequently take place in secluded environments such as backwoods and rural fields should not be ignored. With no audience to applaud or disapprove of conduct, hunter and anglers' actions are largely dictated by their consciences as opposed to onlookers (Leopold 1968, Warchol and Johnson 2011).

Attitudes that encourage breaking game laws are often formed at an early age (Hall et al. 1989, Enticott 2011), and it can be difficult to change people's attitudes, especially if there appears to be an abundance of wildlife. Hunters and anglers may learn to break game and fish laws at the same time they learn to legally hunt and fish, and can view illegal actions, such as poaching, as acceptable behavior early in their development. As hunters and anglers traditionally learn to hunt and fish from their parents, it is suggested that people who illegally harvest large quantities of game and fish acquired their illegal habits from their parents (Glover and Baskett 1984, Hall et al. 1989, Enticott 2011).

Agencies promote ethical behavior in hunters and anglers with enforcement of harvest regulations by conservation officers (Shelley and Crow 2009). Common repercussions for illegal behavior include fees, revocation of hunting and fishing privileges, and even incarceration. Agencies also work to shape hunter and angler ethics through other means, such as stressing the importance of conservation behavior to parents of young hunters and anglers and through hunter-safety courses. Ethics and responsibility remain a staple of nearly every hunter-education course (Hilaire et al. 1998). However, some argue that ethics and responsibility should be a given a greater importance in these courses, perhaps even elevated to the same level of importance as safety (Vitali 1990, Gilbert 2000). Many agencies work directly with hunter-education instructors in developing solutions to promote better ethics, and work with hunters and anglers to develop conservation plans. Collaborative efforts among agencies, hunters, and anglers increase conservation behavior among hunters and anglers (Arlinghaus 2006*b*, Hawkins 2017).

CONCLUSIONS

Effective harvest management requires a firm understanding of the social and the ecological segments of the system. We highlight important considerations for human dimensions that will allow further assessment and integration of human behavior into social-ecological systems. Agencies should be concerned with making sure that hunter and angler behavior is integrated in harvest management for *two* primary reasons. First, the current funding model for wildlife conservation in North America relies on the users for license sales and the excise taxes collected on sporting goods. Declining participation rates should concern agencies about future financial resources and spur managers to better understand the *who, where, when, and why* of their constituents. Second, an understanding of human behavior is required to understand how hunters and anglers will respond to changes in policy and regulations so that agencies can predict effects on the population that is being managed. Agencies can have the best grasp on the ecology of the system, but hunter and angler behavior may have the potential to confound predicted effects of regulations on harvested populations (Conroy 2021 [Chapter 1]). Human dimensions is well suited to inform policy and management of the social–ecological systems.

The dilemma of a declining user base is a serious concern to fish and wildlife conservation without a change in funding mechanisms. Despite considerable and well-intended efforts by agencies and nongovernmental organizations such as Delta Waterfowl, Pheasants and Quail Forever, and Backcountry Hunters & Anglers, programs for recruitment, retention, and reactivations have not brought the panacea of new participants that were anticipated. The future of harvest management must include a fundamental shift in the perspective of who is a potential stakeholder in fish and wildlife conservation to build relevancy of conservation and inclusion of a more diverse group of stakeholders. Understanding the values, attitudes, and preferences of women and black and Indigenous people, as well as people of color in general, will help develop the foundation to build a greater diversity of stakeholders in fish and wildlife management. In addition, recruitment, retention, and reactivation efforts only focused on "low hanging fruit" such as provision of equipment or skills training may not have the ability to increase participation in wildlife-based recreation. Rather, fundamental or "big shifts" may be required by expanding or altering traditional management approaches such as changing regulation structures to be more inclusive of new users (Hinrichs 2019). This will be no easy task. Further, careful balances among strategies should be made to ensure an optimal number of hunters and anglers, aided by model-based predictions of sportsperson populations (Graham et al. 2021 [Chapter 6]). The heterogeneity of hunters and anglers indicates that not all management approaches will appeal or influence the different segments of the population similarly and should be carefully considered in evaluating alternative management scenarios.

We must reshape how we view the social components of hunting and fishing in the context of harvest management, and also more broadly in fish and wildlife management. Management of social components has primarily been reactive in nature to the problem at hand (Allen et al. 2011, Pope et al. 2016). We contend, despite merit of reactionary approaches, there may be greater benefits from management approaches that allow wildlife agencies to anticipate changes in a proactive or predictive manner (Runge 2021 [Chapter 7]). Models that integrate both human behavior and the ecology in socio-ecological systems should be used to predict the outcomes of alternative management scenarios, such as different regulation structures and policies. We call for an understanding that hunters and anglers should be managed just as we manage habitat or game species, which requires a firm understanding of hunters and angler behavior and at a larger spatiotemporal perspective. Proactive, effective management of sportspersons through the evaluation and monitoring of the behaviors of hunters and anglers will be essential in the new paradigm of harvest management.

ACKNOWLEDGMENTS

M. Gruntorad was supported by Federal Aid in Wildlife Restoration W-130-R, administered by Nebraska Game and Parks Commission. C. Chizinski was supported by Hatch funds through the Agricultural Research Division at the University of Nebraska-Lincoln and from Federal Aid in Wildlife Restoration project W-120-T, administered by the Nebraska Game and Parks Commission. The authors thank the two reviewers of this chapter who greatly improved clarity and direction.

5 Expanding Participation in Recreational Fishing through Multiple License Markets

Richard T. Melstrom

CONTENTS

INTRODUCTION

State wildlife agencies depend heavily on the funds generated by license sales to pay the costs of fish and wildlife management (Hiller et al. 2021*b* [Chapter 2]). On average, one-third of state agency funds come from license sales (Lueck 2000). This means that agencies are often obligated to raise license prices when costs exceed revenue. However, raising prices can be controversial. Although academic research and public opinion polls generally show that hunters and anglers support modest price increases, some users and policymakers inevitably oppose such changes (Sutton et al. 2001, Whitehead et al. 2001, Hayes 2018), and state and federal policies can work to keep prices low. For example, the amount of federal funds agencies receive is tied to hunting and fishing license sales (U.S. Code, volume 16, section 777c), so prices need to be low to maximize sales and hence federal funding. A state wildlife agency is also often responsible to legislators and a group of commissioners (Hiller et al. 2021*b* [Chapter 2]) with their own beliefs about prices and regulations. For example, legislators may ideologically oppose price increases, or oppose increases because they fear the economic consequences of less spending on hunting and fishing equipment when participation declines. Thus, there is a critical need for state wildlife agencies to understand how license prices affect sales and revenue.

Agencies can use license demand models to explain the relationship between price, sales and revenue. The *quantity demanded* in a demand model is the total sales of a single good—for example, resident annual fishing licenses. Casual observation or graphical analysis will show that sometimes sales fall substantially, and other times hardly budge, after a price change. This does not mean that price has no consistent effect, but that other factors—for example, individual employment status and spring temperatures, to name only two—which can vary significantly from year to year, also have an important and potentially large effect on sales. License demand models are useful because they disentangle the effect of price on sales from other factors.

Here, I show how license demand models can be used to price recreational fishing licenses while taking sales and revenue into account, and how analysts can develop license demand models by statistically analyzing price and sales data. I assume maximizing sales is the objective of most state wildlife agencies. I therefore focus on pricing licenses to maximize sales when the agency has separated the market for recreational fishing between different types of anglers, specifically resident and nonresident anglers. Every state has separated its market—for example, by establishing licenses for residents versus nonresidents, seniors versus nonseniors—which provides opportunities for additional sales and revenue. However, an agency needs to know something about demand in different markets to set the right prices; otherwise, the state's price and licensing strategy may needlessly reduce participation in the fishery.[1]

The first of three sections in this chapter examines current price statistics and several historical time series of resident annual fishing license prices. The second section analyzes optimal prices in a single license market; the third considers two license markets. My goal was to use basic economic principles to generate insights into optimal prices that would be accessible to readers regardless of background.

Information presented here compliments several published papers as well as a number of unpublished reports that develop license demand models. The consensus in this literature is that increasing the price of a license will increase revenue but reduce sales (Beardmore and Harris 2016). This is certainly true for resident annual fishing licenses across the country (Bennear et al. 2005). Thus, for example, increasing resident annual license prices by 10% will increase revenue by an amount somewhat less than 10%. One corollary of this finding is that agencies are not maximizing revenue from their resident annual licenses. Although states probably do not intend to maximize license revenue, this finding is important because it indicates that modest budget deficits at state wildlife agencies can be addressed by increasing prices. However, this point should not be overgeneralized. Several studies show that some licenses are overpriced in terms of revenue maximization; in other words, there can be situations when raising the price will decrease revenue. Teisl et al. (1999), after studying over a dozen Maine licenses, determined that the resident combination licenses, as well as the nonresident annual, one-day and three-day fishing licenses, were overpriced. Melstrom (2018) determined that Pennsylvania overpriced their nonresident three-day fishing license. So raising the price of resident licenses is the most definite way to increase revenue. However, I will demonstrate that states can benefit when they coordinate prices across different licenses rather than gaging licenses individually.

Price is important, but it is just one among many factors affecting the demand for fishing licenses. A large literature documents the effect of travel costs, time costs, and catch-related factors on participation in recreational fishing (Oh et al. 2005, Dabrowska et al. 2014, Hunt et al. 2019). State policies and regulations that set bag limits (Knoche and Lupi 2016, Cha and Melstrom 2018), gear restrictions (Beardmore et al. 2011*a*), and consumption advisories (Jakus et al. 1997, MacDonald and Boyle 1997, Breffle et al. 1999) also have significant effects on the demand for fishing. In addition, there are often opportunities for substitution across licenses—for example, between single-day and annual licenses—which means that sales depend on the price and availability of substitutes. All these factors affect sales and revenue, and add to the challenge of optimally pricing licenses.

Many state wildlife agencies are experiencing declining license sales and therefore revenue and federal funds tied to participation. True, the total number of fishing license holders in the United States of America stabilized at about 30 million between 2000 and 2019, but the trend has been uneven geographically. Although some states have experienced modest sales growth, several others have experienced definite declines, averaging in many cases about –1% per year since 2000 and in certain years –5% or more. Conservation officials and administrators are understandably concerned by these trends. Analyzing these trends is beyond the scope of this chapter, but I want to make clear that prices are not the culprit. For the states that I collected long-run price data there is no evidence of a general increase in real prices (prices have tended to just keep up with inflation in the long run), so other factors must explain the historical sales decline. I examine these price trends in the next section.

CURRENT LICENSE PRICES AND TRENDS

I collected 2019 price data for resident annual and nonresident annual fishing license prices nationally.[2] Every state separates fishing licenses for residents and nonresidents, and these two license types are among the oldest and largest-revenue generators for agencies. The prices of these licenses can vary substantially. For the resident annual license, most prices range between US$11 and $30, with an average of $23.72, a minimum of $6.00 (Hawaii), and a maximum of $49.94 (California). Most annual nonresident fishing licenses are priced between $41 and $80, with an average of $59.38, a minimum of $13.00 (Washington D.C.), and a maximum of $145.00 (Alaska; Fig. 5.1). The markup from the resident to the nonresident annual license exceeds 100% in almost every state.

In addition to annual fishing licenses, most states offer some type of day or multi-day license. These licenses are less standard in their terms across states, but at least 33 states offer one-day licenses, ten offer three-day, three offer seven-day, eight offer other varying day limits (such as two-day, ten-day, and 14-day), and four offer no day or multi-day license. Some states have formally separated these multi-day licenses between residents and nonresidents, some are non-specific, and some have informally but essentially separated these licenses by charging more for the day license than the resident annual license. For the states that explicitly list a resident one-day license (this includes states that offer a one-day license only to residents), the average price is $9.97. For states that explicitly list a nonresident one-day license, the average price is $14.12.

I paired the 2019 resident annual license data with prices that I recorded for 2016 (Fig. 5.2). No state decreased their price during this period. Price increased on average $1.66, or 7.5%. This increase slightly exceeds the amount of inflation over the same period, which was about 6%, but this length of time is too short to draw conclusions about long-run price trends.

To analyze long-run trends, I collected historical price data going back to 1929 for Minnesota, Ohio, and Pennsylvania. Current prices in these states lie in the middle of the national price distribution, so their experience is likely indicative of the trends that occurred in many other states (Fig. 5.3). Two features stand out in the series. First, all three states, although especially Ohio and

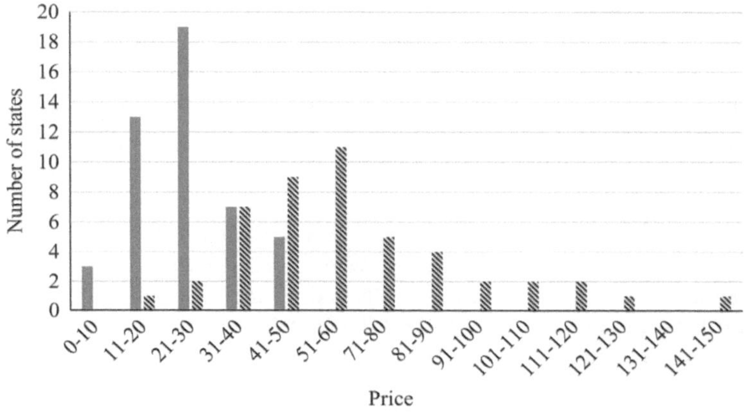

FIGURE 5.1 The distribution of resident (solid bars) and nonresident (hashed bars) annual fishing license prices (U.S. dollars) in the United States of America during 2019. Data collected by the author.

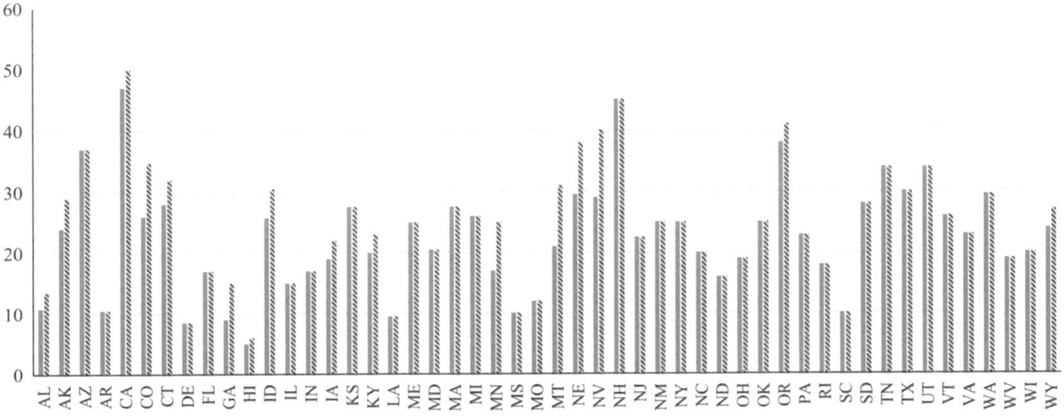

FIGURE 5.2 Prices (U.S. dollars) of resident annual licenses among states in the United States of America, ordered alphabetically, during 2016 (solid bars) and 2019 (hashed bars).

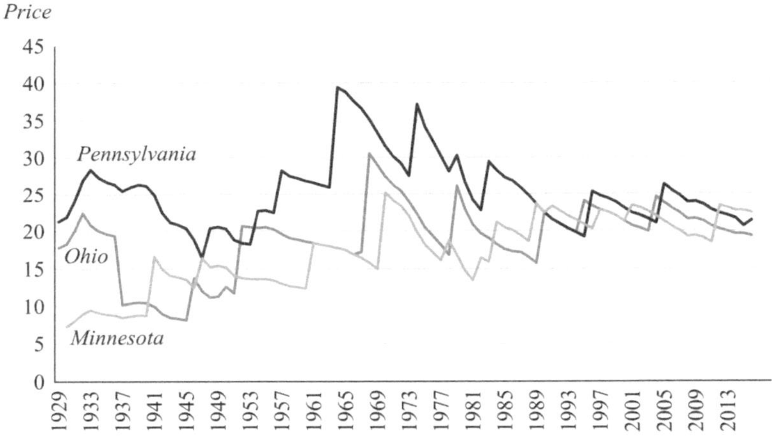

FIGURE 5.3 Historical prices (U.S. dollars) of resident annual licenses in Minnesota, Ohio, and Pennsylvania, adjusting for inflation.

Pennsylvania, experienced substantial price swings from the 1930s to the 1970s. There is an unmistakable pattern of price increases (in both real and nominal terms) followed by erosion in the real price due to inflation. In the 1960s and 1970s, the rate of inflation meant that the real value of a license could decrease by as much as $10 before the state raised the price again, which is quite a substantial sum when prices are between $15 and $30. However, inflation has lowered considerably since the 1980s, and this almost certainly explains why there have been fewer extreme adjustments in the past 30 years. Second, when averaged within decades, the real price in each state is essentially unchanged since the 1980s and, in fact, prices in Ohio and Pennsylvania today are almost identical to prices during 1929. Minnesota's price has increased over the long run, but most of this increase occurred before 1975.

Sometimes agencies notice that sales can immediately drop but ultimately rebound after a large price increase. Although some anglers may contribute to the sales drop by lapsing for a year or two to *protest* the price change, sales can also fall because the increase has real effects on the affordability of fishing, and rebound because inflation erodes the cost of a license over time (assuming, of course, the price is not indexed to inflation). For example, suppose a state raises its resident annual license price 10%, from \$20 to \$22. Given inflation is typically 2% annually, and assuming the price does not change once it is \$22, after five years inflation will have brought the real price back to \$20. Inflation will always have a significant effect on sales by changing the affordability of a license every year.

PRICING LICENSES IN ONE MARKET

This section analyzes the demand for a single license. In the first part, I discuss the objectives of a state wildlife agency with respect to sales and revenue using a theoretical demand model. In the second part, I present a demand model estimated using actual price and sales data.

Conceptual Model

To examine the issue of optimal pricing, I must first recognize an agency's objective with respect to the demand for fishing. Is it to maximize net income (the difference between total revenue and total cost), as a business would, or maximize participation? A state wildlife agency is not unlike a business in the sense that it must pay for much of its own operation. On the other hand, an agency often has a constitutional or statutory obligation or mission to manage natural resources for the benefit of current and future generations. I interpret this obligation as maximizing opportunities for the public to fish, hunt, and generally enjoy the outdoors. I therefore believe that an agency's (and the state's) objective with respect to demand is to maximize participation, which will correlate closely with license sales depending on which individuals are required to purchase a license. Revenue matters only insofar as it is needed to pay the costs of an agency's operations. An agency still faces the budget constraint that total revenue must equal or exceed total cost. Repeated budget deficits are not sustainable.[3]

A license demand model describes the relationship between sales, the price of a license, and other variables that affect sales. The simplest demand model, Model 1, expresses the relationship between sales and price only:

$$Q = \alpha - \beta \times P, \tag{5.1}$$

where Q is the number of licenses sold, P is the license price in dollars, and α and β are intercept and slope parameters, respectively. In a two-dimensional graph, Equation (5.1) is a downward-sloping line, with one intercept determined by α and the slope of the line determined by β (Fig. 5.4b). We can see that the demand line intercepts the horizontal axis at the coordinates $P = 0$, $Q = 1000$ ($Q = 1000 - 13 \times 0 \Rightarrow Q = 1000$). The horizontal intercept indicates the number of licenses sold when price is zero. If Q is measured in thousands, then $\alpha = 1000$ means that one million licenses are sold when the price is zero. Demand intercepts the vertical axis at $P = 77$, $Q = 0$ ($0 = 1000 - 13 \times P \Rightarrow P = 77$). The vertical intercept is the lowest price with zero sales; and at least one person would buy a license at a slightly lower price. Economists refer to this price as the *choke price*, which is written in terms of the parameters as α/β. In this example, the choke price $\alpha/\beta = 77$ means that no one will buy a license if the price is raised to \$77.

Equation (5.1) is a demand model in which only price and no other factors affect sales. However, in the real world sales trend upward or downward irrespective of price. A demand model with trending sales, Model 2, is

$$Q = \alpha - \beta \times P + \gamma \times T + \delta \times T^2, \tag{5.2}$$

where T is a measure of time—for example, the number of years prior to the current year, so that $T = 0$ in the current year, $T = 1$ in the prior year, and so on—and γ and δ are additional parameters. The time

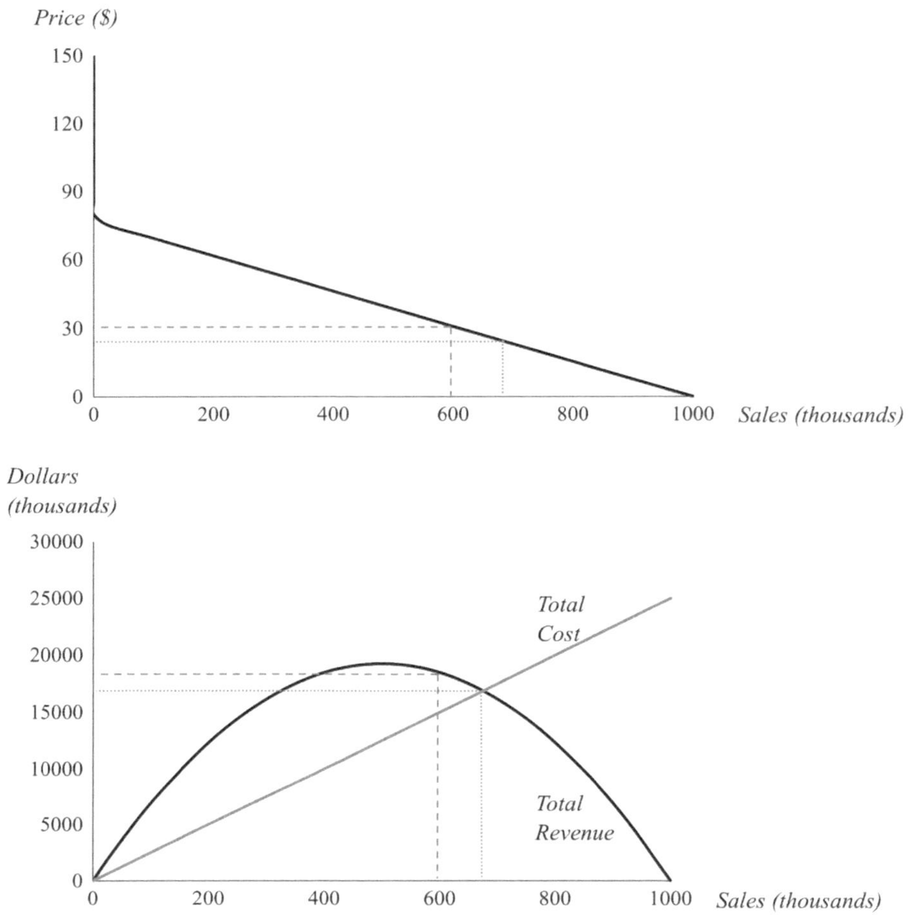

FIGURE 5.4 (a) The relationship between price (U.S. dollars) and sales in a license demand model, when α = 1000 and β = 13. (b) The relationship between revenue and sales. The dashed and dotted lines illustrate different price, sale, and revenue combinations. The dotted line shows the maximum sustainable sales level, in which total revenue equals total cost.

trend in Equation (5.2) is quadratic, which allows sales in the model to trend nonlinearly and, in my experience, closely approximate actual, long-run sales trajectories. Suppose T measures years prior to the current year. If γ and δ are both positive, then sales are trending lower. If γ and δ are both negative, then sales are trending higher. If γ is positive and δ is negative, then sales have a humped-shape trend. I will return to this model when I analyze a demand model with real-world parameters in the next section.

We can use the demand model (Fig. 5.4) to measure sales and revenue at any price. In the example, the agency sells 600 licenses if the price is \$30, and revenue from license sales is $P \times Q$ = \$18,000 (Fig. 5.4 panel a shows revenue as a box tucked between the two axes, below the price and to the left of the sales amount outlined by dashed lines; panel b illustrates every possible revenue and sales combination as the curve labeled *Total Revenue*). The revenue curve has an inverted-U shape because I assume revenue is zero when the agency sells zero licenses, and zero if it sells licenses for \$0. To keep the demonstration as simple as possible, I do not include any Sport Fish Restoration Program funds that are allocated on a sales basis; if I did include Sport Fish

Restoration Program funds, the revenue formula would be $P \times Q + F \times Q$, where F is the apportionment per license. In this example, revenue is greatest when the price is \$38.50 and the agency sells 500 licenses. When the price is lower than \$38.50, the agency will sell more than 500 licenses; note that in this region of the revenue curve, increasing the price will increase revenue. When the price is higher than \$38.50, the agency will sell less than 500 licenses, and in this region of demand, increasing the price will decrease revenue. The dashed lines in the figure show that revenue is \$18,000 when the agency sells 600 licenses (price = \$30). In this case, the agency could earn more revenue by slightly increasing the price but it should lower the price if it wants to maximize participation. Total cost is an upward sloping line (Fig. 5.4b) because costs tend to increase with sales and thus participation; when there are more anglers, an agency inherently needs to hire more conservation officers, stock more fish, and install more access points if it wants to maintain the same level of fishing quality (although in practice agencies do not explicitly tie spending to sales).

Pricing the license to maximize total revenue is not optimal because it does not maximize sales, and it generates more revenue than the agency needs. In the example (Fig. 5.4), the agency maximizes revenue by charging \$38.50 per license and selling 500 licenses, generating \$19,000 in revenue but only \$12,500 in cost. Thus, at \$38.50 per license, the agency generates far more revenue than it needs because total cost is less than total revenue at that use level. The agency could sell substantially more licenses without compromising its budget by lowering the price. Note that a price of \$30 is still too much because then the agency sells 600 licenses, total revenue is \$18,000 and total cost is \$15,000, for a positive net income of \$3000.

The optimal price should be low enough that the agency earns just enough revenue to pay its costs, no more, no less (Fig. 5.4b, dotted lines). Net income is zero where the total revenue curve intersects the total cost line, which occurs when the agency sells 675 licenses, total revenue is about \$17,000, and total cost is about \$17,000. The agency will sell 675 licenses when the price is \$25. Thus, the optimal price is \$25.

The demand for fishing *and* the cost of managing the fishery determine the optimal price. We could not have used the license demand model and, by extension, the total revenue curve to solve for the optimal price without knowing the location of the total cost line. To demonstrate this more concretely, suppose that the agency's total cost is $C \times Q$, where C is the cost that the agency has to spend to manage the fishery for each angler that enters the fishery (I am assuming that there are no fixed costs of managing the fishery; my claim is valid regardless of this simplification). Economists refer to C as the *constant marginal cost*. Balancing total revenue and total cost means that the agency must satisfy $P \times Q = C \times Q$, which implies that the optimal price equals the constant marginal cost, $P = C$. The slope of the cost curve (Fig. 5.4) is 25, so $C = 25$, which explains why the optimal price in the example worked out to \$25. Of course, in the real world, $C \times Q$ may not approximate an agency's operating costs, so we cannot conclude that the optimal price is definitively $P = C$. However, it is definitely true that the general formula $P \times Q = \textit{Total Cost}$ determines the optimal price, and that therefore we need to understand license demand Q as well as cost to determine the optimal price.

An Application

I now present a license demand model developed from real price and sales data. With price and sales data, I can use statistics to estimate the parameters α and β in Model 1 (Equation (5.1)), or α, β, γ, and δ in Model 2 (Equation (5.2)). I can develop even richer models by incorporating additional data on substitute prices, demographics, macroeconomic conditions, climate variables, and catch-related factors. Despite this, I find that Model 2 can predict historical sales remarkably well. Nevertheless, it cannot be understated that Model 2 omits important variables, some of which are controlled by agencies and state policymakers. Agencies that worry about anglers substituting between licenses due to price differences—for example, anglers that switch from the annual license to a lifetime license—understand the importance of substitute prices. Unfortunately, it difficult to identify substitution effects in real-world license demand data because states usually increase the prices of different licenses simultaneously, which creates problems with multicollinearity.

The data come from resident annual licenses sold in Pennsylvania from 1929 to 2016 (Fig. 5.5a, solid line). Sales peaked during 1990 at 1,015,134 licenses, out of a population at the time of 11.9 million. My impression is that sales peaked in other states around the same time. For example, sales of resident annual licenses peaked in Ohio during 1987 and in Minnesota during 1991. Interestingly, sales did not consistently increase between 1929 and 1990. After growing for nearly three decades, sales fell year over year between 1956 and 1964, before growing again for the next 20 years. Sales of the resident annual license in Pennsylvania have declined in a more-or-less steady fashion since 1990, and during 2016 the state sold 690,777 licenses, out of a population of 12.8 million.

I estimated the parameters of Model 2 (Equation (5.2), Table 5.A1) for Pennsylvania's resident annual fishing licenses using ordinary least squares. I adjusted prices for inflation, denominated

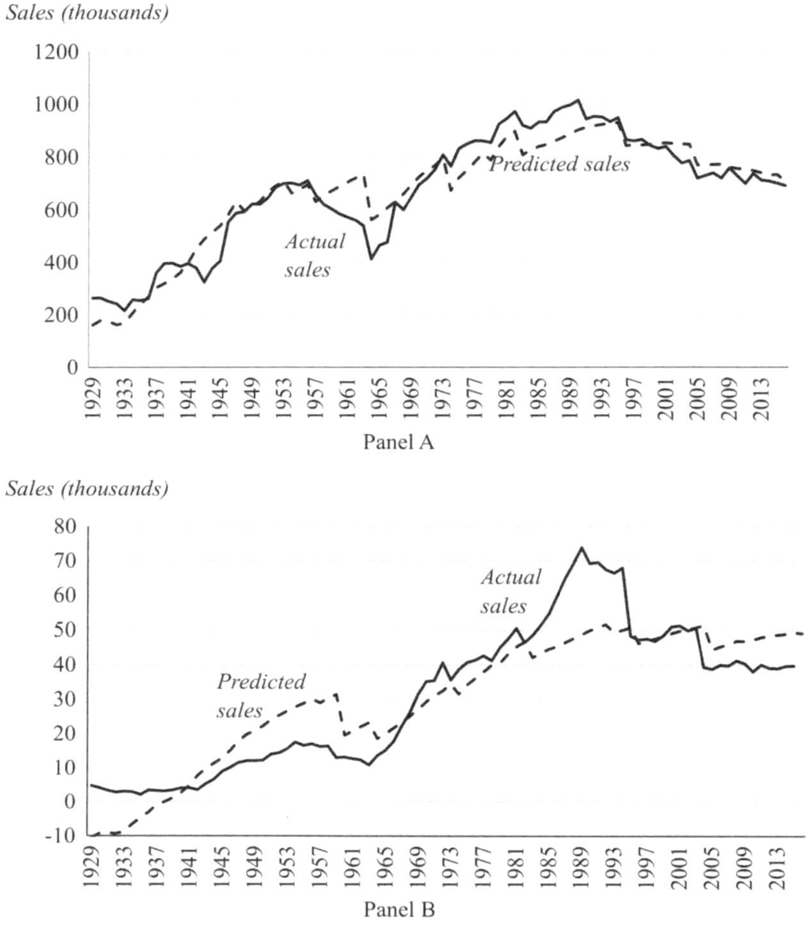

FIGURE 5.5 (a) Actual (solid line) and predicted (dotted line) resident annual license sales (thousands of U.S. dollars) in Pennsylvania. (b) Actual (solid line) and predicted (dotted line) nonresident annual license sales in Pennsylvania.

sales in units of 1000, and indexed the time trend to 2016 so that T measures years before 2016.[4] The model with the estimated parameters for Pennsylvania is

$$Q = 996.2 - 13.5 \times P + 14.1 \times T - 0.2 \times T^2. \tag{5.3}$$

All parameters are precisely estimated, with p-values < 0.001.[5] The appendix to this chapter presents more models statistics, including standard errors. The demand for resident annual licenses in other states is likely to be qualitatively similar. For example, I show in the appendix that the model of Ohio's resident annual license sales has parameters with similar magnitudes.[6] The model in Equation (5.3) predicts historical sales remarkably well, with a correlation coefficient between actual and predicted sales of 0.94 (Fig. 5.5a).

Now examine each parameter in Equation (5.3). The first term on the right-hand side implies that the agency would sell 996,200 licenses during 2016 if the price was \$0. This is close to the historical peak of 1,015,100 licenses during 1990. Nevertheless, the model clearly indicates that sales near one million are a thing of the past. We probably could not expect that level of participation in Pennsylvania during 2016 even if the state gave its resident annual license away for free. This is because external factors have caused sales to shift lower relative to 1990. The positive and negative parameters on T and T^2, respectively, explain the hump-shaped trajectory in historical sales, and indicate that sales will decline into the future if price remains unchanged. What is the effect of a price change? The price parameter indicates that the state sells 13,500 fewer licenses for every \$1 increase in price. During 2016, the price of Pennsylvania's resident annual license was \$21, and the Pennsylvania Fish & Boat Commission sold 690,100 resident annual licenses, generating \$14.5 million in revenue. The model predicts that if the price had been raised to \$22 during 2016, the agency would have sold 676,600 licenses, and generated \$14.9 million in revenue. On the other hand, the model predicts that if the price had been lowered to \$20 during 2016, the agency would have sold 703,600, and generated \$14.1 million in revenue.

The revenue effect of a \$1 price change indicates that Pennsylvania has priced its resident annual license so that it can raise additional revenue if the price goes up, ceteris paribus. This finding provides additional evidence for the claim in the introduction that increasing license prices will usually increase revenue.[7] The model estimates that a 10% increase in price, which is about \$2 in Pennsylvania and a typical adjustment amount, reduces sales by 2%. This sales reduction is not large enough to offset the additional revenue from the higher price. To lose revenue, a 10% increase in price would have to reduce sales by more than 10%. The estimate also provides clear evidence that sales are trending downward due to factors other than price. The effect of price on sales is definitely negative, but too small to affect the trend over the long run. Recall that over the last few decades the price of a resident annual license has just about kept up with inflation.

PRICING LICENSES IN TWO MARKETS

CONCEPTUAL MODEL

State officials have a great degree of power in licensing recreational fishing. They set license prices, determine who needs a license and who is exempt, and can separate its markets by selling different types of licenses to different individuals. For example, all agencies separate licenses by residents and nonresidents, and most separate multi-day and annual users, and seniors and non-seniors. We see similar market separation at work in private industry, for example, students discounted movie tickets and separate Disney World passes for Florida residents and nonresidents. Like industry, state agencies sometimes separate their markets to address the demand of less avid users, for example when they offer a single-day license, but concerns about fairness and equity can also motivate offering different licenses. Whatever the reason, agencies clearly exercise their power to sell licenses to different groups at different prices.

Acknowledging different user groups begs the question: Is there a particular group that the state wants to maximize sales and participation? States may appear to have prioritized resident access over nonresidents by pricing resident licenses lower than nonresident licenses, suggesting that states principally care about sales to residents. This outcome makes sense because prices are usually approved by state legislators representing or protecting the interests of residents. From this perspective, states should price their licenses to maximize participation of residents. However, states may also be motivated to maximize total sales, including both residents and nonresidents, which can lead to a similar disparity in prices (I demonstrate this below). A number of state wildlife agencies have a mission to manage their states' natural resources in the interests of the general public—for example, the Ohio Department of Natural Resources' mission states that the agency protects "natural resources for the benefit of all." Furthermore, Sport Fish Restoration Program funds are apportioned based on all licenses sales, not just residents. Thus, a state could conceivably intend to price its licenses to maximize participation and sales across residents and nonresidents.

I will now expand the model developed earlier to show how separating licenses is useful. I frame the problem as selling resident and nonresident annual licenses, but note that the logic holds in any situation where the agency can categorize people by individual characteristics, such as age, military service, and student status. The two demands are

$$Q_R = \alpha_R - \beta_R \times P_R \tag{5.4}$$

for residents and

$$Q_{NR} = \alpha_{NR} - \beta_{NR} \times P_{NR} \tag{5.5}$$

for nonresidents. I illustrate resident and nonresident demand when $\alpha_R = 1000$, $\beta_R = 13$, $\alpha_{NR} = 450$, and $\beta_{NR} = 3$ (Fig. 5.6a). The figure also shows total demand when residents and nonresidents are charged the same price, $P_R = P_{NR} = P$. Total demand is the horizontal summation of the two individual demand curves, which is why total demand is equivalent to nonresident demand when the price exceeds residents' choke price of \$77 (in this example, nonresident demand has a higher choke price than resident demand). Now, consider a scenario when the agency charges residents and nonresidents the same price. If the agency cannot discriminate between resident and nonresident demand, then the best possible outcome is to find the price that balances total revenue and total cost, $P \times (Q_R + Q_{NR}) = \textit{Total Cost}$. In this example, total revenue equals total cost when the agency sells 1050 licenses (Fig. 5.6b).

Charging residents and nonresidents the same price is not optimal if the state wants to maximize total sales. When the price is \$25, residents buy 675 licenses and nonresidents buy 375 licenses, for total sales of 1050 (Fig. 5.6, dashed line). Yet the state can increase sales by pricing one group's license higher and the other group's license lower. Demand for the resident license is flatter than demand for the nonresident license, which indicates that the agency will sell more licenses if the state lowers the price of the resident license rather than the nonresident license, and lose fewer sales if the state increases the price of the nonresident license rather than the resident license. For example, the state could lower the price of the resident license to \$20 and raise the price of the nonresident license to \$40. Then residents would buy 740 licenses and nonresidents 330 licenses, for total sales of 1070, which is 20 more than when both licenses cost \$25. Total cost is about \$27,000 when the agency sells 1070 licenses (Fig. 5.6b). Does the agency earn enough revenue at these prices to cover its costs? A little more than enough, in fact. Total revenue is \$28,000 (\$20 × 740 + \$40 × 33). Thus, the state can increase total sales without compromising the agency's balance sheet by pricing resident and nonresident licenses differently.

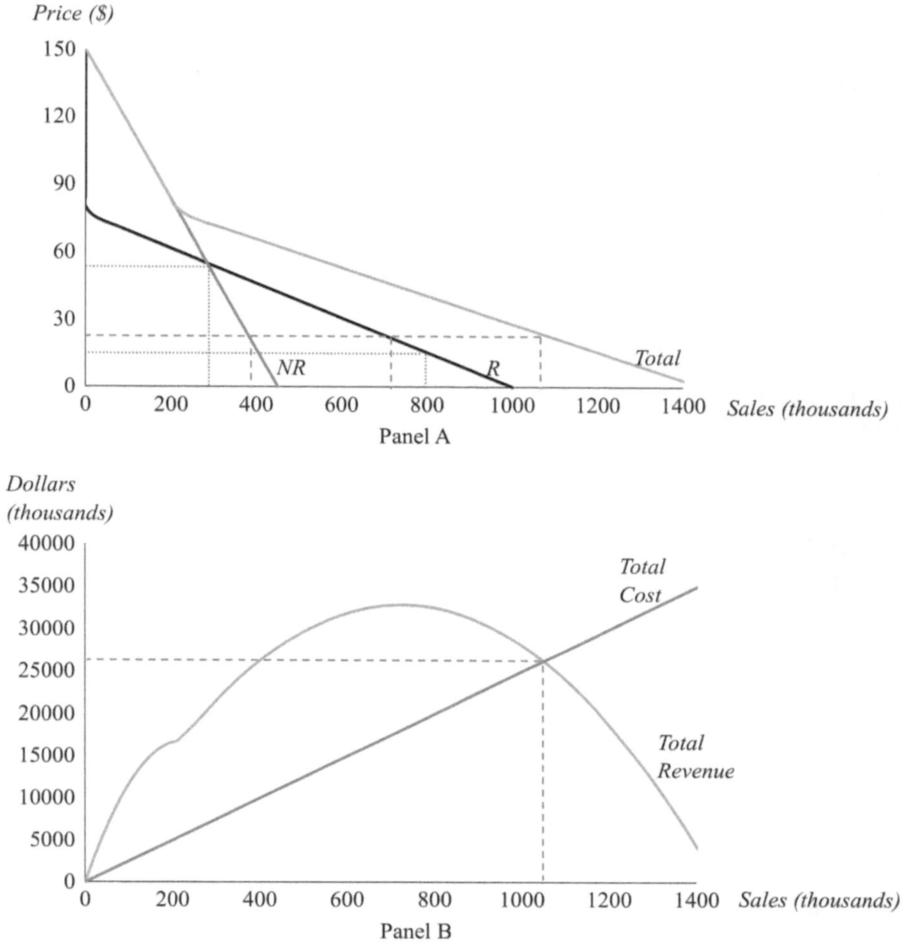

FIGURE 5.6 The relationship between price (U.S. dollars), sales, and revenue across two license demand models, when $\alpha_R = 1000$, $\beta_R = 13$, $\alpha_{NR} = 450$, and $\beta_{NR} = 3$. (a) Resident (R) and nonresident (NR) demands, as well as total demand when the license and price are the same for residents and nonresidents. (b) Total revenue when the agency sells the same license to resident and nonresidents. The dashed line shows the maximum sustainable sales level when selling a single license. The dotted lines show the optimal price and sales levels when selling residents and nonresidents separate licenses.

To be more precise about the optimal solution, the state should price resident and nonresident licenses to satisfy the conditions:

$$P_R Q_R + P_{NR} Q_{NR} = Total\ Cost \tag{5.6}$$

and

$$MR_R = MR_{NR}. \tag{5.7}$$

The first condition says that total revenue from resident and nonresident licenses must equal total cost, which is obvious. The second condition says that the marginal revenue from residents must equal the marginal revenue from nonresidents, which is not so obvious. Marginal revenue is the additional revenue from selling one more license. Thus, Equation (5.7) says that the agency should set its prices so that the revenue from selling one more resident license equals the revenue from selling one more nonresident license. This does not mean that the price of the resident license should equal the price of the nonresident license because the state faces downward sloping demand curves, so it can only sell more if it reduces the price of a license. Thus, the revenue from selling one more license (i.e., the marginal revenue) will always be less than the current price, because the state would have to lower the current price to sell it.[8] I show in the Appendix 5.B how I derived Equation (5.7).

Economists are familiar with Equation (5.7) because it is an important condition in monopoly behavior when the monopolist can set different prices for different groups. So why does the condition appear in a chapter about pricing fishing licenses? State wildlife agencies may not maximize profits, but like monopolists they care about sales and have monopoly-like pricing power. Equation (5.7) shows that license prices should bear some semblance to the prices that monopolists' charge for the same product to different consumers. For example, monopolists will separate their markets between consumers who are price sensitive and consumers who are price insensitive; and they will set a high price for the price-insensitive group and a low price for the price-sensitive group. States should pursue a similar pricing strategy to maximize sales.

Equation (5.7) can be written more intuitively. In terms of the demand models found in Equations (5.4) and (5.5), Equation (5.7) implies that

$$P_R - P_{NR} = \frac{1}{2}\left(\frac{\alpha_R}{\beta_R} - \frac{\alpha_{NR}}{\beta_{NR}}\right). \tag{5.8}$$

Equation (5.8) says that the difference between the optimal prices P_R and P_{NR} equals one-half the difference between the resident and nonresident choke prices. If the agency has separated its market by selling two different licenses and knows that models 5.4 and 5.5 closely approximate actual demand, then it can use Equation (5.8) to measure the optimal difference between the two license prices. One simply has to note where the choke prices are and then divide the difference in half. For example, suppose the resident and nonresident choke prices are \$77 and \$150, respectively (Fig. 5.6a); putting these values in Equation (5.8) yields $P_R - P_{NR} = \frac{1}{2}(77 - 150)$ and therefore $P_R - P_{NR} = -36.5$, which means that the nonresident license should be \$36.50 more than the resident license.

We can solve for the optimal prices rather than just note their difference when we know the equation for total cost. For example, suppose that the hypothetical agency in our example has a constant marginal cost of \$25, so that total cost is $25 \times (Q_R + Q_{NR})$. Then, we can use this information and the information that $P_R - P_{NR} = -36.5$ to solve for the optimal prices. I find that P_R is about \$15.25 and P_{NR} is about \$51.75, with total sales just shy of 1100 (consisting of about 800 residents and 300 nonresidents). Recall that sales were 1050 when the price was \$25 for residents and nonresidents. At optimal price–sales combinations (Fig. 5.6, dotted lines), the agency is just able to cover its costs, with about \$27,500 in both revenue and cost. This example shows that an agency can sell more by offering multiple licenses at different prices than it could if it offered only a single license.

This demonstration assumed that maximizing sales to residents and nonresidents was the objective, but what if the state wanted to maximize sales to residents only? In general, when a state wants to maximize sales across two groups, it will have negative net income (a loss) from one group and positive net income (profit) from the other; this is a natural outcome of keeping prices low for the most price-sensitive group. For example, an agency will sell the nonresident license at a high price to generate positive net income from nonresidents while selling the resident license at a low price and earning negative net income from residents. A state that wants to maximize participation among residents only will maximize net income from nonresidents and use the proceeds to lower the price of the resident license as much as possible while still covering its costs. In the previous example, the hypothetical agency earned positive net income from nonresidents but it was not quite maximizing net income from nonresidents. Based on nonresident demand (Fig. 5.6a), the net income-maximizing nonresident license price is \$87.50, with sales of 188 licenses. This is a substantially higher price than the agency would offer nonresidents if it wanted to maximize total sales. But by maximizing net income from nonresidents, the agency can now "afford" to sell 854 resident licenses. Naturally, total sales when the state maximizes sales to residents is less than total sales when the state maximizes total sales overall (1040 vs. 1100), but sacrificing nonresident sales allows the state to sell more licenses to residents (854 vs. 800).

An Application

I now estimate a demand model for nonresident annual licenses using the Pennsylvania price and sales data. I can use the parameters from this and the resident annual license demand model presented earlier to get insights into optimal price difference in Pennsylvania.

As with the resident annual license demand model, I estimate the parameters α, β, γ and δ for the nonresident demand model using ordinary least squares. I adjusted prices for inflation, denominated sales in units of 1000, and indexed the time trend to 2016. The estimated model is

$$Q = 68.4 - 0.4 \times P + 0.07 \times T - 0.01 \times T^2 \qquad (5.9)$$

All of the parameters have p-values < 0.001, except for the parameter on T, which has a p-value = 0.669.[9] The appendix to this chapter presents complete model statistics. Sales to nonresidents peaked in the late 1980s (Fig. 5.5b). The model does not predict this peak with precision, and it overestimates sales in the 1940s and 1950s, underestimates sales in the 1980s, and overestimates sales again in the 2010s. Nevertheless, predicted sales track the overall trajectory of actual sales, and the two series have a correlation coefficient of 0.86.

The first term on the right-hand side of Equation (5.9) implies that the agency would sell 68,400 nonresident annual licenses during 2016 if the price was \$0. The positive and negative parameters on T and T^2, respectively, suggest that nonresident annual license sales have followed a hump-shaped trajectory since 1929. The price parameter indicates that increasing the price \$1 reduces sales by 400. This level of change means that the state can increase agency revenue by charging nonresidents more for an annual license. During 2016, the price of the nonresident annual license in Pennsylvania was \$51, and the Pennsylvania Fish and Boat Commission sold about 39,700 licenses, so the model implies that a 10% increase in price (a \$5 increase) reduces sales by 5% (2000 fewer license sold out of 39,700). In contrast, the model of resident license demand implies that a 10% increase in price reduces sales by 2%.

What are the optimal prices for residents and nonresidents in Pennsylvania? We can plug the parameters reported in Equations (5.3) and (5.9)—that is, use these parameters to calculate the choke prices—into Equation (5.8) to see what the difference should be to maximize total sales. I find the price difference is $P_R - P_{NR} = -48.6$. The agency could therefore sell more licenses if it could find two prices \$48.60 apart and where total revenue equals total cost. In particular, based on the estimated models, I find that lowering the price for residents to \$20.20 and raising the price for nonresidents to \$68.80 would sell 724,000 resident licenses and 41,000 nonresident licenses for a total of 765,000 licenses, which is greater than predicted total sales of 761,000 at the actual prices of \$21 for residents and \$51 for nonresidents.[10] Total revenue is slightly more when the prices are \$20.20 and \$68.80 for residents and nonresidents, respectively, than when the prices are \$21 and \$51. This means that the agency can increase participation without lowering revenue by slightly lowering the price of the resident annual license, and offset less revenue from sales to residents by substantially raising the price of the nonresident annual license.

We should not, however, be too critical of Pennsylvania's actual prices. The difference ($30) is less than the optimal difference ($48.60), but clearly the right order of magnitude. Moreover, seemingly small and realistic changes to the estimated parameters show that the optimal difference could very well be close to $30. For example, I find $P_R - P_{NR} = -31.5$ if the price parameter in Equation (5.9) is 0.5 rather than 0.4, which is unquestionably possible![11] Therefore, we cannot reject the possibility that Pennsylvania's prices are optimal, based on the estimated model.

It is important to remember that the optimal prices reported here are based on models of Pennsylvania's resident and nonresident annual fishing license sales. The prices should not be interpreted as the optimal prices for those same licenses in other states or for other license types. License demand in other states will be different more or less from Pennsylvania. Even when the differences in demand are small, the optimal prices could be much more or less than $20.20 and $68.80 for a resident and nonresident annual license, respectively, because small differences in the demand parameters can lead to large differences in optimal prices.

That said, if we take the results for Pennsylvania as a rough benchmark, then the actual price differences in other states appear to be close to the optimal levels. Though most states are probably not maximizing sales precisely, because of the lack of prior research on optimal pricing, the price differences in states similar to Pennsylvania are the same order of magnitude as the optimal difference in Pennsylvania. The difference between resident and nonresident annual license prices range from a whopping $116 in Alaska to just $3 in Washington D.C., but most of the differences lie in the $30–$40 range (the average difference is $36). Moreover, the unique structure of demand in various locations could explain variation across states even if all of states want the same thing—namely, to maximize sales. For example, I find that the optimal price difference is $30 in Ohio; although the actual price difference in the last year of the data is $21, Ohio subsequently raised the price of its nonresident annual license so that the difference became $32 during 2019. States similar to Pennsylvania in terms of license sales and demographics have similar differences. For example, New York prices their resident and nonresident annual licenses $25 apart. Without more research I cannot definitively say that these states have optimally priced their licenses but, intentionally or not, most of the actual price differences are within a range that is defensible based on the models estimated in this chapter.

OTHER PRICE AND LICENSING STRATEGIES

States can also separate anglers by type when it is impossible to identify type at the time of sale. For example, some anglers are avid whereas others fish infrequently, and some target bass while others target trout. An agency can separate these anglers by selling different licenses in a manner similar to separating residents from nonresidents, and thus increase sales through optimal pricing. Separating these anglers can be challenging and present risks for the agency when there are opportunities for anglers to substitute between licenses. For example, an agency losses revenue when individuals who would have bought the annual license in the absence of the one-day license switch to the one-day license; the lower the price of the one-day license, the more individuals will switch, although offering a low-priced day license can help recruit new anglers and thus increase total sales. Agencies can price these licenses optimally to limit substitution effects. Private industry addresses substitution effects by designing price–quantity combinations that give individuals an incentive to spend more money by self-selecting into the small or large quantity group.[12]

Another licensing strategy is to offer licenses with species restrictions. Agencies can use these restrictions to monitor and regulate fishing pressure of large game fish, but they can also use them to increase the affordability of fishing for—and thus access to—smaller panfish. For example, agencies can increase participation if they offer low-priced licenses for species targeted by individuals with low demand (or low willingness to pay) and high-priced licenses for species targeted by those with high demand (or high willingness to pay).

CONCLUSION

Participation in recreational fishing may be increased by selling different types of fishing licenses at different prices. Although every state currently separates their licenses by resident and nonresident, dividing the market further would likely increase sales even more; for example, by selling single day and annual licenses (which most states do), or by selling licenses with species restrictions (which a few states do). I showed that for several states real prices have changed little over the long run, with nominal prices keeping up with inflation. I then showed that raising prices will generally increase revenue, with an application to Pennsylvania, and discussed why optimal prices do not maximize revenue. I then showed how states can increase total sales by separating their licenses between residents and nonresidents. States are currently attempting to increase significantly participation in fishing (Hiller et al. 2021*b* [Chapter 2], Graham et al. 2021 [Chapter 6], Gruntorad and Chizinski 2021 [Chapter 4]). Therefore, licensing agencies should plan to earn positive net income from one group, which can then be used to offset negative net income from another group. The optimal difference between the resident and nonresident license price depends on the difference in choke prices between residents and nonresidents, which I demonstrated for Pennsylvania. The funding paradigm for harvest management is critical for the future, and understanding the nature of demand is essential for optimally pricing licenses.

ACKNOWLEDGMENTS

The author thanks C. Braswell for providing valuable research assistance. This work was supported by the Illinois-Indiana Sea Grant College Program, grant number NA18OAR4170082.

APPENDIX 5A

Here, I present the Ohio resident annual license demand model with a quadratic trend. I estimated the parameters using ordinary least squares. The available data covers 1929 to 2015. I adjusted prices for inflation, denominated sales in units of 1000, and indexed the time trend to 2016 so that T measures years before 2016. The model for Ohio is

$$Q = 958.5 - 21.2 \times P + 24.1 \times T - 0.3 \times T^2, \tag{5.A1}$$

all parameters are precisely estimated, with p-values < 0.001 (Table 5.A1). The Ohio model predicts historical resident annual license sales in Ohio just as accurately as the Pennsylvania model for sales in Pennsylvania. The correlation coefficient between actual and predicted sales is 0.95 (Fig. 5.A1). Note that Ohio sales jumped during 1937 after the state slashed its license price from \$1.10 to \$0.60 (these prices are unadjusted for inflation; see Fig. 5.3 for the amounts in real prices).

The first term on the right-hand side of Equation (5.A1) implies that the agency would sell 958,500 licenses during 2015 if the price was \$0. The historical peak number of sales is 1,082,500 during 1987, so the model predicts that Ohio would not experience same amount of participation during 2015 even if the state gave its resident annual license away for free. The positive parameter on T and the negative parameter on T^2 indicate that factors other than price have given sales a hump shape since 1929, and imply that sales will continue to decline into the foreseeable future. The price parameter indicates that increasing the price by \$1 reduces sales by 21,200. During 2015, the price of the resident annual license in Ohio was \$19, and the Ohio Department of Natural Resources sold 636,900 resident annual licenses, which generated \$12.1 million in revenue. The model therefore predicts that if the price was raised to \$20 during 2015, the agency would have sold 615,700 licenses, and generated \$12.3 million in revenue.

TABLE 5.A1
Ordinary Least Square Estimates (and Standard Error, SE) for Parameters from Model 2 (Equation (5.2)) for Three State-Resident Types of Annual Licenses

License Type	Variable Name	Parameter Symbol	Parameter	SE
Pennsylvania, resident	Constant	A	996.210	43.338
	Price	B	–13.472	1.819
	T	Γ	14.083	1.365
	T^2	Δ	–0.234	0.015
	Observations	88		
	F-statistic	227		
	R^2	0.886		
Pennsylvania, nonresident	Constant	A	68.370	5.501
	Price	B	–0.370	0.087
	T	Γ	0.069	0.160
	T^2	Δ	–0.009	0.002
	Observations	88		
	F-statistic	111		
	R^2	0.798		
Ohio, resident	Constant	A	958.468	48.609
	Price	B	–21.172	1.964
	T	Γ	24.074	1.272
	T^2	Δ	–0.345	0.014
	Observations	87		
	F-statistic	276		
	R^2	0.909		

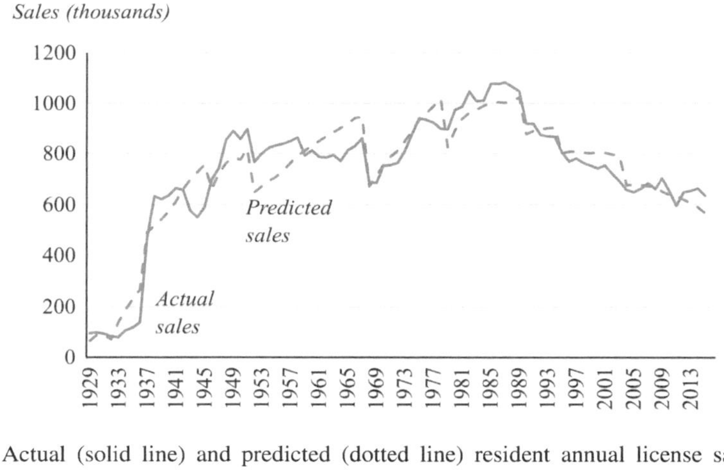

FIGURE 5.A1 Actual (solid line) and predicted (dotted line) resident annual license sales (US dollars) in Ohio.

APPENDIX 5B

The objective of a state that wants to maximize sales including residents and nonresidents can be written as

$$\max Q_R + Q_{NR}, \tag{5.B1}$$

subject to the budget constraint

$$P_R(Q_R)Q_R + P_{NR}(Q_{NR})Q_{NR} = TC(Q_R + Q_{NR}). \tag{5.B2}$$

Here I write price as a function of sales, that is, $P_R(Q_R) = \frac{\alpha_R - Q_R}{\beta_R}$ and $P_{NR}(Q_R) = \frac{\alpha_{NR} - Q_{NR}}{\beta_{NR}}$. I solve this maximization problem using the method of Lagrange multipliers, where the Lagrangian is

$$\mathcal{L} = Q_R + Q_{NR} + \lambda[TC(Q_R + Q_{NR}) - P_R(Q_R)Q_R - P_{NR}(Q_{NR})Q_{NR}]. \tag{5.B3}$$

The solution is found by solving the conditions

$$\mathcal{L}_R = 1 + \lambda[TC' - P_R(Q_R) - Q_R P_R'] = 0 \tag{5.B4}$$

$$\mathcal{L}_{NR} = 1 + \lambda[TC' - P_{NR}(Q_{NR}) - Q_{NR} P_{NR}'] = 0 \tag{5.B5}$$

$$\mathcal{L}_\lambda = TC(Q_R + Q_{NR}) - P_R(Q_R)Q_R - P_{NR}(Q_{NR})Q_{NR} = 0 \tag{5.B6}$$

where a′ indicates a derivative. Equations (5.B5) and (5.B6) yield

$$\lambda = \frac{1}{TC' - P_R(Q_R) - Q_R P_R'} = \frac{1}{TC' - P_{NR}(Q_{NR}) - Q_{NR} P_{NR}'}, \tag{5.B7}$$

which implies that $P_R(Q_R) + Q_R P_R' = P_{NR}(Q_{NR}) + Q_{NR} P_{NR}'$. This expression says that the optimal pricing strategy equates the marginal revenue from residents with the marginal revenue of nonresidents, or $MR_R = MR_{NR}$ (Equation (5.7)). Marginal revenue is the revenue from selling an additional license. Note that this condition is written as a function of license sales, that is, the quantity of licenses. Alternatively, I can express the condition in terms of license prices, $P_R - (\alpha_R - \beta_R P_R)/\beta_R = P_{NR} - (\alpha_{NR} - \beta_{NR} P_{NR})/\beta_{NR}$, which simplifies to $2P_R - \alpha_R/\beta_R = 2P_{NR} - \alpha_{NR}/\beta_{NR}$ and therefore $P_R - P_{NR} = \alpha_R/2\beta_R - \alpha_{NR}/2\beta_{NR}$. Equation (5.B6) simply restates the budget condition 5.B2, which says that revenue from resident and nonresident licenses should equal the total costs of the agency, or $P_R Q_R + P_{NR} Q_{NR} = \textit{Total Cost}$.

NOTES

1 Most state wildlife agencies do not have the authority to create licenses or set prices. Instead, these powers are held by the state legislature, although some state wildlife commissions can authorize price increases.
2 These prices exclude saltwater fishing licenses in states that separate saltwater and freshwater fishing licenses.
3 Maximizing sales is consistent with the hypothesis that bureaucracies and non-profits are service maximizers (Hewitt and Brown 2000). Alternatively, one could argue that bureaucracies are budget maximizers (Blais and Dion 1990).
4 We can therefore interpret the intercept parameter α in this model as the number of licenses (in thousands) that the agency could sell during 2016 if the price was zero.
5 I also estimated the price parameter β after differencing the model, and found the parameter to be similar

to the ordinary least squares estimate. Differencing removes problems with time series data, including series with a high degree of autocorrelation or a unit root, but precludes estimating the parameter α in the demand model. However, we can use first differencing to check the sensitivity of the β to the modeling assumptions. The parameter after differencing is –9.3 ($p < 0.001$), which is significantly lower than the ordinary least squares estimate, but still quite negative. See Melstrom (2018) for research on license demand models that uses regression methods developed for time series data.

6 The constant should vary with the resident population and thus the pool of potential license purchasers.

7 Note that if we illustrated Pennsylvania's demand and revenue curves (as in Fig. 5.4), then license sales would appear in the vicinity of the dashed and dotted lines.

8 Marginal revenue is approximately equal to $(TR_1 - TR_0)/(Q_1 - Q_0)$, where TR is total revenue, Q is quantity, and the subscript 1s and 0s indicate the amount before and after a price change, respectively. For example, we know that Pennsylvania could sell 13,500 more licenses but lose \$–400,000 (\$14,100,000–\$14,500,000) in revenue by lowering the price from \$21 to \$20, so marginal revenue is \$–400,000/13,500 = \$–29.6.

9 However, T and T^2 are highly correlated, and a joint test of significance has a p-value < 0.001.

10 The model overestimates sales slightly, because Pennsylvania actually sold 731,000 resident and non-resident annual licenses during 2016.

11 The true parameter β could be 0.5 rather than the estimated parameter 0.4. I fail to reject the hypothesis that β is 0.5 at the 95% confidence level ($p = 0.139$).

12 The optimal price–quantity combinations can depend on the seller's willingness and ability to divide up consumers. For example, Disney World sells one-day through ten-day licenses, each with a park hopper option, several half-day passes, as well as several different annual passes to its theme and water parks. The day passes are priced so that visiting for an additional day costs less than the previous day, that is each day has a progressively lower price. Users who buy the additional day passes spend more money than they would if the day pass was always the same price per day, and users who buy the annual pass spend more money than they would if Disney offered only day passes. Note that the basic Disney World theme parks annual pass is more than \$1,000, while a basic day pass is \$109. In contrast, an annual pass to Disney's water parks is \$139, and a day pass is \$69. So in the first case the annual pass is ten times more than the day pass, and in the second case it is two times more. Like Disney and its water park passes, states charge on average twice as much for an annual fishing license compared to a single-day license.

6 Marketing and Ecological Models to Predict Permit Purchasing Behavior of Sportspersons

Katherine A. Graham, Nathaniel B. Price, Valerie K. Jones, Joseph J. Fontaine, and Christopher J. Chizinski

CONTENTS

INTRODUCTION

Agencies monitor participation in recreational hunting and fishing to anticipate future capacity for fish and wildlife management and to inform decision making (Riley et al. 2003, Zinn 2003, Schorr et al. 2014). The number of game animals harvested is closely tied to the number of individuals participating in hunting and fishing (i.e., sportspersons, Gruntorad and Chizinski 2021 [Chapter 4]), and one of the key uncertainties in harvest management is sportsperson behavior (Conroy 2021 [Chapter 1], Runge 2021 [Chapter 7]). For example, the size of harvest may become disconnected from sportsperson satisfaction, which is dependent on multiple factors such as cost and access (Gruntorad and Chizinski 2021 [Chapter 4]). Therefore, understanding the behavior of sportspersons (Fig. 6.1) is essential to understanding where, when, and what fish and wildlife species are sought and harvested. Additionally, information on the behavior of sportspersons is essential for the management of sportsperson populations. Similar to how mangers manipulate wildlife populations, managers could seek to influence sportsperson populations (Martin and Pope 2011). Many examples of sportsperson management can be found in the recruitment, retention, and reactivation (referred to as *R3*) literature (Larson et al. 2014, Metcalf et al. 2015, Council to Advance Hunting and the Shooting Sports 2016, Price Tack et al. 2018), though not all agencies explicitly manage sportspersons or even conceptualize sportspersons as a group that can be managed.

Traditionally, social surveys (e.g., mail, in-person, web) of sportspersons (Kuehn et al. 2013, Laborde et al. 2014, Quartuch et al. 2017) have been used to greatly improve our knowledge of participation, and are critical for garnering support of agency programs and management efforts (Price Tack et al. 2018). Yet the costs associated with regularly surveying sportspersons brings into question the long-term viability of *only* using survey efforts (Roberts 2007) to assess sportsperson behavior. Sportspersons are often burdened with answering multiple surveys throughout the year, which can lead

FIGURE 6.1 Understanding the behavior of sportspersons is essential to understanding where, when, and what fish and wildlife species are sought and harvested. Here, a newly recruited waterfowl hunter shows off her first waterfowl harvest in the field (photo provided by the authors).

to survey fatigue and result in reduced response rates (Callegaro and Yang 2018) and careless responses (Meade and Craig 2012, Huang et al. 2015). Therefore, it is important for fish and wildlife agencies to develop new approaches to understand sportsperson participation (Galea and Tracy 2007).

Most state and provincial fish and wildlife agencies have access to important information about patterns in sportsperson participation through their electronic-licensing databases. Electronic-licensing databases hold unique identifiers, names and addresses, dates of purchase, demographic information, and types of stamps, licenses, and permits purchased during any given year. License databases provide fish and wildlife agencies with the ability to trace sportsperson transactions and behaviors over time. Further, license databases house a trove of information on sportsperson demographics and important covariates that influence the relationships between license purchasers and any number of conditions that affect purchasing behavior (Hinrichs et al. 2020, Gruntorad and Chizinski 2021 [Chapter 4]). Traditionally, license databases have been analyzed in hindsight (i.e., looking back at previous purchasing patterns) to determine how license sales respond to various factors (e.g., spring temperatures, change in license or permit costs, pandemics). Although these analyses provide important information to wildlife managers, these analyses can result in a reactionary-management approach (Allen et al. 2011, Pope et al. 2016). We contend, despite the potential merit of reactionary approaches, that there may be impactful benefits from management approaches that use forward projection of sportspersons, such that allow wildlife agencies to anticipate changes in a proactive or predictive manner (Runge 2021 [Chapter 7]).

The disciplines of ecology and marketing may seem quite disparate disciplines of science, yet both have placed an importance on tracking groups of individuals, whether that be animal populations or customers. Further, both disciplines have developed tools that can be used in a predictive capacity. Increasingly, companies apply data science methodology and customer relationship management to customer data to better understand their customers and predict future behavior. Analysis of purchase data has been common in marketing since the 1960s, progressing from simple descriptive statistics to advanced predictive models of future purchases (Fader and Hardie 2009, Kumar and Petersen 2012, Kumar et al. 2019). Over the same time, ecologists have also made great advances on a similar class of predictive statistical models, using the presence of marked individuals in a population (i.e., capture-mark-recapture) to assess the dynamics of wild populations. More recently, capture-mark-recapture models have been applied to purchase data in electronic-licensing systems (Gude et al. 2012, Schorr et al. 2014, Graham 2019). However, perhaps due to limited communication among disciplines (Pennington et al. 2013), there has been limited exchange of ideas regarding these two similar, specialized classes of statistical models, as well as limited application to understanding sportsperson behavior.

Herein, we describe the adaptation of marketing techniques and capture-mark-recapture models to license databases, as an approach to shift from reactionary management of sportspersons to a predictive and proactive approach of assessing and evaluating sportsperson populations. Although we do not explicitly discuss analyses in this chapter, we have provided data and analysis code (written in R [R Core Team 2020]) publically archived at: https://github.com/chrischizinski/angler_predictive_models) to illustrate these techniques. The overarching goals of this chapter are to (1) highlight the importance that electronic-licensing databases have in managing sportspersons; (2) provide a conceptualization of the sportsperson population in relation to important parameters derived from predictive models; and (3) describe several approaches for predictive modeling of sportsperson populations. Ultimately, these approaches will allow fish and wildlife managers to better forecast the abundance and dynamics of sportspersons and will provide the social information required to understand harvested populations.

SPORTSPERSONS AS A POPULATION: CONCEPTUALIZATION OF SPORTSPERSON PURCHASES

The conceptual description of an individual's transition through the purchasing process is essential to applying statistical models to a sportsperson population. We conceptualize the purchasing process as consisting of multiple latent (i.e., hidden) states characterized by differences in participation (Fig. 6.2). The distinction between sportsperson states is inherently probabilistic because we are attempting to differentiate between a state with zero probability of purchase (i.e., non-participant or inactive) and a state with nonzero probability of purchase (i.e., active). The problem requires that we consider the spectrum of situations between one extreme, where active sportspersons purchase at every given opportunity (100% probability of subsequent purchases), and the other extreme, where customers only purchase once (0% probability of subsequent purchase).

For the purpose of this conceptualization, participation in hunting or fishing is defined by purchasing the appropriate stamp, permit, license, or combination thereof (hereafter all referred to as license), to legally engage in the activity. As such, our conceptualization ignores illegal behaviors like hunting without the required license (e.g., poaching), because we use license purchases as a proxy for participation. We fully understand that purchasing the appropriate license does not fully equate to participation in the activity, but for the majority of sportspersons a purchase typically reflects whether the activity occurred. For simplicity, we describe our conceptualization in discrete time and define purchase histories (akin to encounter histories in capture-mark-recapture) as binary strings indicating the presence (1) or absence (0) of a purchase within the discrete-time intervals (e.g., 00100110 representing yearly purchases). Licenses may be purchased, in practice, for a variety of reasons (e.g., other recreational activities, weather, finances) throughout a calendar year (Martin 2016), but we disregard that level of detail in purchase time by summarizing in annual fashion to simplify the conceptualization.

At the base of our framework are two groups: participants and non-participants (Fig. 6.2). Participants can exist as one of two states: *active* or *temporarily inactive*. The active state occurs when the appropriate license has been purchased during the given time frame of interest. Individuals who are repeatedly active can be considered *frequent participants* and will have purchase histories with a large proportion of ones (e.g., 11110111). The temporarily inactive state

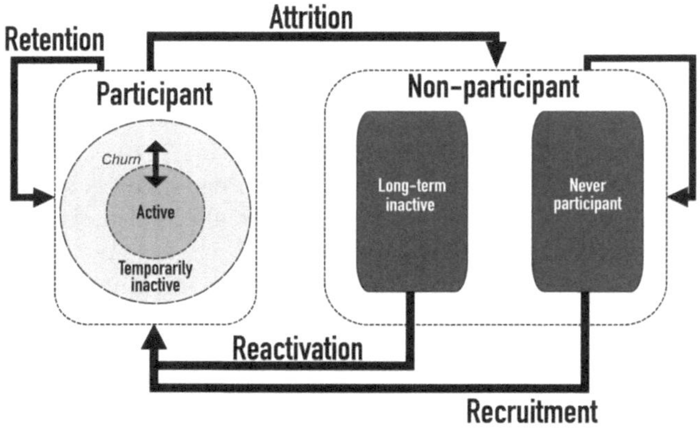

FIGURE 6.2 Conceptual model of the sportsperson purchasing process. Sportspersons exist at any point in time as participants or non-participants. Participants (left-panel) can exist as one of two states: *active* or *temporarily inactive*. Active participants have purchased a license during a given time frame. Individuals who repeatedly purchase licenses, or are repeatedly active, can be considered *frequent participants* and will have purchase histories (represented in similar fashion as encounter histories used in mark-recapture with 1 s and 0 s) with a large proportion of 1 s. Individuals who purchase the appropriate licenses semi-frequently (1 s interspersed with 0 s) fall into the temporally inactive state, termed *sporadic participants*. Retention describes the process of individuals continuing to be participants in subsequent time steps, and attrition occurs when individuals move from the participant to non-participant group (right panel). Non-participants are represented by *long-term inactive* or *never participants* (i.e., individuals who have never participated in the activity before). Individuals may move from the non-participant population to the participant population in two ways: recruitment (from never participant status) or reactivation (from long-term inactive status).

occurs when appropriate licenses are purchased semi-frequently (relative to the time frame of interest). Individuals who fall into the temporary inactive state are considered *sporadic participants* and will have purchase histories of ones interspersed with zeros (e.g., 10100101). *Churn* is the process of individuals moving from the active state to the temporarily inactive state, and is often described as a rate of individuals who participated in time step $t + 1$ that also participated in time step t. Churn rates can also be measured in time steps other than one, but it is important that the churn rate is adequately described (e.g., two-year churn rate, four-year churn rate). Our definition of a participant does not require license purchase every year (i.e., be a frequent participant), which is crucial to interpreting parameters from predictive statistical models. Further, the literature on recreational identity indicates that identifying as a hunter or angler does not require regular participation (Jun and Kyle 2012, Schroeder et al. 2013), as participants may still identify as hunters and anglers even though they purchase licenses infrequently. Thus, our statistical definition of participation fits with this concept of leisure identity.

The process of retention describes those individuals who continue to be participants in subsequent time steps, and is often exhibited as a rate, as a probability, or as the number of unique individuals (i.e., total individuals at t) retained from one year to the next (Council to Advance Hunting and the Shooting Sports 2016, Schorr et al. 2014, Price et al. 2020). Retention rates vary spatiotemporally as well as by harvest activity. For example, in Nebraska retention rates of hunters tend to be greater than 67%, whereas retention rates of anglers are closer to 57% (Hinrichs et al. 2020). The opposite process of retention is attrition, which occurs when individuals move from the participant to non-participant group. The attrition process can be a punctual (participant to non-participant in a single step) or gradual (active to temporarily inactive to long-term inactive) process. Individuals who become temporarily inactive often have a greater probability of becoming inactive for longer periods (Hinrichs et al. 2020). For example, Hinrichs et al. (2020), using discrete-time Markov chain model (Winston and Goldberg 2004) analysis, indicated that female anglers had a 0.39 probability of becoming inactive for one year, with this probability increasing to 0.75 after another year of inactivity (eventually approaching >0.90 after two more years). Generally, with five years of inactivity sportspersons can be considered long-term inactive with little chance (<0.05 probability) of becoming active again.

The other group at the base of our framework is non-participants. Non-participants also exist as one of two states *long-term inactive* or *never participants* (i.e., individuals who have never participated in the activity before; Fig. 6.2). In the United States of America, the population size of non-participants is large, much larger than the pool of participants (e.g., nationwide angling non-participants in 2016 was 86% [U.S. Fish and Wildlife Service and U.S. Census Bureau 2016]). Certified license sales (U.S. Fish and Wildlife Service 2017*a*, *b*) and national surveys of wildlife-based recreation (U.S. Fish and Wildlife Service and U.S. Census Bureau 2016) have indicated that the pool of non-participants continues to grow for a multitude of socio-demographic reasons (Winkler and Warnke 2013, Larson et al. 2014, Price Tack et al. 2018). In our conceptualization, we represent the non-participants as long-term inactive participants or never participated. In reality, it can be difficult (if not impossible) to identify these groups in a license database. Never participants do not show up in the database because they have never purchased a license. Truncated license data (see below) also excludes long-term inactive participants because they purchased licenses outside the time frame that data were retained.

Individuals may move from the non-participant population to the participant population in two ways: recruitment or reactivation. Recruitment is the process from which never participants become participants, and can be expressed as the number of unique individuals or as a rate (Council to Advance Hunting and the Shooting Sports 2016, Price et al. 2020). State fish and wildlife agencies, as well as non-governmental organizations, have expended considerable effort in recruiting new individuals into the participant pool. However, the recruitment process is complex and has multiple hierarchical forces that influence whether the individual becomes a participant (Larson et al. 2014). In many cases the increased efforts to recruit many new sportspersons has not been met. Recruitment rates also vary temporarily and by activity. In Nebraska, recruitment rates for spring turkey hunters are such that the ratio of recruits to the previous year's active participants is approximately 0.40, but the recruitment rate for waterfowl hunters is about half of the level of recruitment for spring turkey hunters (M. P. Vrtiska, Nebraska Game and Parks, personal communication). Reactivation is the process by which long-term inactive participants begin participating once again and can also be expressed as the number of unique individuals or as a rate (Council to Advance Hunting and the Shooting Sports 2016, Price et al. 2020). Reactivation can be difficult to assess and is largely dependent on the extent of the license database (see truncated data below). In a study of hunters and anglers in Nebraska, Hinrichs et al. (2020) estimated that the probability of long-term (more than five years) inactive male hunters becoming active again was <0.01.

When applying our conceptualization, it is important to consider the timespan of available license data because data availability can influence the interpretation of parameters from predictive models (Hinrichs et al. 2020). We distinguish between the effects of truncated purchase histories and the hidden nature of customer states. If there are active customers at the end of the available license data, or if customers are active prior to the available license data, then repeat-purchase data are truncated (i.e., we have incomplete information on full purchase histories). The data available from most state agencies likely fall into this situation as many fish and wildlife agencies' licensing databases have only been in existence for a decade or less. Further, many state agencies have changed license vendors or changed their databases, often resulting in customer identification codes being disconnected (i.e., broken) and limiting the temporal extent of purchase histories. If management agencies plan to better leverage their license databases, it should be a top priority to maintain the integrity of customer identification codes during transitions between vendors and database systems.

When repeat-purchase data are truncated, then we may not know (1) whether the last recorded purchase is a final purchase, or (2) if the first recorded purchase is truly an initial purchase (Fig. 6.3). The truncated purchase histories can influence the context and interpretation of the parameters from modeling approaches. For example, many sportspersons do not purchase a permit every year and instead purchase certain permits every other year or every couple of years (Hinrichs et al. 2020). Such sporadic purchasing behavior could appear to be attrition when viewed at the end of the time frame, whereas if the time frame of the data availability was expanded, the sportsperson would have the potential to become active again. In addition, we assume that recruitment coincides with initial purchase. However, this does not preclude the possibility of a segment of the sportsperson population remaining unseen during a study due to data censoring. For example, a customer recruited before the time frame of available data might not purchase during the period of the study because either they transitioned to inactive or the length of study is short relative to their purchase probability. In addition

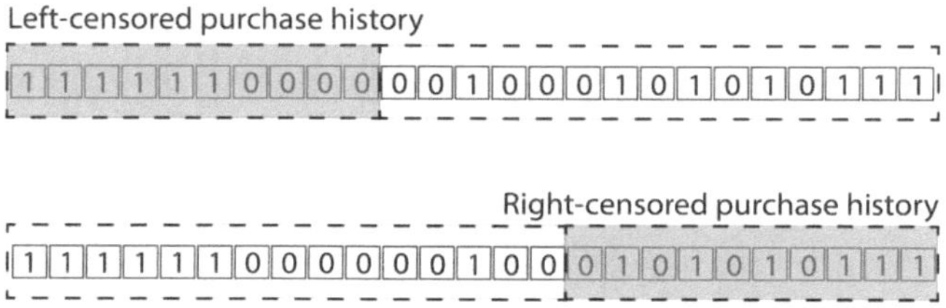

FIGURE 6.3 Illustration of truncation on a 25-year hypothetical customer purchase history. The gray boxes represent the yearly purchases that would be missing due to a truncation. The open boxes represent the data that exist in the license database. The illustration of the left-censored purchase history shows how the customer could be interpreted as a customer who did not recruit until year 11, despite already being a frequent participant in year 1. The illustration of the right-censored purchase history shows how the customer could be interpreted as someone who has lapsed in year 14, despite returning as a customer later in time.

to considering context to interpret the potential biases from short-term data strings, managers can evaluate the associated errors estimated for parameter estimates as a measure of the uncertainty in recruitment, retention, and reactivation rates.

MARKETING MODELS: BUY 'TIL YOU DIE AND REPEAT-PURCHASE DATA

Fader and Hardie (2009) reviewed specialized models in the marketing literature developed for the analysis of repeat-purchase data. Mixture models are typically used, structured with probability distributions that describe heterogeneity in purchase rates and attrition rates (see e.g., Ehrenberg 1959, Grahn 1969, Chatfield and Goodhardt 1970, Schmittlein et al. 1987, Fader et al. 2005, 2010). Marketing models have several similarities amongst themselves: (1) they often model repeat-purchase data based on recency (i.e., the timing of the last purchase) and frequency (i.e., how many purchases were made) information, as opposed to a purchase history; (2) follow a cohort of customers after initial purchase; and (3) model variation using probability distributions alone as opposed to functions of covariates (Fader and Hardie 2009). The class of specialized mixture models developed for analyzing repeat-purchase data are often referred to as *buy 'til you die* models (Dziurzynski et al. 2014, Platzer 2016), which are based on survival models, and model two processes (Fader et al. 2010). First, they model the repeated purchase process that explains how frequently customers make purchases while they "survive." Second, they model the attrition process, which describes how likely a customer is to become inactive during a period. It should be noted that though the buy 'til you die family of models has shown itself helpful in predicting future purchase behavior, they are a narrow subset of broader customer relationship management approaches to quantitatively evaluate customers (e.g., customer lifetime value, residual lifetime value, discounted expected-residual transactions). Unlike the buy 'til you die models, most empirical marketing models (e.g., Hidden Markov models, non-parametric Bayesian models, machine learning models) do not have an absorbing death state. However, there is less connection between these empirical marketing models and our conceptualization as well as the ecological models described later. As such, we focus in this chapter only on the buy 'til you die models because of their close conceptualization to the sportsperson purchase process. Conceptually it makes sense to think a company's customers act as a population in that they are recruited, they are retained for some period of time (i.e., survival), and eventually they are lost through attrition (i.e., death). Further, their parameters and interpretation are similar to those vital rates that are estimated from ecological models.

There are several types of customer models based on the buy 'til you die survival framework (Platzer 2016). One of the relatively simpler models that fits with the sportsperson-licensing context is the betageometric/betaBernoulli (referred to in literature as *BG/BB*) model that describes buyer behavior as described by Fader et al. (2010). Such an approach can be used to model customer transaction probability p (i.e., probability to purchase during a given period) and the dropout probability θ (i.e., probability to not purchase during a given period). The betageometric/betaBernoulli model can be used to model data in discrete, non-contractual settings (i.e., those settings consistent with purchasing yearly hunting and fishing licenses), and the parameters can be interpreted in a sportsperson recruitment, retention, and reactivation context. The betageometric/betaBernoulli model is related to the more widely used *Pareto/NBD* (negative binomial distribution) model that analyzes non-discrete (i.e., continuous) purchases. The betageometric/betaBernoulli has four parameters: α and β, which are used to calculate the transaction probability, and δ and γ, which are used to estimate the dropout probability. We were unable to find any applications of the betageometric/betaBernoulli models used in a sportsperson or, more generally, outdoor-recreation context in the peer-reviewed literature. However, Fader et al. (2010) used the betageometric/betaBernoulli approach to model donations to a non-profit organization between 1995 and 2000, and the authors indicated similarities between donor behavior and our conceptualization of sportsperson purchasing behaviors (Fig. 6.2). For example, the models indicated that even among frequent donors, it was unlikely that most donors would donate every year; rather, the mean was 3.75 donations in five years. Further, donors who were completely absent after their first donation were very unlikely to donate again (0.7 times in five years). However, Fader et al. (2010) noted that the size of the non-repeat donor population was much larger than the repeat donors (similar to observations among non-participants in hunting and fishing), and the individually infrequent donations were cumulatively substantial.

ECOLOGICAL MODELS: MARK-RECAPTURE AND REPEAT-PURCHASE DATA

When using capture-mark-recapture methods, animals are typically captured, marked, and later recaptured to draw inferences about the size of a population and the vital rates (i.e., survival, recruitment, population growth rate) that influence maintenance, decline, or expansion of the population (Amstrup et al. 2005). Sampled individuals from the population of animals of interest are repeatedly captured and marked with a tag or other device that can identify either individuals or batches. Successive sampling of the population and further markings of newly caught individuals provide counts of the number of marked and unmarked animals. The basic Cormack-Jolly-Seber model (Cormack 1964, Jolly 1965, Seber 1982), and its subsequent modifications and extensions, is structured such that each animal in the population has two possible states, alive or dead, and two possible outcomes, captured or not captured. In this framework, the capture frequencies of individuals are described in discrete time and defined as encounter histories (akin to sportsperson purchase histories described earlier). The encounter histories are binary strings indicating the presence (1) or absence (0) of a capture within the discrete-time intervals (e.g., 00100110 representing eight periods for potential capture). The statistical models that have been developed to analyze capture-mark-recapture data provide a tool to determine probabilities to describe whether an animal not recaptured has perished since the last sampling event or simply avoided capture (Powell and Gale 2015). Therefore, the state of the animal (i.e., alive or dead) is a latent variable that is inferred from the observed encounter history. For example, a capture history of 1100000 may strongly indicate that the individual has perished if other individuals' capture histories show evidence of high probability of capture while animals are alive, whereas if the other capture histories suggest low probability of capture (e.g., a capture history of 1100001, which indicates that the individual survived, but eluded capture four times), then an individual with a capture history of 1100000 would have a higher likelihood of being alive. The probability of survival, Φ, is an important metric for modeling and forecasting populations. Further, we can calculate the capture probability, p, which provides an estimate of the probability that given an individual survives, it is captured. Several factors can affect the probability of capture (Nichols et al. 1984*b*), including those introduced by the researcher and those learned by the animal. Another important population vital rate that can be calculated from this framework is the recruitment rate, f, which describes the proportion of new individuals who have joined the population through birth or immigration. The well-established vital rates provided by the capture-mark-recapture framework (White and Burnham 1999, Williams et al. 2002) are the basis for answers concerning fundamental questions for fisheries and wildlife management, ecology, and evolution.

Given our prior conceptualization, it is possible to extend the Cormack-Jolly-Seber framework to a sportsperson population. Sportspersons exist with two possible states, participant or non-participant, and two possible outcomes during each of a series of discrete intervals, purchased or did not purchase. Like the animal population, an important challenge is identifying when an individual does not purchase a license, whether they are still a participant or if they have ceased to be a participant (i.e., due to the non-contractual nature of fish and wildlife licenses). Therefore, the state (i.e., participant or non-participant) of the sportsperson is also a latent variable that can be inferred from observed purchase histories. However, we must "translate" the parameters calculated through capture-mark-recapture analyses into a social context. We interpret the survival parameter as an estimate of sportsperson retention (retention = 1 − attrition), or the probability that an individual is retained in the sportsperson population from one year to the next (Schorr et al. 2014). Although individuals can literally die (no longer purchasing a license), survival in this model structure can also be influenced by the multitude of factors that influence if and when sportspersons purchase permits. The interpretation of detection probability, as obtained from the capture-mark-recapture model applied to permit purchases, is a little trickier. We know that to legally participate in hunting or fishing, a sportsperson must buy the appropriate license and thus, we theoretically have 100% detection of sportsperson purchases. Therefore, we interpret the detection probability in the sportsperson context as the parameterization of the probability that the sportsperson is active in a given year, given that they have also been retained. Thus, a person with a purchase history of 11000001 is less active than a person with a purchase history of 1101111, although both have been retained or "survived" through the study period. The concept of recruitment is conceptually similar to that in the animal population. The recruitment rate is the proportion of individuals at $t+1$ (relative to the number of individuals at t) who have moved from the non-participants to the participant group (Fig. 6.2; Schorr et al. 2014, Graham 2019).

The first published example of an application of capture-mark-recapture techniques to sportsperson-license data was Gude et al. (2012), who analyzed purchasing habits of Montana residents for deer and elk licenses from 2002 to 2007. Using a multi-state model in a capture-mark-recapture framework (Nichols and Kendall 1995, Pradel 2005), Gude et al. (2012) estimated hunter retention, detection probability, and age-class transition rates. Sex and age class of residents were included in the models as predictor variables to determine the influence these factors had on deer and elk-hunter parameters. Gude et al. (2012) found that males had greater recruitment rates, retention rates, and license purchasing probabilities than females, and young adults had lesser license purchasing probabilities than the other age classes. The impact of this approach was the ability to use predictions to suggest that programs focused on increasing recruitment, retention, and license-purchase probabilities of older-age males may have greater influences on license sales than programs solely focused on youth recruitment.

Schorr et al. (2014) extended the work of Gude et al. (2012) by using a Pradel model structure (Pradel 1996) in a capture-mark-recapture framework to estimate Montana deer- and elk-hunter retention, recruitment rates, and hunter-population growth rate. They also used the population-growth-rate estimates to forecast future hunter populations. The Pradel model, an extension of the Cormack-Jolly-Seber model, is a temporal symmetry model that uses a forward-time model to estimate survival and a reverse-time model to estimate recruitment. Schorr et al. (2014) used covariates of gender, age, residency, license price, cohort, and the number of deer and elk antlerless-licenses sold (proxy for hunting opportunity) to estimate hunter retention and recruitment. The authors found that millennial hunters increased during the 11-year analysis period, driven by high recruitment rates of young hunters, especially young women. Additionally, the authors found that because baby boomers constitute such a large proportion of the hunting population, any decreases in retention of this cohort resulted in large declines in the Montana hunter population. Finally, the authors projected that the Montana elk- and deer-hunter population could decrease from approximately 190,000 during 2012 to between 165,000 and 171,000 by 2021 (>10% decline). Such a decline in the elk- and deer-hunter base is critical for managers to recognize and adapt to as they consider potential effects of regulations of the elk and deer populations, as well as losses in future funding.

More recently, Graham (2019) extended the methods used by Schorr et al. (2014) to waterfowl hunters in the Central and Mississippi Flyways. Graham (2019) used the Pradel model to estimate waterfowl hunter recruitment rates, retention probabilities, and population growth rate, and investigated the role of several socio-demographic covariates including Ducks Unlimited membership on recruitment, retention, and license-purchase probability. Similar to Montana deer and elk hunters, female waterfowl hunters were recruited at greater rates, though male waterfowl hunters had greater retention rates than females. Additionally, those in younger generations were less likely to be retained, but had greater recruitment rates as compared to baby boomers. However, covariates including hunting opportunity and rurality had only marginal effects on recruitment, retention, and purchase probability. One important factor influencing retention and purchase probability was Ducks Unlimited membership, which positively influenced both. Further, Ducks Unlimited members, who also volunteered for Ducks Unlimited, had even greater retention and license-purchase probabilities than those hunters that were just members who did not volunteer. Graham (2019) concluded that increased collaboration between state agencies and conservation organizations will be beneficial for increasing waterfowl-hunter participation, and potentially participation in other hunting activities.

MOVING TO A PREDICTIVE FRAMEWORK

Research fields related to repeat-purchase analysis and capture-mark-recapture analysis are well developed with new papers on applications and methodology extensions published every year. In addition, numerous statistical software packages are capable of analyzing repeat-purchase or capture-mark-recapture data. Thus, there is sound theory and established statistical procedures that allow for the forward projection and proactive management of sportsperson populations.

The parameter estimates obtained from the previously mentioned models are valuable for many reasons in the management of sportspersons. First, parameter estimates allow for the evaluation of recruitment, retention, and reactivation programs and efforts. Given limited resources (time, money, and personnel) constrain our natural-resource agencies, it is imperative that recruitment, retention, and reactivation programs are evaluated for effectiveness and adjusted to meet agency objectives. Models described in this chapter can provide a basis for the comparison. Second, calculating sportsperson population vital rates provides a much-needed foundational understanding of the dynamics of sportsperson populations. For example, current economic models assume homogeneity of purchasers (Melstrom 2021 [Chapter 5]), despite considerable evidence to the contrary in the human dimensions literature (Gruntorad and Chizinski 2021 [Chapter 4]). Including covariates like gender, generation, rurality, and hunting opportunity into predictive models allows for a clearer picture of who are active participants, and which participants are more likely to become inactive. By taking advantage of this information, segments of the sportsperson population can be more effectively targeted with specific programs and marketing, allowing for them to be managed (i.e., influencing where, when, what, and how often various permits are purchased) in a proactive manner.

Lastly, parameter estimates from these models can be used to simulate potential management alternatives (Gangl 2021 [Chapter 24], Morina et al. 2021 [Chapter 19], Runge 2021 [Chapter 7]). Estimates of the sportsperson population vital rates can be used to parameterize approaches like the stage-based, stochastic model proposed by Price Tack et al. (2018) or to model space-time effects of harvest pressure on local populations (Kaemingk et al. 2021 [Chapter 3]). These projection models can be used in conjunction with decision analysis to predict how sportsperson populations will respond to different management alternatives (Price Tack et al. 2018) and ultimately result in more effective management decisions (Conroy 2021 [Chapter 1], Runge 2021 [Chapter 7]).

CONCLUSION

The literature on ecological and marketing models is vast. This chapter demonstrates the usefulness of capture-mark-recapture models from ecology and buy 'til you die models from marketing in the context of sportsperson-license data; each approach has strengths and weakness. For example, capture-mark-recapture methods are widespread and well developed in modern wildlife management, which may make the interpretations easier and more familiar to managers making management decisions concerning hunters and anglers. Furthermore, capture-mark-recapture methods allow managers to incorporate individual-level and landscape-level covariates to determine what may influence sportsperson populations, which lends itself to developing marketing and strategies for recruitment, retention, and reactivation. However, we note that balanced, model-selection techniques can require enhanced computing power when many covariates are of interest; in our experience, even when using supercomputers, the models may take weeks or months to run.

Buy 'til you die models and the other customer-relationship-management models also have a long history and established theory. Used by companies all over the world, this approach provides the basis for management decisions in the private marketplace. Some of the simpler models (like betageometric/betaBernoulli) do not readily include covariates, making it difficult to identify important factors influencing purchasing trends. More complex models are currently in use, such as those used by big-tech companies (e.g., Google, Amazon) to predict consumer behavior and dramatically improve marketing to customers.

We believe that a component of a new paradigm for harvest management is for agencies to accept the perspectives offered by data-driven forward-looking management approaches, through the application of ecological models or customer-based marketing approaches. Certainly, sportsperson management has a history of relying on tradition and emotion in the hope that inactive or non-participant hunters and anglers will return to purchase a traditional product. Although an appeal to tradition and emotion has had limited success, it is not a sustainable way to sell hunting and fishing permits. In this chapter, we provided a conceptualization that can act as the foundation for further exploration into modeling the dynamics of sportsperson populations. In the next decade, human dimensions research will contribute to a new paradigm in harvest management through advancements in these model structures to enable managers to better understand sportsperson behavior. The result should be more effective solutions and proactive management for sustainable harvest.

ACKNOWLEDGMENTS

K. Graham was supported by Federal Aid in Sport Fish Restoration and Federal Aid in Wildlife Restoration project W-123-R, which was administered by the Nebraska Game and Parks Commission. C. Chizinski was supported by Hatch funds through the Agricultural Research Division at the University of Nebraska-Lincoln and from Federal Aid in Wildlife Restoration project W-120-T, administered by the Nebraska Game and Parks Commission. N. Price was supported by Federal Aid in Sport Fish Restoration and Federal Aid in Wildlife Restoration project FW-23-R, which was administered by the Nebraska Game and Parks Commission.

Section IB

Harvest Management Decision Processes

A young angler participates in a public meeting of the Florida Fish and Wildlife Conservation Commission. Photo by Tim Donovan, Florida Fish and Wildlife Conservation Commission.

7 A Decision Analytical Framework for Developing Harvest Regulations

Michael C. Runge

CONTENTS

INTRODUCTION

The design and implementation of fish and wildlife harvest regulations are decisions. Like many other natural resource management decisions, these decisions are generally made by a government agency on behalf of (and perhaps in consultation with) the public it serves (Hiller et al. 2021b [Chapter 2]) and meant to achieve some desired outcomes. These decisions can be complex and contentious: *complex*, because the system being managed is a coupled social–ecological system, the dynamics of which we do not understand well; and *contentious*, because the outcomes desired by various stakeholder groups differ and can compete with one another. How can a management agency successfully navigate this complexity and contention?

Mike Conroy, in the opening chapter of this book (Conroy 2021 [Chapter 1]), describes the underlying paradigm of modern harvest management, which might be described briefly as dynamic decision making in the face of uncertainty to achieve a sustainable harvest and meet other social objectives. It is valuable to note that the principle of sustainable yield provides the ecological foundation of this paradigm. However, modern development nests this principle within the social context of setting harvest regulations and recognizes regulations as a decision, allowing us to borrow tools from decision analysis, operations research, and other fields in the search for solutions.

In this chapter, I elaborate the decision analytical structure of fish and wildlife harvest regulations, both to describe modern approaches to harvest regulation under a single, common framework and as a process for development or revision of harvest regulations. Decision analysts recognize that all decisions have a similar set of components, captured by the PrOACT framework (Fig. 7.1; Hammond et al. 1999): a ***Problem*** context, which identifies the decision maker and frames the decision; a set of ***Objectives***, which define the long-term outcomes desired by the decision maker; a choice among a set of ***Alternatives***; an analysis of the predicted ***Consequences*** of each alternative in terms of each objective; and some way of identifying the preferred alternative, by navigating the ***Tradeoffs*** across objectives. These elements provide a way to decompose and analyze a decision, and the remainder of the chapter discusses the details of these elements in the context of harvest management.

The PrOACT framework can also serve as a process, the sequence of which is important. The decision analysis begins with an understanding of the identity of the decision maker and their authority to act; this context provides the basis for the rest of the framing. The early identification of fundamental objectives is a central tenet of value-focused thinking (Keeney 1992)—any decision is an attempt to achieve something we value, which we cannot do well unless we actually specify what we are aiming to achieve. An understanding of the decision maker's authority and objectives allows the creative generation of alternative approaches to flow naturally. The key role of science is to harness existing knowledge to make predictions about the consequences—How well will the proposed alternatives achieve the desired objectives? From a decision-making point of view, that scientific task is grounded in the value-based framing of the decision. Finally, once an evaluation of the consequences is available, a variety of tools from the field of decision analysis

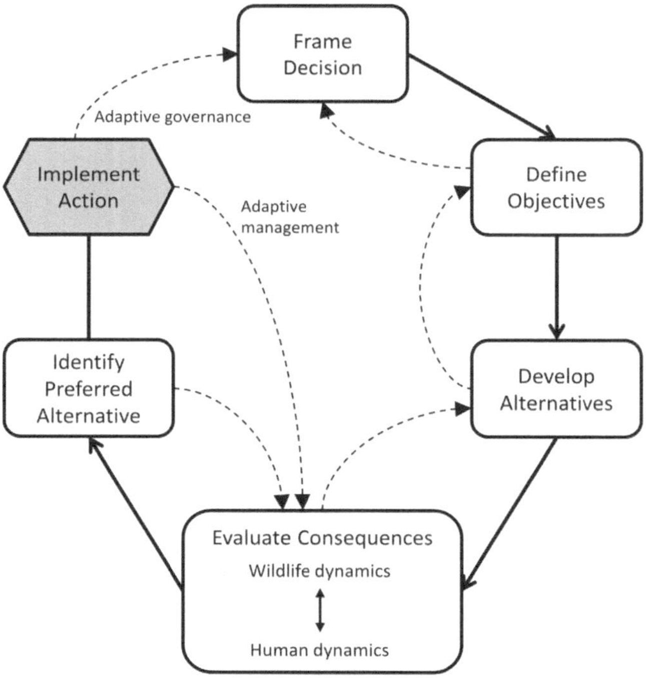

FIGURE 7.1 Decision analytical framework for developing and choosing harvest regulations. Modified and adapted from Hammond et al. (1999).

can help the decision maker navigate the challenging tradeoffs, risks, and uncertainties that may be inherent in the decision. As a process, the PrOACT sequence is not linear; there can be frequent loops to earlier steps (dashed lines in Fig. 7.1). For instance, an initial analysis of consequences might generate insights about new alternatives to consider, or even new objectives that were not explicit at an earlier stage. Two of the feedback steps have become central to the modern understanding and practice of wildlife and fisheries management: *adaptive management*, to allow learning about the system to occur in the process of management, with responsive adjustment of future actions (Williams et al. 2007); and *adaptive governance*, as experience managing a system generates insights about the very framing of the problem, leading perhaps to changes in objectives, alternatives, or even the institutions in charge of the decision (Pahl-Wostl 2009). Federal, Tribal, State, and Provincial agencies across the world are increasingly recognizing the value of structured decision-making approaches for establishing and modifying harvest regulations, as a way to engage stakeholders, incorporate science, and be transparent to the public.

DECISION CONTEXT

Hunting, trapping, and fishing all occur in a cultural context, which can take many forms, including subsistence harvest in indigenous communities, landowner-based rights found in Western Europe, and shared public resource settings familiar to those in North America. Many of the underlying principles discussed by Conroy (2021 [Chapter 1]) hold across all these contexts (Runge et al. 2004), but the decision-making processes can be quite different, because the decision authorities are very different. Modern developments in harvest management, however, have arguably been led by North American agencies and scientists, and the contents of this book reflect an emphasis on the North American setting. Harvest is a central concept in the North American Model of Wildlife Conservation (Geist et al. 2001), under which wildlife and fisheries are treated, both legally and culturally, as shared community resources, and where the desire for sustainable harvest motivates conservation. The opportunity for (or at least the promise of) access to harvest by all citizens has generated such public interest (and also such public contention) that there have been enough resources for the science and practice of harvest management to advance in rich ways. In the sections that follow, the North American Model strongly influences the framing of harvest management decisions. I undertake this narrowed focus without apology, but do note that this is not a universal treatment of the decision context for harvest management.

THE DECISION MAKERS AND THEIR AUTHORITY

Under the North American Model, the management of harvest is strongly influenced by the Public Trust Doctrine (Geist and Organ 2004, Hiller et al. 2021b [Chapter 2]), the notion that wildlife and fish are resources held in trust by government entities for public use. Thus, the decision makers who set harvest regulations are doing so on behalf of the public whom they serve, and in modern democracies, the authority for the decision makers to act is granted by the public, often through legislation enacted by elected representatives. In North America, the decision makers are typically Federal, Tribal, State, or Provincial agencies; it is rare that the jurisdiction for harvest management falls to more local authorities. The authority for decision making varies across taxa, depending on the governing statutes. Harvest management of migratory birds in Canada, the United States of America, and Mexico is under the authority of the respective Federal governments (Dahlgren et al. 2021 [Chapter 21], Vrtiska 2021 [Chapter 20]), largely because of treaties that recognized the necessity of cross-border coordination (enacted in U.S. law as the Migratory Bird Treaty Act of 1918 [MBTA], 16 U.S.C. §§703–712). Authority for harvest management of terrestrial mammals (including ungulates and furbearers) typically rests with Tribes, States, or Provinces (Hiller et al. 2021b [Chapter 2], Diefenbach et al. 2021 [Chapter 22]), unless the species is listed under the U.S. Endangered Species Act of 1973 (16 U.S.C. §1531 et seq.) or the Canadian Species at Risk Act

(S.C. 2002, c. 29), although circumstances when a listed species can be harvested are rare (e.g., subsistence harvest, incidental take, or conservation measures). Harvest of marine mammals in North America, where allowed, is generally managed at the Federal level: the U.S. Marine Mammal Protection Act of 1972 (MMPA, 16 U.S.C. §1361 et seq.) prohibits take of marine mammals except in a few situations, notably involving Alaska Native communities or as take incidental to otherwise lawful activities; Canada's Marine Mammal Regulations (SOR/93-56, under the Fisheries Act, R.S.C., 1985, c. F-14) allow harvest of cetaceans and seals under limited conditions. Management of harvest of marine fisheries in North America, which includes a substantial proportion of commercial harvest, occurs at both Federal and international levels, with considerable input from States and Provinces. Management of inland fisheries, by contrast, tends to occur at the Tribal, State, and Provincial levels (Gangl 2021 [Chapter 24]), except in the case of anadromous fish, where coordination with Federal agencies is sometimes required. The theme here is that the starting point for harvest management is the statutory authority granted by a governing body—the statutory authority identifies the decision maker, specifies the scope of their authority, and often outlines the nature of the process for setting regulations.

Stakeholders

In the spirit of the Public Trust Doctrine, if resources are managed in trust for the benefit of the public, then involving the affected public in the decision making makes eminent sense. Indeed, harvest management in North America typically includes considerable stakeholder involvement. From a decision analytical standpoint, stakeholders and decision makers are not equivalent; stakeholders do not have the authority to make the decision. But stakeholders, as the public for whom the resources are held in trust, can be incredibly helpful to the decision makers in at least three ways. First, stakeholders provide a test of the public trust, affirming the legitimacy of the agencies to act. Second, stakeholders can provide valuable input about the fundamental objectives being sought through the decisions. Third, because harvest regulations are actions taken on a coupled social–ecological system (Kaemingk et al. 2021 [Chapter 3]), stakeholders can provide insights on the efficacy of proposed regulations.

Examples of stakeholder engagement in harvest regulations cross all taxa and are documented in many of the chapters in this book. The process of setting migratory bird hunting regulations in the United States of America involves the U.S. Fish and Wildlife Service (the Federal agency with delegated jurisdiction under the MBTA for the overall harvest framework), which works closely with four Flyway Councils. Each Council is a consortium of State, Tribal, and non-governmental agencies who represent stakeholder interests (Vrtiska 2021 [Chapter 20]). Marine commercial fishing regulations are established by the U.S. Department of Commerce, National Oceanic and Atmospheric Administration, National Marine Fisheries Service, which receives recommendations from the regional fishery management councils, eight stakeholder councils established under the Magnuson-Stevens Fishery Conservation and Management Act of 1976 (16 U.S.C. §1801 et seq.). State agencies have a variety of ways of engaging stakeholders in setting harvest regulations for terrestrial mammals, non-migratory game birds, and stepped-down regulations for waterfowl (e.g., Fuller et al. 2021 [Chapter 8]). Perhaps some of the most important modern advances in harvest management in the late twentieth and early twenty-first centuries have been the sophisticated ways of involving stakeholders in the decision-making process (Conroy 2021 [Chapter 1], Hiller et al. 2021*b* [Chapter 2], Kaemingk et al. 2021 [Chapter 3]).

MANAGEMENT OBJECTIVES

Decisions about fish and wildlife harvest management can be contentious because the public has competing objectives for the resources; understanding and clearly articulating the objectives embodied in the enabling legislation and the stakeholder desires is the central task of value-focused decision making (Keeney 1992). One of the ways that harvest management has substantially advanced in the last two decades is through a more profound understanding of the complexity of the objectives being sought.

The foundational notion of sustainable harvest (Conroy 2021 [Chapter 1]), which we can trace to Leopold (1933), embeds two important fundamental objectives: the desire for *extraction* (say, to maximize annual yield) and the desire for *conservation* (to maintain the population, as well as the opportunity for harvest, in perpetuity). The tension between these objectives and the tradeoffs needed to balance them is the hallmark of sustainable harvest. Although those objectives describe two critical aspects of harvest management, they only begin to capture the range of outcomes that we desire from harvest management systems. I have proposed a generic objectives hierarchy for harvest management—an attempt to describe the full range of outcomes we seek, at least in caricature (Table 7.1). At the higher level, the hierarchy is broken into six categories of objectives: extraction, conservation, damage control, ethical treatment, human dimensions, and economic outcomes. At the lower level, the *fundamental objectives* are described, at least in general terms. The use of this hierarchy in practice would require the decision makers and stakeholders to identify which objectives were important and to describe them in detail appropriate for the specific context.

Under the category of extraction is the foundational desire to maximize annual take (Table 7.1). This objective is easy to state, easy to represent mathematically, and seems to intuitively capture a central purpose of harvest management. But the motivations for this objective may be complex and may reflect aspects of other objectives (Table 7.1). Maximizing annual harvest contributes to economic benefits, be they subsistence harvest or commercial value. Maximizing annual harvest also acts as a proxy for harvest opportunity or other elements that motivate recreational hunters, trappers, and anglers. Unpacking why we would want to maximize annual take leads to consideration of this range of other objectives.

TABLE 7.1
A Generic Objectives Hierarchy for Evaluating Harvest Regulations

Category	Fundamental Objectives
Extraction	Maximize annual take
Conservation	Maintain the target population in perpetuity
	Minimize impacts on non-target species (i.e., incidental take)
Damage control	Minimize the effects of overabundant or nuisance species on ecosystems, ecosystem services, or other outcomes valuable to humans
Ethical treatment	Maximize ethical treatment of animals, including humane handling and fair chase, and minimize unretrieved take
Human dimensions	Provide equitable access, opportunity, and allocation for interested hunters, fishers, and trappers
	Maximize participation by hunters, fishers, and trappers, through recruitment, retention, and re-activation
	Maximize satisfaction of hunters, fishers, and trappers
	Maximize non-consumptive recreational opportunity associated with the target species (e.g., birdwatching, wildlife watching)
Economic outcomes	Provide opportunity for subsistence harvest, in appropriate circumstances
	Maximize commercial value of the harvest
	Maximize the indirect economic value of the harvest activity (through gear sales, guiding industry, and other benefits to local economies)

When harvest pressure is low relative to the capacity of a wild population to sustain it, conservation objectives do not rise to a level of concern. But, worldwide over the last four centuries, we have had to learn time and again that wild populations are not inexhaustible resources: large whales in the 1800s (Dolin 2008); beaver (*Castor canadensis*), waterfowl, and wading birds in the early 1900s (Dunlap 1988, Organ et al. 1998); pelagic fisheries in the 1970s (Myers et al. 1997)—these examples and others have taught us the importance of embedding conservation objectives in our harvest management systems. We have also learned that harvest can have a conservation impact on non-target species (by-catch, incidental take). Many of the provisions of the MMPA, for example, are designed to address by-catch of marine mammals in commercial fisheries (16 U.S.C. §1387).

On the other hand, we have also begun to see harvest management as a tool to manage damage from overabundant species (Paukert et al. 2021 [Chapter 18]). In ecosystems that have become degraded, through habitat modification, loss of predators, or introduction of non-native species, some fish and wildlife populations can grow, possibly exceeding our social tolerance for them, if they have effects on other resources of value to us. For instance, beavers have increased in population in much of the eastern United States of America over the past 30 years, because of absence of natural predators and a decrease in interest in recreational trapping (Organ et al. 1998). It is likely beavers are only returning to levels that would have been common 400 years ago. In the meantime, however, human land use has greatly expanded, and the impact of beavers on low-lying agricultural fields, county roads, and other infrastructure can be substantial. Similar examples abound in terrestrial (e.g., white-tailed deer, *Odocoileus virginianus*), wetland (double-crested cormorant, *Phalacrocorax auratus*), freshwater (grass carp, *Ctenopharyngodon idella*, and related species), estuarine (blue catfish, *Ictalurus furcatus*), and marine habitats. In some cases, reduction of overabundant populations is sought through harvest management (Paukert et al. 2021 [Chapter 18]).

Other objectives associated with harvest management concern the ethical treatment of animals. Often these values are reflected in the practices that are allowed and the expectations for how hunters, trappers, and anglers will pursue their prey. For example, regulations for terrestrial hunting often forbid the use of baiting stations (Virginia Department of Game and Inland Fisheries 2014); one of the explanations for this might be an ethical objective associated with fair chase (Posewitz 1994). Trapping regulations have seen significant changes in the last several decades in an effort to reduce suffering of trapped or snared animals (Hiller et al. 2021a [Chapter 23]). Ethical treatment objectives can arise from the fishing, hunting, or trapping community, but can also arise from broader public concern about animal welfare and can be important in justifying harvest of wild animals to stakeholders who are opposed to the very concept.

Other objectives for harvest management focus on the human experience, rather than the effect on the animal population. These objectives include desires for equitable access to and allocation of harvest; maximizing opportunity (Fuller et al. 2021 [Chapter 8]); maximizing participation through recruitment, retention, and reactivation (Gruntorad and Chizinski 2021 [Chapter 4]); and maximizing hunter, trapper, or angler satisfaction (Robinson et al. 2021 [Chapter 9]). In addition, agencies are also beginning to consider the importance of non-consumptive recreational activity associated with harvest species (e.g., the value of waterfowl to birdwatchers; Devers et al. 2017). Through stakeholder forums and surveys, agencies seek to better understand what humans are seeking through their interactions with wildlife (Gruntorad and Chizinski 2021 [Chapter 4]), and to consider how those objectives might influence harvest regulations.

Finally, there are a set of objectives for harvest management that are associated with economic outcomes. In some cases, a pertinent economic outcome is subsistence harvest, and the provision of satisfactory opportunity for such harvest is important (Natcher 2009). In other cases, maximizing the sustained commercial value of the harvest is important, for example, in furbearer management (Hiller et al. 2021a [Chapter 2]) or in many fisheries. Another economic objective might involve indirect economic value, for example, through gear, guiding services, damage control, or other benefits to the local economy (Munn et al. 2010).

From a process perspective, early and clear articulation of the fundamental objectives being sought through the harvest regulations invites stakeholder engagement, provides transparency to the public, and helps to guide the evaluation of alternatives. A first step is simply the realization that the set of objectives being sought is often quite diverse and nuanced; a generic objectives hierarchy (like Table 7.1) can help to generate thoughtful discussion about what objectives might be germane in a particular setting.

MANAGEMENT ALTERNATIVES

An agency in the midst of setting harvest regulations is essentially making a choice from among the universe of possible harvest regulations that could have been set. What does that universe of options look like? Harvest regulations tend to be fairly complex: they have elements that are fixed (e.g., methods of take) and elements that can vary from year to year (e.g., season length or quota); there are temporal and spatial details; they might be state-dependent (i.e., the regulations in a particular year depend on the status of the harvested population); there are licensing requirements. It is useful to think of a set of harvest regulations as a portfolio of items, much like a portfolio of dishes that you order from a menu. There are many ways you can combine a choice of appetizer, a choice of salad, a choice of entrée, a choice of dessert, and a choice of beverage. The menu provides the set of options in each category; the set of alternatives is all the possible ways of combining those choices. In that spirit, I propose a menu of the items that might be under consideration as part of a harvest regulations package (Table 7.2).

The timing and length of the hunting, trapping, or angling season is one of the important elements of a harvest regulations package, and one that is often used to adjust the harvest in response to its status or to environmental conditions. The timing of the season often coincides with some aspect of the life history of the species (e.g., avoiding harvest during the breeding season, or focusing harvest during migration). Season length affects the degree of opportunity and is often highly correlated with harvest rate. In developing regulatory alternatives for consideration, agencies sometimes specify season length as a continuous variable and sometimes as a discrete variable. As an example of the latter, season length options for mid-continent mallard (*Anas platyrhynchos*) in the U.S. Mississippi Flyway are currently restrictive (30 days), moderate (45 days), or liberal (60 days, U.S. Fish and Wildlife Service 2019a).

Another set of options that agencies use in setting harvest regulations is the spatial areas in which harvest can occur. These spatial designations can apply to all participants, for example, hunting zones that have specific rules associated with them. But, in some harvest settings, the spatial zones are associated with individual harvesters, for example, traplines for furbearers that are licensed to only one trapper (Hiller et al. 2021a [Chapter 23]). Spatial zones allow agencies to focus higher harvest in areas that can sustain it and decrease harvest elsewhere.

Along with season length, one of the primary ways that agencies try to control the magnitude of harvest is through daily, possession, or seasonal harvest limits. For example, again in the Mississippi Flyway, the daily bag limit for mallard is two birds in restrictive seasons and four in moderate or liberal seasons (U.S. Fish and Wildlife Service 2019a). These limits frequently apply to individuals, but in some cases, the limit is for the aggregate across all harvesters. For example, in 2020, the Puget Sound Pacific halibut (*Hippoglossus stenolepis*) recreational fishery had a quota of 77,500 pounds; anglers report their catch on a daily basis, and the season closes when the quota is reached (Washington Department of Fish and Wildlife 2020a).

Some of the most complicated elements of harvest regulations are aspects that aim to focus harvest on specific segments of the population, often through differential harvest limits. These harvest limits can differ by species (e.g., different bag limits for various duck species, Vrtiska 2021 [Chapter 20]), by size (e.g., minimum, maximum, or slot lengths for fish; Gangl 2021 [Chapter 24]), by size- or age-related features (like antler size in ungulates; Morina et al. 2021 [Chapter 19]), or by sex or reproductive condition (e.g., male-only harvest; Diefenbach et al. 2021 [Chapter 22])).

TABLE 7.2
Potential Elements of Harvest Regulations Packages

Regulation Category	Regulation Element
Season dates and lengths	–
Spatial zones	As applied to all participants (e.g., hunting zones) or to individuals (e.g., licensed traplines)
Daily, possession, and seasonal harvest limits	By species
	By size (min/max/slot), age, or related features (e.g., antlers)
	By sex or reproductive condition
	As applied to individual hunters, fishers, or trappers, or to the aggregate of them (e.g., an aggregate quota)
	Conditional opportunities (e.g., "earn-a-buck")
Other aspects	Shooting hours
Participation incentives	–
Licensing	Eligibility (residency, training requirements, age, etc.)
	Fees
	License structure
	Compliance requirements: permits, tags, reporting
	Monitoring requirements (especially for commercial fisheries)
Allowed methods of take	Allowed gear (types of gun, ammunition, tackle, traps)
	Allowed methods of finding and attracting targets (bait, dogs, calling, lures, etc.)
	Lethal removal versus catch and release
Commercial regulations	Can the harvested animals be sold commercially?
Public access	--
Population augmentation	Species introduction or reintroduction
	Stocking
	Habitat management
Other population control methods	Conservation (depredation) orders
	Sharpshooters
	Electrofishing
	Sterilization
	Introduction of predators
	Pesticides and piscicides (e.g., rotenone)

Conditional harvest limits are also used, such as an *earn-a-buc*k scheme whereby a hunter cannot harvest an antlered deer until they have first harvested an antlerless one (Diefenbach et al. 2021 [Chapter 22]).

Increasingly, as agencies and non-government organizations have seen a decline in the number of people participating in recreational hunting, trapping, and fishing, there has been interest in creating participation incentives, designed to attract new participants or reactivate previous participants (Gruntorad and Chizinski 2021 [Chapter 4]). Sometimes these incentives might be variations on other elements proposed (Table 7.2), like special season dates for youth or special site access. For example, Maryland offers Junior Hunt Days for deer, turkey, and waterfowl, for hunters 16 and younger accompanied by a mentor (Maryland Department of Natural Resources 2020). Other incentives might be unique elements of a harvest management program, such as a reverse mentoring program for older fly anglers, or a fishing daughters club (Bylander 2016).

The details of the licensing process are also part of the harvest regulations package. Eligibility for a license, including residence, training, and age requirements, can help to manage the pool of participants. Where demand for opportunity greatly outstrips the capacity of the wild population to sustain harvest, limited-entry schemes such as lotteries and waiting lists can be used to identify who is eligible for a license in each season. Fees and license structure can be used both to limit and to incentivize participation; agencies typically have different types of licenses with different prices, segmented by factors like residency (resident, non-resident), age (youth, regular, senior), and duration (one-day, three-day, seasonal, lifetime), and how they design and price these options can affect participation and revenue (Melstrom 2021 [Chapter 5]). License tiering could also be an option, say, having several classes of license that differ in the regulations associated with them (Vrtiska 2021 [Chapter 20]). Other elements of licensing include compliance and monitoring requirements, for example, tags that need to be attached to a hide, pelt, or fish to verify it was caught legally; reporting requirements; and, in the case of commercial fisheries, requirements to have independent monitors on board to document the catch and by-catch.

Harvest regulations also need to specify the allowed methods of take. This includes the allowed gear, whether it be the type of hunting implement (rifle, shotgun, bow), the type of ammunition (especially, lead or lead-free), the type of fishing tackle, or the type of trap or snare. The methods of take can also include how the prey is found and attracted, for example, whether bait is allowed (often prohibited for terrestrial mammals like deer, but normal practice for fishing), whether dogs are allowed (as in fox, bear, or cougar hunting), and a range of other practices. The allowed methods of take can also govern the disposition of prey once caught, notably whether a recreational fishery is catch and release (which is typically coupled with a requirement for barbless hooks). It is worth noting that the allowable methods of take often interact with the seasonal timing, so that, for instance, bear hunting with dogs may have a separate season than bear hunting with bait and archery, muzzleloader, and gun hunting of deer may all have separate seasons.

As noted earlier, the statutory setting will often describe whether commercial harvest is permitted or not. When permitted, the harvest management strategy can also include elements that are specific to the commercial aspects that are not otherwise specified by other elements proposed (Table 7.2). These regulations could apply to commercial landings of fish, for example, specifications about how the provenance of the catch is meant to be verified. Regulations might also apply to settings where the commercial aspect is not the hunting or fishing, per se, but rather the guide or charter service.

Access to land or water on which to hunt, trap, or fish is a critical aspect of opportunity, and one that can be part of the harvest management strategy. Private lands can provide access for owners, shareowners (e.g., hunt clubs), guests, lessors, or clients, but under the North American Model, fair access to public lands is a central component (Organ et al. 2012). Agencies have many tools for apportioning access to public lands and waters, while also balancing conservation objectives, such as lotteries, permits, residency requirements, and seasonal restrictions. Agencies also have tools for promoting access, like information sharing, private–land partnerships, and land acquisition.

The last two elements of a harvest regulation strategy concern other methods of managing the harvested population, including both augmentation and control. On the augmentation side, public agencies and non-governmental organizations can seek to increase the population available to be harvested by introducing or reintroducing species into suitable habitat, as in the case of the efforts of the National Wild Turkey Federation and the Rocky Mountain Elk Foundation. Stocking is often a common practice, especially in inland fisheries, where the intent is not to establish a self-sustaining population, but rather to provide a continuous supply of new stock from a hatchery or other source (Gangl 2021 [Chapter 24]). One of the most important ways to augment populations that can be sustainably harvested is through habitat management. Although this topic is beyond the scope of this chapter and, indeed, this book, habitat management is often a central component of a harvest management strategy (Runge et al. 2006), particularly for wildlife.

Finally, for overabundant species that cannot be managed through harvest alone, there are a variety of methods for population control. Conservation orders can be established that promote high removal rates (Paukert et al. 2021 [Chapter 18]). Trained sharpshooters can be brought in to remove animals (Diefenbach et al. 2021 [Chapter 22]). Electrofishing can be used to stun fish, then selectively remove the overabundant or undesirable ones (Paukert et al. 2021 [Chapter 18], Sylvia et al. 2021 [Chapter 17]). Sterilization methods can be used to reduce the reproductive capacity of a population, without having to remove animals; this is most common in deer and wild horse population management (Diefenbach et al. 2021 [Chapter 22]). Sometimes sterilization is coupled with stocking, for example, introducing sterile triploid trout to provide harvest opportunity without the risk of population escape (Dillon et al. 2000). Introduction of predators can help to reduce overabundant populations. Finally, in extreme situations, pesticides and piscicides (e.g., rotenone) can be used to reduce or eliminate unwanted populations (Sylvia et al. 2021 [Chapter 17]).

There is a rich and complex array of elements (Table 7.2) that can be considered as part of a regulatory portfolio. From a process perspective, articulating all the regulatory portfolios that are being considered is valuable, because rather than just presenting a single option, the public sees (and perhaps comments on) several alternative portfolios. Further, articulation of multiple portfolios allows those alternative regulations packages to be transparently analyzed. Fuller et al. (2021 [Chapter 8]) and Robinson et al. (2021 [Chapter 9]) show examples of how multiple alternatives for harvest regulations were designed. One notable value in designing multiple alternatives is that it allows substantive, creative input from stakeholders at an early point in the process, before analysis takes place.

ANALYSIS OF CONSEQUENCES

At the consequence evaluation stage of a decision analysis, the goal is to predict how well each of the alternatives will achieve the fundamental objectives. If the framing of the decision context is done well, then the consequence analysis should be only a scientific task, in which the best available science is examined to evaluate the alternatives. The scientific assessment provides the results that allow for an informed debate about the tradeoffs inherent in selection of an alternative for implementation (see *Choosing the Best Regulations* below). Note that the central scientific task is a predictive task: the decision maker needs scientists to forecast what would happen if a given regulatory strategy was implemented.

The science of harvest assessment advanced considerably in the late twentieth and early twenty-first centuries, driven by the realization that the system being managed is not a wildlife population, but a complex, coupled social–ecological system consisting of both humans and wildlife (Kaemingk et al. 2021 [Chapter 3]). In the early and mid-twentieth century, as the quantitative science of harvest management developed, the paradigm focused on how harvest regulations affected harvest rate, which in turn affected the wildlife population dynamics. The profound contribution of the late twentieth century, however, was to change this paradigm, and acknowledge that harvest regulations work by acting through humans. That is, regulations work by changing the behavior of hunters, trappers, and fishers, which, in turn, affects the wild populations (Fig. 7.2). Season length, daily limit, gear restrictions, and access rules—all these elements of a harvest management alternative—act by changing the participation and effort of humans; the effect of regulations on harvest rate is only indirectly through changes in human behavior. In the early twenty-first century, this understanding is advancing even further to consider the interactions between the human and wild systems, and how they affect the dynamics that are relevant to harvest management (Gruntorad and Chizinski 2021 [Chapter 4], Kaemingk et al. 2021 [Chapter 3]).

PREDICTING POPULATION DYNAMICS

The classic, quantitative harvest analysis of the twentieth century focused on the dynamics of the wild population under the influence of harvest (Fig. 7.2, gray box on left side). A sophisticated

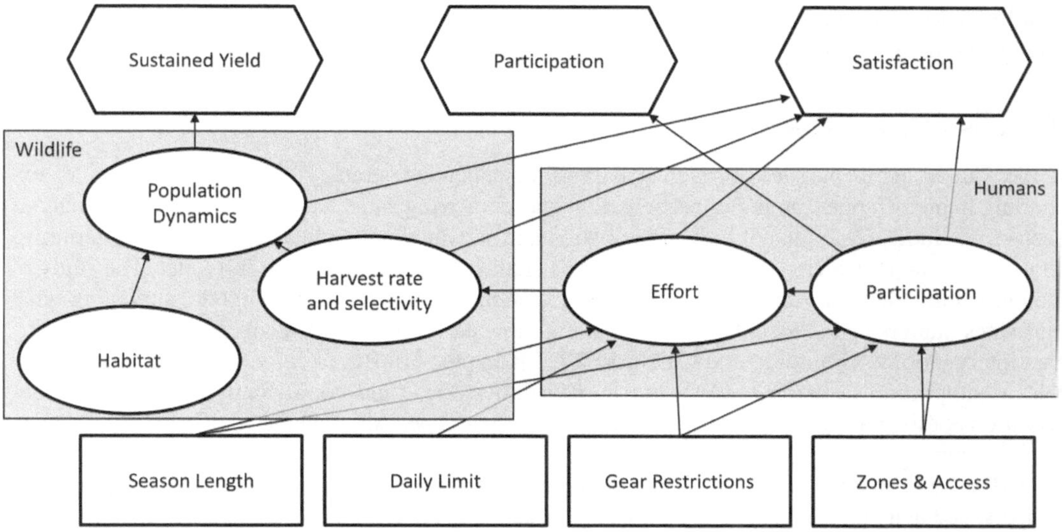

FIGURE 7.2 Schematic diagram of the coupled social–ecological system that underlies the development of harvest regulations. A subset of fundamental objectives is shown as hexagons at top. A subset of elements of the regulatory alternatives is shown as rectangles at bottom. Intermediate variables are shown as ovals.

mathematical treatment of these dynamics has been developed and provides the basis for analysis of sustainable harvesting in all taxa (Quinn and Deriso 1999, Reynolds et al. 2001). The central concept is maximum sustainable yield (Conroy 2021 [Chapter 1]), a property that arises because of internal feedbacks in the dynamics of the wild population, including negative density dependence, in which survival or reproductive rates increase as the population size decreases. Thus, part of the predictive task at the core of an analysis of alternative harvest management strategies is to forecast the feedbacks in the population dynamics as a function of harvest, tracing these forecasts to metrics that reflect the fundamental objectives associated with yield and conservation. This necessarily requires an understanding of density-dependent processes in the population, both their magnitude and their functional form (Runge and Johnson 2002). In addition, there may be other compensatory mechanisms in the dynamics that mitigate the effect of harvest, for example harvest selectivity in conjunction with individual heterogeneity (Arnold 2021 [Chapter 14]); advances are still being made in understanding these types of mechanisms.

The harvest analyses of the twentieth century made a hidden assumption—that the wild population dynamics were stationary. This does not mean that the dynamics were assumed to be deterministic, but rather that the stochastic processes describing the population dynamics were not changing over time. Land-use and climate change effects have challenged the appropriateness of this assumption. There is growing realization, then, that the predictions needed to assess harvest management also need to forecast how the habitat may change over time and, in turn, how that habitat change can affect the harvest dynamics (Nichols et al. 2011). For example, harvest strategies for polar bears (*Ursus maritimus*) may be quite sensitive to loss of sea ice and how it affects polar bear population dynamics (Regehr et al. 2017).

Harvest can have direct effects on the target population, but it can also have indirect effects on both the target population and other populations of interest (Harms and Dinsmore 2021 [Chapter 15]). Thus, the predictive task may include forecasting the effects of harvest on survival, reproduction, and movement of the target population; physiology and stress of the target population; and dynamics and conservation of non-target populations.

Given the complexity of the population dynamics of harvested species, the range of scientific predictions needed to understand the effects of harvest on wildlife and fish population dynamics could be wide. But the scope of the assessment arises from the decision framing and reflects the specific needs of the decision maker to discern among the alternative strategies being considered.

Predicting Human Dynamics

If the values being pursued through a harvest management strategy include, for example, expanding hunter, trapper, or angler participation and improving satisfaction, then a decision analysis needs to include predictions of how participation and satisfaction would differ under the alternative strategies being considered. To some extent, this challenge has not yet been fully met. The study of the human dimensions of harvest has done an excellent job in a descriptive mode, surveying what motivates hunters and anglers, and understanding how participation or satisfaction was affected by past interventions (Gruntorad and Chizinski 2021 [Chapter 4]). But to fully embrace its predictive role in supporting decision makers, the human dimensions field could focus more on how observations of the past inform predictions about the effectiveness of strategies that have not yet been implemented.

At the heart of the predictions about humans is an understanding of the behavior of hunters, trappers, and anglers, and how their behavior is affected by regulations (Graham et al. 2021 [Chapter 6], Gruntorad and Chizinski 2021 [Chapter 4], Sylvia et al. 2021 [Chapter 17]). How is the average number of days afield affected by the season length? How many hunters would avail themselves of an earn-a-buck opportunity? What will the compliance rate be for gear restrictions meant to reduce by-catch? Is hunter satisfaction increased with simpler regulations? Continued development in this field could help to move from descriptive statistics to predictive forecasts, using modeling, expert judgment, non-market valuation, and other tools for predicting human behavior.

Predictions about human dynamics are also needed to understand the economic consequences of harvest management. How would different regulatory alternatives, acting through changes in human behavior interacting with wildlife dynamics, affect the provision of subsistence harvest, the growth of local economic benefits, the commercial profit generated by harvest, and the income taken in by wildlife agencies for use in conservation? These types of economic analyses are most common in commercial fishery settings (Clark 2006), but would be relevant in many other harvest management contexts.

Predicting the Interaction of Population and Human Dynamics

At the cutting edge of current understanding of the effects of harvest management is the recognition that the dynamics of human behavior and wild populations are tightly intertwined, and that novel dynamics can arise from that interaction. There is not yet a full theory for the nature of these interactions, but many tantalizing examples of the relevance exist. Changes in angler behavior in response to regulations may produce compensatory effects in the harvest dynamics (Feiner et al. 2021 [Chapter 16]). Evolution of the target population can arise through harvest selection (Festa-Bianchet and Arlinghaus 2021 [Chapter 12]). Harvest management strategies may need to account for ecoevolutionary dynamics (Wszola and Fontaine 2021 [Chapter 13]). As for the other aspects of consequence analysis, the prediction need arises from the decision context: What dynamics need to be represented to fairly evaluate the alternative strategies being considered against the fundamental objectives that have been identified?

CHOOSING THE BEST REGULATIONS

Sometimes, the clear articulation of objectives and a detailed description of a set of alternatives, coupled with a cursory consequence analysis, is enough to identify a preferred alternative. Other times, a detailed consequence analysis is needed, but once completed, the preferred alternative is evident without much contention. But there are many cases when further steps are needed after the consequence analysis before a preferred alternative can be identified. It is at this stage (*Identify Preferred Alternative*; Fig. 7.1) that several tools from the field of decision analysis can be deployed to aid the decision maker and stakeholders in deliberating the merits of the various alternatives. The appropriate tools arise from considering the impediments to identifying a preferred alternative.

Balancing Multiple Objectives

In harvest management, as in most natural resource management decisions, one of the primary impediments is often that there are multiple objectives, the objectives compete with one another, and it is difficult to know how to balance these objectives in selecting a preferred alternative. Even the classical twentieth-century approach to harvest management (maximum sustained yield) contains an embedded tradeoff between two objectives: the desire to maximize annual harvest and the desire to conserve the population in perpetuity. In this case, the tradeoff was solved by designing an objective function that combined both objectives: maximizing the sum of current and (possibly discounted) future harvests. But as we expand the number of objectives being sought, it becomes more difficult to find a simple mathematical description that naturally captures the tradeoffs. Instead, the decision maker and stakeholders really need to grapple with how much they care about the individual objectives, how much they would trade achievement of one objective for achievement of another. This is the realm of multi-criteria decision analysis (MCDA, Keeney and Raiffa 1976), a set of techniques for understanding the tradeoffs, eliciting values preferences from the decision makers and stakeholders, and combining those preferences with the consequence analysis to identify a preferred alternative. A commonly used method is the SMART method (Simple Multi-attribute Rating Technique, Goodwin and Wright 2004); examples of this method are provided in the next three chapters of this book (Cummings and Bernier 2021 [Chapter 10], Fuller et al. 2021 [Chapter 8], Robinson et al. 2021 [Chapter 9]).

Managing Uncontrolled Variation

Harvest management strategies are often recurrent—that is, a similar decision is made each year. One of the advantages of being able to adjust the regulations each year is the opportunity to respond to variation that is outside the manager's control, such as stochastic variation in the weather affecting survival or reproductive rates in the target population or participation by the human population. The challenge in designing a harvest strategy that is responsive to uncontrolled variation is knowing how much to correct. There is a branch of engineering that studies optimization methods for such settings (so-called Markov decision problems), and methods like stochastic dynamic programming have been widely used for designing harvest strategies that are responsive to environmental variation (Johnson et al. 1997).

Managing Uncertainty

The coupled social–ecological system that is being affected through harvest regulations is complex (Fig. 7.2), and there are many aspects of it that we do not understand well. Sometimes, this uncertainty is a factor in choosing a preferred alternative—which harvest strategy best meets our objectives may depend on critical aspects of the system dynamics that we do not yet understand. How do we make choices in the face of uncertainty? It is valuable to distinguish two kinds of uncertainty: reducible uncertainty, which we can resolve through research or monitoring; and irreducible uncertainty, which we cannot resolve, either because it is not knowable before we make a decision or because we do not have the resources to resolve it (Bolam et al. 2019).

In the case of uncertainty that is irreducible or practically irreducible, the decision maker is faced with risk—the chance that the best outcome will not be achieved, owing to uncertainty.

The degree of that risk may differ among the alternatives being considered, so the decision maker sometimes needs to weigh the expected outcome against the risk of being wrong. There is a rich set of tools from the field of risk analysis that address this type of decision problem, which focuses on estimating the probabilities of the outcomes, eliciting the decision maker's risk tolerance, and combining those elements into an assessment that balances risk against reward (Runge and Converse 2020). In some cases, the probabilities of the outcomes cannot be estimated, but a robustness analysis can be performed; Tyre and Tenhumberg (2021 [Chapter 11]) demonstrate the use of info-gap decision analysis for management of harvest.

In the case of uncertainty that can be reduced, either prior to making a decision or in the course of implementing a recurrent set of decisions, the question for the decision maker is whether resolution of the uncertainty would change the choice of harvest strategy. That is, does the uncertainty matter *to the decision maker*? The expected value of perfect information (EVPI) and related methods provide a way to calculate how much the expected outcomes of the decision could be improved by resolving uncertainty prior to implementing the decision (Runge et al. 2011). These methods have been used to evaluate the importance of uncertainty in harvest management contexts (e.g., Frederick and Peterman 1995, Johnson et al. 2014).

FEEDBACKS

In many cases, there is an opportunity to adjust future decisions in response to new information because the harvest management strategy is implemented in a recurrent manner. Some of these feedback loops are shown in Fig. 7.1. Several purposes of monitoring arise from anticipating these feedback opportunities, and a well-designed monitoring system will take these opportunities into account.

State-dependent Decisions and the Need for Monitoring

For harvest management strategies that are meant to respond to changes in the state of the system (see *Managing uncontrolled variation* above), a monitoring system is needed to track the relevant system variables. This is common practice in terrestrial and aquatic harvested systems, where variables like the population size, the harvest or catch, and relevant habitat variables are estimated on an annual basis. For example, harvest regulations for mid-continent mallard are set annually using two state variables: the estimated breeding population size and a measure of the quality of breeding habitat (Johnson et al. 1997).

Epistemic Uncertainty and the Need for Monitoring

When uncertainty about how the harvested system works (so-called *epistemic uncertainty*) matters to a decision maker (i.e., when the value of information is high, and when that uncertainty is reducible), the decision maker can improve management outcomes over time by being responsive to learning. This requires a monitoring system that is designed to provide information that reduces the critical uncertainty. Adaptive management is defined by this concept: reduction of critical uncertainty during management, with responsive adjustment to the strategy implemented (Walters 1986). There are several examples of adaptive harvest management, the most notable, again, being management of mid-continent mallard (Johnson et al. 1997).

Adaptive Governance and the Need for Monitoring

Another way in which decision makers and stakeholders learn over time is simply through implementing management, which can provide insights about the initial framing of the decision, and such insights may lead to future adjustments. For instance, over time, the decision maker may realize that there are other objectives that were not originally considered, but the absence of which became apparent as experience with the harvest strategy accrued. Or, stakeholders may devise some new, creative alternatives based on insights gleaned from implementation. In the most profound cases, the decision maker and stakeholders may even realize that the institutional structures being used to manage the resource need to change. These are examples of double- and triple-loop learning (Pahl-Wostl 2009). The flexibility to respond to this kind of learning arises from the practice of adaptive governance—structuring of institutions and management processes so they can be responsive to system change, learning, and insights (Folke et al. 2005). Such adaptation is enhanced by well-designed monitoring systems that provide feedback to decision makers and stakeholders about the relevant outcomes of management.

CONCLUSION

The development of a harvest management strategy is a decision—a choice among myriad alternative strategies, designed to achieve outcomes valued by the decision maker and stakeholders. The field of decision analysis provides a rich set of tools both for structuring and analyzing decisions. Many of these tools have been applied in the context of harvest management, as all the chapters in this book attest. Modern advances in harvest management have arisen, in part, through deeper understanding of the decision contexts in which harvest management takes place, the multifaceted objectives that stakeholders desire from harvested systems, and the complex nature of the coupled social–ecological system being managed. More widespread use of decision analysis in the development of harvest regulations provides the opportunity for greater transparency, public engagement, and long-term support.

What steps can we take in the next two decades to continue to advance the quality of decision making in harvest management? First, we could benefit from increased training in decision analysis in undergraduate and graduate wildlife management programs to raise consciousness about the centrality of decision making to harvest management. Second, we could benefit from continued recognition that population dynamics and human dimensions are not two separate fields, they are coupled dynamics that cannot be fully understood separately. Third, the human dimensions field could shift from a descriptive focus to a predictive focus to provide harvest managers with predictions of how humans will respond to harvest regulations. Fourth, investment in monitoring and adaptive management could be more effective if we focused on the role of learning in improving decisions, rather than just increasing knowledge. A decision analytical focus, with the public trust doctrine as central guidance, will continue to ground the integrity and advance the relevance of harvest management in North American and beyond.

8 Engaging Hunters in Selecting Duck Season Dates Using Decision Science: Problem Framing, Objective Setting, and Devising Management Alternatives

Angela K. Fuller, Joshua C. Stiller, William F. Siemer, and Kelly A. Perkins

CONTENTS

HARVEST MANAGEMENT CONTEXT

Migratory bird hunting season setting in the United States of America is an annual process (Blohm 1989), which falls under the Migratory Bird Treaty Act of 1918 and establishes the earliest and latest dates that states can set their hunting seasons and specifies that season lengths cannot exceed 107 days (Vrtiska 2021 [Chapter 20]). The U.S. Fish and Wildlife Service works in partnership with four Flyway Councils (Atlantic, Mississippi, Central, and Pacific) consisting of officials from state and provincial wildlife management agencies. The season lengths and bag limits vary by flyway, as a result of factors including differences in bird abundance and number of hunters. Within states, hunting zones and split seasons are designed to maximize hunting opportunities throughout states according to temporal, geographic, and demographic variability in duck populations (Department of Interior, Fish and Wildlife Service 2019). In 1990, the U.S. Fish and Wildlife Service placed limits on the number of zones or splits a state may have to ensure equitable hunting opportunities throughout the flyways. States with more than three zones in 1990 (e.g., New York) were grandfathered into the new frameworks for zone and boundaries. Any significant changes in grandfathered states would need to comply with the current frameworks. Every five years, the U.S. Fish and Wildlife Service allows states to make changes to waterfowl hunting zone boundaries or the number of zones.

Each Flyway Council has a technical committee that reviews and evaluates species population status, harvest, and hunter participation data and makes recommendations to the migratory bird program of the U.S. Fish and Wildlife Service. Recommendations are based on an adaptive harvest management program (U.S. Fish and Wildlife Service 2019*a*) that explicitly accounts for uncertainty (Walters 1986) and includes monitoring (Cummings and Bernier 2021 [Chapter 10]) programs (e.g., aerial waterfowl surveys, preseason banding, waterfowl harvest information program) and factors including population size, nesting success, habitat condition, harvest rates, and hunter participation. The Harvest Management Working Group includes representatives from the U.S. Fish and Wildlife Service, the U.S. Geological Survey, the Canadian Wildlife Service, and the Flyway Councils. The working group works to develop optimal regulatory strategies that use harvest management objectives, an evaluation of regulatory alternatives (i.e., flyway-specific season lengths, bag limits, and framework dates), and population models that describe hypotheses about the effects of harvest and environmental factors on waterfowl abundance, and the relative weights or reliability of each population model (U.S. Fish and Wildlife Service 2019*a*). The Division of Migratory Bird Management reviews the recommendations from the Flyway Councils and considers current regulatory policies and species biology before making recommendations to the U.S. Fish and Wildlife Service Regulations Committee. The Regulations Committee considers the recommendations from the Flyway Councils and the Migratory Bird Program recommendations, and annual decisions are forwarded for approval to the U.S. Department of Interior, Director of the U.S. Fish and Wildlife Service and the Assistant Secretary of Interior for Fish, Wildlife, and Parks. The approved regulations are published in the Federal Register and open for public comment before the final ruling is published. Once the federal regulations are finalized, individual state natural resource agencies can then select waterfowl hunting season dates that fit within these guidelines (Vrtiska 2021 [Chapter 20]).

Harvest management objectives typically include economic, social, and ecological values (McDaniels et al. 2006, Hiller et al. 2021*b* [Chapter 2], Robinson et al. 2021 [Chapter 9], Runge 2021 [Chapter 7]), but given that the federal framework includes biological objectives, the season setting process left to the states is largely centered on social objectives. Primary among those objectives is setting seasons that satisfy the desires of waterfowl hunters. Waterfowl hunters constituted 11% of all hunters in the United States of America in 2011 (~1.5 million people) and have an important economic impact on local, state, and national economies (U.S. Fish and Wildlife Service 2015*a*). Duck hunters are arguably the stakeholders most affected by decisions about waterfowl hunting season dates.

Anyone who is affected by, or can affect, a wildlife management decision is a stakeholder in that decision (Susskind and Cruikshank 1987, Decker et al. 1996). Wildlife managers engage stakeholders in the processes of making, implementing, and evaluating wildlife management decisions (Lauber et al. 2012). When conducted effectively, engaging stakeholder perspectives in the decision-making process can yield multiple benefits including helping managers understand the needs, beliefs, values, and behaviors of those affected by a decision, informing judgment, and influencing stakeholders' beliefs, attitudes, and behaviors in ways that improve the effectiveness and durability of wildlife management decisions (Leong et al. 2012). Including stakeholder perspectives can also provide valuable local knowledge to policymakers, as well as increase the transparency and legitimacy of the decision-making process, which can increase the likelihood of social acceptance of policy implementation and increase public trust in decision makers (De Marchi 2003, Newig 2007, Reed 2008, Dietz and Stern 2008, Kahneman and Tversky 2000, Grimmelikhuijsen 2010, Robinson and Fuller 2017). Eliciting stakeholder concerns and opinions with the intent to incorporate public values in the decision-making process can be seen as a form of *participatory democracy* (Escobar 2017), often viewed as a right of citizens. Effective stakeholder engagement in decision making can ultimately lead to greater trust and satisfaction in agency management of natural resources (Decker et al. 2012).

CASE STUDY: SELECTING DUCK HUNTING SEASON DATES IN NEW YORK

In this chapter, we describe the problem framing, objectives, and alternatives steps of the decision-making process that was used by the New York State Department of Environmental Conservation (NYSDEC) to develop duck season dates within the federal framework. Our aim is to demonstrate how hunters can be integrated into a decision science approach. We focus on the process itself instead of the specifics of the modeling or the results because the problem framing, objectives, and alternatives steps of the decision process are those that engage hunters.

New York has more complexity than other states when selecting season dates. Due to a grandfathered zoning agreement with the U.S. Fish and Wildlife Service resulting from the high degree of habitat diversity across the state (e.g., significant coastal habitats on Long Island, numerous mountainous areas, lake plains marshes, and two Great Lakes), New York has more waterfowl zones than any other state in the United States of America. The existence of five waterfowl zones allows for an equitable distribution of harvest opportunity for hunters across New York (Fig. 8.1). Season dates in four of the five zones are established by the NYSDEC, and the fifth (Lake Champlain) is shared with Vermont and selected by the Vermont Fish and Wildlife board.

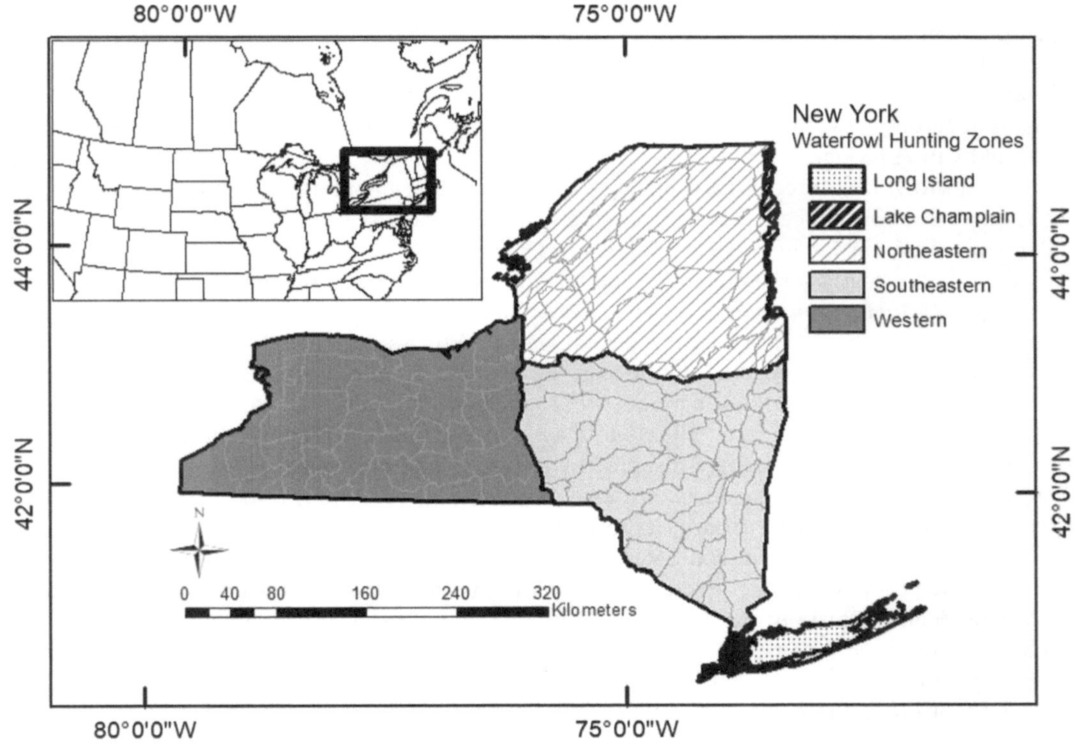

FIGURE 8.1 The five waterfowl hunting zones in New York. Each zone was evaluated separately for the decision problem of selecting duck season dates.

Complexities Associated with Setting Duck Hunting Season Dates

The NYSDEC has used various approaches over time to set season dates. In the 1990s, the NYSDEC used public meetings to obtain stakeholder input to help inform season date recommendations by NYSDEC managers (Odell 2000). After several restrictive, 30-day duck seasons in the early 1990s, hunters in the Western Zone of the state expressed frustration and dissatisfaction with the decision-making process. Hunters believed that the NYSDEC was not appropriately representing the interests of duck hunters in their zone and believed that the selected dates did not maximize duck hunting opportunity. In 1996, the NYSDEC developed and implemented a waterfowl hunter task force in the Western Zone to address this concern. The process was modeled after white-tailed deer (*Odocoileus virginianus*) citizen task forces (Hall 1991) to include the managed community directly in the decision-making process. Participants were tasked with gathering feedback from other hunters in the zone on season date preferences for duck hunting. Since 1997, the NYSDEC has implemented waterfowl hunter task forces in all waterfowl zones, with the most recent waterfowl hunter task force established on Long Island in 2006. The waterfowl hunter task forces are comprised of avid duck hunters in each zone, most often representatives of organized, long-standing waterfowl hunting organizations, as well as the New York State Conservation Council in each NYSDEC region, county sportsmen's federations, waterfowl interest groups, sportsmen's organizations with waterfowl hunting interests, and some *at-large* members to provide input for under-represented areas within the zone. Task force members are selected as representatives for all hunters within the zone. The duties of task force members are to actively solicit the opinions and desires of hunters in their zone and to be available to all hunters to receive their opinions. Task force members must be willing to meet with other task force members to discuss issues, negotiate, and ultimately, compromise in order to recommend season dates that all task force members can live with and that are satisfying to hunters in the zone. The task force in each zone meets once annually to develop season date recommendations for NYSDEC's consideration. The final decision regarding season dates and regulations are made by the NYSDEC Bureau of Wildlife Chief and Commissioner.

The waterfowl task force process engages the most avid segment of the regulated community in the season setting process. However, criticisms and concerns have been expressed, such as under-representation for unaffiliated hunters, lack of clearly defined objectives, lack of transparency of input to task force members, and lack of objectiveness of the process (Enck and Van Den Berg 2007). Hunters, especially those not affiliated with organizations, expressed concern that they were not being fairly represented (Odell 2000). Using representatives from organizations to represent the general public can be seen as a pluralist model of policymaking (Konisky and Beierle 2001) and may not truly represent the values of society at large. Some hunters expressed concern that although they could contact waterfowl hunter task force members directly, if the task force disagreed with their values or opinions, their input was not heard or considered during the decision-making meeting. Odell (2000) also noted the difficulty of getting waterfowl hunters from a large geographic area, with diverse backgrounds and interests, to agree on optimal season dates. The lack of agreement is likely driven by the complexity of the problem and the diversity of motivations for individual hunters pursuing ducks. Consensus decision making can result in inferior policy choices if a premature consensus is reached where some participants values may have been suppressed and important concerns ignored, which may be the result of influence by personalities of the participants, leading to conformity or *group think* (Gregory et al. 2001). Professional facilitation is one means to reduce influential power dynamics (Reed et al. 2018), but care must still be taken to ensure that all stakeholder views are represented.

An additional complexity for setting duck season dates is the lack of migration and abundance data at the appropriate scale. When waterfowl task force members discussed their recommendations for season dates, they often were forced to revert to anecdotal observations from a handful of locations or limited information from waterfowl counts on wildlife refuges that may or may not adequately represent migration chronology throughout the diverse habitats of each zone. Unfortunately, this strategy relies heavily on a relatively small sample size from specific locations and is rarely quantifiable. Nevertheless, understanding migration chronology of important species was identified as very important to task force members during a 2007 workshop (Enck and Van Den Berg 2007). Lack of relevant data has been identified as one of the challenges to effective agency decision making (Cummings and Bernier 2021 [Chapter 10]). Even so, this should not represent a stumbling block to decision making because models may be derived from conceptual, quantitative, or empirical field or laboratory data, or informed by expert opinion (Fuller et al. 2020).

Lastly, the lack of clearly stated objectives and goals was another major concern expressed by both the waterfowl task force and the NYSDEC. The waterfowl hunter task forces were originally asked to develop dates that maximize hunter satisfaction; however, satisfaction was not clearly defined. Hunter satisfaction is multifaceted, and understanding which type of satisfaction the agency and hunters are trying to accomplish is vital. Stakeholders may hold strong, but imprecise opinions resulting in their values and objectives being constructed rather than revealed in a decision-making process, which argues for decision aiding approaches that help stakeholders identify and define their values (Hammond et al. 1999, Gregory et al. 2012). Hunters may know the aspects of waterfowl hunting that contribute toward their satisfaction; however, they may not have an understanding of how those values align with the selection of the best waterfowl season dates, largely because it is difficult to evaluate the consequences and tradeoffs associated with different season dates without a formal quantitative process. The lack of formal methods for incorporation of stakeholder values and relevant science to generate viable management options to address a particular problem is seen as a primary failing of most stakeholder processes (Wilson and Arvai 2006). Articulation of objectives is particularly important because a wise decision is one that is focused on the achievement of objectives and follows a process that increases the odds of achieving the stated objectives (Hammond et al. 1999).

Is a New Decision Process for Season Setting Warranted?

Given feedback from the task forces and hunters, the NYSDEC recognized the need for evaluating the potential for a new process that would be responsive to the needs of the regulated community. The agency conducted an internal decision-making process in 2016 to evaluate if engaging in a new process for setting duck seasons would be warranted. The process was facilitated by someone external to the agency. The NYSDEC used a structured decision-making approach (Gregory et al. 2012, Runge 2021) [Chapter 7] to evaluate which method for season setting would best meet their objectives. Members of the migratory game bird group of the NYSDEC participated in the decision-making process. The problem statement was to select duck season dates through an objective, clear, and transparent decision-making process that maximizes duck hunter inclusion and is scientifically defensible and data driven. The fundamental objectives (the ends NYSDEC was trying to achieve) that addressed the concerns of the NYSDEC and were important relative to a process for arriving at season dates included:

1. Maximize hunter satisfaction with, and inclusion in the decision-making process
2. Ensure that the decision-making process is scientifically defensible and data driven
3. Maximize transparency of the decision-making process

There were two means objectives for achieving hunter satisfaction: (a) maximizing the representation of waterfowl hunter values and (b) maximizing the incorporation of hunter values into the final decision.

The NYSDEC generated decision-making process alternatives representing different ways to achieve the stated fundamental objectives:

1. Status quo (annual task force meetings)
2. Modified task force process #1 (biennial or triennial meetings and decisions; add additional task force members; add additional NYSDEC representation)
3. Modified task force process #2 (modified task force process #1 + define task force membership application or approval process, clearly redefine or reinforce roles and responsibilities of representatives, hire new professional facilitators)
4. Modified task force process #3 (modified task force process #2 + increased educational outreach by increasing the number of meetings prior to final season setting meeting to share data, one-year maximum term for task force members, social survey to define and evaluate hunter preferences, and geographic representation)
5. Structured decision-making process without task force input
6. Structured decision-making process with task force input on objectives and alternatives steps of the decision

To evaluate the consequences and tradeoffs, the NYSDEC used a simple multi-attribute ranking tradeoff technique where consequences (the outcome of taking a particular alternative on achievement of an objective) were elicited directly from managers and biologists using Likert scales. Objectives were weighted by the managers and biologists, where maximizing hunter satisfaction received the greatest weight (0.47), followed by ensuring a data-driven decision-making process (0.31), and maximizing transparency (0.22). The alternative with the highest expected utility value was the structured decision-making process with task force input (Fig. 8.2), which is the approach that the NYSDEC implemented. This alternative specified a formal structured decision-making framework where the waterfowl task force members were engaged in the objective setting phase and in the selection of the alternative season dates, and hunters at large would be surveyed to better understand the relative importance of hunter satisfaction objectives.

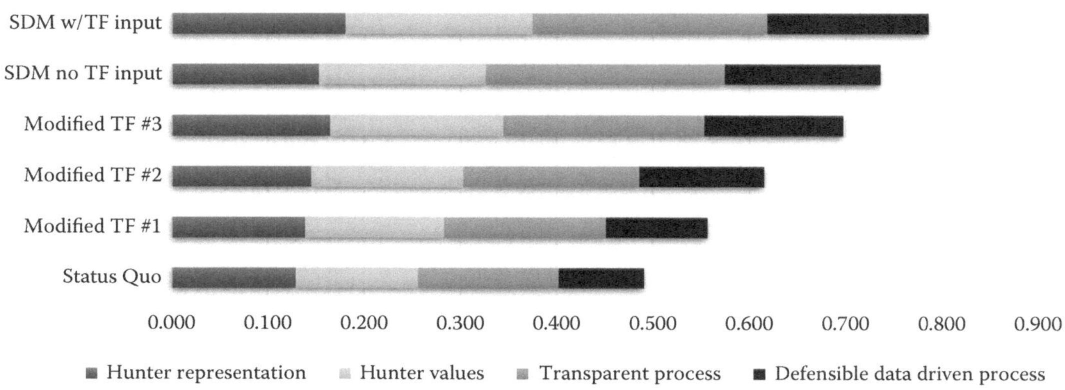

FIGURE 8.2 Decomposition of the expected utility value for the waterfowl season date decision by the New York State Department of Environmental Conservation. Each shade represents the value of the consequence of each alternative on each fundamental objective, multiplied by the weight on the fundamental objective. The total length of each bar corresponds to the expected utility value for each alternative. TF = task force. Hunter representation and hunter values are combined into the single fundamental objective, to maximize hunter satisfaction, with a weight of 0.47. The objective to ensure that the decision-making process is scientifically defensible and data driven had a weight of 0.31 and maximize the transparency of the decision-making process had a weight of 0.22.

Implementing Structured Decision Making for Problem Framing, Objectives, and Alternatives Regarding Selection of Duck Season Dates

After evaluating six possible processes for selecting duck season dates (described above), the NYSDEC selected and implemented a structured decision-making process with task force input for selecting duck season dates. The new process allowed for a greater number of duck hunters to be engaged (waterfowl hunter task force members and the general public through hunter surveys) and incorporated representative regional data on duck abundance over time (using eBird data described below). Here we outline the problem, objectives, and alternatives steps of the decision-making process.

Problem

The problem statement was to select duck season dates for each waterfowl hunting zone using an objective, clear, and transparent decision-making process that maximizes duck hunter inclusion and is scientifically defensible and data driven. The scale of the problem is statewide, but decisions can vary by waterfowl zone (Fig. 8.1), given the varying habitat conditions across the state and stakeholder views that can vary spatially (Leong et al. 2012). The decision will be revisited every five years, barring any changes to the number of days allowed by the federal framework (i.e., if the season length is shortened). Stakeholders included the waterfowl hunter task forces in each zone and a stratified random sample of duck hunters surveyed from across the state. The structured decision-making process will identify the preferred season date structure that maximizes elements of waterfowl hunter satisfaction.

Objectives

Objectives are statements that describe the values and preferences of those involved in the decision process (Runge 2021 [Chapter 7]). The NYSDEC conducted a one-day workshop with the four waterfowl task forces to identify the fundamental objectives of duck hunters across the state. Twenty of the 45 waterfowl hunter task force members participated in the workshop. The meeting was facilitated by an independent third-party facilitator to ensure that participants were free to communicate their opinions and values without perceived or actual influence from NYSDEC managers. After a brief presentation explaining the structured decision-making process (Runge 2021 [Chapter 7]) and the task force role in the process, participants were split into smaller breakout groups of six to seven participants. Each group was asked to generate objectives related to duck hunter satisfaction and a NYSDEC note taker kept a list of all objectives that were articulated. Breakout groups were then brought together, and a complete list of all objectives was shared. Task force members identified 47 potential unique objectives to maximize duck hunter satisfaction. Then, NYSDEC and the third-party facilitator led a discussion to refine the list of potential objectives to elucidate the fundamental objectives (Gregory et al. 2012) that led to hunter satisfaction with duck season dates.

The focus of the discussion was on the 13 species of ducks that are frequently harvested by hunters in New York, including wood duck (*Aix sponsa*), green-winged teal (*Anas crecca*), blue-winged teal (*Spatula discors*), northern pintail (*Anas acuta*), American widgeon (*Mareca americana*), mallard (*Anas platyrhynchos*), American black duck (*Anas rubripes*), lesser scaup (*Aythya affinis*), redhead (*Aythya americana*), greater scaup (*Aythya marila*), canvasback (*Aythya valisineria*), bufflehead (*Bucephala alebola*), and common goldeneye (*Bucephala clangula*). The six fundamental objectives as identified by the task force included:

1. Maximize the opportunity to see and shoot wood duck and teal species (i.e., being able to hunt when wood duck and teal species are most abundant).
2. Maximize the opportunity to see and shoot mallard and black duck (i.e., being able to hunt when mallard and black duck are most abundant or most susceptible to decoying).

3. Maximize the opportunity to see and shoot diving ducks (e.g., scaup, redhead, common goldeneye) (i.e., being able to hunt when diving ducks are most abundant or most susceptible to decoying).
4. Maximize the opportunity to see and shoot ANY ducks, regardless of species (i.e., being able to hunt when abundance of ducks [any species] is highest, or when the variety of duck species is greatest).
5. Maximize the opportunity to go duck hunting (i.e., including the most weekend days and holidays in the season; holidays were defined as Columbus Day, Thanksgiving Day, Christmas Eve, Christmas Day, New Year's Eve, and New Year's Day).
6. Minimize the overlap of waterfowl and deer hunting seasons (i.e., minimize the duck season overlap with the first seven days of the firearm deer season and avoid opening duck season on the same day as youth big-game hunting season).

Several of the objectives include subobjectives (Fig. 8.3). For example, maximizing the opportunity to see and shoot mallard and black duck requires that birds occur in high numbers (opportunity to *see* = abundance) and that the birds are susceptible to harvest by hunters (i.e., opportunity to *shoot* = maximize newly arriving migrating ducks via immigration). Mallard and black duck migrate later in the fall. In contrast, species like teal and wood duck have nearly reached their peak migration early in the season. Therefore, there is no way to maximize the opportunity to shoot teal and wood duck by capturing their migration; thus, we can only capture the periods when the species are most abundant.

Estimates of relative abundance for each species in each zone were derived from spatiotemporal exploratory models using eBird data from 2004 through 2016 (Fink et al. 2019), with relative abundance predicted for the year 2016. Estimated relative abundance for a species was defined as the expected number of birds encountered on a one-hour, 1-km eBird checklist beginning at 0700 hours. To measure the consequences of each alternative related to species abundance, we applied the mean relative abundance to the three days preceding and posterior to the ordinal date of the estimate (e.g., an estimate from ordinal date 286 would be applied to date 283 through 289). To estimate the period when most species are abundant, we normalized the weekly relative abundance estimates for each species individually (i.e., a value of 0 represents the lowest relative abundance and 1 represents peak abundance) and summed the normalized values for each week. We applied this value to individual days of the calendar year similar to the abundance estimates previously described. We used migration data (i.e., newly arriving migrating and immigrating ducks) to measure susceptibility to harvest, calculated as the percent change relative to the maximum relative abundance, between the weekly relative abundance estimates (e.g., the percent change between ordinal date 286 and 293). If the value indicated a positive change, it was applied to all dates between the two point estimates (e.g., day 287 through 293); if the value was ≤ zero, each date received a 0% migration value. For the any duck objective, to measure the variety of any duck

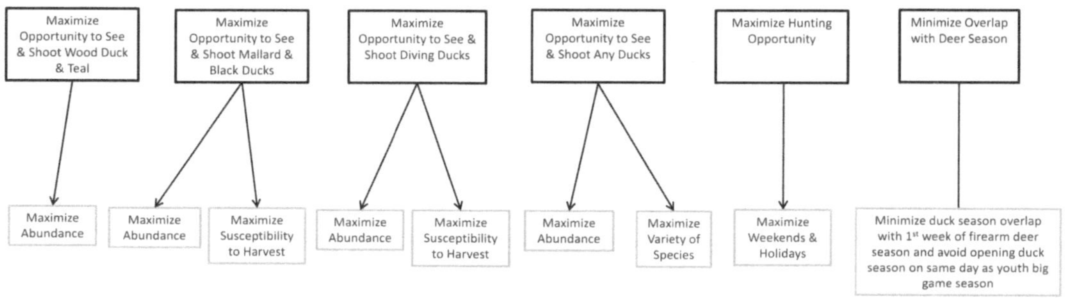

FIGURE 8.3 Objectives hierarchy for deciding dates for waterfowl harvest season.

species, we measured species diversity using a Shannon evenness index (Shannon 1948) using the relative abundance of all species. For richness, we calculated the number of species present on a given date at a relative abundance ≥ 0.99.

To understand how important each of the objectives were to the waterfowl hunting community, we conducted a statewide survey with a stratified random sample of 6000 waterfowl hunters (Siemer and Stedman 2018). We drew the sample from the pool of 2017 Harvest Information Program registrants that were ≥18 years old and indicated they hunted ducks in New York State during the 2016–2017 hunting season ($n = 18{,}207$). We drew 30% of the sample ($n = 1800$) from hunters who provided an email address to the NYSDEC and the remaining 70% ($n = 4200$) was drawn from duck hunters who did not provide an email address. This approach yielded a sample that mirrors the proportion of 2016–2017 waterfowl hunters who provided an email address (i.e., 30% of the hunter population provided an email address to NYSDEC during 2016–2017 and 70% did not). Hunters with email addresses received an online survey and those without email addresses received a mail survey. The survey instrument included rating and ranking questions that allowed us to quantify the relative weights that hunters placed on objectives as well as measurable attributes related to duck–hunter satisfaction. Each respondent was asked to rate how important 12 specific factors were as reasons for their preferences about when the waterfowl hunting season should be open in the zone that was most important to them and to rank the importance of the six fundamental objectives as reasons for their season date preferences. Weights were calculated for each survey respondent, and the weights were averaged across all respondents within each zone, resulting in a set of weights for each zone. The survey response rate was 47% ($n = 2791$ hunters). The rank-order centroid method was used to calculate weights on the fundamental objectives (Edwards and Barron 1994; Goodwin and Wright 2009). The Cornell University Office of Research Integrity and Assurance reviewed and approved the questionnaire (Institutional Review Board for Human Participants, Protocol ID# 1006001472).

Seeing and shooting mallard and black duck was ranked in all four waterfowl hunting zones as the most important influence on satisfaction with waterfowl hunting season dates (range 0.251–0.270) in the zone of most importance to the respondent. Seeing and shooting diving ducks (range 0.065–0.115) and minimizing overlap of waterfowl and deer seasons (range 0.065–0.092) were ranked as having the least influence on satisfaction with waterfowl hunting season dates in the zone of most importance to the respondent (Siemer and Stedman 2018; Fig. 8.4).

Alternatives

Alternatives are the management choices or means by which decision makers can achieve their stated objectives (Gregory et al. 2012). The NYSDEC worked with the waterfowl hunter task forces to develop season date alternatives for each waterfowl zone that were focused on achieving the fundamental objectives. Every zone included the status quo alternative, describing the most frequently used season dates over the past ten years. Consensus was not required from the task force, which allowed for inclusion of multiple versions of season dates that potentially satisfied the same objective. The NYSDEC provided task force members with simplified graphs of relative abundance for each species group throughout the entire time window of the federal frameworks (e.g., the Saturday closest to 24 September through 31 January), which allowed task force members to visualize season dates when duck abundance was greatest and when the greatest increase in relative abundance occurred (Fig. 8.5).

The waterfowl hunter task forces were asked to develop a season date alternative for each objective separately, and then to devise season dates that could simultaneously accomplish multiple objectives. For example, in the Western Zone, a season from November 23 to January 21 was chosen to maximize opportunity for seeing and shooting diving ducks, whereas a season from October 5 to December 3 was chosen to maximize seeing and shooting wood duck and teal. A split season of October 19 to November 10 and November 30 to January 5 was chosen to maximize seeing and shooting any duck (i.e., to maximize the number of ducks and the number of duck

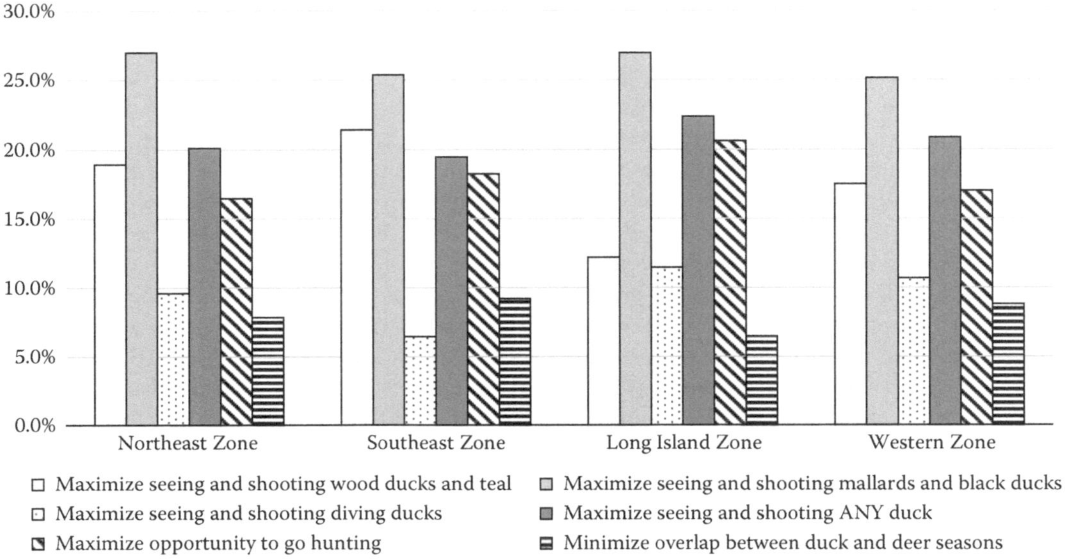

FIGURE 8.4 Mean rank order centroid weights for six fundamental objectives that influence satisfaction with duck hunting season dates in New York.

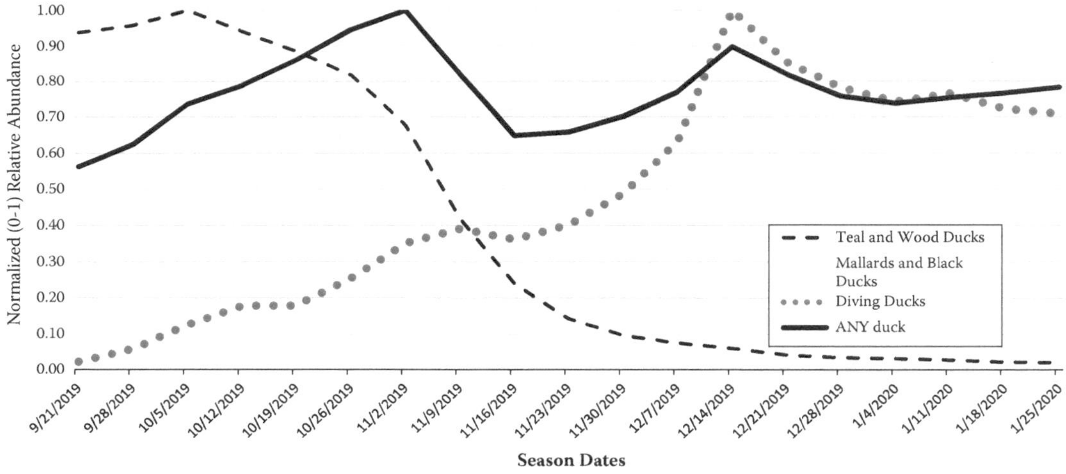

FIGURE 8.5 Normalized estimates of relative abundance (0–1) for teal and wood duck, mallard and black duck, diving ducks, and any ducks in the Western Zone of New York, derived from eBird spatiotemporal models (Fink et al. 2019). Relative abundance data presented are the median mid-point estimates of each week based on ordinal date. We normalized weekly relative abundance estimates for each species individually (i.e., a value of 0 represents the lowest relative abundance and 1 represents peak abundance) and summed the normalized values for each week.

species present). Each zone generated up to ten season date alternatives that were then evaluated to assess how well each alternative did at achieving all of the objectives. The alternative that best achieved all of the objectives was chosen for each zone using multi-criteria decision analysis whereby an expected utility value (weighted average) was calculated that incorporated the predicted outcomes and relative weights for each objective to quantify the overall performance of each alternative (Gregory et al. 2012, New York State Department of Environmental Conservation 2019). Final selection of season dates for each waterfowl zone is described in detail and available online (New York State Department of Environmental Conservation 2019). The season date structure for each zone was implemented for a five-year period (2019–2020 season through the 2023–2024 season), provided there are no changes to the number of days allowed under the federal framework. The five-year period allows hunters to plan ahead and the time frame is short enough that duck migration timing is unlikely to change significantly. After the five-year period, the NYSDEC will work with the task force to update the process by incorporating the latest duck migration and abundance data and will re-evaluate the season date decision using the updated information.

DISCUSSION

Selecting duck hunting season dates within the federal framework is largely a decision focused on objectives related to duck hunter satisfaction and maximizing the opportunity to participate. Ensuring hunter satisfaction is important to agencies because it can increase the likelihood of continued hunter participation (Schroeder et al. 2006). The management objective for mid-continent mallard that drove the federal regulations until 2018 was to maximize long-term cumulative (i.e., sustainable) harvest, while maintaining the population size above the goal of the North American Waterfowl Management Plan (Johnson et al. 2015). Although this objective maximizes harvest, and is an objective of importance to duck hunters, it is not a surrogate for maximizing hunter satisfaction, an objective of interest to wildlife managers and hunters alike (Johnson and Case 2000). The newly developed multiple species (i.e., multi-stock) adaptive harvest management framework that was implemented in 2018 considers the habitat requirements and distribution of multiple duck species in the Atlantic Flyway. In addition to sustaining duck populations at levels that meet the demands for hunters, the new framework also includes an objective of maximizing hunter satisfaction (e.g., no closed seasons, maximize percentage of satisfied hunters, simple regulations, hunters see a lot of ducks, minimize year to year changes in regulations) with harvest opportunity and regulations (Padding et al. 2018).

There are many components of hunter satisfaction, and those components and how they lead to acceptability of regulations generally are not well understood (Witter et al. 2006, Johnson et al. 2015). Even among the waterfowl task force group, comprised of representatives of waterfowl organizations, there was not an *a priori* definition or articulation of the important components of hunter satisfaction. The important components or objectives describing hunter satisfaction only became apparent after engaging in our structured decision-making process by working with task force members to identify their values and to express them as objectives. Waterfowl hunters are not a homogenous group (Schroeder et al. 2006), and thus the "average" waterfowl hunter is not easily characterized. The objectives setting stage of our structured decision-making process allowed us to better understand the different dimensions of duck hunter satisfaction regarding season dates and allowed us to characterize values in different waterfowl zones across the state (New York State Department of Environmental Conservation 2019). Consistent with previous studies (Heberlein and Kuentzel 2002, Brunke and Hunt 2008, Bradshaw et al. 2019), seeing and bagging game was a primary factor influencing hunter satisfaction in all waterfowl zones. The structured decision-making process allowed the waterfowl task force to be clear and transparent regarding the important components of satisfaction regarding season dates and the process allowed for a full articulation of the aspects of waterfowl hunting that were important and that might vary depending on timing of the season.

In the season setting structured decision-making process in our case study, we were able to evaluate season dates that would likely satisfy hunters in each zone based on species they are interested in pursuing (e.g., wood duck, mallard, canvasback). In this way, even for hunters not on the task force, hunters could find a season date that aligned with their hunting objectives and also understand how that objective fared in achieving all of the objectives as compared with the season date alternative that was ultimately selected. Our approach of pairing social considerations (i.e., hunter satisfaction) with biological data (migration chronology and duck abundance using eBird data) was a way for the task force to understand how each season date alternative fared with respect to their objectives. The previous task force process and other surveys typically only informally incorporated social values and used anecdotal observations regarding migration chronology. Acceptability of regulations by hunters is an important consideration to state natural resource agencies, and acceptability can be highly variable among hunters (Johnson et al. 2015). The structured decision-making process is transparent and reproducible and easily explained, which can lead to greater acceptability of the final decision.

Inclusion of stakeholders in decision making can lead to greater trust and satisfaction in the decision-making process (Lauber and Knuth 1997, Decker et al. 2012), which can increase acceptability of management decisions to a wider variety of stakeholders. In the New York case study, inclusion of waterfowl hunters via the hunter survey allowed hunter voices to be heard while also not alienating the already established waterfowl hunter task force. The state agency was able to be involved in helping to guide meetings with the waterfowl hunter task force, deepening connections and maintaining effective communication, which is associated with building trust (Young et al. 2013). This two-way knowledge exchange or co-productive model engages stakeholders in discussion, seeking their input regarding the decision, with the final authority to implement the decision resting with the agency (Rowe et al. 2005, Reed et al. 2018). Engagement with stakeholders can lead to attitudinal change among participants that may allow stakeholders to better understand the decision problem; attitudinal change is more likely when stakeholders are engaged throughout the entire decision-making process, stakeholder values are integrated in the process, and when there is transparency of the decision-making process (Sterling et al. 2017).

Although state wildlife agencies routinely engage duck hunters when making decisions about duck hunting season dates, approaches to hunter engagement vary by state. To better understand the processes that state natural resource agencies use when setting season dates, we searched state agency websites in the Atlantic Flyway for information describing how agencies make season date decisions. Information on the process that each state uses to set season dates within the allowable federal framework is not always transparent or made publicly available through the internet. However, we highlight several states for which information is provided either on their agency website or in press releases. Decisions made by state agencies regarding waterfowl season setting may include input from the public and various stakeholders by means of a task force, a committee of interested parties, unsolicited hunter input (e.g., phone calls, email), or solicited hunter input (e.g., public comment periods, quantitative hunter surveys) (Smith 2009). New Jersey sets season dates with input from a committee of sportsmen comprised of members of various waterfowl hunting organizations such as the New Jersey Waterfowl Association and Ducks Unlimited. In Massachusetts, season dates are set by the Fisheries and Wildlife Board, a governor appointed committee that oversees the Division of Fisheries and Wildlife. New Hampshire held public meetings in 2014 and 2019 to gather public input, which was used by the New Hampshire Fish and Game Department to make season date decisions. Maryland Department of Natural Resources conducted extensive public outreach regarding the 2019–2020 migratory game bird hunting season. This was used by the Wildlife and Heritage Service to help set season dates in consultation with the Governor's Wildlife Advisory Commission and the Migratory Game Bird Advisory Committee. Virginia conducted waterfowl hunter surveys (2000, 2004, 2010) to help the Virginia Department of Game and Inland Fisheries set season dates. To gain a broader understanding of public opinion, in 2011, North Carolina conducted its first statewide waterfowl hunter survey in 20 years. In 2011, Pennsylvania also completed its first statewide waterfowl hunter survey conducted by the Pennsylvania game commission. For the states in the Atlantic Flyway for which information was available on their state agency website—although many did incorporate hunter input when developing waterfowl season dates—we found no evidence of the use of a formal decision science approach to setting waterfowl season dates (with the exception being New York which we highlight in this chapter). Decision science is particularly useful because it clearly frames management problems as decisions and utilizes specific criteria (i.e., objectives) for making a management decision.

Structured decision making provided a transparent process that allowed for an explicit description of the steps involved in the season date decision and helped in facilitating communication (Gregory and Keeney 2002) among the NYSDEC and stakeholders. Engaging the waterfowl hunter task force in developing objectives and conducting a statewide survey of hunter values allowed stakeholders to better understand how their concerns were incorporated into the decision. Many wildlife agencies routinely survey their stakeholders, but there are few examples of how agencies engage stakeholders directly in decision-making processes (but see Robinson et al. 2017, 2021 [Chapter 9], Runge 2021 [Chapter 7]). The waterfowl season date decision is the third formal decision science framework that incorporates hunter values that has been used by the NYSDEC, demonstrating the agency's continued interest in and satisfaction with the process. Involving stakeholders in decision-making processes of wildlife management agencies can be challenging if agencies lack the capacity or capability associated with the social science element of stakeholder engagement (Fuller et al. 2020). If agencies lack capacity, but wish to incorporate a diversity of stakeholder interests, partnering with academic institutions or consultants could be one approach to begin the integration of social values in decision-making processes. Stakeholder groups are more likely to support the use of an inclusive process, such as structured decision making, than decisions made with a process that is less transparent (Decker et al. 2012, Riley and Gregory 2012) or a process that lacks a formal evaluation of alternatives. Thus, inclusive processes benefit agencies because of resulting decisions that are durable (Fuller et al. 2020).

ACKNOWLEDGMENTS

The authors appreciate input of the New York State waterfowl hunters who participated in the survey portion of this work, and the waterfowl hunter task force members who participated in the structured decision making process. The authors are thankful for the input, coordination, and expertise of biologists C. Adams, T. Bell, J. Eckler, K. Hamilton, S. Heerkens, J. Hurst, J. O'Connor, K. Parker, M. Schiavone, and M. Wasilco throughout the development of this important process for selecting duck seasons. The authors thank T. Auer, D. Fink, and C. Wood for their time and expertise to utilize the eBird Citizen Science data to model duck relative abundance. The authors thank J. Cummings and P. Devers for helpful comments. Funding for this work was provided by New York Federal Aid in Wildlife Restoration Grant WE-173-G.

9 Using Structured Decision Making to Incorporate Ecological and Social Values into Harvest Decisions

Case Studies of White-tailed Deer and Walleye

Kelly F. Robinson, Angela K. Fuller, and Michael L. Jones

CONTENTS

INTRODUCTION

Determining regulations to ensure sustainable harvest of valued species is a common goal of fisheries and wildlife management agencies, and there are many common challenges. Specifically, decision makers often must account for the values and objectives of multiple stakeholder groups, human and wildlife population- or community-level responses to management actions and associated uncertainties, and constraints related to funding for implementation of regulations. Stakeholder groups generally do not all share the same values, often instead having contradictory objectives, which likely have not been fully clarified (Keeney 2004, Gregory et al. 2012). In addition, the list of solutions or management actions that could be taken to achieve the objectives and make the decision might be incomplete (Gregory and Keeney 2002). Finally, multiple jurisdictions or agencies might be involved in setting harvest regulations, and these groups may or may not work together to manage the resource (Tyre and Michaels 2011, Robinson et al. 2021). Each of these challenges of fisheries and wildlife harvest management can make management decisions particularly difficult and possibly lead to regulations being called into question.

Decision analysis (i.e., structured decision making and adaptive management) provides a framework for multi-objective decision making to aid management agencies in setting harvest regulations (see Runge 2021 [Chapter 7]). This framework provides a defensible and values-based method for making difficult decisions that include contested stakeholder values and technical complexity (Gregory et al. 2012, Robinson and Fuller 2017). The decision analytic framework is designed to incorporate both social and ecological science into a decision-aiding process (Peterman and Peters 1998, Robinson et al. 2019). The steps include framing the problem, describing the objectives that must be achieved and how they will be measured, determining a set of management actions that could be implemented to achieve the objectives, predicting the effects of each management action on each objective, and finally evaluating tradeoffs among the objectives, based on the predicted outcomes (Hammond et al. 1999, Gregory et al. 2012, Runge 2021 [Chapter 7]).

The incorporation of both ecological and social science is an integral part of the use of decision analysis for harvest management (Fuller et al. 2021 [Chapter 8]), and indeed for natural resource decision making as a whole (Bennett et al. 2017, Robinson et al. 2019; Fuller et al. 2020). Without both of these components, decisions that could appear to be the best choice might not be supported by stakeholders or might misrepresent the dynamics of the harvested population, ultimately hindering conservation and limiting access to the resource. For example, an early 2000s exercise in decision analysis for walleye (*Sander vitreus*) harvest management in Lake Erie included fisheries managers and biologists from the agencies that surround the lake, as well as modelers, but notably did not include stakeholders (e.g., anglers). Although the resulting harvest policy from the process was arguably suitable, given the stated goals for the fishery, it was not supported by fishery stakeholders (Jones et al. 2016). This example highlights the importance of a comprehensive representation of stakeholder values and transparency in the decision-making process. However, not all harvest management decisions require the integration of social and ecological objectives. Different problems and stakeholder groups likely will lead to the use of different methods.

METHODS FOR THE INTEGRATION OF SOCIAL AND ECOLOGICAL SCIENCE IN HARVEST MANAGEMENT DECISIONS

Decision analysts can use a variety of methods to integrate social and ecological science into the decision analytic process for harvest management. We provide a description of the methods that are frequently used or that we suggest would be most helpful for harvest management decisions. We suggest that including both social and ecological scientists in the decision-making process, as well as the tools that these scientists provide, engaging stakeholders fully in decision-making processes, and using participatory modeling strategies can enhance the integration of these disciplines (Bennett et al. 2017, Robinson et al. 2019). We describe these methods in more detail and then provide two case studies that are examples of ways that these methods have been used for harvest management for fisheries and wildlife.

Include Both Social and Ecological Scientists

Harvest management decisions have both social and ecological components, so it stands to reason that social and ecological scientists should be involved in aiding the decision analysis process.

Each decision analysis step can benefit from social and ecological thought, and incorporating the necessary experts and stakeholders early can allow each step to be more fully explored (Robinson et al. 2019), ultimately providing the decision makers with a more holistic view of the decision problem and objectives. For example, when framing the problem, different stakeholders likely view the problem differently, which should be considered in crafting the problem statement (Robinson et al. 2019). When considering the ecological effects of a dam removal, for instance, different stakeholders might inherently envision different post-removal time frames for the riverine ecosystem, leading to temporal mismatch in the problem framing step (Lin et al. 2019). Similarly, the scale of the problem will rely both on social factors in a region, such as cultural values or local demography, as well as ecological factors like climate or land use (Robinson et al. 2016*a*, 2017, Lin et al. 2019, Fuller et al. 2021 [Chapter 8]).

The objectives and tradeoffs steps, in particular, involve both ecological and social values related to the harvested resource. Setting objectives requires understanding diverse values of stakeholders that could be affected by harvest management decisions. Social scientists can aid stakeholders in considering values that are important to them, but social scientists may not be familiar with the process of decision analysis (Robinson et al. 2019). Some social scientists have spent a great deal of time studying what constitutes a satisfying hunting or fishing experience, including size and age of animal harvested, hunting or fishing opportunity, perception of an abundant resource, and conflict with other hunters or fishers (Siemer et al. 2014, 2015, Robinson et al. 2016*a*, 2017). In addition, social data, such as information collected from large-scale surveys, are often necessary to understand how stakeholders in a region value the achievement of multiple objectives, which then can be used for evaluating tradeoffs. By including social scientists in the objectives-setting step, more time is afforded for creation of relevant survey instruments (Robinson et al. 2019, Fuller et al. 2021 [Chapter 8]).

The multi-objective nature of most harvest management decision problems also requires predicting ecological and social consequences of the alternatives, again necessitating scientists from these different disciplines. Although predicting consequences has largely involved ecological models, there are methods and protocols available for predicting and measuring social values, as well (Robinson et al. 2019, Gruntorad and Chizinski 2021 [Chapter 4]). The first step toward measuring social values is to create a well-formed set of objectives that fully describes the nuances of stakeholder values (Stedman 2003, Robinson et al. 2019). Making full use of natural, constructed, and proxy attributes to measure the achievement of objectives can allow groups to define both the meaning of objectives more clearly and exactly how they will be measured (Keeney 1992). Natural attributes, such as measuring cost in terms of dollars or population size in terms of number of fish, are most easily interpreted and should be used if such an attribute exists. However, natural attributes are often difficult to define for social objectives. In these cases, constructed and proxy attributes can be particularly useful. Constructed attributes allow stakeholders to create a scale of measure specific to the objective, such as the cultural and spiritual quality of a river (Failing et al. 2013) or respecting non-human life (Runge et al. 2011). Indirect measures of objective achievement, or proxy attributes, can also be employed for objectives for which natural or constructed attributes are difficult to identify (Keeney 1992). The goal in creating these measures is for all attributes to meet five criteria: that they are unambiguous, comprehensive, direct, operational, and understandable (Keeney and Gregory 2005). In all cases, integrating social and ecological thought into the formation of these measures can lead to more well-informed and defensible predictions of consequences on stakeholder values.

Engage Stakeholders in All Stages of the Decision Process

Incorporating stakeholder concerns into harvest management decisions requires more than simply holding public forums to allow citizens to voice their concerns. These types of engagements rely on breadth and quantity and do not lead to organized sets of concerns that are well articulated by stakeholders or easily understood by decision makers (Gregory 2017). By contrast, fully engaging stakeholders throughout the decision-making process provides those stakeholders with an opportunity to articulate their value sets based on relevant science, weigh those values, and importantly, adjust their values or weights based on the stated values of other stakeholders and predicted outcomes of management actions (Jones et al. 2016, Gregory 2017). In addition, by including stakeholders fully in the decision-making process, decision makers can proactively work within the social construct of stakeholder influence (Hiller et al. 2021*b* [Chapter 2]). Ultimately, including stakeholders in a meaningful way can lead to greater buy-in and trust for the eventual decision (Decker et al. 2012).

In-depth stakeholder engagement can be accomplished through multiple processes. Most often, this means engaging a small group of stakeholders longitudinally in a series of workshops (Failing et al. 2004, Jones et al. 2016, Robinson et al. 2021). Other methods for meaningful engagement include meetings of multiple groups across a large spatial area with centralized coordination (Gregory 2017) or using a hybrid approach that includes small-group discussions and large-scale surveys (Robinson et al. 2016*a*, 2017, Gregory 2017, Fuller et al. 2021 [Chapter 8]). Within all of these strategies, the process of structured decision making provides a framework for stakeholders to articulate their values in a well-formed objectives hierarchy, create alternatives to achieve those objectives, understand the predicted consequences of those actions on their objectives, and make tradeoffs that incorporate the shared concerns of the group (Gregory 2017).

Use Participatory Modeling Strategies

The consequences step of the structured decision-making process often involves building models for prediction. These models might be qualitative, such as an influence diagram that depicts relationships among objectives and alternatives, or they might be fully quantitative. In all cases, the goal of these models is to predict the ability of each alternative to achieve each of the stated objectives (Conroy and Peterson 2013). These objectives are often diverse, representing the multiple needs and values of different stakeholder groups. As such, participatory modeling to build predictive models for the objectives step provides a method for incorporating the diversity of stakeholders and their relevant knowledge (Röckmann et al. 2012, Robinson and Fuller 2017). The process of participatory modeling requires the stakeholder group to work together to build predictive models specific to the decision. This process can be challenging, as many stakeholders will have little to no background in model building, but it can also allow for participants to better integrate and understand the steps of the structured decision-making process, provide necessary data for the model (Irwin et al. 2011), and feel as though the model describes the system accurately (Robinson and Fuller 2017). Most importantly, the act of collaboratively building a model for structured decision making can result in more trust in these models, which in turn leads to less difficulty in implementation of the models for management purposes (Moore and Runge 2012, Jones et al. 2016).

CASE STUDIES FOR INTEGRATION OF SOCIAL AND ECOLOGICAL SCIENCE

We provide two case studies that make use of the methods described above to integrate social and ecological science into decisions for harvest management. These studies take different approaches in application of these methods, including how stakeholders are engaged throughout the process and how complex models are built and described. The details of each of these projects, such as the inclusion of both recreational and commercial fishers, necessitated these different approaches and provide examples of when to make use of these different techniques.

White-tailed Deer Buck Harvest Management in New York State

White-tailed deer (*Odocoileus virginianus*) is an important game species in much of eastern North America (U.S. Fish and Wildlife Service and U.S. Census Bureau 2016), including New York State, where more than 500,000 people hunt deer each year (New York State Department of Environmental Conservation 2020). The New York State Department of Environmental Conservation (NYSDEC) manages harvest of white-tailed deer to balance the interests of stakeholders with the ecological constraints of population growth (New York State Department of Environmental Conservation 2011). In 2011, NYSDEC received requests from hunting groups to change harvest regulations for white-tailed deer bucks to a system that included mandatory antler restrictions. This new regulation, which would restrict harvest of bucks to those with a certain number of antler points, would reduce the harvest of yearling (1.5-years old) bucks and increase the number of larger-antlered bucks available for harvest (see Morina et al. 2021 [Chapter 19]). However, many hunters were concerned that the implementation of mandatory antler restrictions would reduce their ability and freedom to harvest a deer during the hunting season.

A statewide survey of deer hunters indicated that although 57% of surveyed hunters supported the implementation of mandatory antler restrictions, there was variation in support throughout the state, and 34% of hunters opposed the measure (Enck et al. 2011). Furthermore, implementation of voluntary antler restrictions, in which hunters simply would choose not to harvest smaller-antlered bucks, was supported by 54% of hunters (Enck et al. 2011). In addition, NYSDEC implemented pilot areas for mandatory antler restrictions in the southeastern part of the state, and as with the statewide survey, there were conflicting opinions after implementation. In particular, a majority of hunters in these areas supported the continued implementation of the pilot program although a majority also reported unmet expectations and dissatisfaction with the outcomes of the program (Enck et al. 2011). Based on the lack of clarity from hunter survey results, NYSDEC decided to require that a *super majority* (i.e., ≥ 67%) of hunters in a region support mandatory antler restrictions implementation with *strong opposition* not to exceed 20%. Ultimately, this rule was unsatisfactory to stakeholders and managers, and seemed to lead to greater political activism by stakeholder groups (Robinson et al. 2016*b*).

The rules for mandatory antler restrictions implementation represented a desire from NYSDEC managers to be responsive to stakeholder desires, but the process lacked a clear set of objectives and potential actions to achieve those objectives. In addition, NYSDEC's 2012 Deer Management Plan stated that the agency could use various measures to reduce the harvest of yearling bucks, but did not provide a list of options for accomplishing this task (New York State Department of Environmental Conservation 2011). Stakeholders, however, viewed mandatory antler restrictions as the objective, rather than one management action that might achieve multiple objectives. There was also a suite of uncertainties in the ecological and social realms related to effectiveness of and satisfaction with potential regulatory packages. And finally, there was concern from the management agency that lobbying efforts by vocal parties might misrepresent the desires of the hunting public (Robinson et al. 2016*b*). These concerns led to the implementation of a structured decision-making process for white-tailed deer buck harvest management that included a team of biologists and managers from the state agency, social scientists, decision analysts, and ecologists, hereafter referred to as the *working group*.

Decision Framework and Problem Statement

The goal of this structured decision-making process was to aid the Commissioner of NYSDEC, who was the decision maker, in setting regulations for white-tailed deer harvest in the state. Although the Commissioner was the ultimate decision maker, NYSDEC biologists and managers had delegated authority for this process. The white-tailed deer harvest decision required consideration of multiple objectives and tradeoffs at different levels. The agency's objectives and tradeoffs were related to hunter satisfaction, white-tailed deer population management, and management costs. In turn, hunters had objectives related to the size, age, and sex of deer that they would prefer to harvest, as well as other aspects of the hunting experience that might be affected by regulation changes. These aspects included ability to be in the field, the number of bucks that can be harvested, and ease of adhering to regulations. The harvest decision also included uncertainties related to hunter preference and white-tailed deer population demographics. Finally, hunter values and deer population demographics varied spatially throughout the state, which led to the creation of seven buck management zones (i.e., Adirondack, Lake Plains, Mohawk, Northwestern, Southeastern, Southern Tier, and Westchester/Suffolk) that aggregated portions of the state with similar deer population characteristics and environmental characteristics like snow cover and land use (Kelly and Hurst 2016, Robinson et al. 2016*a*). Based on this decision context and the direction provided by the 2012 Deer Management Plan, the problem statement, as defined by the working group, was "*to develop a decision framework that uses objective criteria to evaluate optimal strategies for reducing harvest of yearling bucks, including mandatory antler restrictions*" (Robinson et al. 2016*a*,*b*).

Objectives

The objectives hierarchy created by the working group for this decision problem incorporated the economic, ecological, and social concerns of agency staff and hunters. The inclusion of social scientists in the working group provided the needed expertise to develop objectives that described deer hunters' concerns about the components of satisfaction with hunting (Robinson et al. 2016*a*,*b*). The objectives related to hunter satisfaction were defined based on previous research by the social scientists in the group that described what constitutes a satisfying hunting experience. These values included not just number of deer harvested, but also ability to be afield, complexity of regulations, and the freedom to choose a deer for harvest. Ultimately, the objectives hierarchy was composed of a set of fundamental and means objectives that described the agency's needs related to costs, hunter desires, and deer population management, as well as a series of measurable attributes for each of these objectives (Fig. 9.1; Robinson et al. 2016*a*).

Alternatives

Six alternatives were considered for this decision problem. These alternatives were created from previous hunter input and the results of relevant statewide hunter surveys with questions related to harvest regulations (Enck and Brown 2008*a*, Enck et al. 2011). The group worked to be creative in describing a set of actions that could in some way reduce the harvest of yearling bucks. For example, an alternative to implement mandatory antler restrictions in the first part of the regular firearms season was created. This regulation was created to both limit yearling harvest and maintain some portion of the season when hunters have more freedom of choice in buck harvest. The list of alternatives (Table 9.1) ranged from maintaining status quo, to mandatory antler restrictions for all or part of the hunting season, to voluntarily passing up bucks with small antlers.

Consequences

The consequences step for this decision problem involved making predictions about both ecological and social outcomes. In addition, the group needed to account for uncertainties related to deer population demographics, effects of novel harvest alternatives (e.g., partial mandatory antler restrictions and voluntary restraint; Table 9.1), and predicting hunter actions under the suite of alternatives. A combination of participatory modeling, expert elicitation, hunter surveys, and empirical data was used to predict the consequences of the six alternatives (Table 9.1) on each of the objectives (Fig. 9.1). This combination of social and ecological tools allowed us to create a robust set of predictions for all objectives in our decision problem (Robinson et al. 2019).

We used a stochastic simulation model of the white-tailed deer population (Collier and Krementz 2007, Robinson et al. 2014) to predict how changes in harvest rates under different alternatives would influence measurable attributes related to the deer population. Through the process of participatory modeling, the group modified the model to predict the necessary outcomes,

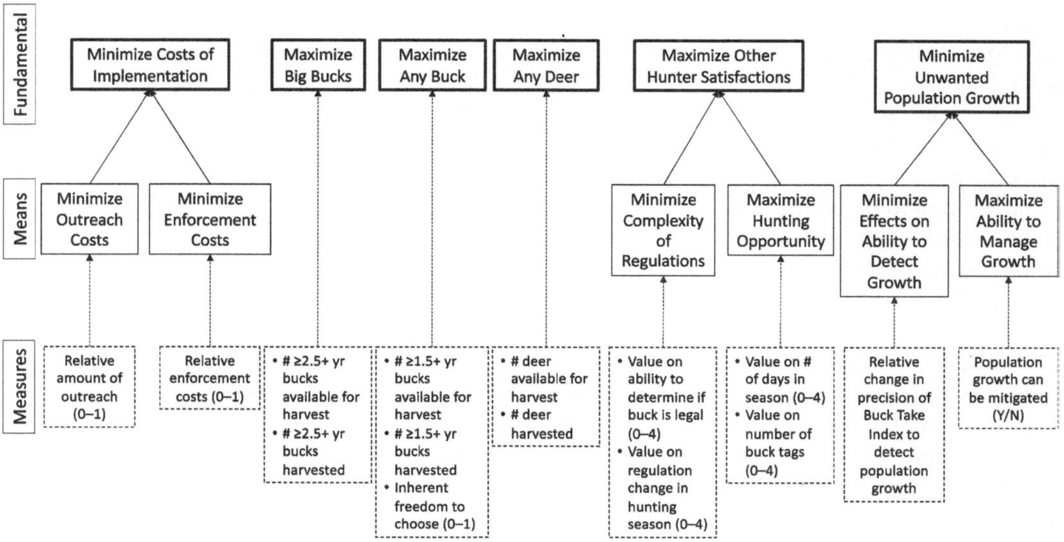

FIGURE 9.1 Objectives hierarchy for the decision problem of setting harvest regulations for white-tailed deer buck harvest in New York State. The top row represents the fundamental objectives. Solid arrows connect means objectives (row 2) to the fundamental objectives, and dashed arrows connect the measurable attributes (row 3) to their associated objectives. Big bucks are ≥2.5 years old, any buck includes bucks ≥1.5 years old, and any deer includes antlerless deer and all bucks.

TABLE 9.1
Harvest alternatives for the white-tailed deer buck harvest management decision problem in New York State[1]

Alternative	Description
Status quo	Continue with the current regulations in which a legal buck had an antler at least 7.5 cm (3 in.) long, with the possibility of harvesting up to two bucks in a season
One-buck	Decrease the bag limit for bucks from two to one
Mandatory antler restrictions	Legally harvestable bucks must have three or four points on one antler, depending on the zone. Points must be ≥ 2.5 cm (1 in.)
Partial mandatory antler restrictions	Mandatory antler restrictions in place from archery season through the first one to two weeks of the firearms season, depending on the zone
Shorter season	Reduce the firearms season by one to two weeks, depending on the zone
Voluntary restraint	Educational outreach to encourage hunters to voluntarily pass up harvest of yearling bucks

Note
[1] Modified from Robinson et al. (2016*a*).

gather data and expert opinion to parameterize the model for each of the seven buck management zones, and determine the relevant uncertainties that should be included. For each harvest alternative in each zone, the model predicted population growth, age- and sex-specific harvest, and age- and sex-specific number of deer available for harvest after five years (Fig. 9.1). The model included data collected by NYSDEC, data from other state agencies, data from relevant publications, and expert knowledge from NYSDEC biologists (see Robinson et al. 2014 model description). In particular, some of the harvest alternatives considered had not been implemented previously in any state. Therefore, we used expert elicitation techniques to determine the range of harvest rates that might be expected under these alternatives (Robinson et al. 2016*a*). Using the strategies of participatory modeling allowed the group to understand and verify the model inputs and outputs and provide needed data and expert knowledge.

In addition to predictions about the status of the deer population, we needed to predict aspects of hunter satisfaction and management costs. By including social scientists in the working group, we were able to develop a self-administered questionnaire designed specifically to understand how hunters value different aspects of satisfaction with white-tailed deer hunting in New York State. This survey instrument used a series of rating and ranking questions that were tied directly to the fundamental objectives of the decision framework. For example, hunters were asked to rank the seven dimensions of satisfaction that constituted the objectives hierarchy (Fig. 9.1). Hunters responded using a scale of 1–5, how important it is to "see more bucks with big antlers than I have seen in the last five years" and "keep the regular firearms season at least as long as it is now in the zone I hunt" (Siemer et al. 2015). The survey was administered statewide to 7000 registered hunters throughout New York State (Siemer et al. 2015, Robinson et al. 2016*a*). The results of the survey provided information about hunters' willingness to voluntarily pass up shots at yearling bucks, which was translated into a harvest rate for the population model. The survey also provided a means to predict hunters' relative values for aspects of complexity and hunting opportunity under each alternative (Robinson et al. 2016*a*). Finally, to predict the costs associated with implementation of each harvest alternative, we used the direct rating method (Goodwin and Wright 2009) to elicit the relative costs associated with outreach and education from NYSDEC biologists and the relative change in compliance and enforcement from NYSDEC law enforcement staff.

Tradeoffs

The hierarchical nature of the fundamental objectives meant that tradeoffs needed to be made at the level of the management agency and among hunters themselves. The NYSDEC biologists and managers had delegated authority from the agency Commissioner to place weights on the three overarching objectives for the state: costs of management (0.10), deer population control (0.15), and the grouping of four objectives that made up hunter satisfaction (0.75), as well as the means objectives associated with costs and population control (Fig. 9.1). We used the results of the rating and ranking questions from our statewide hunter survey to calculate weights associated with the fundamental objectives of hunter satisfaction for each buck management zone. These survey data allowed us to gain the spatial resolution necessary to understand how New York hunters value harvest outcomes and hunting opportunities throughout the state. We used the weights associated with all objectives, along with the predicted outcomes, normalized to a 0–1 scale, to calculate the expected utility value (i.e., a weighted average) for each alternative (see Robinson et al. 2016*a* for a full description of the weighting process).

Decision and Importance of Integration of Social and Ecological Science

The tradeoffs analysis indicated that the optimal decision for each of the seven buck management zones was the status quo alternative, with the voluntary restraint, shorter season, or mandatory antler restrictions alternatives ranked second, depending on the zone in question. For example, mandatory antler restrictions was ranked second in the Lake Plains zone, but voluntary restraint was ranked second in the Adirondack zone (Fig. 9.2; Robinson et al. 2016*a*). Voluntary restraint was never ranked less than third among zones. Ultimately, the agency decided not to continue with the status quo alternative. They acknowledged that voluntary restraint was less optimal only because it performed poorly in terms of the cost objective, which they decided could be accounted for in their budget and through partnerships with hunting groups. Therefore, NYSDEC staff recommended to the Commissioner, and the Commissioner decided, that educational efforts to promote voluntary restraint be implemented. Since the completion of the decision analysis, NYSDEC has been promoting voluntary restraint through their website and associated educational efforts, themed as "*let young bucks go and watch them grow.*"

This structured decision-making case study highlights the importance of understanding not only the ecological consequences of implementing a particular harvest management policy, but also the social consequences. The management agency placed 75% of their weight on the objectives related to hunter satisfaction, reflecting their view that the decision was mostly social, as described in the problem statement (Robinson et al. 2016*a*,*b*). Although the quantitative aspects of the decision analytic process were potentially difficult to convey to the public, NYSDEC

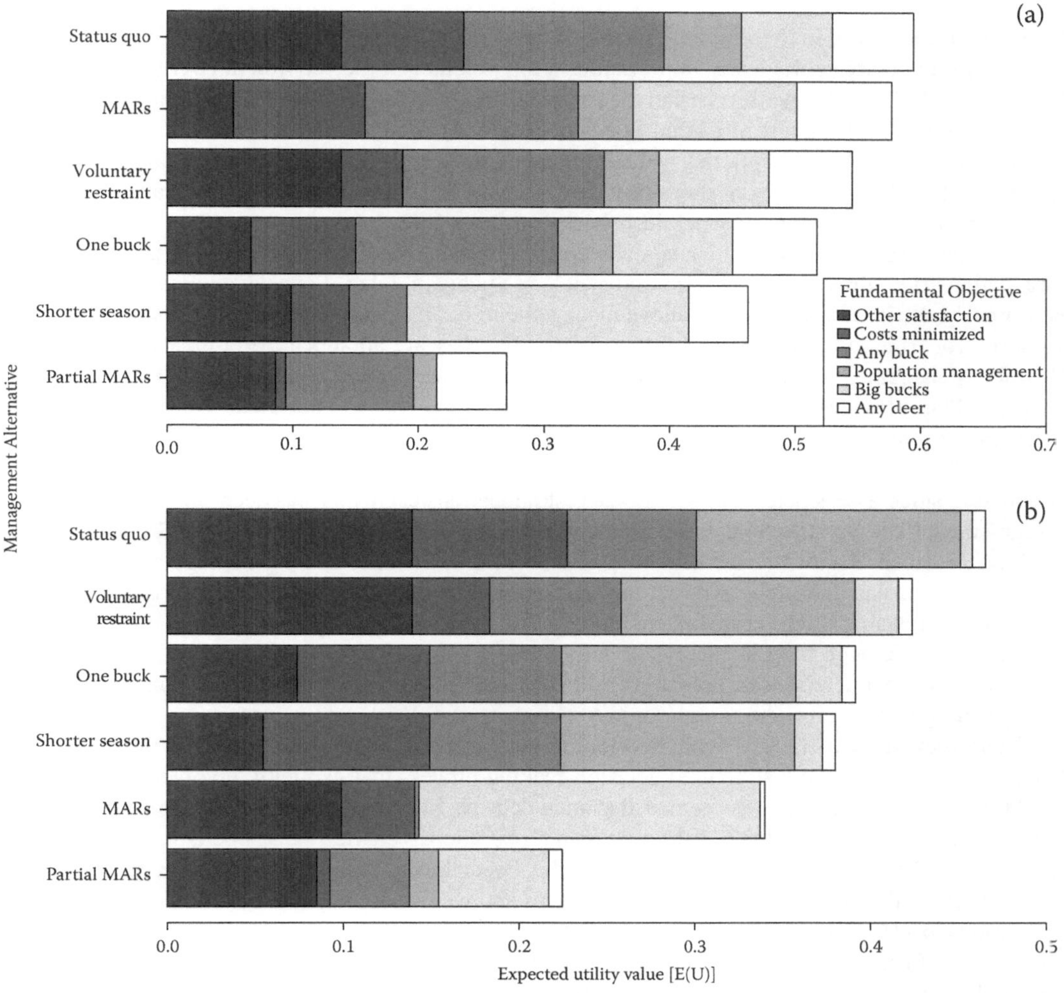

FIGURE 9.2 Tradeoffs analysis for evaluating the management alternatives for setting harvest regulations for white-tailed deer buck harvest in two regions of New York State: (a) the Lake Plains region, (b) the Adirondack region (modified from Robinson et al. 2016*a*). Each bar represents the expected utility score for an alternative, and the colors within a bar represent the weighted score for the fundamental objectives. The alternative with the greatest utility score is the optimal alternative. Descriptions of fundamental objectives are provided elsewhere (Fig 9.1).

managers described the results of this process at public forums throughout the state, including figures for each zone (similar to Fig. 9.2), and the reasons for choices related to objective weights. We also created a pamphlet that was distributed to hunters and available on the NYSDEC website (https://www.dec.ny.gov/docs/wildlife_pdf/buckmgmtsdm2016.pdf) that described the process and outcomes. The hunter survey results indicated that in all buck management zones, hunters were divided in terms of the age and sex of deer that they preferred to harvest (Siemer et al. 2015). As a result, the transparency that the structured decision-making process lent to the decision of how best to regulate harvest of white-tailed deer bucks was paramount, as hunters could see how the decision was made, even though the hunting public had diverse opinions and values about buck harvest.

Management of Commercial and Recreational Harvest of Walleye in Lake Erie

Walleye in Lake Erie, in the Laurentian Great Lakes, are of enormous socio-economic importance to the region and are ecologically important as a top predator in the ecosystem (Berger et al. 2012, Kayle et al. 2015). The walleye fishery comprises commercial (100% of Canadian harvest) and recreational (97% of harvest in United States of America) components (Berger et al. 2012). These two fishing sectors have conflicting values for the fishery and have been at odds since the 1970s (Jones et al. 2016). In addition, there has been an historical lack of trust among stakeholders for management agencies regarding percid fisheries in Lake Erie.

The Lake Erie Committee, which is made up of representatives from each of the five management jurisdictions (i.e., Province of Ontario and states of Michigan, Ohio, Pennsylvania, and New York) surrounding the lake, coordinates walleye management. Each of these agencies has decision-making authority for fisheries management in their waters, but they are all signatories to the Joint Strategic Plan for Management of Great Lakes Fisheries (Great Lakes Fishery Commission 2007). The Joint Strategic Plan provides a means for all agencies around the lake to coordinate the management of fisheries and encourages the use of consensus for setting lake-wide total allowable catches (Jones et al. 2016). Although the agencies have worked together historically to set harvest regulations, conducting stock assessments and setting harvest allocations each year has been difficult because of disagreements among stakeholders regarding their values, between stakeholders and managers regarding the science and management methods used, and at times among managers (Jones et al. 2016). In particular, the process of setting the annual recommended allowable harvests and associated total allowable catches for walleye in Lake Erie has lacked transparency from the viewpoint of stakeholders. Although the Lake Erie Committee had previously made use of the tenets of decision analysis to create a new harvest policy in the lake, stakeholders were not invited to join this process, again leading to concerns about transparency (Jones et al. 2016).

Those previous experiences with developing harvest policies for Lake Erie led the Lake Erie Committee to embark on a structured decision-making process that explicitly involved stakeholders in 2009. A working group was formed that included the Lake Erie Committee members (i.e., the managers), commercial fishers, fish processing operations, charter boat operators, recreational fishers, Sea Grant personnel, and advisors to the Great Lakes Fishery Commission (Kayle et al. 2015). These stakeholders represented larger groups and were encouraged to communicate the efforts and results back to their groups throughout the decision analysis. In addition, members of the Quantitative Fisheries Center at Michigan State University served as third-party facilitators. This group, the Lake Erie Percid Management Advisory Group, met for a series of workshops over the course of three years, working through the steps of evaluating management objectives and uncertainties, refining stock assessment and prediction models, and eventually recommending a harvest control rule for Lake Erie walleye to the Lake Erie Committee (Jones et al. 2016).

Decision Framework and Problem Statement

Lake Erie is the shallowest and most productive of the Laurentian Great Lakes. It comprises three separate basins that differ in ecological and physical characteristics. Walleye spawn in late March and early April, mainly in the shallow western basin of the lake. Juvenile and adult walleye then migrate to the deeper central and eastern basins during the summer. The western and central basins were included in the decision-making process for walleye harvest, as relatively little harvest occurs in the eastern basin (Jones et al. 2016). Walleye reproductive success and subsequent recruitment to the fishery is affected by a combination of factors that can affect fecundity (e.g., maternal temperature, resource abundance) and egg and larval survival (e.g., spring warming rate, abundance of predators and food resources; Hartman 1972, Hokanson 1977, Shuter and Koonce 1977, Malison et al. 1994, Madenjian et al. 1996, Hansen et al. 2015*a*). The *periodic* life-history strategy of walleye, low recruitment punctuated by rare larger year classes (Winemiller and Rose 1992), creates challenges for management of these types of species.

Lake Erie walleye harvest is managed by setting a recommended allowable harvest each year and allocating that recommended allowable harvest as total allowable catch between the recreational and commercial sectors. The total allowable catches are allocated based on the relative surface area of Lake Erie in U.S. and Canadian waters of the western and central basins of Lake Erie, such that the United States of America receives 56.9% of the catch and Canada receives 43.1%. The Lake Erie Committee uses the results of a stock assessment each year to set the recommended allowable harvest and allocate the total allowable catches. As previously mentioned, stakeholders in the walleye fishery were dissatisfied with the lack of transparency involved in setting the recommended allowable harvest and total allowable catches each year. In addition, there were concerns among stakeholders about the effects of each sector on the fishery.

The Lake Erie Percid Management Advisory Group was formed as an advisory group to the ultimate decision makers, the members of the Lake Erie Committee. Through the creation of a *terms of reference* to specify the purpose for the advisory group, the Lake Erie Committee agreed to (1) explicitly consider all management recommendations from the Lake Erie Percid Management Advisory Group, (2) ensure transparency and accountability when setting total allowable catch, and (3) evaluate outcomes of the recommendations through consultation with the advisory group (Kayle et al. 2015). Overall, the decision problem for walleye in Lake Erie was to determine the best method to manage a large, binational freshwater fishery to maximize returns for commercial and recreational fishers in Lake Erie (Jones et al. 2016).

Objectives

Through the series of Lake Erie Percid Management Advisory Group workshops, the group voiced a set of values, including fishery sustainability, addressing recent declines in the fishery, ensuring economic viability for the commercial sector, accounting for risk and uncertainty, transparency, and more even distribution of benefits from the fishery (Kayle et al. 2015). Overall, the fundamental objectives of this decision process were to (1) maximize commercial fishery returns, (2) maximize recreational fishery catch rates, and (3) minimize risk of low population abundance. The set of objectives was created to ensure that fishery benefits reached both the commercial and recreational sectors and that the fishery is sustainable and can recover from the recent declines observed. The measurable attributes for these objectives included the mean abundance of walleye, catch per effort for the recreational fishery, and harvest and yield for the commercial fishery. Thresholds for commercial yield and recreational catch per effort were also included to account for risks to industry viability (Jones et al. 2016).

Alternatives

Although other actions could potentially be taken to achieve some of the fundamental objectives, the nature of fisheries management in Lake Erie required that the alternatives be constrained to recommended allowable harvest and total allowable catch. As such, the group considered a set of harvest control rules that described aspects of fishing mortality and biomass. The feedback portion of these policies meant that the harvest control rules dynamically set the fishing mortality rates each year, based on the most recent stock assessment. The harvest control rules used two reference points, a target related to the fishing rate at maximum sustainable yield, and a limit related to the assessed biomass of the stock relative to the unfished spawning stock biomass (Jones et al. 2016). The harvest control rule also included a probabilistic portion to account for uncertainty in the stock assessment. In short, these probabilistic harvest control rules had an *a priori* probability (P^*) associated with violation of the limit reference point, with the goal of not exceeding P^* (Jones et al. 2016). Finally, the group decided to evaluate rules that included a cap on the change in the total allowable catch that might occur year to year. Overall, the group evaluated 96 different probabilistic biomass feedback policies that differed in terms of fishing mortality, limit reference points, P^* values, and annual total allowable catch changes (see Jones et al. 2016 for a full description of the harvest control rules evaluated).

Consequences

The consequences step for walleye management in Lake Erie required the use of a stock assessment framework to evaluate the historic and current state of the stock and the integration of the stock assessment into a management strategy evaluation (Punt et al. 2016) to predict the effects of the harvest control rules on the measurable attributes described for the stakeholders' objectives. Through a series of workshops and the use of participatory modeling, the members of the Lake Erie Percid Management Advisory Group worked to modify an existing statistical catch-at-age model and create the management strategy evaluation framework for predictive modeling.

Statistical catch-at-age models have been used to assess walleye stocks in Lake Erie for almost 30 years (Berger et al. 2012, Jones et al. 2016). Catch-at-age models use a population model and likelihood methods to infer demographic rates and fishery parameters such as catchability of fishing gear from fishery-dependent (catch, effort) and fishery-independent (surveys) data (Fournier and Archibald 1982). The Lake Erie walleye model used age-structured data from harvest (catch and effort from Ohio and Michigan recreational fisheries, Ontario commercial fishery), catch-per-unit-effort data from gillnet surveys in both countries, bottom trawl surveys targeting age-0 walleye, and age composition of all catches to fit cohorts of fish forward in time to estimate the yearly abundance of each age group (Berger et al. 2012, Jones et al. 2016). Through the participatory modeling process, the model used by the Lake Erie Committee was refined for the structured decision-making process for walleye (see Jones et al. 2016 for a full description of the models). Importantly, although many of the stakeholders in the group were unfamiliar with quantitative fisheries models, the participatory modeling process allowed them to see the types of data that fed into the model and the reasons decisions were made during model building. In addition, because of the gap in technical knowledge among stakeholders, the model was evaluated by an outside panel of walleye and stock assessment experts. The resulting model was unanimously supported by Lake Erie Percid Management Advisory Group members and led to an increase in trust in the decision-making process (Jones et al. 2016).

The results of the stock assessment modeling work, as well as a stock-recruitment model fit to walleye data, were then incorporated into a forecasting phase through the management strategy evaluation process. Management strategy evaluation provides a framework to simulate the effects of different harvest control rules, as well as associated uncertainty, into the future (Polacheck et al. 1999, Deroba and Bence 2008). The rules were simulated forward in time in a stochastic age-structured operating model, providing a means of evaluating how robust harvest control rules are to uncertainty. The management strategy evaluation allowed for comparisons of the aforementioned harvest control rules across a range of states of nature, which in turn provided information for making tradeoffs among the different objectives related to the fish population and fishery economics (Deroba and Bence 2008, Jones et al. 2016). In the case of Lake Erie walleye, the management strategy evaluation provided information relevant to the objectives of Lake Erie Percid Management Advisory Group members, in the form of the measurable attributes (e.g., recreational catch per hour, biomass of commercial catch; Jones et al. 2016). These models are a decision-aiding tool, as different harvest control rules will differentially achieve the objectives of different stakeholder groups. Therefore, the group then formally analyzed the tradeoffs to understand how specific harvest control rules would affect the commercial and recreational walleye fisheries and the set of fundamental objectives.

Tradeoffs

Making tradeoffs among the evaluated harvest control rules proved to be a challenge for the Lake Erie Percid Management Advisory Group because of the technical nature of the results produced by the management strategy evaluation process. However, the combination of trust built through the structured decision-making and participatory modeling process and the ability of the facilitators and modelers to present the results as a series of boxplots, tables, and summary statistics led to substantive discussions about the tradeoffs related to these different policies (Jones et al. 2016). The group's main focus was on the tradeoff between ensuring economic stability for commercial fishers through an adequate total allowable catch while maintaining acceptable levels of catch for recreational fishers. Tradeoff plots provided a succinct view for the group of the relative risk of the different harvest control rules to the commercial and recreational stakeholders (Fig. 9.3; Jones et al. 2016). These plots showed that there was a non-linear pattern in the effects of different levels of fishing mortality on risk for the two sectors, with greater (but still moderate) levels of fishing mortality providing lower risks to commercial stakeholders with relatively little effect on the risk to the recreational fisheries (Jones et al. 2016). As a result of the participatory nature of the structured decision-making process, recreational stakeholders were willing to accept a decision that included less conservative fishing mortality rates because they could see the benefits to the commercial stakeholders. In addition, commercial fishers were willing to accept a more conservative policy than they had initially requested (Jones et al. 2016).

Decision and Importance of Integration of Social and Ecological Science

Based on the analysis of the tradeoffs, stakeholders recommended, and the Lake Erie Committee adopted, a harvest control rule that accounted for the objectives and relevant tradeoffs among stakeholders (Jones et al. 2016). The structured decision-making process was viewed by all stakeholders as providing a more transparent process for setting recommended allowable harvest and total allowable catch each year. In addition, stakeholders also stated that the management decisions were derived from sound science (Jones et al. 2016). Overall, the explicit integration of stakeholders into the harvest management process for walleye through a series of workshops and participatory modeling led to a more complete and acceptable decision. More recently, the Lake Erie Percid Management Advisory Group has completed a similar process for yellow perch (*Perca flavescens*) harvest management on Lake Erie (Belore et al. 2020).

CONCLUSIONS

Decisions for fisheries and wildlife harvest management require balancing multiple objectives related to a diversity of stakeholder values. These objectives generally fall into broad categories of ecological, social, and economic values (McDaniels et al. 2006). Structured decision making

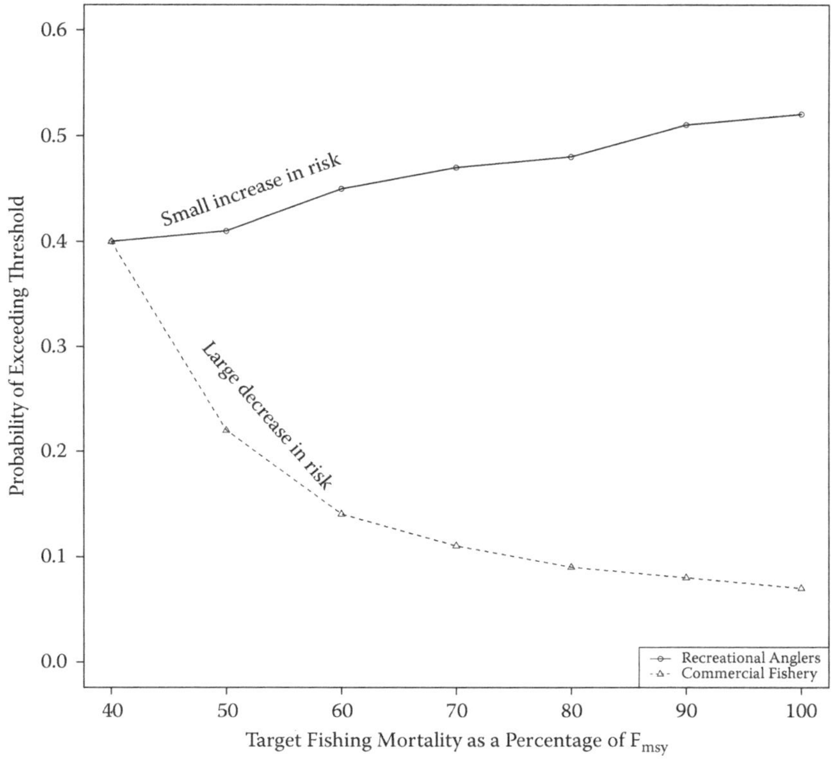

FIGURE 9.3 Tradeoffs analysis for evaluating levels of target fishing mortality rate for walleye harvest in Lake Erie. Y-axis represents a measure of risk: the probability of failing to meet a threshold outcome judged by the respective stakeholder group as indicative of a successful fishery (total harvest greater than 1.8 million kg/year for commercial harvest, catch-rates per hour greater than 0.4 for recreational anglers). At low to moderate target fishing rates, the decline in risk for the commercial stakeholders is much more rapid than the corresponding increase in risk for recreational stakeholders.

provides a clear framework for accounting for stakeholders' objectives, determining actions that could help achieve those objectives, predicting the consequences of each action on each objective, and making necessary tradeoffs among objectives. We argue that the integration of both social and ecological science into structured decision making for harvest management strengthens the resulting decisions and engenders trust in the process as a whole (Decker et al. 2012). This integration of disciplines can be accomplished by a number of means, including the incorporation of ecological and social scientists throughout the process, engaging in meaningful collaborative processes with stakeholders, and using participatory modeling to build predictive models in the consequences step.

The above case studies of deer and walleye harvest management provide examples of this integration, making use of different aspects of these methods, tailored to the decision at hand. In the decision for white-tailed deer buck harvest management, the statewide survey designed around the structured decision-making process provided insight into a stakeholder base that was divided in terms of their hunting preferences. The ecological modeling aided decision makers in understanding how the competing values of stakeholders could or could not be achieved, providing them with the necessary transparency in the decision-making process. In the case of walleye management, stakeholder desires were also at odds, but with an added component of economic stability required for the commercial fishing sector. The small group setting, which included representatives of key stakeholder groups and multiple meetings, increased trust both among participants and in the highly quantitative management strategy evaluation process. These case studies are just two examples of the ways to more fully integrate social and ecological science into a decision analytic process for harvest management. The commonalities are embedded in the need to account for multiple objectives, large spatial scales, and stakeholders with different opinions and values. Ultimately, challenges still exist for full integration of social and ecological science into structured decision making (Robinson et al. 2019), but the resulting decisions are more transparent, robust, and satisfying to decision makers and stakeholders alike.

ACKNOWLEDGMENTS

The authors thank C. McGowan, L. Powell, and M. Runge for helpful comments on this chapter. The authors also thank all participants in the two case studies for their work on those projects. This is contribution number 2021-01 of the Quantitative Fisheries Center.

10 Structured Decision Making Provides Insight When Selecting Population Monitoring Programs

Jonathan W. Cummings and Chris Bernier

CONTENTS

INTRODUCTION

A hallmark of monitoring programs is estimating the state of a population to inform management decisions (Williams et al. 2002, Lyons et al. 2008). State variables for population monitoring commonly include occupancy, abundance, or population trend. A harvest management monitoring program typically includes the activities, data collection, and accompanying estimation methods used to assess and track the state variables over time to inform management decisions (Williams et al. 2002). Data collection in such a monitoring program can range from monitoring total annual or daily harvest, conducting necropsies or other assessments to determine age, sex, and health of each harvested individual, implementing surveys to assess daily effort of sportspeople and harvest rates, and conducting surveys of marked, individual animals to determine demographic rates, to direct observations of individuals who support population estimation methods. The design of a monitoring program—the choice of data collection method and estimation technique—affects the achievement of monitoring objectives.

The main tradeoff to consider when designing a monitoring program involves minimizing the cost of the monitoring program and maximizing the information obtained. Managers may have multiple additional objectives when designing a monitoring program however, and these objectives—as well as their relative importance—may vary from agency to agency (Conroy 2021 [Chapter 1]). Accounting for all these objectives simultaneously requires going beyond evaluation of bias and precision (i.e., estimator performance) of a given estimation method (e.g., Roseberry and Woolf 1991, Davis et al. 2007, Millspaugh et al. 2009, Skalski et al. 2012).

Value of information (Moore and Runge 2012, Conroy and Peterson 2013) analyses can provide a framework for determining whether specific information is worth its cost, although these analyses have rarely occurred for harvested wildlife management (but see Johnson et al. 2014, Roberts et al. 2018). Multiple criteria decision analyses (Keeney and Raiffa 1976) can be combined with estimation error evaluation methods to account for cost and other objectives in the evaluation of monitoring programs. These analyses can be formal components of the structured decision-making approach to decision analysis (Runge 2021 [Chapter 7]).

The question of which monitoring program to employ is relevant to fisher (*Pekania pennanti*) management in Vermont, United States of America. The primary objectives for management—identified through consultation with the Vermont Fish and Wildlife Department (hereafter, Department)—are to maintain a sustainable population and to maximize constituent (i.e., trappers) and public support for the appropriate management of this species. Presently, the Department uses harvest and necropsy data to assess the state of the fisher population on an annual basis. Citizen science volunteers including trappers, students, researchers, and the general public engage in carcass processing providing the Department with a harvested individual's age, sex, and condition. Trappers are also surveyed on an annual basis to engage these key constituents and determine yearly fisher catch per unit effort estimates. Several estimation methods can be used to estimate population size and trend based on these data sources. For example, with the harvest totals alone, an index-based estimation method can be used. The sex-age-kill method (Eberhardt 1982) can be employed without the benefit of necropsy if trappers can accurately identify the sex of the animal and classify animals into age groups (first-year breeders vs. older breeders). Some of age-based population reconstruction methods such as Fry (1949) and Downing (1980) are available for use in situations when it is possible to collect necropsy data.

Different estimation methods require different sources of information and vary in their cost and level of citizen engagement, leading to the question of which will best meet overall needs. The Department requires a monitoring program that provides unbiased and precise estimates of the fisher population at low cost, while maximizing stakeholder and citizen support. In this chapter, we present a *simple multi-attribute rating technique* (often referred to as *SMART methods*; Edwards 1971, von Winterfeldt and Edwards 1986, Goodwin and Wright 2009) decision tool for selecting a monitoring program for fisher (*Pekania pennanti*) management in Vermont, United States of America.

Our objectives were to (1) describe the decision problem of monitoring program selection for Vermont furbearer management, (2) describe our simulation of that problem and its analysis, (3) determine the monitoring program that will best meet the objectives for Vermont fisher management, and (4) compare results with the selection of a monitoring program based solely on estimator performance.

METHODS

FISHER

Female fishers produce their first litter at 24 months and then produce one litter per year thereafter, with males becoming reproductively active at age 1 year (Powell and Zielinski 1994). Offspring sex ratio has not been found to differ from 50:50 (Buskirk et al. 1994). Annual birth rates average between 1.18 and 2.16 offspring per breeding female across studies in the northeastern United

States of America (Buskirk et al. 1994, Paragi et al. 1994, Koen et al. 2007). The estimated life span of a fisher is ten years (Powell et al. 2003). Annual survival probabilities depend on age and the level of trapping; without trapping mortality, survival ranges between 0.7 and 0.8 for juveniles and is greater than 0.9 for adults (Buskirk et al. 1994). With trapping, annual survival probability can diminish to as low as 0.34 (Buskirk et al. 1994).

Problem

The primary question for the Department is, what monitoring program will best achieve Department objectives by providing the most cost-effective information for guiding fisher management in Vermont? This question was triggered by the cost of the current necropsy-based data collection procedure. Vermont trappers are required to record, tag, and submit all carcasses for processing within 48 hours of the close of the trapping season. Trapper records provide information on the number of animals harvested per trapping season, while necropsies are conducted on every harvested fisher to obtain age and sex data and collect samples for toxicity such as anticoagulant rodenticides, disease such as canine distemper (Needle et al. 2019), nutrition, and other body characteristics. The Department benefits from an informed citizenry, including the trappers who directly participate in the harvest of fishers, academic institutions who help conduct the research necessary for fully assessing the status of the species, and the general public who influence the management of wildlife through the public input process. Such an informed set of stakeholders is aware of the management and monitoring efforts undertaken for game species management. Conducting necropsies and trapper surveys uses warden time and Department staff time and equipment. Further, due to the mandatory reporting and carcass delivery process, volunteers are required to assist with the necropsy process itself. Therefore, managers questioned whether the benefits of the necropsy approach justify the costs.

Objectives

The eight objectives included for evaluating alternative fisher monitoring programs fall into three categories (Table 10.1): (1) maximizing information about the state of the population (objectives 1–5), which included minimizing the bias in population growth rate (lambda) estimates, maximizing the precision in lambda estimates, minimizing the bias in abundance estimates, maximizing the precision of abundance estimates, and maximizing the probability of detecting disease or toxins in the population; (2) minimizing cost (objective 6), and (3) maximizing public knowledge and engagement (objectives 7 and 8), which included separate objectives for trappers, as well as public and academic institutions.

To quantify the assessment and achievement of these objectives, measurable attributes were identified, and importance scores were given to each of the objectives (Table 10.1). These objectives and measurable attributes were developed via in close consultation with the Department's furbearer management team, which consists of two wildlife biologists, a game warden, and the Department's wildlife management program director.

Objective weights, representing the relative importance of the objectives, were elicited from the lead fisher biologist with a swing weighting exercise (von Winterfeldt and Edwards 1986, Clemen 1996) and placed on a 100-point scale. Most of the weight was placed on the estimation performance (68%), 12% each to disease and cost objectives, and the remaining 8% on engagement objectives (Table 10.1).

Alternatives

Estimation Methods

We evaluated four methods available for converting the fisher monitoring data into estimates of abundance: an index method, two virtual population reconstruction methods, and the sex-age-kill

TABLE 10.1
List of Fisher Monitoring Program Objective Categories, Objectives, the Desired Direction for the Objectives, How the Objectives Were Measured, and Their Relative Importance Used to Assess the Fisher Monitoring Program in Vermont, United States of America[1]

Category	Objective	Direction	Measurable Attribute	Weight (Relative Importance)
Information on population state	1. Precision of lambda estimate	Maximize	5th to 95th percentile range of median	20
	2. Precision of abundance estimate	Maximize	5th to 95th percentile range of median	18
	3. Bias in lambda estimate	Minimize	Median error in lambda estimate	16
	4. Bias in abundance estimate	Minimize	Median error in abundance estimate	14
	5. Disease detection	Maximize	Probability of detection	12
Cost	6. Cost	Minimize	Thousands of U.S. dollars	12
Citizen engagement	7. Trapper knowledge and engagement	Maximize	Constructed 0–10 scale	5
	8. Academic and public knowledge and engagement	Maximize	Constructed 0–10 scale	3

Note
[1] Relative importance weights were elicited during interviews with lead fisher biologist using a swing weighting exercise.

method. Each of these estimators has unique data input requirements, resulting in different implementation costs.

The *index method* takes the annual harvest data (either total or by age and sex) and an estimated harvest rate model as inputs (Petrides 1949, Eberhardt 1982). Harvest rate is normally based on repeated removal sampling or expert opinion and is used to extrapolate from the observed count to the total population size. Although the number of harvested individuals provides an index of the total population size, variability in effort and harvest success result in this being an imperfect indicator of the total abundance (Wilberg et al. 2009). Regardless, fisheries management often uses index methods for stock assessment (National Oceanic and Atmospheric Administration 2020*b*).

The estimators introduced by Fry (1949) and Downing (1980) are known as *population reconstruction methods*. These methods use back-calculation to produce year-, age-, and sex-specific abundance tables from age- and sex-specific harvest data. Summing over the age classes produces annual abundance estimates. The Fry method estimates the minimum population size necessary to produce the observed harvest without accounting for additional sources of mortality. In contrast, the Downing reconstruction method uses a weak proxy of adult mortality to reconstruct the pre-hunt population by backward addition of known mortality and a minimal assumption of unaccounted-for mortality (Downing 1980) in an effort to produce a less biased estimate of abundance. These population reconstruction methods have been used extensively in fisheries where the Fry method is termed Virtual Population Analysis (National Oceanic and Atmospheric Administration 2020*b*). Downing (1980) was the first to apply these methods for use and adoption by management agencies for management of terrestrial wildlife (Davis et al. 2007).

The *sex-age-kill method* is a life-history-based method that uses harvest data by age group, as well as estimates of the proportional mortality due to harvest, and the adult female birth rate to estimate the abundance of a population (Eberhardt 1982, Roseberry and Woolf 1991, Millspaugh et al. 2009). The exact age of harvested individuals is not needed; however, animals must be aged by group, defined here as young (age 0 year), juveniles (age 1 year, pre-breeding), recruits (age 1 year, post-breeding), and adults (age 2+ years). A modified version of this method has been applied to management of white-tailed deer in Pennsylvania (Norton et al. 2013). Starting from the adult male harvest rate, the abundance of adult males is estimated from the adult male harvest rate and the proportion of the total mortality that is due to harvest. The adult female abundance is derived from the male abundance estimate and the adult sex ratio. The juvenile abundance from the female estimate and the birth rate which are added to the estimate of adult male abundance to produce the total abundance estimate.

Monitoring Programs

We considered six alternative monitoring programs, consisting of a combination of the four estimation methods (index, Fry, Downing, and sex-age-kill) and two data collection process (with and without necropsies). When necropsies are conducted, all four estimation methods are available for use, resulting in four alternative with-necropsy monitoring programs. However, age data are not available and sex data are likely to be more error prone without necropsy information. The population reconstruction methods (Fry and Downing) require aging to year; thus, without the availability of necropsies, these methods lack sufficient data inputs. Therefore, the index and the sex-age-kill methods remain as the two *without-necropsy* monitoring program alternatives. Annual trapper surveys continue for all six alternatives because the benefits of maintaining trapper engagement in fisher management compensate for the survey costs.

Consequences

Bias and Precision of Abundance and Lambda Estimates (Decision Objectives 1–4)

We simulated a fisher population, its harvest, and data collection (both with and without necropsy) to evaluate each of the six monitoring program alternatives with respect to bias and precision of abundance and lambda estimates (Table 10.1). The model was a modified age- and sex-based Leslie matrix model (Leslie 1945), in which the annual cycle was further decomposed into discrete periods.

The annual cycle consisted an instantaneous census, followed by a 31-day harvest period, a four-month pre-breeding period, an instantaneous birthday, and a seven-month post-breeding survival period which culminated in the next year's census (Fig. 10.1). We simulated conditions similar to the Vermont fisher population over a 30-year period arbitrarily labeled as 1901 to 1930. The population model was parameterized with an initial population size (seed) of 4277 individuals, allocated by age and sex as a stable-age distribution for the first-year census period as follows (age 0–10+ years):

$$\text{Males} = [535, 267, 191, 154, 129, 112, 99, 88, 78, 69, 59]$$

$$\text{Females} = [813, 380, 267, 212, 178, 153, 132, 114, 99, 83, 65].$$

We assumed that the age at first reproduction was one year for males and two years for females, and the final age class represented a composite age class of ten years and more. The harvest season was assumed to be 31 days beginning on December 1 of each year, with all ages susceptible to additive harvest (harvest was 100% additive to natural mortality).

Demographic rates (birth, survival, and harvest) were simulated from a series of logistic-link and log-link linear models (Table 10.2). We modeled an expected birth rate of 1.68 offspring per

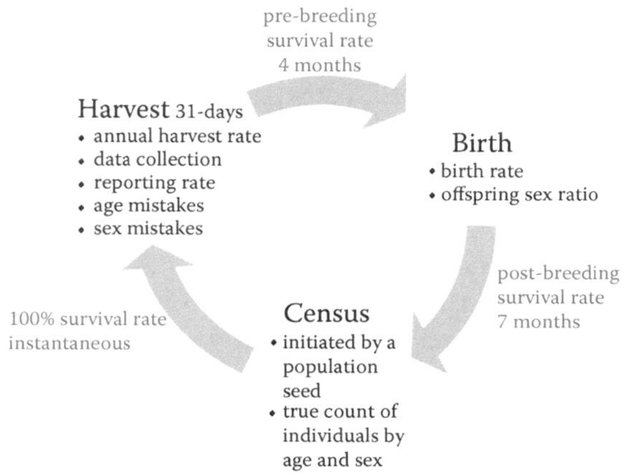

FIGURE 10.1 Simulated life cycle for fishes in Vermont, United States of America. Beginning with the census, all individuals survive until the harvest period. Harvest is determined by the simulated harvest rate, with simulated observation error. Following the harvest, individuals who survive the pre-breeding period advance in age and reproduce based on the birth rate. Individuals who then survive a post-breeding period are censused, producing the *true* abundance for that year.

reproductive female allocated to sex assuming a 50:50 offspring sex ratio. The expected pre-breeding and post-breeding survival rates were modeled to approximate the reported survival rates without trapping. Our expected juvenile pre-breeding and post-breeding survival rates were 0.73 for both sexes, and adult post-breeding rates were 0.86 for males and 0.90 for females. The baseline harvest rate was a function of simulated annual harvest effort, which ranged between 11.3 and 19.2 thousand trap nights per year (C. Bernier, personal communication). This resulted in an expected annual harvest rate of 0.15 across ages and sexes, with higher harvest rates for young and old individuals and slightly lower rates for three- to seven-year-olds. Apart from the post-breeding survival rate, all rates (birth rate, pre-breeding survival, and harvest effort) include stochasticity in the model intercept term (β_0), with values drawn from a bounded uniform distribution for each simulation. The annual number of harvested individuals was simulated using a binomial distribution where the number of trials was the census abundance at each age and the probability of success was the simulated annual harvest rate. We simulated 250 30-year population trends and recorded population size and lambda at each time step with resultant lambda values that ranged from 0.96 to 1.04 (Fig. 10.2). The simulation was coded in both Microsoft Excel and R (R Core Team 2019); R functions can be found at https://code.usgs.gov/vtcfwru/ampop.

Each stochastic simulation produced annual harvest totals by age, sex, and year (a true value), to which we applied a *harvest data error*. This error accounts for the cumulative probabilities of reporting error (the probability that a harvested animal was not reported), sexing error (the probability that a harvested male or female was incorrectly classified by sex), and aging error (the probability that a harvested animal of age *f* was not classified into the correct age group). With necropsies, trappers must submit carcasses to the state for processing, so we assumed that all trapped fishers were reported, sexed, and age correctly for the necropsy data set. Without necropsies, the likelihood of reporting a harvested individual likely remains high as pelts still require tagging. However, without carcasses sex and age group determination would occur in the field by trappers or wardens, which we have observed to increase the likelihood of error. We modeled a 95% reporting rate (5% reporting error), 10% aging error, and 10% sexing error rate for the without-necropsy monitoring alternatives. Age errors were modeled with a distance-to-and-from age class formula:

TABLE 10.2
List of Models Used to Simulate Fisher Population Rates in Vermont, United States of America, the Linear Transformation Applied (if any), and the Model's Input Parameter Values Used for Simulating Abundance through the Annual Life Cycle

Modeled Demographic Rate	Equation	Stochasticity	Beta Parameter Values
Birth rate	$\log(rate) = \beta_0 + \beta_1 * age + \beta_2 * age^2$	simulation	β_0 ~ Uniform (0.245, 0.375) $\beta_1 = 0.090$ $\beta_2 = -0.008$
Pre-breeding female survival rate	$\text{logit}(rate) = \beta_0 + \beta_1 * age + \beta_2 * age^2$	simulation	β_0 ~ Uniform (0.875, 1.110) $\beta_1 = 1.650$ $\beta_2 = -0.160$
Pre-breeding male survival rate	$\text{logit}(rate) = \beta_0 + \beta_1 * age + \beta_2 * age^2$	simulation	β_0 ~ Uniform (0.875, 1.110) $\beta_1 = 1.500$ $\beta_2 = -0.125$
Post-breeding female survival rate	$\text{logit}(rate) = \beta_0 + \beta_1 * age + \beta_2 * age^2$	No	$\beta_0 = 1.850$ $\beta_1 = 0.450$ $\beta_2 = -0.028$
Post-breeding male survival rate	$\text{logit}(rate) = \beta_0 + \beta_1 * age + \beta_2 * age^2$	No	$\beta_0 = 2.250$ $\beta_1 = 0.400$ $\beta_2 = -0.020$
Annual female harvest rate	$\text{logit}(rate) = \beta_0 + \beta_1 * age + \beta_2 * age^2 + \beta_3 * effort$	annual	$\beta_0 = -1.777$ $\beta_1 = -0.325$ $\beta_2 = 0.030$ $\beta_3 = 0.0325$
Annual male harvest rate	$\text{logit}(rate) = \beta_0 + \beta_1 * age + \beta_2 * age^2 + \beta_3 * effort$ $effort$~Uniform (11.3, 19.2)	annual	$\beta_0 = -1.900$ $\beta_1 = -0.250$ $\beta_2 = 0.020$ $\beta_3 = 0.0325$

$$A_t = \sum_{f=1}^{F} A_f * \frac{e^{f+(f-t)^2 * E}}{\sum_{i=1}^{T} e^{f+(f-i)^2 * E}} \tag{10.1}$$

where A_t, *age to* is the number of individuals in the resulting data count for age t, A_f, *age from* is the number of truly harvested individuals for age f, and E the age error parameter, was set to –25 to produce no error for the necropsy monitoring program alternatives and to –2.791 to produce 10% error rates for the without-necropsy alternatives. Anecdotally, these represent the expected errors in the data collection process from the two monitoring programs. Harvest data sets were then analyzed to estimate the true population size.

We estimated the fisher population abundance and lambda from the simulated annual harvest and necropsy data with each of the four estimation methods estimators. For the index method, we used a constant value of 0.15385 (the average harvest rate across age and sex for the stable population trajectory simulation) as the harvest rate input in all evaluations. Using the simulated

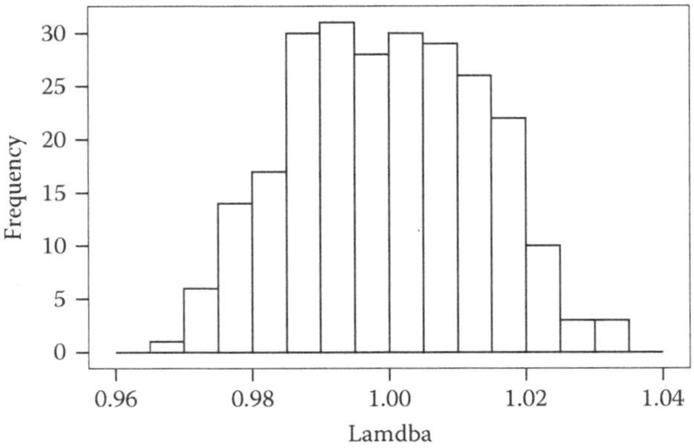

FIGURE 10.2 Simulated population growth rates (lambda) from 250 simulations of fisher population models used to assess monitoring programs in Vermont, United States of America. Lambda values are summarized in 0.005 increment bins from 0.960 to 1.040.

true harvest rate removes the primary source of bias in the abundance estimate, but bias resulting from fluctuations in the harvest rate or age and sex distribution remains.

For the sex-age-kill method, as with the index method, we used constant values—the *true* values on average from a stable population trajectory—of 0.4702 for the harvest mortality ratio and 1.128 for birth rate inputs (we note that levels of precision in model parameters varied as required by the model, and the high level of precision in the index method and sex-age kill method input values ensured the population estimates did not contain any error resulting from rounding to less precise estimates). Thus, our methods produced the best possible expected performance from the sex-age-kill method.

The Fry and Downing methods take annual sex and age at harvest data as model inputs. In addition, the Downing method requires a grouping age, the age above which all individuals are summed into a single count, as well. We used age 4 years as the grouping age.

We calculated the median bias in abundance and lambda estimates under each monitoring program, determining the median difference between the estimated and true values divided by the true values, $([\widehat{N_t} - N_t]/N_t)$ for abundance and $\left(\frac{[\hat{\lambda}_t - \lambda_t]}{\lambda_t}\right)$ for lambda, where lambda is $(\lambda_t = N_{t+1}/N_t)$. We then report the median of these median values over the 250 simulations. The precision results were reported as the range in the 5th to 95th percentile median values over the 250 simulations.

To ease interpretation, we converted the bias and lambda measures into units representing the numbers of individuals by which the estimation methods would have erred relative to a true population size of 5000 individuals. For example, a percent bias in abundance of 0.5 would be synonymous with an error of 2500 individuals. In the case of lambda, we used the resulting difference in population size over a ten-year period assuming the true population was stable; a bias in lambda value of 0.001 would equate to an error of 50 individuals over a ten-year period with an initial population size of 5000 individuals. We used the absolute value of these measures such that negative and positive bias (e.g., an error of ±100 individuals) are both treated in the same way (an error of 100 individuals) and given the same performance score.

The performance of the abundance estimates and the lambda estimates was also assessed with the coefficient of error summary statistic used by Millspaugh et al (2009). The coefficient of error in abundance expressed as a percentage is calculated as:

$$coefficient\ of\ error = \frac{\sqrt{MSE}}{\left(\frac{\sum_{i=j}^{n}\sum_{j=1}^{y} N_{ij}}{ny}\right)} * 100, \quad (10.2)$$

where

$$\widehat{MSE} = \frac{1}{n}\sum_{i=1}^{n}\left[\frac{\sum_{j=1}^{y}\left(\hat{N}_{ij} - N_{ij}\right)^2}{(y-1)}\right], \quad (10.3)$$

where y is the number of years being compared ($y = 40$), n is the number of simulations ($n = 100$), N_{ij} is the true population for simulation i and year j, and $\hat{N}_{ij}$ is the associated abundance estimate. The coefficient of error in lambda is calculated the same way, but substituting $\hat{\lambda}_{ij}$ and λ_{ij} for $\hat{N}_{ij}$ and N_{ij}:

Disease Detection, Cost, and Public Engagement (Decision Objectives 5–8)

We developed estimates of the monitoring program impacts on disease detection, costs, and public engagement through interviews of the Department's furbearer management team. Based on responses, the probability of detecting disease with a necropsy monitoring program was 75% and 20% without necropsies, the cost of the monitoring programs was estimated to be $120,000 with and $70,000 without necropsy, and the engagement scores were set to 10 (0–10 scale) with necropsy and 6 without necropsy. A typical necropsy session involves two to four state employees, up to four volunteers and additional data are frequently collected by one or more academic institutions per session. Not including state employees, as many as 25 individual volunteers and researchers participate in the five or six necropsy sessions conducted each year. Such a level of participation involves the public and presents an opportunity for communication between game species managers and select members of the public.

Tradeoffs and Decision Analysis

We organized our consequences in a consequence table (Table 10.3) to allow comparison of the predicted outcome of each alternative for each of the objectives (e.g., Mitchell et al. 2013).

Once the consequences were determined, the six monitoring program alternatives were scored across all objectives, given their weights, using a simple multi-attribute rating technique analysis (Edwards 1971, von Winterfeldt and Edwards 1986, Goodwin and Wright 2009). There are three steps in this process: normalizing consequences, applying objective weights, and summing the weighted results (Goodwin and Wright 2009). The alternative with the highest total score was that which fulfills the objectives for the decision to the greatest degree.

RESULTS

Estimator Performance

Considering only decision objectives 1–4, we found that the with-necropsy monitoring program alternatives performed better than the without-necropsy alternatives. However, within the with-necropsy alternatives, which estimator performed best depended on the objective. The sex-age-kill method when used with necropsy data performed best for the bias in abundance estimate objective (Table 10.3, objective 4). The Fry method was best for the precision in abundance objective

TABLE 10.3
Consequence Table with the Predicted Outcome of Each Alternative Monitoring Program, the Six Combinations of Estimation Method with or without Necropsy for Each Objective of a Fisher Monitoring Program in Vermont, United States of America[1]

	Alternatives					
Objective	Index with Necropsy	Fry with Necropsy	Downing with Necropsy	Sex-age-kill with Necropsy	Index without Necropsy	Sex-age-kill without Necropsy
1. Precision of lambda estimate	5733	4992	4762	13,019	5471	10,340
2. Precision of abundance estimate	2914	1563	1747	3606	2815	3247
3. Bias in lambda estimate	124	46	37	94	67	57
4. Bias in abundance estimate	–67	–2545	–2338	–5	–201	–67
5. Disease detection	75%	75%	75%	75%	20%	20%
6. Cost (U.S. dollars)	$120,000	$120,000	$120,000	$120,000	$70,000	$70,000
7. Trapper knowledge and engagement	10	10	10	10	6	6
8. Academic and public knowledge and engagement	10	10	10	10	6	6

Note
[1] Precision and bias units have been converted to represent the numbers of individuals by which the estimation methods would have erred relative to a true population size of 5000 individuals. Values for objectives 5–8 were developed through interviews of Vermont's furbearer management team. Disease detection is the probability of detecting disease under the given monitoring alternative. Units for objectives 7–8 are relative to a zero to ten-point scale.

(Table 10.3, objective 2), and the Downing method produced the least biased and most precise estimates of lambda (Table 10.3, objectives 1 and 3). The Fry and Downing estimates were often quite similar, performing best or second best for the precision in abundance, bias in lambda, and precision in lambda objectives.

The Downing and Fry had the lowest coefficients of error scores for lambda of 2.9 and 2.7 (Fig. 10.3, lower panel), but the highest coefficients of error scores for abundance of 52.1 and 56.1, respectively (Fig. 10.3, upper panel). The index method, without necropsy, resulted in the best coefficient of error score for abundance of 22.1 but a coefficient of variation score of 5.8 for its lambda estimation performance. Thus, the selection of the best estimation method depends heavily on the weighting (importance) of each of the four monitoring objectives related to bias and precision. However, the remaining four objectives must also be considered to identify the optimal monitoring program.

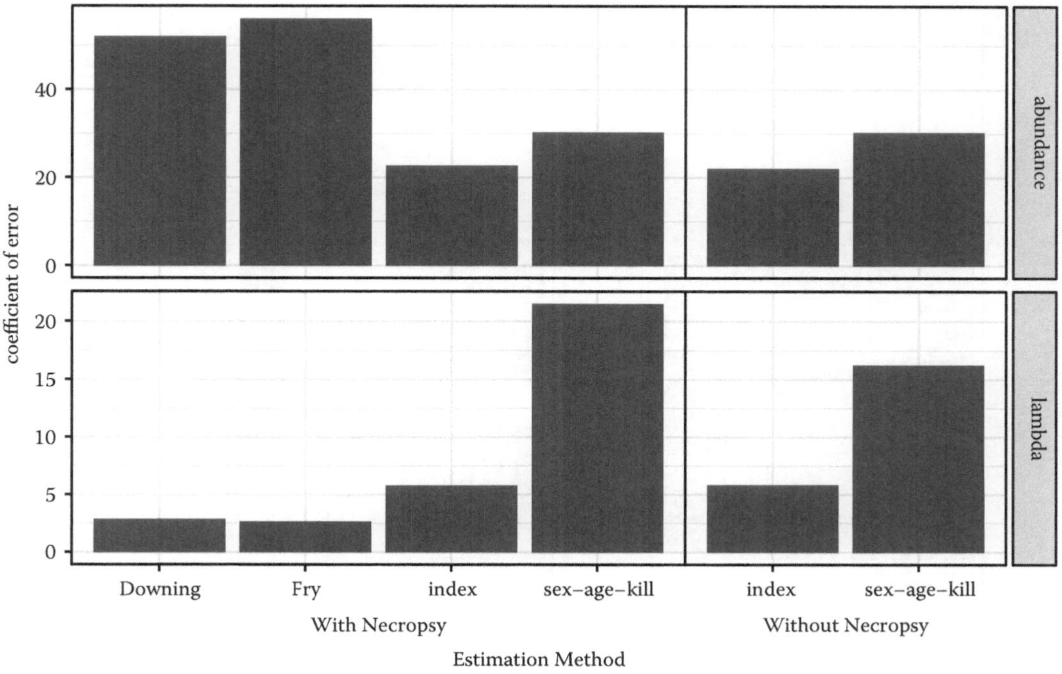

FIGURE 10.3 Coefficient of error in abundance (top panel) and lambda (bottom panel) for each of six monitoring programs (estimation method and necropsy program combination) considered in evaluation for fisher harvest management in Vermont, United States of America. Lower coefficient of error scores indicate better performance.

Decision Analysis

Given the full set of objectives and their weights, we found that the necropsy supported Downing estimation method was the best monitoring program with a weighted performance score of 73.5, followed closely by the necropsy supported Fry monitoring program with a score of 71.4 (Fig. 10.4). The Downing method performed well for most of the objectives, ranking in the top two methods for all but the bias in abundance and cost objectives where it performed poorly. The sex-age-kill with-necropsy program had the lowest weighted performance score of the six monitoring programs with a score of 39.5.

Although we found some unexpected benefits of the without-necropsy programs in terms of estimator performance, necropsy provided numerous benefits that resulted in necropsy-based alternatives performing best overall. The main benefit of the without-necropsy alternatives was their reduction in cost, which contributed 12 weighted-performance points. However, the 12 points from the increased probability of disease detection with necropsy compensated for the reduction in cost (Table 10.1). The benefits of citizen engagement and the improved estimator performance accounted for the increased score of the with-necropsy Downing and Fry alternatives relative to the without-necropsy alternatives.

Sensitivity Analysis

We assessed how changing the weights for each objective affected the relative ranking of the alternative monitoring programs. We found that the with-necropsy Downing alternative was robust

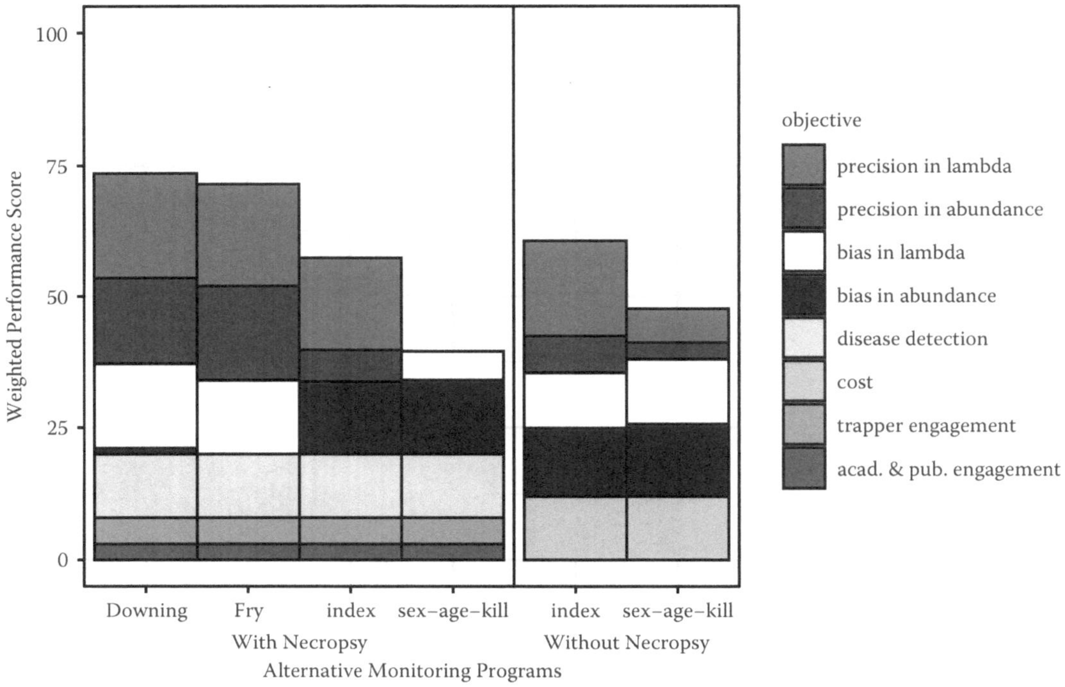

FIGURE 10.4 Results from simple multi-attribute rating technique analysis in the form of weighted performance scores for each monitoring program (estimation method and necropsy program combination) considered during an evaluation for fisher harvest management in Vermont, United States of America. The stacked bars display the contribution from each objective (objectives 1 to 8 from top to bottom of each stack) with a higher score indicating better performance.

across a range of objective weights. While keeping the other seven weights constant (at the weights in Table 10.1), weight on the bias in abundance objective needed to be increased to an importance score (weight) of at least 20 for the without-necropsy index alternative, or at least 35 for the without-necropsy sex-age-kill alternative, and the weight of the precision in abundance objective to 42 for the necropsy-supported Fry alternative to outscore the necropsy-supported Downing alternative. For the without-necropsy alternatives to be favored without altering the weights on estimator performance the weight on cost reduction needed to be greater than 20. Therefore, the robustness of the necropsy-supported Downing alternative was due to (1) the disease detection and engagement benefits of necropsy monitoring programs outweighing their costs, and (2) the high performance of the estimation method when all estimation performance measures were considered. The combined benefits from the with-necropsy Downing method's precision and bias in lambda performance outweighed the benefits of improved bias in abundance performance from the index or sex-age-kill approaches.

DISCUSSION

We demonstrated an approach that combined aspects of monitoring design (Reynolds et al. 2016), structured decision making, and value of information analysis (Moore and Runge 2012) to evaluate alternative monitoring programs, given multiple objectives to a monitoring program selection case study for fisher management in Vermont. Although multiple criteria decision analyses like this one have been applied to many natural resource problems (Mendoza and Martins 2006, Huang et al.

2011, Davies et al. 2013), with several applications to fish and wildlife management (Peterson and Evans 2003, Converse et al. 2011, 2013, Irwin et al. 2011), such analyses have been less commonly applied to common decisions in harvested species management (excepting Robinson et al. 2017, and Mitchell et al. 2018) or monitoring program design (excepting Neckles et al. 2015 and Hauser et al. 2019). Our approach addresses monitoring program design in the context of harvest management at the state agency level, which provides a unique model for future work.

We found that Vermont's furbearer management team will best meet their objectives with continued use of the necropsy program and use of the Downing estimation method. This result demonstrated robustness to objective weights, as it typically required doubling the importance of objectives favoring another alternative to change the ranking of the alternatives. Therefore, managers can be confident that the cost of conducting necropsies is warranted, particularly given that the disease detection and citizen engagement benefits are perceived to provide greater value than the reduction in cost from cessation of necropsies. These citizen engagement efforts have provided benefits to researchers as well. For example, a single year worth of necropsy analysis allowed Johnson State College to develop a library of *Escherichia coli* strains by species through digestive system samples for subsequent use in identifying game species present in a watershed. Similarly, Green Mountain College had the opportunity to collect tissue samples to conduct population genetic analysis and test for canine parvovirus (Bernier and Adler 2012). Hauser et al (2019) also found benefits to the inclusion of citizen science in monitoring programs, and citizen engagement can help shape management (see Fuller et al. 2021 [Chapter 8], Robinson et al. 2021 [Chapter 9]).

The selection of the Downing method from the necropsy supported estimation methods is largely due to its more consistent estimates across simulations. This was likely due to the pooling of adults and the calculation of abundance using a cohort approach that dampen the impacts of data from an unusual year. Although the Downing estimation method performed best overall, its estimates of abundance were strongly negatively biased, and the method is more inaccurate at estimating population status in the most recent years (Davis et al. 2007). The Department may wish to consider alternatives beyond those examined here such as using the sex-age-kill estimation method to estimate abundance along with the Downing methods estimates for lambda to obtain a better combination of estimates. Alternatively, while the Downing abundance estimate is strongly (negatively) biased, the precision is good; thus, it may be predictably biased in a way that enables adjustment to obtain less biased outputs.

The confidence in selection of the Downing estimation method found here would not result from an evaluation limited to estimator performance. Although the coefficient of error evaluation revealed a strong tradeoff between the ability to estimate abundance and the ability to estimate lambda, this evaluation does not weight the relative importance of that error to management and presumes that all information is equally valuable. We note that a weighted evaluation that solely considered the estimator performance objectives would have resulted in a similar ranking of the monitoring programs. However, a cost analysis was required to clarify the benefits of necropsies, which outweigh their cost solely due to the disease and toxicity detection and citizen engagement they provide.

The use of a structured decision-making approach in natural resource management can be challenging. There can be additional upfront costs due to the time invested by management agencies to reflect on the goals and their values as well as determining how to measure and weight them. Many performance measures are available for evaluating population estimator performance, such as mean square error, root mean square error, absolute error, raw error, r-squared, and Akaike Information Criterion (AIC), and the outcome of the evaluation can depend on what measures are selected (Cummings et al. 2011). The Department in our case study could have decided to measure the citizen engagement or disease detection objectives differently, such as conducting citizen and management surveys to assess how necropsy participation influences knowledge and perception of management, or a direct experiment to determine the likelihood of disease and toxicity detection.

Predicting the consequences can also be time consuming and technical, particularly to develop and report on simulation studies.

Structured decision making is a flexible framework in terms of the time commitment and complexity of its application though, and our case study highlights the benefits of careful problem framing and objective selection. Structured decision making has been applied to problems ranging from personal career decisions (e.g., Katz and Sterrett 2019) to complex multiple year, multiple stakeholder, and multiple decision maker decisions (e.g., McGowan et al. 2015). The process can be adapted to the circumstances and complexity necessary to overcome a problem's decision impediment, with a primary benefit being the clarity and assistance to transparent documentation that clarity can provide (Runge et al. 2020).

The level of complexity in the decision framework and model simulation used here supported the decision to maintain a necropsy program. The degree of fluctuation in monitoring results to date has not triggered state-based changes to management actions, and therefore the Department did not feel a need to simulate alternative harvest regulations. Given the past stability of the fisher population and the relatively low pressure from trapping the Department is not actively considering alternative harvest-rate management plans at this time. Therefore, the current monitoring program is mainly precautionary in nature to assess the magnitude of a change in state needed to trigger alternative trapping regulations.

The framework provided here could be extended to support decisions about alternative harvest regulations and adaptive management if the Department desires an evaluation of alternative harvest regulations in the future, perhaps triggered by a change in population status. For example, a closed loop simulation of alternative trapping management actions such as a management strategy evaluation (Bunnefeld et al. 2011*a*) would enable an evaluation of alternative regulations, while accounting for non-linear utility to determine the benefits of regulatory changes from improved estimation performance would support an expanded value of information analysis.

CONCLUSION

Our approach was applied to necropsy as the form of data collection. However, the value of information or benefits of monitoring programs that account for the social complexity of management (e.g., Hiller et al. 2021*b* [Chapter 2], Kaemingk et al. 2021 [Chapter 3]) is a common concern in harvest management with a similar decision structure. The decision to collect data on harvested individuals age, length (a proxy for age), and sex is common in fish and wildlife management. Indeed, an aging technique for aging our study species, fishers, using teeth is similar to methods for otolith aging in fisheries. The evaluation of monitoring benefits through value of information type analyses is also relevant to other methods of abundance estimation such as transect counts and mark recapture, as well as population projection efforts using demographic rates, such a mortality rate and reproductive success studies.

A decision analytical approach to monitoring program selection can provide clarity where typical evaluation procedures are less conclusive, as the coefficient of error evaluation was here. Structured decision-making approaches also provide transparency to the process, as well as a tool for managers to validate and communicate why a particular monitoring program was chosen. The Department has now a documented rational for their costs, and a framework that future estimation techniques or alternative harvest regulations can be readily added to and evaluated with. Such a decision analysis can be repeated with other species, population demographics, harvest rates, data collection regimes, and estimators to provide insight into many wildlife management decisions. We provide a model for the assessment of alternative monitoring programs, and we believe such assessments will be critical for effective harvest management in the future.

ACKNOWLEDGMENTS

The authors thank J. Pontius, R. Mickey, and K. Rinehart for their valuable feedback, and the Vermont Fish and Wildlife Department for providing this research opportunity. The authors thank J. Murdoch, D. Smith, and L. Powell for their reviews of this chapter. T. Donovan advised this research, assisting with conceptualization, methodological development, and documentation of this work. The Vermont Cooperative Fish and Wildlife Research Unit is jointly supported by the U.S. Geological Survey, University of Vermont, Vermont Fish and Wildlife Department, and Wildlife Management Institute.

11 Making Harvest Management Decisions Robust to Uncertainty

Andrew J. Tyre and Brigitte Tenhumberg

CONTENTS

INTRODUCTION

Wildlife and fisheries managers regularly make decisions about harvest in the face of uncertainty about vital rates such as survival and fecundity. Perhaps more important are structural uncertainties such as the degree of density dependence (e.g., compensatory mortality) in vital rates, and social dynamics creating uncertainty in implementation. Managers must satisfy many conflicting stakeholder positions, often with little guidance from policymakers on how to make the inevitable tradeoffs. Decision theory offers a reproducible set of methods for making harvest management decisions in the face of competing objectives (Smith 2020, Runge 2021 [Chapter 7]). The general paradigm for harvest management is optimizing, usually maximizing, a set of objectives such as population size or hunting opportunities (Conroy 2021 [Chapter 1]).

In a typical decision theory analysis, the analyst chooses an objective, or a weighted combination of objectives, and evaluates alternative decisions to identify the best that can be achieved. However, alternatives identified this way can be *brittle* toward uncertainty. Brittle in this case means if the analyst is wrong about one or more details of the model (parameters, structure, etc.), then the best decision might be different. An alternative that gives a very good outcome if the model is exactly right may be terrible if the model is just a little bit wrong. In contrast, a robustness analysis seeks to identify alternatives that are not brittle, but rather perform well across a broad range of uncertainty. This changes the paradigm from maximization of an objective to guaranteeing some minimum performance over a broad range of uncertainty.

It is important to make a critical distinction about the meaning of the term *uncertainty*. To scientists, this term loosely encompasses many things, including statistical variation in an estimate, a poor or incomplete understanding about a process, and inherent variation arising from natural processes such as birth, death, and environmental variability. However, the term uncertainty can also refer to events for which no probability can be calculated—a complete lack of understanding about a process. Thus, when a scientist claims "uncertainty" exists about a natural process, such as the extinction of biodiversity or the changing of the climate, stakeholders may hear "no probability," which may be quite far from the actual state of scientific knowledge.

There are many events for which the probability is not estimable—pandemic diseases fall in this category. Consider the SARS-CoV-2 pandemic; epidemiologists can describe the spread of the disease in a population with a fair degree of accuracy (e.g., Kucharski et al. 2020). However, science cannot give a probability for the emergence of such a virus in any particular year, beyond saying > 0 and increasing (Allen et al. 2017). Not all catastrophic events are uncertain in this sense. Consider hurricanes. Meteorologists have many years of observations of hurricanes, enough to calculate the probability of, say, six Category 4 or greater storms occurring in one year (National Oceanic and Atmospheric Administration 2020c). Meteorologists can even calculate the probability of a hurricane hitting any particular place on the coast. There may be other types of events that decision makers are currently unaware of—the infamous "unknown unknowns"—decision makers are *ignorant* of these. It is impossible to give examples of these types of events or processes, because if an example was available, then decision makers would not be ignorant of the possibility, merely uncertain about the probability that it could occur!

Analysts and decision makers could evaluate the effect of uncertainty by using Monte Carlo approaches, but even there, assumptions must be made about the distributions of uncertain parameters. Ideally the analysis makes as few assumptions as possible about the nature of reality. Models for assessing each alternative are only ever approximations.

To evaluate the robustness of an alternative, rather than choosing to maximize our objective(s), decision makers could state the worst performance they find acceptable. Then the analyst looks for alternatives that are guaranteed to achieve that minimum over the largest range of uncertainty. Consider the mid-continental mallard (*Anas platyrhynchos*) harvest management decision (Conroy 2021 [Chapter 1]); the goal is to maximize the utility of long-term harvest, given a discrete set of possible regulatory combinations. The utility of harvest in each year varies with population size in such a way as to devalue harvest that yields low adult population sizes. To analyze the robustness of these alternatives, decision makers would choose lower bounds for both harvest and total population size. The analyst would then calculate how wrong the models could be and still *guarantee* that alternative would produce harvest and population sizes larger than the bounds.

This chapter focuses on uncertainty where the probability distribution of an outcome cannot be calculated. Decisions must still be made. Broadly, this uncertainty takes at least three forms. First, the analyst could be uncertain about the actual growth rate of a population in the absence of competition, or any other vital rate. Second, and perhaps worse, an analyst could be uncertain about the structural details of a population model. Is this population strongly regulated at its current size, or weakly (Runge and Johnson 2002)? Third, human behavior introduces inevitable uncertainty about how human stakeholders will react to management actions (Tyre and Michaels 2011). All of these are important, and all could be addressed with the robustness approach introduced here. For the sake of brevity, this chapter focuses on the first form of uncertainty only. First, we show a very simple example to demonstrate the overall concept of robustness to uncertainty. Second, we demonstrate the calculation of robust strategies for (st)age-structured, density-independent populations.

ROBUST HARVEST LIMITS FOR NEBRASKA'S MOUNTAIN LIONS

Mountain lions (*Puma concolor*) recolonized Nebraska's northwest sometime in the early 2000s. There is now a sizable breeding population (>30 animals) in the Pine Ridge, and evidence of mountain lion breeding in other parts of the state. A small hunting season took place in 2014, leading to significant controversy and some threat of political action in the state unicameral (single house) legislature. There were two short seasons, and no more than two males could be taken in each season. If a female was taken, the entire hunting season would close. As it turned out, three lions (two males, one female) were harvested in total out of an estimated population of 22 (Nebraska Game and Parks Commission 2020). Despite ongoing controversy, a small hunting season resumed in 2019. Up to eight animals could be taken, and ultimately four were harvested.

The overall goal for management (Nebraska Game and Parks Commission 2020) is vague:

A resilient and healthy mountain lion population is one that: (1) maintains a reasonable proportion of older age animals, (2) maintains a sufficient number of breeding females to recover from mortality events, (3) has healthy individuals with minimal burden from disease or malnutrition, (4) is in balance with available prey, and (5) maintains genetic variability and connectivity to other populations.

Of those objectives, only (2) is measurable given the current monitoring. Nebraska Game and Parks Commission estimates mountain lion populations every two years using scat detection dogs and DNA analysis, but currently the estimated population is not broken down by sex because the genetic estimate data are too sparse. Elsewhere the goal is stated as five to seven mountain lions per 100 km^2 (Haag 2020). The goal ignores the effect of mountain lions on livestock in the area.

The 2017 maximum likelihood estimate by the Nebraska Game and Parks Commission was 59 individuals with a lower 95% confidence limit of 34 (Nebraska Game and Parks Commission 2020). The upper confidence limit was so high that it called into question the validity of the analysis, and so was ignored. If the objective is to avoid local extinction, this is entirely reasonable, but it means that analysts and decision makers have abandoned any pretense of calculating the probability of different population sizes.

Michael Runge and colleagues from the U.S. Geological Survey developed a formula for determining *prescribed take level* for a species (Runge et al. 2009). This formula is intended to guide the setting of limits for *take*, which is mortality or disturbance usually incidental to some other regulated activity like mining, oil and gas exploration, or road construction. The formula is derived from the logistic model of population growth, and is an extension of the *Potential Biological Removal* formula written into the Marine Mammal Protection Act. Recall that the maximum sustainable yield for a population growing according to the logistic equation is

$$MSY = \frac{Kr_{max}}{4}. \tag{11.1}$$

where K is the carrying capacity or equilibrium population size in the absence of harvest and r_{max} is the intrinsic per capita rate of population growth (Powell 2020, Conroy 2021 [Chapter 1]). Harvest, or take, less than maximum sustainable yield will be sustainable in the sense of having a positive equilibrium population size $K/2 < N^* < K$ (Fig. 11.1). Runge et al. introduced a new *management objective variable*, F_0 (mean annual fecundity, chicks hatched per pair), which ranges between 0 and 2. The prescribed take level in year t is then

$$PTL_t = F_0 \frac{\tilde{r}_{max}}{2} \tilde{N}_t. \tag{11.2}$$

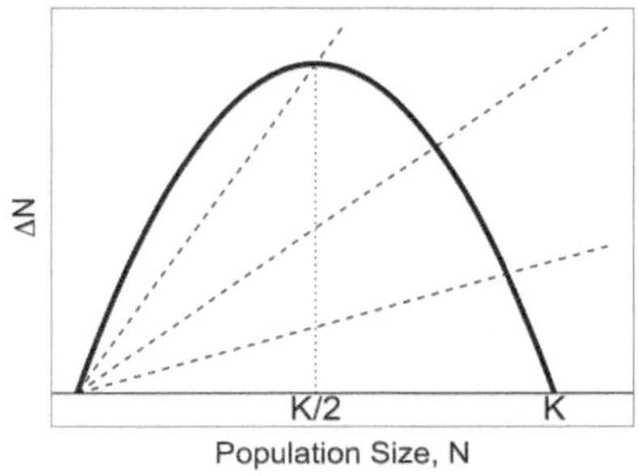

FIGURE 11.1 Proportional harvest of a population experiencing logistic growth. The dark curve is the change in population size as a function of population. The dashed lines are different levels of proportional harvest. Equilibrium population size N* for a given harvest is at the intersection of the line with the curve.

where the $\tilde{}$ over r and N_t indicate that these quantities are not known. The analyst may have information on the probability distributions of these quantities, but not necessarily. A management objective $F_0 = 0$ corresponds to no allowed take, and the population will eventually stabilize at K. $F_0 = 1$ is the maximum-sustainable-yield harvest level. When $F_0 > 1$ the population will stabilize at a level less than K/2. A value of $F_0 > 2$ will lead to extinction. The advantage of specifying take in this manner is that we do not have to estimate K. All the analyst needs to guess or estimate is the current population size, and the species potential for population growth at low population sizes.

Another way to think of F_0 is as the fraction of K at the equilibrium: $N^* = K\left(1 - \frac{F_0}{2}\right)$.

It is instructive to work backward from harvest quotas set by management agencies (i.e., prescribed take level) using agency estimates of $\tilde{N}_t$ and $\tilde{r}_{max}$ to infer what sort of management objective is in place. Ignoring uncertainty, and using $\tilde{r}_{max} = 0.28$ (Logan et al. 2001) and $N_{2014} = 22$, $PTL_{2014} = 4$, rearrange (1) to get $F_0 = \frac{2PTL_t}{\tilde{r}_{max}\tilde{N}_{2014}} = 1.3$. This is larger than a maximum-sustainable-yield harvest, and the expectation is that the population will stabilize below half of the carrying capacity. The actual take of three lions yields $F_0 = 0.97$, approximately a maximum-sustainable-yield harvest with N^* around $K/2$.

This is limited information, and thus the conclusion comes with large caveats. First, the prescribed-take-level model ignores sex structure of the population and harvest, and that could matter for this species. Second, the estimate of $\tilde{r}_{max}$ comes from a remote area of Arizona. Nebraska lions experience other sources of anthropogenic mortality, including roadkill, responses to human safety, and cattle depredation. These reduce annual survival rates compared to more pristine wilderness areas, and would lead to lower values of $\tilde{r}_{max}$. A third reason to regard these calculations with skepticism comes from the small number of animals involved. Small populations experience stochastic fluctuations simply due to demographic stochasticity.

For all those reasons, casting this problem in a robustness framework is attractive. What follows here is based on the approach to uncertainty outlined in Ben-Haim (2001). The management action is the total allowed take, PTL. The decision maker must specify a *minimum performance* for the system. In this case, it seems reasonable to specify that as a desired fraction of the carrying capacity, K. This does not mean K is known or even estimated, only that whatever equilibrium is reached will be this fraction of the unknown K. This value is not something management is *trying* to achieve, rather it is the minimum level that decision makers are prepared to accept. The analysis will *guarantee* that the equilibrium population size is at least this big, as long as parameters are wrong by less than some amount. The performance guarantee is $F_0 = 1$, corresponding to $K/2$.

There are two things that are unknown: N, the estimated population size, and r_{max}, the per capita growth rate of the population in the absence of competition. For any combination of those two things, plus the specified take PTL, the value of F_0, or the fraction of K at equilibrium $1 - F_0/2$ can be calculated.

Plotted as a surface (Fig. 11.2), it is clear that as $\tilde{N}$ and $\tilde{r}_{max}$ increase, F_0 decreases. In contrast, as $\tilde{N}$ and $\tilde{r}_{max}$ decrease toward the origin of the plot, F_0 increases, and eventually $F_0 > 2$, above which the equilibrium population size is extinction.

Choose any particular point as the initial "guess" for $\tilde{N}$ and $\tilde{r}_{max}$. Now imagine a box of uncertainty expanding out from that initial guess. At each corner of the box is a possible value of F_0. One of the corners (the one closest to the origin in this case) will have a higher value than all the others. In fact, this corner has a higher F_0 than any other point on the edge of our box of uncertainty. Call the distance from the initial point to the edge of the box ϵ. F_0 is guaranteed smaller than the value at the corner closest to the origin. That corner represents the worst possible outcome for errors that are equal to or less than ϵ. There is a box, rather than a circle, of uncertainty because each of the errors can be as large as ϵ. The *worst case* corner is therefore $\sqrt{2}\epsilon$ away from the nominal point. If perturbations or errors are modeled as the total perturbation, then a circle of radius ϵ represents the uncertainty. That makes the math below harder so this example sticks with the box.

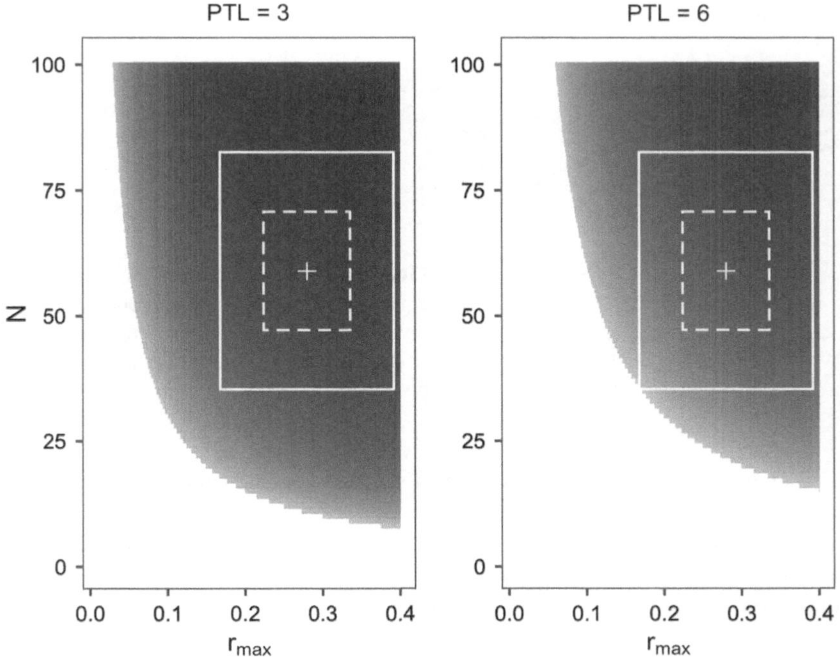

FIGURE 11.2 Management goal F (annual fecundity) as a function of N (population size) and r (growth rate) for a total take of 3 or 6. Black is $F_0 = 0$; light gray is $F_0 = 2$. The white portion is where the harvest would lead to extinction. The crosses are the nominal point of N = 59 and r = 0.28. The boxes are increasing levels of uncertainty.

The bigger ϵ can be before intersecting any particular level of F_0, the more *robust* the management alternative is to error. Clearly, if more individuals are taken from the population, our worst case corner will intersect the desired performance guarantee more quickly. However, it is still useful to be able to quantify the tradeoff between the robustness measured by ϵ and the level of take. Calculate this by re-arranging the prescribed-take-level equation to solve for F_0, and then multiplying each of the uncertain parameters by the fractional error $1 - \epsilon$. When $\epsilon = 0$, the parameters are known exactly:

$$
\begin{aligned}
PTL &= F_0 \frac{\tilde{r}_{max}}{2} \tilde{N} \\
F_0 &= \frac{2PTL}{\tilde{r}_{max} \tilde{N}} \\
F_0(\epsilon) &= \frac{2PTL}{\tilde{r}_{max}(1-\epsilon)\tilde{N}(1-\epsilon)} \\
&= \frac{2PTL}{\tilde{r}_{max}\tilde{N}} \frac{1}{(1-2\epsilon+\epsilon^2)}
\end{aligned}
\tag{11.3}
$$

Note that this equation only works for $\epsilon < 1$, which is OK, because more than 100% error is a pretty terrible situation.

As expected, increasing error decreases the *worst case* fraction of K at equilibrium for any level of take (Fig. 11.3). Increasing the take (PLT value) decreases the fraction of K for the same level of error. If six animals are harvested, errors less than 15% *guarantee* that the equilibrium will be greater than $K/2$ (equilibrium population size for a maximum-sustainable-yield harvest). Note that this is not saying that the population will not reach $K/2$, or even that the population will decrease from 59. It states that errors bigger than 15% in both parameters *may* lead to

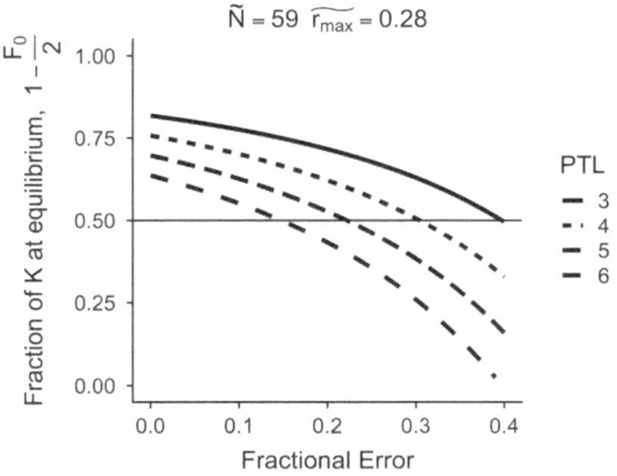

FIGURE 11.3 Fraction of K (carrying capacity) at equilibrium for different take options (prescribed-take-level values) and fractional errors. Above the horizontal line, the equilibrium population is guaranteed to be greater than K/2.

equilibrium sizes less than $K/2$. Whatever equilibrium is reached may be higher or lower than 59, because K is unknown.

Reducing take to four animals this year means that the fractional error can increase to just over 30% and still guarantee that the ultimate equilibrium is at least $K/2$. So, reducing the allowed take by two animals doubles the robustness to uncertainty.

STRUCTURED, DENSITY-INDEPENDENT POPULATIONS

Age- or stage-structured models are very common in fish and wildlife management (Caswell 2001). The parameters are readily interpretable in the context of real life-history data, and complexities like sex and patch structure can also be incorporated. There are some powerful mathematical proofs for square matrices that make robustness analysis quite straightforward. The basic ideas are demonstrated with an age-structured, density-independent model.

Peregrine falcon (*Falco peregrinus*) suffered terribly during the period that the pesticide DDT was in use. After listing as an endangered species, falconers, sportspeople who rear young raptors for hunting, became a critical part of the recovery process. Countless volunteer hours were invested in rescuing chicks, hacking (rearing to independence), and releasing back into the wild. This helped the species recover, because peregrines lay more eggs than the number of chicks they can rear to fledging. This *facultative brood reduction* allows a nesting pair to take advantage of plentiful years without causing too much loss in bad years. By taking some of the chicks from the nest early on, falconers were able to increase the total reproductive output of the remnant wild population.

After the U.S. Fish and Wildlife Service delisted the population in 1999, falconers were keen to obtain permission to resume *taking* chicks for falconry. The U.S. Fish and Wildlife Service needed to choose a threshold number of chicks that could be removed from a wild population, and ensure the population would continue to grow. In 2004, they chose 5% of all the chicks in a statewide population as an upper bound on the number of permits that could be issued by each state agency.

A matrix population model of falcon life history can be used to assess the robustness of this decision (Craig et al. 2004). Imagine that the U.S. Fish and Wildlife Service wants to guarantee that the change in population size is bigger than 1 (population is at least stable). All of the parameters (Table 11.1) are uncertain to some degree. However, it is difficult to visualize the robustness in >2 dimensions, so this example is restricted to just uncertainty in adult survival, S_2, and the proportion of two-year-old birds that breed.

These parameters are combined into a square matrix A that will project a vector of the number of individuals in each stage forward by one year (Caswell 2001)

TABLE 11.1
Parameter values and definitions for the peregrine falcon (*Falco peregrinus*) matrix from Craig et al. (2004)

Parameter	Value	Description
F	1.66	No. of nestlings fledged per pair
R	0.5	Proportion female nestlings
S_0	0.544	Survival of fledgling to age one year
S_1	0.67	Survival of age one to age two years
S_2	0.8	Survival of birds age two years and above
B	0.5	Proportion birds breeding at the second birthday
h	0	Proportion of nestlings taken

$$A = \begin{bmatrix} 0 & (1-h)FRBS_1 & (1-h)FRS_2 \\ S_0 & 0 & 0 \\ 0 & S_1 & S_2 \end{bmatrix}. \quad (11.4)$$

The leading eigenvalue of this matrix, λ is the ratio of next year's total population to this year's total population when the population is at the stable stage structure. The population grows when $\lambda > 1$, and declines when $\lambda < 1$.

It turns out that this boundary is the solution to

$$det(A(B, S_2) - \lambda I) = 0 \quad (11.5)$$

where $A(B, S_2)$ is a matrix conditional on the values of B and S_2 for any value of λ (Deines et al. 2007; Lubben et al. 2008, 2009), and I is the identity matrix with 1 s on the main diagonal and 0 s everywhere else. With two dimensions and $\lambda = 1$, this generates a line that divides the space of B and S_2 into areas where $\lambda > 1$ and $\lambda < 1$. As in the prescribed-take-level example, this line should be as far away from the *nominal point* (assuming zero uncertainty) as possible (Fig. 11.4).

Computational algebra software such as MATLAB can readily solve the equation in more dimensions, but it is difficult to plot multi-dimensional hypersurfaces.

DISCUSSION

Wildlife and fishery managers often must make decisions under extreme uncertainty where the probability distribution of an outcome cannot be calculated. In situations like this decision makers should use a robustness approach that has the goal to *guarantee* some minimum performance level for the greatest range of uncertainty, rather than maximizing a performance criterion. This chapter demonstrated what this looks like for two different simple harvest models. The first example used a simple density-dependent population model to calculate the size of proportional errors in two model parameters that *guarantees* that the equilibrium population size will be greater than K/2 (equilibrium population size for a maximum-sustainable-yield harvest). The second example considered a stage-structured, density-independent population model to calculate the permissible error in two model parameters that still *guarantees* population persistence, that is, the long-term growth rate $\lambda > 1$. The concepts are not limited to harvest decisions. Post van der Burg and Tyre (2011) used this concept together with a simple matrix model to evaluate nest protection programs for ground-nesting birds.

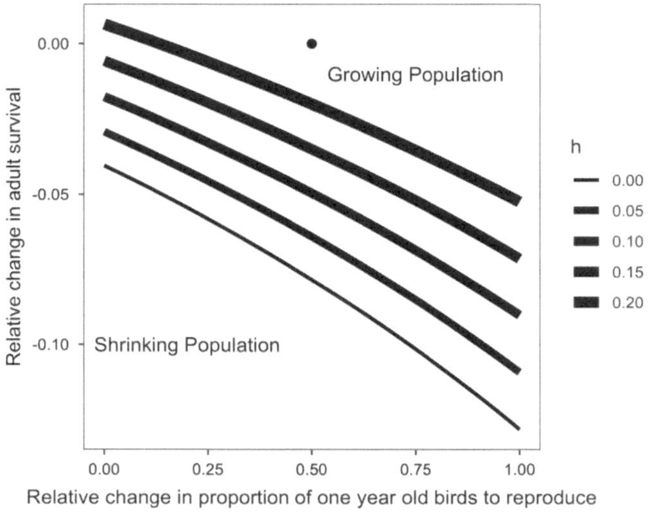

FIGURE 11.4 Contours of lambda = 1 (discrete population growth) as a function of the value of B (proportion of birds breeding) and S_2 (annual survival, age 2+) for different levels of per capita take, h. The solid point is the nominal point (zero uncertainty).

How could robustness be implemented? Fundamentally, this paradigm shift arises from changing how decision makers formulate the overall objective of harvest management. Consider the stated goal of mountain lion management in Nebraska. As stated the objectives are unmeasurable, because of ambiguous words like "reasonable," "sufficient," "healthy," and "balance." Other than population size and age structure, and possibly the "balance with prey," none of stated objectives are affected by the harvest management decision. A robust restating of the objective might be

Maintain at least 30% of adult lion population older than three years, and population density greater than 5 female lions per 100 km^2.

The specific numbers in both of these components would be subject to considerable debate! Nonetheless, stating them as minimum desired, rather than something specific to be achieved, could reduce debate. This simple step would transform management, whether the prescribed-take-level approach or something more sophisticated is used to calculate the effects of different management alternatives. The desire for age and sex specificity would require the use of a structured population model like that illustrated with the Falcon example. It is worth noting that as stated, the alternative that is most robust to uncertainty is no harvest at all. Providing an additional objective, such as provide at least two hunting opportunities per year, would be necessary to logically choose a harvest alternative greater than 0.

The quantitative methods introduced here are not difficult. The prescribed-take-level example can be worked out in a spreadsheet, even without the algebra, and would be broadly applicable to most harvest management. The matrix example requires more quantitative sophistication, but matrix models are introduced to undergraduates, and obtaining the solutions requires only modest programming well within the reach of quantitatively trained ecologists with graduate degrees. The difficult part is getting decision makers to specifically state objectives, and to change their paradigm from maximization to guaranteeing a minimum performance.

The robustness approach presented in in this chapter could be expanded in several ways. For example, analysts could add weights to the uncertainty of different parameter values thereby making some uncertainties "more important" than others. However, this approach risks making assumptions that are perhaps unsupported by the data. Another unexplored direction is how to account for the random changes of parameter values over time. The prescribed-take-level model assumes the population will approach an equilibrium if take values are selected according to the model. But in each subsequent year there is random variation in both r_{max} and N. It is not clear whether this simple equilibrium approach can account for the potential compounding of this between year variation. For example, what if there are three years in a row with low values of r_{max} and N? Does that change the performance guarantee? Finally, our examples did not consider the effect of different types of uncertainty. Epistemic uncertainty is associated with what is known about the world, for instance the error in estimates of mountain lion population size. There was some population size in 2017, and the estimate is just imprecise. In contrast, aleatory uncertainty arises because of actual changes in what is real. So, from one year to the next, the mountain lion population will either grow or decline, influenced by the availability of food, vagaries of weather, demographic stochasticity, and harvest. Epistemic uncertainties do not compound, but aleatory ones do. This issue is not unique to robustness analysis.

The examples here are characterized by an extreme lack of data, which makes the robustness approach attractive. If an analyst has data, a robustness approach to decision making is still useful. Imagine the analyst had sufficient data to characterize probability distributions for all the parameters in the Falcon matrix. They could still switch the perspective from maximizing take to maximizing the probability that $\lambda > 1$. The intermediate case is actually more difficult; what to do when there are data to calculate a probability distribution for some parameters but not others. The analyst has to decide how to weight the parameters with against parameters without probability distributions, and that may end up making assumptions that are untenable. One approach would be to calculate the probability of achieving a minimum performance guarantee conditional on the parameters with extreme lack of data, and then use robustness to guarantee some specific probability of achieving the goal.

Social indeterminism (Tyre and Michaels 2011) arising from the vagaries of human behavior may affect the actual management actions more than the parameters of the matrix. This *implementation uncertainty* (e.g., Conroy 2021 [Chapter 1]) can be difficult to measure, and developing methods for incorporating it into robustness analysis is a fruitful area of work.

We opened by describing how robustness analysis avoids brittle strategies that fail badly if models are a little bit wrong. However, in both examples, the robustness varies smoothly with increasing harvest pressure. Situations with high performance at small errors but worse performance with increasing errors did not occur. These strategies would show crossing robustness curves in Figs. 11.3 and 11.4. Does this mean that harvesting natural populations is inherently robust, or that the models are too simple?

Making decisions that are robust to uncertainty, rather than maximizing some population metric like total harvest, is a paradigm shift with a lot of potential to improve outcomes for fish and wildlife populations.

ACKNOWLEDGMENTS

B. Tenhumberg was financially supported by the USDA-(NIFA Grant 2017-03807) and the National Science Foundation (DEB 1655117). A. Tyre was partially supported by the University of Nebraska Agricultural Research Station.

Section II

Harvest Outcomes

Artist's Impression: Different approaches to managing a natural resource can result in very different tradeoffs for the managed population. Here, the abundance of the resource is the same for both scenarios, but there are drastically different user experiences as well as population characteristics.—Jennifer Clausen

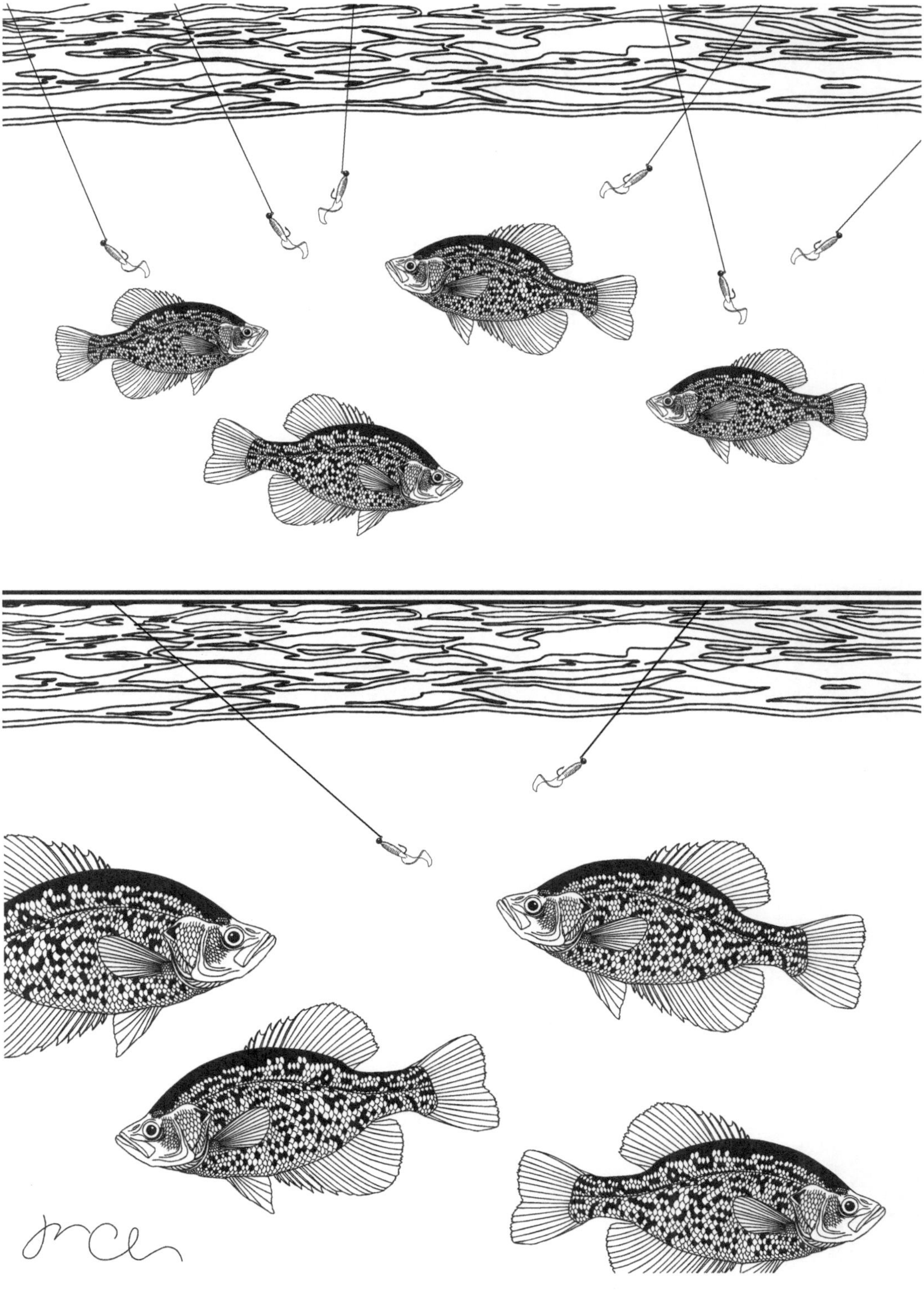

Section IIA

Evolutionary and Population Dynamics for Harvested Species

Fishers along the Columbia River in Oregon, United States of America use a seine net and spears to harvest salmon in c1906. Photo by Benjamin A. Gifford, from the collections of the Library of Congress (https://www.loc.gov/pictures/item/2001698216/).

12 How Regulations Can Affect the Evolutionary Impacts of Recreational Harvests on Fish and Mammals

Marco Festa-Bianchet and Robert Arlinghaus

CONTENTS

INTRODUCTION

Harvest of fish and wildlife by humans is nearly always selective, altering the age-, sex-, or phenotype-specific mortality schedule of exploited populations (Allendorf and Hard 2009). The frequency distribution of harvested individuals often differs from that of the total population in size, sex, age, health status, or various morphological characteristics (Kuparinen and Festa-Bianchet 2017). Selectivity arises from personal preference, gear, bait or weapon type, regulations, access to fishing or hunting sites, and age- and sex-specific differences in vulnerability. In addition, heavy harvests, even when entirely unselective, can create selection pressures favoring the evolution of fast life histories that bet on reproduction early in life to offset the increased mortality relative to natural mortality schedules (Jørgensen and Holt 2013, Dunlop et al. 2018).

This chapter will examine how regulations can affect the probability that recreational harvests of fish and mammals could lead to evolutionary changes in life history and other traits. Fish and mammals differ in many biological traits, such as mortality schedules, lifetime growth, and reproductive patterns. Trait differences could determine the effects of selective harvests in these taxa. Specifically, many fishes have indeterminate growth and their reproductive output can increase substantially with body size (Barneche et al. 2018). In female mammals, large body size increases reproductive success, but usually not more than a few percentage points (Gaillard et al. 2000a). In many hunted ungulates, litter size is fixed at one or varies at most from one to three (Gaillard et al. 2000b). Therefore, from a life-history perspective, intense and trait-selective harvests are more likely to affect reproduction and maturation schedules in fishes than in large mammals. In contrast, selective harvest in mammals is often directed at specific morphological features, such as horn size—a type of selective harvest not found in recreational fishing, where body size is often a key target (Beardmore et al. 2015). Hence, we expect that harvest-induced evolution will mostly affect morphological traits desired by hunters in mammals, whereas a range of traits may evolve in recreational fisheries (Matsumura et al. 2011, Arlinghaus et al. 2017b). In fishes, there is ample experimental and modeling evidence that body size can be affected by intense, persistent, and selective harvests (Conover and Munch 2002, Edeline et al. 2009, Ahti et al. 2020). In both mammals and fishes, selective harvesting can also drive behavioral changes (Arlinghaus et al. 2017b, Leclerc et al. 2017).

Heavy and trait-selective harvest will not necessarily cause evolution. For evolution to occur, the selected trait must be heritable. Harvest-induced selection must also be stronger than any opposing natural or sexual selective forces and persistent in space and time. Often, harvests act against a set of adaptive traits favored by natural selection, such as large size in fishes, which increases fecundity, lowers natural mortality, and may also be affected by sexual selection (Sbragaglia et al. 2019). Hence, natural selection will typically favor large body size, yet large fish are often under heavy harvest pressure. Natural selective forces persist in the environment of harvested populations (Festa-Bianchet and Mysterud 2018) and may create counter-selection gradients weakening harvest selection (Edeline et al. 2007).

Harvests affect the density, sex-age structure, and the ecology of harvested species, and it can be difficult to partition observed changes in phenotype or life history into their ecological, demographic, and evolutionary components (Olsen et al. 2004, Hendry et al. 2008). For example, any substantial selective harvest of ungulates based on horn or antler size, or of fish based on a minimum size, will lead to a demographic increase in the proportion of smaller individuals in older age classes, regardless of evolutionary change. A greater challenge in identifying evolutionary effects is that the reduced density of harvested population leads to life-history changes through greater per capita food availability. Those changes largely overlap with the evolutionary predictions derived from reduced adult life expectancy, including greater reproductive output, early reproduction at smaller size in fishes, earlier first reproduction, and greater maternal care in mammals (Festa-Bianchet 2003, Heino et al. 2015). Thus, observed shifts in maturation age might be caused by either or both genetic and plastic changes, which can be very difficult to tease apart in the absence of pedigrees or genetic data. For fisheries, though theory and artificial selection experiments have produced a coherent body of theory that predicts evolutionary adaptations in a range of life-history traits (Conover and Munch 2002, Heino et al. 2015, Uusi-Heikkila et al. 2015, Therkildsen et al. 2019), there is little empirical evidence of genetic change caused by fisheries-induced evolution in the wild, including recreational fishing (Hutchings and Kuparinen 2020, but see Bowles et al. 2020). Although theory is well developed (Matsumura et al. 2011, Ayllon et al. 2018), and selection caused by recreational fishing on a range of traits has been documented (Sutter et al. 2012, Hessenauer et al. 2015, Alós et al. 2016), there is very limited field evidence of fishery-induced evolution in recreationally exploited stocks (Bowles et al. 2020). That is possibly caused by data limitations, as clear evidence of evolution in the wild is generally difficult to document (Kruuk 2017, Hutchings and Kuparinen 2020).

WHAT ARE THE EVOLUTIONARY IMPACTS OF HARVEST?

Several reviews of the evolutionary impacts of harvests have been published (Heino et al. 2015, Kuparinen and Festa-Bianchet 2017). Here we will summarize the main results, as a background to how regulations may affect evolutionary changes. Most reported harvest-induced evolution concerns life history, behavior or morphological traits, but physiological traits may also be affected (Hessenauer et al. 2015, Hollins et al. 2018). Heavy harvest select for a "faster" life history because the much greater adult mortality may favor individuals who reproduce as early as possible, to ensure some reproductive output before death (Heino et al. 2015). Females are then expected to allocate a greater proportion of their available resources to reproduction (Arlinghaus et al. 2009), because the survival cost of reproduction will be low if, independently of reproductive effort, survival is already very low because of harvest. To date, evidence for harvest-induced changes in life history is mostly for heavily exploited fish in commercial, not recreational fisheries (Heino et al. 2015). These changes may slow recovery after fishing ceases, because life-history strategies that evolved under heavy harvest can be suboptimal when fishing stops and the population is only subjected to natural selection (Swain et al. 2007, Enberg et al. 2009, Uusi-Heikkila et al. 2015, Hutchings and Kuparinen 2020). Reversal of harvest-induced evolution is expected to be slow, because natural selection is generally weaker than the selective pressure imposed by heavy harvests (Conover et al. 2009).

A recent analysis of empirical evidence for commercially harvested marine fishes, however, suggests that many stocks recovered rapidly following a decline in harvest rates in situations where fishery-induced evolution was implied (Hutchings and Kuparinen 2020). Quick recovery suggests that either plasticity, not evolution, was the dominant mechanism of phenotypic change or that evolutionary recovery may be more rapid than predicted by models. One mechanism that may speed up recovery is sexual selection favoring crossing with individuals adapted to natural selection that may disperse from protected areas (Sbragaglia et al. 2019). For mammals, heavy harvests have been suggested to favor a faster life strategy in brown bears (*Ursus arctos*) (Zedrosser et al. 2011) and beavers (*Castor canadensis*) (Boyce 1981). Heavy harvests may also affect the timing of parturition in moose (*Alces alces*) (Kvalnes et al. 2016) and wild boar (*Sus scrofa*) (Gamelon et al. 2011). There have been, however, few attempts to test for possible evolutionary effects in other species, and the required data are available for very few systems.

Behavioral changes have been mostly investigated in fishes, where risk of angling appears to select for "shy" or less aggressive fish that are less likely to take a bait (Arlinghaus et al. 2017*b*). Other behavioral changes may include reduced exploration and activity (Alós et al. 2012) or late "wake-up" times in species with clearly developed chronotypes (Martorelli-Barcelo et al. 2018). Evolution of boldness or other personality types, however, is very much dependent on the angling gear and cannot be generalized across lure types and fisheries (Monk 2019). Other studies found no link between shyness and vulnerability to angling, for example in largemouth bass (*Micropterus salmoides*) (Keiling et al. 2020). In carp (*Cyprinus carpio*), boldness was a key trait under selection by angling (Klefoth et al. 2017). The possibility of fisheries-induced behavioral change favoring shy fish has important implications for population dynamics and community ecology, because shyness may be correlated with lower feeding rates and slower growth (Rhoades et al. 2019). Shy fish are at lower risk of taking baits, but also have lower feeding and growth rates. A population of shy fish is therefore expected to grow more slowly than one of bold fish, given the strong links between feeding rate and fertility. There is, however, little empirical evidence supporting this expectation (Arlinghaus et al. 2017b) and more research is needed at a longer temporal scale, to determine if fish may recover from prolonged starvation with compensatory growth at a later date (Ali et al. 2003). This is another important difference between fish and mammals, as mammals face a risk of dying from starvation, not just a slower growth rate, if they substantially reduce food intake. It has also been suggested that timing of migration in salmonids may be affected by fishing pressure, as early running fish face greater fishing pressure (Tillotson and Quinn 2017, Morita 2019). Among mammals, ungulates that avoid roads and open habitats during the hunting season have higher survival (Ciuti et al. 2012). So far, however, these behavioral changes have not been conclusively tied to evolutionary change, and may partly be due to plasticity or learning. Similarly, some studies that revealed that fish in exploited populations were harder to catch could not unambiguously differentiate between harvest-induced genetic changes to vulnerability to fishing and alteration of hook-avoidance behavior through plasticity and learning (Philipp et al. 2015). Strong evidence of evolutionary change in behavior is provided by experimental studies that selected for fast or slow vulnerability to angling over four generations of largemouth bass (Sutter et al. 2012). Whether and to what degree shyness or boldness may evolve depends not only on the behavioral selectivity of angling, but also on the correlation between behavioral and life-history traits and size selectivity in the fishery (Andersen et al. 2018). Although recreational fisheries are often expected to select for shy fishes, particularly when artificial lures are used, selection by angling can also operate on traits such as stress responsiveness (Koeck et al. 2019), sociability (Louison et al. 2019), exploration (Monk 2019), or habitat choice (Monk and Arlinghaus 2018).

Phenotypic evolutionary changes that have been investigated include mostly maturation timing, growth rates, and body size in fishes (Heino et al. 2015) and weapon size in male ungulates subjected to intense trophy hunting. Conclusive evidence for hunting-induced evolution in the wild exists for one pedigreed population of bighorn sheep (*Ovis canadensis*) (Pigeon et al. 2016). Harvest-induced evolutionary changes also appear likely from several broad-scale investigations of hunting records of wild sheep in areas without a quota system (Hengeveld and Festa-Bianchet 2011, Douhard et al. 2016). Reductions in size and prevalence of tusks in elephants (*Elephas maximus* and *Loxodonta africana*) are consistent with an evolutionary response to human harvests (Kurt et al. 1995, Chiyo et al. 2015). Despite a long tradition of trophy hunting and sometimes intense harvests, however, there is no clear evidence of evolutionary changes in antler size for any species of wild deer (Festa-Bianchet and Mysterud 2018). Investigations in some antelope species have been inconclusive (Crosmary et al. 2013).

WHAT ARE THE ECOLOGICAL DRIVERS OF HARVEST-INDUCED EVOLUTION?

Evolutionary change should favor phenotypes that are most adapted to the selective pressures in their environment, regardless of whether those pressures are natural or artificial. Given the differences in the relationship between body size and reproductive success between fish and mammals, however, it is not surprising that intense harvests can drive faster changes in female reproductive strategy in fish than in mammals. Many fishes have indeterminate growth: they keep increasing in skeletal size through most or all of their lifetime, and their fecundity rises exponentially with mass. In contrast, most mammals have determinate growth: their bones stop growing at, or soon after, reproductive maturity, and litter size has limited variability. In both taxa, larger females generally enjoy higher reproductive success than smaller females, but the size–reproduction relationship is substantially steeper in fishes, where reproductive output often increases exponentially with body size (Barneche et al. 2018). Interindividual and especially interpopulation variability in size at maturity and growth rate are thus much greater in fish than in mammals. For example, in landlocked salmon (*Salmo salar*), mass at maturity for females varies 400-fold and fecundity 150-fold (Hutchings et al. 2019). This large variation has a strong plastic component that may reduce the expected evolutionary changes from selective pressures acting on life-history traits (Eikeset et al. 2013). Nevertheless, for large- and medium-sized mammals, variation in both mass at maturity and yearly fecundity is at most fourfold, reducing the scope for phenotypic responses to harvest relative to fish. A fish that is naturally selected to accumulate somatic resources and not reproduce until age 5 years could instead be selected to direct those resources to reproduction at age 3 years and at less than half the mass by intense fishing mortality. A female mammal naturally selected to not reproduce until age 3 years is unlikely to be able to reproduce at age 1 year even if adult mortality is increased by hunting. Evolution will change phenotypes only where there is scope for change and natural selection favors it, and for life-history traits that scope appears much greater in many fishes than in most mammals or birds, but no comparative studies across taxa exist.

The importance of the artificially selected trait on survival or reproduction will also affect the potential for harvest-induced selection to lead to evolution. Evolution of smaller horns is unlikely even under intense and persistent trophy harvest if horn size is not a major determinant of male reproductive success, for example in chamois (*Rupicapra rupicapra*) or mountain goats (*Oreamnos americanus*; Festa-Bianchet 2017). In fishes, a typical expectation is that both body size and growth rate are selected against through positive size-selective fishing (Conover and Munch 2002). However, decades of research including theoretical models and phenotypic time series have cast doubt on this apparently clear-cut prediction (Enberg et al. 2012). Fish growth rate varies with both energy intake, which is affected by feeding behavior, and allocation to energetically expensive tissues or processes like gonads, soma, brain, or maintenance. Depending on the intensity of selection, natural selection, and the scope for evolution, growth rate can increase or decrease in response to fishing. Perhaps the most consistent response to fishing is changes in maturation timing, which may reduce post maturation growth, but even this response has not been consistently found in laboratory studies of experimental evolution (Uusi-Heikkila et al. 2015). Therefore, many details of selectivity, intensity of selection, and the co-variance of traits determine the ultimate direction of evolution in response to harvesting, and no general predictions are possible without considering the details of each species and harvest regulations (Heino et al. 2015).

THE MANAGEMENT DRIVERS OF HARVEST-INDUCED EVOLUTION AND HOW REGULATIONS CAN LOWER THE RISK OF HARVEST-INDUCED EVOLUTION

Two major drivers of harvest-induced evolutionary change can be modified by regulations: harvest rate and selectivity. As many have pointed out (Eikeset et al. 2013, Heino et al. 2015, Kuparinen and Festa-Bianchet 2017), most issues of harvest-induced evolution can be solved by simply lowering the rate of exploitation. Regulations directly and only addressing size selectivity are necessary and sufficient only in rare cases to complement management tailored at controlling overall mortality.

The main driver of life-history evolution in fishes is elevated harvest rate. Although quotas, or other means to reduce the intensity of harvest such as protected areas, may reduce the evolutionary effects of harvesting, it is unclear whether regulations that specifically address the risk of life-history evolution are necessary. Reductions in harvest rate dictated by a need to maintain ecological sustainability, such as to maintain yield within sustainable limits, may suffice to address harvest-induced evolution (Hutchings and Kuparinen 2020). Usually, harvest rates that maximize ecologically sustainable fishing rates are also those that curtail the most important evolutionary changes (Eikeset et al. 2013). Some level of adaptive change to harvesting is likely inevitable in commercial fisheries (Hutchings 2009, Zimmermann and Jørgensen 2017) and possibly in any fishery (Matsumura et al. 2011). It has also been argued that some regulations could direct harvest-induced changes in directions favoring humans. For example, harvest slot- or maximum-size limits in recreational fisheries may increase, rather than decrease, growth rates and yield (Matsumura et al. 2011). The judgment of fisheries-induced evolution is thus socially constructed and not necessarily negative.

Heavy fishing pressure can select for several types of behavioral changes in fish. When fishing is by passive means (bait, baited traps, gillnets), it can select against active, explorative phenotypes and generally favor low movement rates (Biro and Post 2008, Andersen et al. 2018, Koeck et al. 2019). Those behavioral changes can have important consequences for population growth rate and community ecology because fish with lower appetite may grow more slowly, consume less prey, and have a lower reproductive rate. Such evolution may also lead to hyperdepleted catch rates, where catch rates drop faster than underlying abundance (Alós et al. 2019). Such changes would not only leave behind unhappy anglers, but also constitute a challenge for fishery-dependent stock assessments. It is unclear how regulations may address this selection other than by reducing fishing mortality, rotating gear types and the areas and timing of fishing (Thorbjørnsen et al. 2018), and perhaps stocking with specific phenotypes (Hessenauer et al. 2015). Any form of spatial closure, however, may introduce more selection pressures on spatial behavioral types (Villegas-Rios et al. 2017, Alós et al. 2019). Stocking specific behavioral phenotypes can produce other undesirable genetic changes through hybridization (Le Luyer et al. 2018).

A new area of research in fishing-induced selection is on physiological traits (Hollins et al. 2018). Physiology is the ultimate mechanistic underpinning of behavioral and life-history traits. Behaviors such as aggression and aerobic scope can be linked, so that selection on behavioral traits may affect metabolism and thermal tolerance (Redpath et al. 2010, Hessenauer et al. 2015, Duncan et al. 2019) possibly making fish less able to deal with a warming ocean (Duncan et al. 2019). Currently, there are no studies on how fishing regulations may affect this selection.

Harvest-induced evolution of phenotypic changes in mammals has mostly been studied in the case of trophy hunting (Festa-Bianchet and Mysterud 2018). These changes are induced by very intense selective harvests, and can be avoided or reduced by regulations that decrease either harvest intensity or selectivity. A recent comparison of long-term changes in horn size of shot bighorn sheep rams found strong evidence of phenotypic declines, after adjusting for age and environmental conditions, in the Canadian province of Alberta, but not in most U.S. states (Lasharr et al. 2019). Regulations in Alberta are more liberal than in any American jurisdiction, which typically involve quotas. To weaken the evolutionary impacts of selective harvests, regulations should encourage the harvest of older individuals, so that those with the largest horns can reproduce before they are shot. To prevent harvest of large-horned males before they can breed, regulations must consider the mating system and age-specific horn-growth patterns of each species. For example, the horns of mountain sheep (*Ovis*) grow substantially over the first four to six years of life, before approaching an asymptote, yet large horn size has little effect on male siring success before about seven years of age (Martin et al. 2016). Under the definition of *legal* ram used in most of Alberta (Fig. 12.1), males with rapidly growing horns can be shot at four years of age (Festa-Bianchet et al. 2014). That regulation sets up a very strong selective pressure because large horns

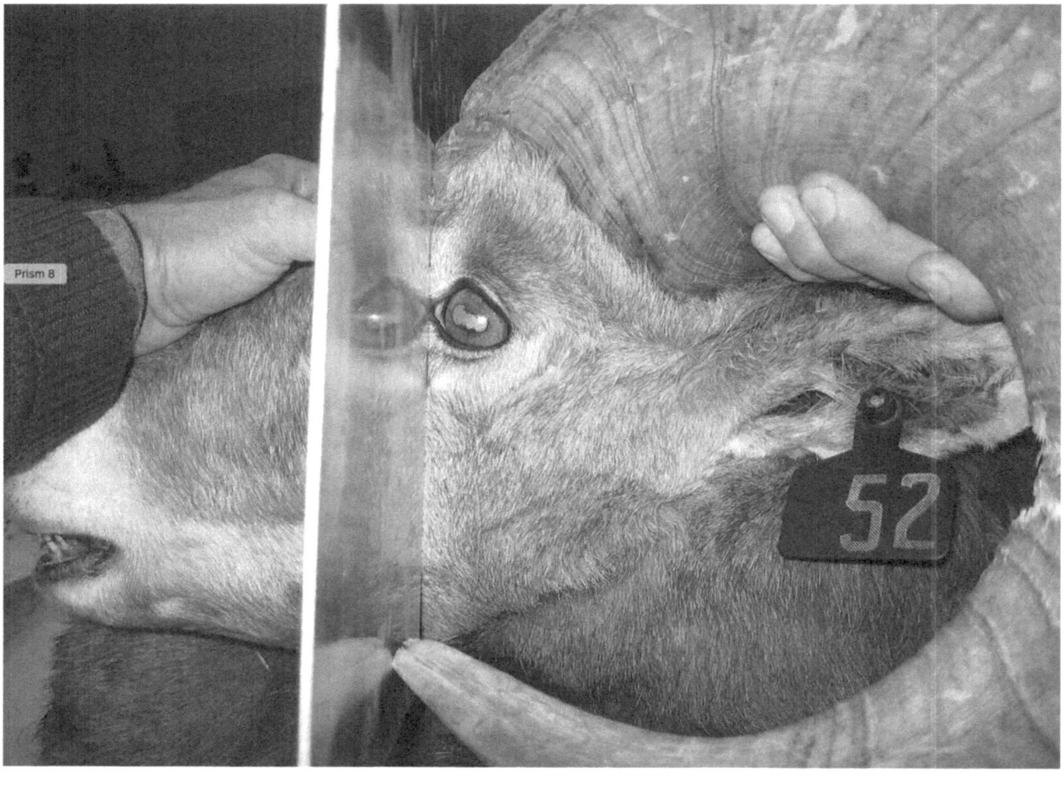

FIGURE 12.1 An exceptionally developed four-year-old bighorn sheep ram shot in Alberta, Canada. Its horns meet the Alberta definition of "legal" ram, or 4/5 of a curl. If a line drawn from the base of the horn to the front of the eye intersects the tip of the horn, the ram can be shot by a person with a *trophy sheep* license. An unlimited number of licenses are available to provincial residents. This ram's horns grew much more rapidly than average, and he was likely shot before it reproduced. Large horns have little effect on the siring success of rams who do not survive to rut as seven-year-olds. Photo: Alberta Fish & Wildlife, used with permission.

are associated with a negative fitness effect (getting shot) at an earlier age than a positive effect (high rank, access to estrous females). For species such as mountain goats, however, asymptotic horn length is reached at about four years (Festa-Bianchet and Côté 2008) and horn size is weakly related to siring success (Mainguy et al. 2009), suggesting that selective hunting of males with large horns would have a weaker evolutionary effect than in mountain sheep (Festa-Bianchet 2017).

A mix of penalties and incentives has been used in various contexts to direct harvest to different types of animals. In North America, various species of deer are often harvested under regulations that specify a minimum number of tines (see Morina et al. 2021 [Chapter 19]). Regulations can have a strong selective effect and change male age structure (Wallingford et al. 2017), but there is no empirical evidence that minimum-tine regulations lead to evolutionary changes in antler size or shape (Festa-Bianchet and Mysterud 2018). In Europe, harvest plans for ungulates often have specific quotas for several sex-age classes, or for males with different horn or antler characteristics (Büntgen et al. 2018). For example, regulations and strong penalties can substantially decrease the harvest of lactating female chamois (Rughetti and Festa-Bianchet 2014), sometimes with perverse effects as the harvest is directed to the most productive age class, young adult females that have not yet reproduced (Rughetti et al. 2017). In British Columbia, Canada, outfitter quotas for Stone sheep (*O. dalli stonei*) are partly affected by the age structure of previous harvests. Outfitters whose clients take mostly young rams may see their quota reduced, and those who harvest older rams may see an increase in allocation (Bill Jex, personal communication). An alternative is a temporal rotation of protected areas to weaken selective pressure and ideally provide a source of unselected immigrants. These rotations are commonly used in Europe, but are rare in North America, and nothing is known about their evolutionary impacts.

At the landscape level, large protected areas may be an effective source of unselected immigrants if those immigrants are not shot during the hunting season. For example, in Alberta there is a spike in the harvest of bighorn rams in the last few days of October, as rams leave National Parks looking for breeding opportunities (Poisson et al. 2020). In fish with long-distance dispersal, marine protected areas may provide a source of genes to replenish heavily harvested areas outside protected areas, although protected areas might also select on spatial behavioral types by favoring fish that do not leave protected areas (Alós et al. 2019). Thus, marine protected areas are only a long-term solution when coupled with reasonable quotas that lower overall fishing mortality and protect much of the spatial behavioral variability within a species (Thorbjørnsen et al. 2018). In terrestrial mammals, females tend to be philopatric, so protected areas are sometimes viewed as a potential source of trophy males. That strategy may depauperate, both demographically and genetically, populations that are supposedly protected (Poisson et al. 2020).

In recreational fisheries, some modeling research has examined the potential for simple size-based regulations to reduce selective pressures induced by angling. Those models suggest that minimum-length limits are the most evolutionary damaging regulation, as they select for reduced post-maturation growth rate (Matsumura et al. 2011). In contrast, harvest slot- and maximum-size limits could reverse that selective pressure and favor rapid growth rate. But no regulation, other than total catch and release with limited hooking mortality rate, can avoid selection altogether. Related work in other model contexts reported similar findings (Zimmermann and Jørgensen 2017, Andersen et al. 2018, Ayllon et al. 2018, Wszola and Fontaine 2021 [Chapter 13]). Therefore, it is doubtful whether size limits alone can avoid fisheries-induced evolution altogether if fishing pressure is high. Any regulation must account for catch-and-release mortality (Johnston et al. 2015b). Also, harvest slot sizes must be narrow enough to allow enough fish to grow past the upper legal limit, otherwise they would effectively be the same as minimum-size limits (Ahrens et al. 2020).

Harvest slot size regulations for mammals are uncommon and mostly rely on antler or horn characteristics. Slot sizes are part of the hunting tradition in parts of Europe, where licenses are for a specific age class that is determined by its horn or antler size (Rughetti and Festa-Bianchet 2010, Büntgen et al. 2018). These regulations usually require the hunter to take some training courses and pass a test—a common practice in Europe that is rare in North America. There are a few recent uses of slot sizes in regulations in North America (see also Morina et al. 2021 [Chapter 19]). In a few areas of British Columbia, male moose can be shot only if they have very small or very large antlers. In some counties in Texas, white-tailed deer can be shot only if they have at least one unbranched antler (referred to by hunters as a *spike*) or a minimum of 33-cm inside spread.

One goal of slot sizes is to reduce the potential evolutionary pressure on large individuals from size-selective harvesting. Those large individuals may also have different behavioral types compared to smaller individuals. Clearly, the effectiveness of these regulations will depend on the proportion of individuals that grow past the maximum legal size in fishes or that survive past the horn or antler size where they are vulnerable to harvest in mammals. We know of no research that has examined the consequences of slot sizes on the potential of harvest-induced evolution in mammals. Slot sizes have been the subject of substantial research in fisheries (Matsumura et al. 2011, Zimmermann and Jørgensen 2017, Ayllon et al. 2018). Slot sizes could in theory reduce the evolutionary pressure on large individuals while maintaining some large and mature individuals in harvested populations (Mysterud and Bischof 2010). The empirical consequences of different slot size regulations are ripe for investigation in wildlife management.

WHY HORN SIZE IS SHRINKING IN BIGHORN RAMS IN ALBERTA AND NOT IN IBEX MALES IN CANTON GRISONS

Two mountain ungulates show how regulations can stimulate or prevent harvest-induced evolutionary changes in horn size (Table 12.1). Horn size is shrinking in bighorn sheep in Alberta, Canada (Festa-Bianchet et al. 2014, Pigeon et al. 2016, Lasharr et al. 2019) but not in ibex (*Capra ibex*) in Canton Grisons, Switzerland (Büntgen et al. 2018). The ibex case has been used to

TABLE 12.1

Hunting regulations for bighorn sheep rams in Alberta, Canada, and Alpine ibex males in Canton Grisons, Switzerland[1]

Regulation	Bighorn Sheep, Alberta	Alpine Ibex, Grisons
Hunting season length	55–65 days	20 days between Oct 5 and Nov 5
Quotas	None for provincial residents, harvest limited by availability of "legal" rams. About 80 permits for nonresidents	Yes, based on annual sex-age classified counts
Other limitations	Hunters cannot buy a license the year after harvesting a ram. Nonresident hunters must hire a guide, have shorter hunting season	Hunters from Grisons can apply once every ten years, must take a female before killing a male
Based on horn size?	Yes, any ram with horns describing a minimum of 4/5 of a curl can be shot.	Permits specify age class, identified by horn length and growth annuli
Age-specific harvest?	No, rams with rapidly growing horns are shot at a younger age	Yes, quotas for five age classes. Hunter's age determines age of ibex that can be shot
Area-specific licenses	No, license is valid anywhere in the province with a trophy sheep hunting season	Yes, license is only valid for one of eight identified ibex colonies

Note

1 Based mostly on Festa-Bianchet et al. 2014 and Büntgen et al. 2018.

argue that selective hunting has limited evolutionary consequences (Reznick et al. 2019), yet ibex hunting regulations in Grisons specifically seek to reduce selective pressures on males with rapidly growing horns, and largely appear to be successful (Büntgen et al. 2018). In both ibex and bighorn sheep, dominant males serially defend a single estrous female and subordinates attempt alternative mating tactics, which can at times be successful (Hogg and Forbes 1997, Willisch et al. 2012). The species differ in male survival patterns. Annual survival of male ibex aged two to six years is over 98% (Toïgo et al. 2007), whereas for bighorn rams of the same age it is 84%–92% (Loison et al. 1999). Hunting regulations in Alberta combine an absence of quotas with a morphology-based definition of "legal" male and a two-month hunting season. Those regulations strongly select against rams with rapidly growing horns, which become legal for harvest at four to six years of age, before large horns have a positive effect on their siring success (Martin et al. 2016). In contrast, male ibex hunting includes strict quotas with allocation of permits to five age classes based on horn size. Hunters must take an identification course and can only hunt for 20 days. They must first harvest a female before being allowed to harvest a male (Büntgen et al. 2018).

A WORD ABOUT SOCIAL ISSUES

Consistently, opinion surveys show that "hunting" is viewed more favorably than "trophy hunting" and biologists debate the role and ethics of trophy hunting in the conservation of biodiversity (Batavia et al. 2018, Dickman et al. 2019). Public acceptance of harvest hinges on sustainability. The concept of evolutionary sustainability is relatively new, compared to ecological sustainability. It is not well established and is considered unnecessary by some in the wildlife establishment (Boyce et al. 2019). Claims in popular media that any kind of selective harvest leads to undesirable evolutionary consequences contribute to the negative perception of recreational harvests. Emphasis on harvesting large trophies was initially promoted to reduce harvests (Monteith et al. 2013), but it can turn into a marketing ploy, leading to semi-captive breeding of *varieties* or simply of large trophies destined for slaughter (Darimont and Child 2014). We suggest that the commercialization of trophy hunting and a strong emphasis on trophy size as a measure of success (Knox 2011) worsen the social perception of recreational hunting. Given that recreational hunting can be a strong component of biodiversity conservation programs (Di Minin et al. 2016, Naidoo et al. 2016), a negative perception is a problem (Hampton and Teh-White 2019). Negative public perception is less of an issue in fisheries because many trophy fisheries are now based on a catch-and-release ethic. Catch and release can often be conducted with limited impact on the individual, reducing or eliminating any selective harvest. In some countries, however, voluntary catch and release, particularly the specific targeting of large fish, is strongly morally questioned as "*playing with food for no good reason*" (Arlinghaus 2007). Germany is a prime case where it is considered morally superior to harvest fish, also large fishes, than to catch and release them. Under such situations, a policy of aggressive harvesting would reinforce fisheries-induced evolution.

CONCLUSION

Harvest-induced evolution is likely to occur when recreational harvests are very high, selective, or both. Yet despite well-established theory and increasing experimental evidence, there is limited evidence for this process in the wild. The limited empirical evidence may be due to the slowness of the evolutionary process in some systems, regulations that limit its effectiveness, or the inability of available genetic tools to document ongoing evolution. In addition, these effects have been properly investigated in only a few systems. Although harvest-induced evolution, especially in fishes, can decrease population growth, other conservation issues may be more important, such as habitat deterioration and fragmentation, climate change, introduced species, and exotic diseases. In fish, stocking has likely had a stronger genetic effect than fisheries-induced evolution over time. That said, fish and wildlife managers should strive to curtail and prevent harvest-induced evolution as much as possible, for biological, economic, and ethical reasons. Our top recommendation is to reduce harvest mortality to a degree that is both ecologically and evolutionarily sustainable. For mammals, aiming the harvest to individuals that have already reproduced would be a good start. That aim requires knowledge of age-specific male reproductive success, which is not available for most species. In recreational fishing, minimum-length limits, particularly when set very high, are the most damaging tools from an evolutionary perspective, whereas slots-length limits may be the most efficient strategy (for a case pro harvest slots, Matsumura et al. 2011; for a case pro protected slots, Wszola and Fontaine 2021 [Chapter 13]). Currently, empirical research is insufficient to provide generalizable insights from one species to another. Consequently, most solutions will be dictated by the life history and growth traits of each harvested species and the specific gear used in different systems. Long-term records kept by fish and wildlife agencies, combined with temporal or spatial changes in regulations, provide fertile grounds for future research, especially in combination with experimental changes in regulations driven by specific research objectives. Multi-species or multi-jurisdiction comparisons, taking advantage of different and long-established management schemes, are another approach that is likely to provide insights useful both for our understanding of the role of evolution in harvested species and for improved conservation of those species.

ACKNOWLEDGMENTS

The authors are grateful to L. Powell and K. Pope for the invitation to write this chapter and for their editorial help. The authors thank C. Hasler and J. Heffelfinger for their review comments. M. Festa-Bianchet's long-term research on large herbivores is supported by the Natural Sciences and Engineering Research Council of Canada and the Alberta Conservation Association. R. Arlinghaus acknowledges funding through the European Union (European Maritime and Fisheries Fund) and the State of Mecklenburg-Vorpommern (Germany) (Grant MV-I.18-LM-004, B 730117000069).

13 Harvest Regulations in Evolving Fisheries

Lyndsie S. Wszola and Joseph J. Fontaine

CONTENTS

INTRODUCTION

Harvest regulations in freshwater recreational fisheries address multiple ecological and social objectives. For example, regulations may seek to increase or maintain population abundance (Hansen et al. 2015b), alter age composition, improve yields (Minnesota Department of Natural Resources 2018), and facilitate harvest of fish in a preferred size group (Oele et al. 2016). Managers and policymakers depend on scientific insights into the relationships between fish, people, and environment to craft effective regulations. Though processes such as fish growth and survival may be modeled with the understanding that they vary in relation to environmental variables like food availability, water clarity, or temperature, the models built to inform harvest regulations generally assume that such relationships will be constant over time. A growing body of evidence demonstrates that fisheries harvest may change the relationships between fish and their environments, especially for large fish, by increasing a fish population's mortality rate even when anglers are not intentionally harvesting the largest fish in a population.

To understand why harvest, especially of large fish, may lead to changes in how fish populations grow and interact with their environments, we must briefly explore some core concepts from life-history theory. A life history is the sequence of events an organism does or experiences in their life and the times at which they do them (Stearns 1989, Charnov 1993). For example, a sockeye salmon's (*Oncorhynchus nerka*) life history may be defined by the ages and sizes at which it migrates to the ocean, returns to its natal stream or lake, and reproduces (Burgner 1991). The ages and sizes at which the salmon undergoes each of these changes, and how well suited the timing of these events is to its environment, will influence its fecundity, the number of gametes it produces, and ultimately its fitness, the number of offspring that inherit its genes. We introduce the concept of life histories here because they are both essential to shaping the nature and timing of interactions between fish and their environments and because they are under intense evolutionary pressure. The core principle behind life-history evolution is that the prevalence of a life-history strategy is correlated with the fitness of individuals enacting that strategy (Charnov 1993). Organisms must make tradeoffs between allocating energy to maintenance, growth, and reproduction throughout their lives, presenting fish with a life-history conundrum. For example, the larger a female fish grows, the more eggs it can produce and the greater its fitness will be (Barneche et al. 2018). However, if the female fish lives in an environment where it has a high probability of dying before reproducing, it is less likely to benefit from growing a large body that produces lots of eggs (Thorson et al. 2017). It may therefore be more advantageous for a fish living in a high-risk environment to mature small and reproduce early, even if it means producing fewer eggs over the course of its lifetime, than to wait and reproduce at a larger size it may never achieve. Of course, this is not a decision individual fish consciously make, but a product of genes and environments we may observe and interpret as a life history. This central question of how big individual fish grow and when they mature motivates the discussion around the evolutionary effects of fishing. When anglers artificially increase mortality by harvesting fish, they may be creating a high-risk environment where it is adaptive to reproduce early and small, rather than late and large.

Though modern fisheries are largely sustainable and benefit from a long tradition of scientific management, it is increasingly apparent that addressing widespread life-history shifts toward maturation at smaller sizes and younger ages among harvested fish populations is necessary to help maintain fisheries sustainability (Law 2000, De Roos et al. 2006, Swain et al. 2007, Allendorf et al. 2008, Heino and Dieckmann 2008, Allendorf and Hard 2009, Pukk et al. 2013, Heino et al. 2015, Kuparinen et al. 2016). Fishing-induced reductions in size and age at maturity have been documented in diverse fish species (Kuparinen and Festa-Bianchet 2016, Kindsvater et al. 2017, Dunlop et al. 2018). In addition to imposing selection on size, fishing may also select for otherwise maladaptive behaviors, such as reduced nest guarding or increased timidity, that further impact individual fitness (Uusi-Heikkilä et al. 2008, Arlinghaus et al. 2017b, Festa-Blanchet and Arlinghaus 2021 [Chapter 12]) and population resilience to environmental change.

The individual fitness and fishery sustainability consequences of changing fish life histories may translate to altered community dynamics. Smaller fish face higher risk of predation, further reducing survival and lifetime reproduction, and changes to reproductive and migratory behavior may destabilize foodweb dynamics (Nilsson and Bronmark 2000, Hutchings 2005, Urban 2007, Barneche et al. 2018). Fishing-induced changes in life-history expression coupled with changes in abundance can cause trophic cascades by reducing the number and gape size of large predatory fish (Andersen and Pedersen 2010, Palkovacs 2011, Palkovacs et al. 2011, DeLong and Luhring 2018). Ecological and evolutionary changes in fish body size may be maintained by newly density-dependent ecological interactions and predator–prey role reversal (Fauchald 2010, Svedäng and Hornborg 2014, Eikeset et al. 2016). For example, fishing-induced changes in the abundance and body size distribution of northern pike (*Esox lucius*) led to a reduction in North Sea stickleback (*Gasterosteus aculeatus*) mortality, increasing the size of the stickleback population and subsequently increasing mortality of pike eggs with corresponding negative effects on pike recruitment (Nilsson et al. 2019). Thus, fishing-induced evolution has the potential to affect the fitness of individual fish with unexpected consequences for the population and community, as well as the food, economic, and recreational values that fisheries provide humans (Blanchard et al. 2009, Eikeset et al. 2013, Zimmerman 2015, Lindmark et al. 2019).

Efforts to manage harvest-induced evolution have shown that harvest-induced life-history changes may be extremely difficult to reverse, especially for large, long-lived, late-maturing species (Walsh et al. 2006, Kuparinen and Merilä 2007, Árnason et al. 2009, Enberg et al. 2009, Hutchings 2009, Kuparinen and Hutchings 2012, 2017, Dunlop et al. 2015). However, there are also an increasing number of successes (Conover et al. 2009). For example, yellow perch (*Perca flavescens*) in the Laurentian Great Lakes exhibited significant evolutionary recovery of size and size at age following fishing moratoria, likely due in part to perch expressing relatively short generation times that enable them to evolve quickly (Feiner et al. 2015).

Fisheries managers, especially those responsible for freshwater recreational fisheries, face several challenges when seeking to make evidence-based management decisions that account for potential evolutionary impacts of harvest. Fish stock responses to fishing are defined by eco-evolutionary feedbacks between angler behavior and fish genetic, population, and community dynamics, so eco-evolutionary models must also account for social variation in harvest patterns if they to be applicable to conservation challenges (Johnston et al. 2012). Many models of fishing-induced evolution are consequently system specific; thus, the potential for fishing-induced evolution in freshwater systems has been historically understudied. Though extensive work has occurred in the marine policy realm concerning fishing-induced evolution, freshwater systems face different challenges and opportunities than do marine systems. Managers of marine systems must design policies and regulations for systems defined by commercial fishers who harvest large quantities of fish with commercial gear. Managers of freshwater fisheries, in contrast, largely craft regulations for recreational anglers who make harvest decisions for individual fish. It is therefore largely unclear to what extent fishing-induced evolution affects freshwater fisheries and what management actions would be appropriate to mitigate any deleterious life-history evolution. Herein, we have built a simple and flexible eco-evolutionary fisheries harvest model to explore how evolution affects population responses to different regulations in a simulated freshwater recreational fishery. We assess the extent to which evolution may be expected to influence the results of management actions by modeling the response of a simulated fish population to different harvest regulations and levels of exploitation with and without a genetic contribution to maximum size, growth coefficient, and age and size at maturity. Our objectives were to model the eco-evolutionary effects of regulations and fishing mortality on population growth, age and size at maturity, and yields.

METHODS

Model Overview

We built an individual-based eco-evolutionary model that simulated the response of a fish population to varying regulations and levels of exploitation under the assumption that asymptotic length, growth coefficient, and age and size at reproduction were either not heritable or partially heritable. The simulated fish population was based on walleye (*Sander vitreus*), a socially and ecologically important freshwater species across much of northern North America. Walleye populations have extremely diverse life histories across their range and have thus formed the basis of numerous basic and applied studies of fish life history (Charnov 1993, Lester et al. 2014, Bowles et al. 2020). The simulated fish population was based on a population of walleye with a spring spawning season and summer to winter fishing season encompassing both open water and ice fishing seasons.

Fisheries managers often seek to increase the frequency of large reproductive females in walleye populations to make populations more self-sustaining and allow anglers to catch more large fish (Minnesota Department of Natural Resources 2018). This objective is expressed through many different regulations, including *minimum length limits*, under which anglers may harvest only fish above a minimum size threshold; *harvest slot limits,* under which anglers may harvest only fish above a minimum size threshold and below a maximum size; and *protected slot limits*, under which anglers may harvest only fish below a lower size threshold or above an upper size threshold (Fig. 13.1; Gangl 2021 [Chapter 24]). We investigated the eco-evolutionary dynamics of fish harvested according to all three regulation types. We first simulated the effects of a walleye minimum size limit, 381 mm (Minnesota's regulation from Lake Superior and tributaries; Minnesota Department of Natural Resources 2020) under high (60% of vulnerable fish caught and all legal caught fish harvested) and low (10% of vulnerable fish caught and all legal caught fish harvested) catch conditions. We also modeled the effects of a walleye harvest slot limit from Wisconsin: walleye greater than 381 mm and less than 508 mm may be harvested (Wisconsin Department of Natural Resources 2020). The harvest slot limit overlapped the average starting

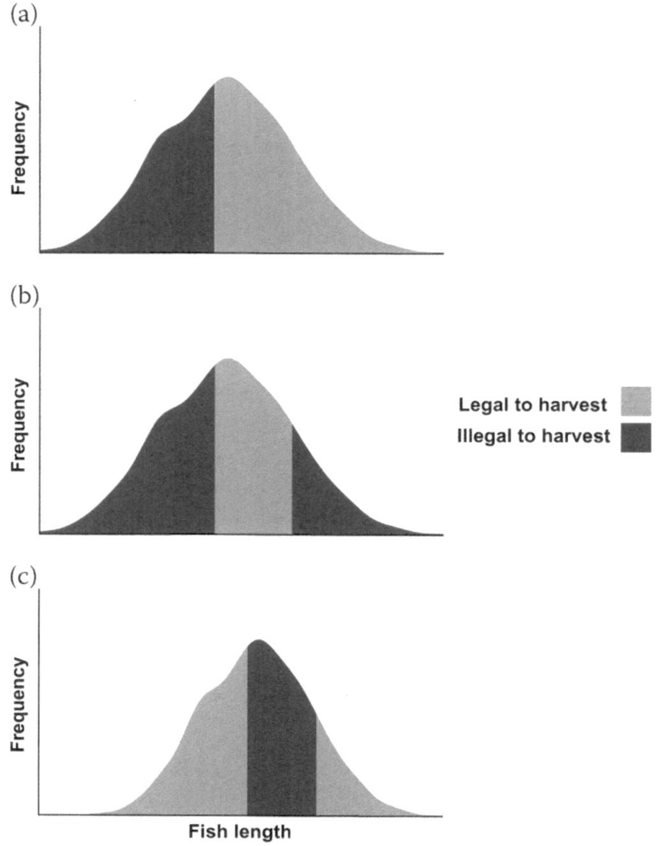

FIGURE 13.1 Minimum length limit harvest regulations (a) allow anglers to harvest only fish greater than or equal to a specified minimum length. Harvest slot regulations (b) stipulate that anglers may harvest fish larger than a minimum length or smaller than a maximum length. Protected slot regulations, the inverse of harvest slot regulations, permit anglers to harvest fish smaller than a lower length threshold or larger than a higher length threshold.

length at maturity in the simulated populations, so we finally modeled an experimental protected slot limit that was the inverse of the harvest slot limit (walleye greater than 381 mm and less than 508 mm must be released). We ran three iterations of each harvest regime for 100 time steps each under the assumption that asymptotic length (the fish's theoretical maximum length), and thus and age and length at maturity were either not heritable ($h^2 = 0$) or partially heritable ($h^2 = 0.3$; Bowles et al. 2020).

Individual fish in the model grew according to a logistic individual-growth function ((13.1)), the starting asymptotic length parameter value of which was drawn from a normal distribution based on an average northern Wisconsin growth trajectory (Beverton and Holt 1957, Tsehaye et al. 2016). In the growth function, fish length at a given age (L_t) was defined by the fish's asymptotic length (L_∞; starting mean = 700, SD = 100) multiplied by the quantity one minus e (Euler's number; about 2.72 and the base of the natural logarithm) raised to the product of the negative growth coefficient (k; starting mean = 0.13) multiplied by the difference between the fish's age (t) and the theoretical age at which the fish's size would equal 0 (t_0; –1.23):

$$L_t = L_\infty * (1 - e^{-k(t-t_0)}) \tag{13.1}$$

Model Functions

Population Initiation

Each harvest simulation was initiated by creating a starting cohort of 1000 individuals per year for 12 years and projecting the surviving number of individuals per age group as a function of the specified natural mortality rate (0.3) for a period of 12 years. All fish in the model were assumed to be female for computational efficiency. Each fish in the starting population was assigned an asymptotic length drawn from a starting distribution with mean equal to 700 mm and standard deviation equal to 100 mm. The model assumed that an increase or decrease in asymptotic size evoked tradeoffs in other aspects of the fish's life history because fish must allocate surplus energy between growth and reproduction. Specifically, as asymptotic length increased, the *k* growth coefficient was assumed to decrease via a function of e raised to the log of asymptotic size multiplied by a constant that we calculated using the starting asymptotic length (700) and *k* (0.13) parameters ((13.2); Thorson et al. 2017):

$$k = e^{-0.31 \ln(L_\infty)} \tag{13.2}$$

Fish were therefore initiated into the model with an asymptotic length phenotype that shaped their life history on an axis from large and slow growing to small and faster growing. It is important to note that the exact value of the relationships between parameters was chosen to demonstrate the concept of life-history evolution and not to predict growth in any one walleye population, though they are within the bounds of observed values for walleye (Bozek et al. 2011).

Fish were then advanced to the correct length for their age according to the individual growth function ((13.1)) using their individual asymptotic length and *k* parameters with $t_0(-1.23)$ from the Wisconsin walleye growth model. We assumed that all fish matured at 60% of their asymptotic length, a value within reasonable bounds for walleye, and calculated age and length at reproduction for each fish accordingly (Bozek et al. 2011). After initiation, each model run followed an annual cycle with six steps, repeated for the specified number of iterations or until the fish population became extinct by dropping to 0 individuals. For each simulation run, we monitored the number and biomass of adult and juvenile fish, the number of recruits, the number of fish harvested, the total mass harvested, and mean and standard deviations of asymptotic length, current length and age, length and age at first reproduction, and mean and standard deviation of *k* growth coefficient.

Fish Grow

At the start of each time step (one year in the simulation), all simulated fish advanced their age by one year and updated their length according to their age, *k* growth coefficient, and asymptotic length. Juvenile fish that achieved their length at first reproduction matured to adults. Fish also updated their vulnerability and legality; fish with lengths ≥ 300 mm were considered vulnerable to catch. Those whose lengths fell into the legal harvest size ranges of the simulation's specified regulation were considered vulnerable to catch and harvest (Tsehaye et al. 2016).

Fish Reproduce

Adult fish reproduced as a function of their mass and expected reproductive investment. For simplicity, we assumed that the population was closed and new individuals could not enter except through reproduction from existing individuals. All adult fish made a reproductive effort, and total per-individual reproduction, r_i was equal to the product of individual mass (m), gonadal-somatic index, the proportion of total body mass devoted to eggs (g; 0.24), expected survival rate from egg to first fall (z_r; 0.0001), and a sex-ratio correction of 0.5 (because all simulated fish are female) divided by the mass of one egg (m_e; 0.0001g; Bozek et al. 2011, Lester et al. 2014; (13.3)):

$$r_i = \frac{g * m * z_r}{2m_e} \tag{13.3}$$

Recruit survival was estimated per a Ricker stock recruitment model for Escanaba Lake in Wisconsin (Ricker 1975, Hansen et al. 1998). The number of recruits (R) that survived was defined by the number of adults (S), a density-independent term (a; 3.392) and a density-dependent mortality term (b –0.001176) such that recruit survival increased up to a threshold of adult abundance after which it decreased:

$$R = Se^{a + bS} \tag{13.4}$$

The model therefore sampled the density-dependence adjusted number of surviving recruits per the Escanaba stock recruitment model from the pool of all possible recruits created by adult fecundity.

Surviving recruits were then assigned asymptotic length phenotypes. We assumed that asymptotic length was a quantitative trait with genetic and environmental components and that the environment imposed a constant phenotypic pressure toward the starting values ((13.5)). A genetic asymptotic length value for each new fish was drawn from a normal distribution with mean equal to the parent's mean genetic asymptotic length and standard deviation equal to the population's standard deviation of asymptotic length. Each fish was then assigned a phenotypic asymptotic length ($L_{\infty p}$) according to its asymptotic length "genotype," ($L_{\infty g}$), and a draw from a normal distribution defined by the starting mean and standard deviation of asymptotic lengths (μ_S, σ_S), and the heritability of asymptotic length (h^2), such that:

$$L_{\infty p} = L_{\infty g} * h^2 + N(\mu_s, \sigma_s) * (1 - h^2) \tag{13.5}$$

Fish in a model run with $h^2 = 1$ would therefore derive their asymptotic length exclusively from their inherited value, whereas fish in a model run with $h^2 = 0$ would derive their asymptotic length phenotype exclusively based on a draw from the starting distribution of asymptotic lengths. The k, age at reproduction, and length at reproduction were then calculated for each fish from their asymptotic length phenotype.

Fish Are Harvested

Fish were harvested in the model according to length-specific vulnerability to harvest, catch rate, keep rate of caught fish, regulations, and discard mortality rate ((13.6)). The total expected harvest (F_H) was equal to the product of the number of fish vulnerable to angling in the population (N_v), the catch rate (c; 0.1 in the low-catch scenarios, 0.6 in the high-catch minimum length regime), the proportion of vulnerable fish that were legal to harvest (l), and the probability that an angler would keep a legal fish once caught (p_k; 1):

$$F_H = N_v * c * l * p_k \tag{13.6}$$

We also included discard mortality in the calculation of total fishing mortality because vulnerable fish not legal to harvest may still be caught and therefore subject to some risk of mortality from handling, such that total discard mortality (F_d) was equal to the product of the number of vulnerable fish in the population (N_v), the catch rate of vulnerable fish (c), the proportion of vulnerable fish illegal to harvest (1-l), and the expected mortality of fish released after capture (Z_d; 0.1):

$$F_d = N_v * c * (1 - l) * Z_d \tag{13.7}$$

Total fishing mortality, F, was therefore equal to $F_H + F_d$. Walleye experience a "knife's-edge" shift in angling vulnerability from being relatively invulnerable to angling at lengths below 300 mm to being vulnerable to angling at lengths ≥ 300 mm, so fish were considered to be vulnerable to catch at lengths ≥ 300 mm and vulnerability was updated at each reproduction and growth step (Allen et al. 2013). Harvest was considered by default to be additive to natural mortality (for discussion of this assumption, see Pöysä et al. 2004, De Roos et al. 2007, Weber et al. 2016, Sylvia et al. 2021 [Chapter 17]).

Fish Die Naturally

We did not explicitly include non-human predators in the model, but fish that survived density-dependent mortality at the recruit stage and subsequent harvest were subject to density-independent natural mortality. An individual probability of mortality was calculated for each fish in each time step by making a random draw from a binomial distribution with probability equal to 0.3. Fish that drew a *1* were removed from the population. Though we chose for simplicity to use one mortality rate across age groups for all fish that survived to the age-0 recruit phase, we have provided an option in the code to create an age-dependent natural mortality structure where mortality declines as age increases until a threshold age is reached, at which point mortality becomes constant (as estimated in Tsehaye et al. 2016).

RESULTS

All populations with a genetic component of asymptotic size evolved reduced asymptotic size and consequently increased growth coefficient and reduced age and size at maturity when harvested at the higher catch rate (60% of vulnerable fish caught; all legal caught fish harvested; Fig. 13.2 and Table 13.1). In contrast, harvest at the lower catch rate (10% of vulnerable fish caught; all legal caught fish harvested) did not appear to induce evolution under any harvest regulation (Table 13.1). At the higher catch rate, the minimum length limit populations evolved the greatest reduction in asymptotic size, followed by the harvest slot populations, with the protected slot populations exhibiting the least evolutionary change in asymptotic size, growth rate, and age and size at maturity.

At time step 100 in high-catch scenarios, the evolving protected slot populations exhibited a mean asymptotic length of 674 mm (SD = 115 mm) compared to their starting mean of 706 mm (SD = 86 mm). In contrast, the high-catch minimum length limit populations evolved to a smaller mean asymptotic length of 617 mm (SD = 103 mm) from a starting mean of 704 mm (SD = 86 mm) by time step 100 and the high-catch harvest slot populations evolved to an intermediate mean asymptotic length of 629 mm (SD = 108 mm; Figs. 13.2 and 13.3) from a starting mean of 705 mm (SD = 87). Mean length and age at maturity consequently declined in the evolving high-catch scenarios, with the minimum length limit populations declining most sharply followed by the harvest slot and protected slot populations. The *k* growth coefficient likewise increased in the evolving high-catch populations in proportion to the decline in asymptotic length (Table 13.1; Figs. 13.2 and 13.3).

Population sizes and harvest yields increased under all harvest regulations, catch rates, and evolutionary assumptions until time step 50, after which they remained relatively constant. Evolving and non-evolving protected slot populations under high and low catch conditions consistently provided greater yields than did minimum length limit and harvest slot populations despite maintaining consistently smaller population sizes (Fig. 13.4 and Table 13.2). Evolving minimum length limit and harvest slot populations yielded fewer harvested fish per time step on average than did non-evolving minimum length limit and harvest slot populations (Table 13.2). The difference in yields between evolving and non-evolving populations was less apparent in

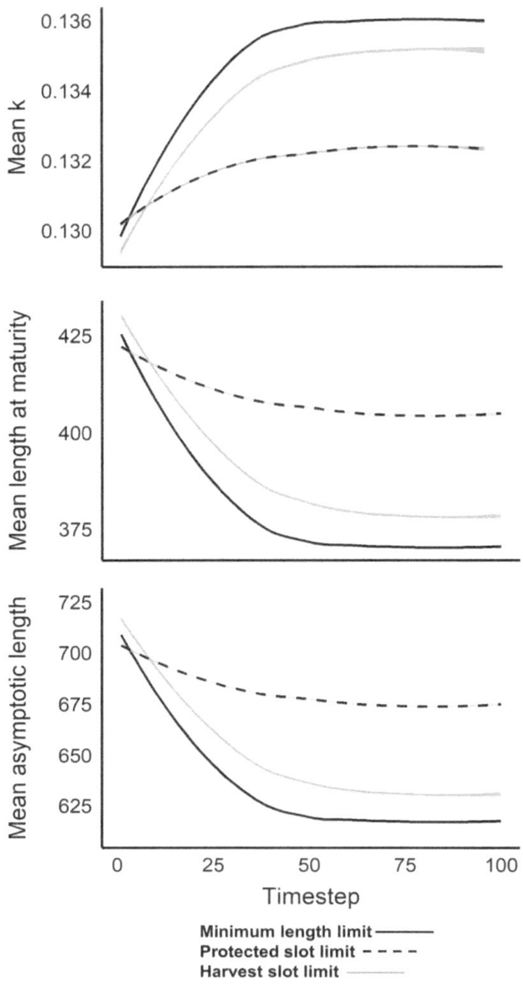

FIGURE 13.2 Under simulated high-catch conditions, evolution toward reduced asymptotic size and consequently increased k growth coefficient and reduced asymptotic length and length at maturity was greatest in minimum length limit populations, intermediate in harvest slot populations, and lowest in protected slot populations. Simulated fish populations were based on walleye (*Sander vitreus*) in Wisconsin, United States of America, and generalized management approaches.

protected slot populations and low-catch scenarios. Population sizes after time step 50 followed a similar pattern; evolving minimum length limit and harvest slot populations averaged smaller population sizes than their non-evolving counterparts under high-catch but not low-catch conditions (Table 13.2). Evolving protected slot populations averaged larger population sizes than did non-evolving protected slot populations under high-catch conditions but similar population sizes under low-catch conditions.

DISCUSSION

Managers of freshwater recreational fisheries are facing an era of rapid environmental change equipped with a long tradition of scientific management and ever-improving scientific insights into the dynamics of social–ecological systems (Arlinghaus et al. 2016a, Ward et al. 2016). Among the many emerging challenges facing fisheries managers is how to plan harvest policies that consider potential selective effects of harvest (Festa-Blanchet and Arlinghaus 2021 [Chapter 12]). Fishing-induced evolution has been implicated in widespread fish life-history changes, but actionable guidance for managing potential evolutionary effects is lacking, particularly in freshwater systems (Heino et al. 2015). We built an eco-evolutionary model to assess how harvest regulations interacted with catch rates to shape the evolution

TABLE 13.1
Means and Standard Deviations of Asymptotic Length (L_∞), k Growth Coefficient, Length-at-Reproduction (L_r), and Age-at-Reproduction (t_r) at the First and Last Step of Each Simulation with and without an Evolutionary Contribution to Asymptotic Size[1]

Regulation	Heritability	Catch Rate	Time Step	Mean L_∞	SD L_∞	Mean k	SD k	Mean L_r	SD L_r	Mean t_r	SD t_r
Minimum	0.3	0.6	1	704	86	0.130	0.005	422	52	5.82	0.27
length limit	0.3	0.6	100	617	103	0.136	0.007	370	62	5.52	0.35
	0.3	0.1	1	706	88	0.130	0.005	424	53	5.83	0.28
	0.3	0.1	100	700	113	0.131	0.007	420	68	5.80	0.36
	0.0	0.6	1	698	100	0.131	0.006	419	60	5.80	0.32
	0.0	0.6	100	694	101	0.131	0.006	416	60	5.78	0.32
	0.0	0.1	1	698	100	0.131	0.006	419	60	5.80	0.32
	0.0	0.1	100	698	100	0.131	0.006	419	60	5.80	0.32
Harvest slot	0.3	0.6	1	705	87	0.130	0.005	423	52	5.82	0.27
limit	0.3	0.6	100	629	108	0.135	0.007	378	65	5.57	0.37
	0.3	0.1	1	705	87	0.130	0.005	423	52	5.82	0.27
	0.3	0.1	100	712	110	0.130	0.007	427	66	5.84	0.35
	0.0	0.6	1	699	101	0.131	0.006	419	60	5.80	0.32
	0.0	0.6	100	695	100	0.131	0.006	417	60	5.79	0.32
	0.0	0.1	1	700	101	0.131	0.006	420	60	5.80	0.32
	0.0	0.1	100	699	100	0.131	0.006	420	60	5.80	0.32
Protected slot	0.3	0.6	1	706	86	0.130	0.005	423	52	5.82	0.27
limit	0.3	0.6	100	674	115	0.132	0.007	404	69	5.71	0.38
	0.3	0.1	1	706	86	0.130	0.005	424	52	5.83	0.27
	0.3	0.1	100	709	111	0.130	0.007	425	67	5.83	0.35
	0.0	0.6	1	700	101	0.131	0.006	420	60	5.80	0.32
	0.0	0.6	100	695	101	0.131	0.006	417	60	5.78	0.32
	0.0	0.1	1	700	100	0.130	0.006	420	60	5.80	0.32
	0.0	0.1	100	700	101	0.130	0.006	419	60	5.80	0.32

Note

1 Simulated fish populations were based on walleye (*Sander vitreus*) in Wisconsin, United States of America, and generalized management approaches.

of fish life history, population sizes, and yields. Fish populations simulated under high-catch conditions evolved reduced asymptotic size, which resulted in reduced age and size at maturity and increased growth rate under all harvest regulations. There were differences in the pace at which evolution proceeded and the outcomes for population dynamics and harvest. Populations harvested under the high-catch minimum-length-limit regime evolved reduced asymptotic size, and consequently reduced size and age at maturity, more quickly and to a greater extent than any of the other populations. Populations harvested according to harvest slot regulations evolved at an intermediate pace, followed by protected slot populations at the slowest pace. Populations harvested at the lower catch rate exhibited little detectable evolution under any regulation.

Our findings demonstrate that evolutionary responses to fishing are defined not only by shifts in the life history of individual fish, but also by changes to a fish's role in a population and ecosystem. That the high-catch minimum length limit populations evolved most rapidly emphasizes a key element of life-history theory that will likely shape the eco-evolutionary effects of fisheries policy. Evolution of

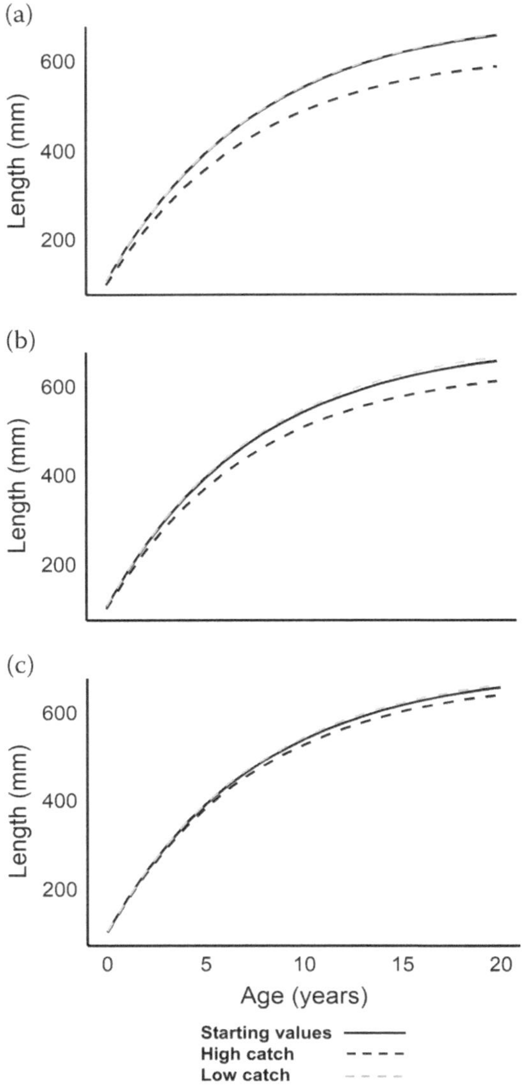

FIGURE 13.3 Changes in growth trajectories of simulated individual fish as a consequence of life history evolution were most pronounced in minimum length limit populations (a), followed by harvest slot populations (b), with protected slot populations (c) exhibiting the least change in growth and life history under high-catch conditions. Populations harvested under low-catch conditions did not exhibit evolutionary changes under any harvest regulation. Simulated fish populations were based on walleye (*Sander vitreus*) in Wisconsin, United States of America, and generalized management approaches.

reduced size and age at maturity can lead to reduced population sizes and harvest yields, as was evident in the smaller population sizes and yields displayed by evolving minimum length limit and harvest slot populations compared to their non-evolving counterparts. Continually removing the largest fish in a population imposes selection toward maturation at earlier ages and smaller sizes because fish that mature later may be harvested before reaching maturity. Raising the risk of mortality significantly at any size can spur selection for early maturation, even at the cost of lifetime reproduction (DeLong and Luhring 2018), underscoring one of the challenges of managing fishing-induced evolution—harvest regulations can affect life-history evolution via multiple mechanisms.

Populations harvested under harvest slot regulations evolved more slowly than did populations harvested under minimum length limits even though they provided similar yields. This difference is likely due to the fact that harvest slot regulations, though they allowed fish to be harvested as they approached reproductive size, also protected the largest fish in the population, conserving some advantage for fish to grow to larger sizes. Protected slot regulations performed even better at

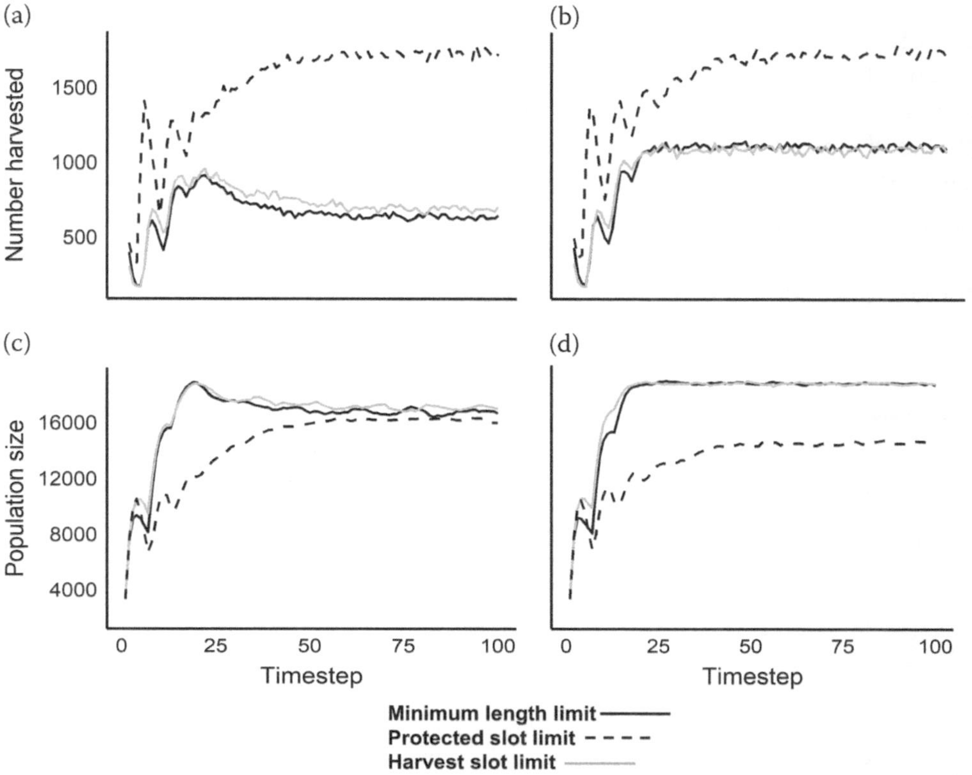

FIGURE 13.4 Under simulated high-catch conditions, evolving minimum length limit and harvest slot populations exhibited reduced yields (a) and population sizes (c) compared to their non-evolving counterparts (b, d). Protected slot populations exhibited smaller differences in yields and population sizes among evolving and non-evolving populations. Though they averaged smaller population sizes overall, protected slot populations consistently yielded the greatest numbers of harvested fish with and without an evolutionary component of asymptotic size. Simulated fish populations were based on walleye (*Sander vitreus*) in Wisconsin, United States of America, and generalized management approaches.

minimizing directional selection on body size while facilitating growth and recruitment and still providing for sustainable harvest. Our findings suggest that protected slots, which are successful at managing the size and age structure effects of recreational fisheries (Ng et al. 2016), may also shape eco-evolutionary dynamics to protect larger-reproducing fish with high reproductive output. The ability of protected slot limits to provide larger sustained harvest over the course of the simulation may be surprising considering previous findings that minimum length limits outperform slot limits at optimizing harvest and age structure objectives (Gwinn et al. 2015). This divergence in regulation performance under different evolutionary assumptions demonstrates that evolutionary impacts on fisheries are a function of feedbacks between social, ecological, and evolutionary dynamics.

By protecting fish as they reached reproductive size, protected slot limits reduced the selective pressure to mature early and allowed higher fecundity fish to survive and pass on their traits. Protected slot limits additionally allowed the harvest of more small fish than did minimum length limit and harvest slot regulations, reducing the likelihood that a given harvested fish would be in the larger legal size group. One limitation to our model is the assumption that anglers would be willing to harvest smaller fish. It would be extremely insightful in future work to explore how varying patterns of angler preference and fish population structure affect the ability of protected slot regulations to limit fishing-induced evolution and maintain sustainable yields. It is important to

TABLE 13.2

Means and standard deviations of population sizes and harvest yields after time step 50 under high and low catch conditions with and without a genetic contribution to asymptotic length. Simulated fish populations were based on walleye (*Sander vitreus*) in Wisconsin, United States of America, and generalized management approaches.

Regulation	Heritability	Catch Rate	Mean Population Size	SD Population Size	Mean Yield	SD Harvest
Minimum length limit	0.3	0.6	16,718	217	639	30
	0.3	0.1	14,795	532	266	20
	0.0	0.6	18,603	91	1047	31
	0.0	0.1	14,617	592	263	19
Harvest slot limit	0.3	0.6	17,058	223	695	32
	0.3	0.1	14,529	453	212	17
	0.0	0.6	18,595	113	1028	32
	0.0	0.1	14,342	407	205	17
Protected slot limit	0.3	0.6	16,170	168	1722	47
	0.3	0.1	14,545	436	335	21
	0.0	0.6	14,397	260	1630	49
	0.0	0.1	14,400	419	326	21

note that conservation of large-maturing fish in the population can be achieved by several means, not just protected slots. Harvest slots and maximum length limits grounded in fish life history have proved effective at limiting fishing-induced evolution while providing for sustainable yields (Ayllón et al. 2018). The finding that protected slot limits and harvest slot limits appeared to limit fishing-induced evolution while providing for sustained harvest is therefore not necessarily an endorsement of any specific regulatory framework, but rather another piece of evidence that conserving the advantage of maturing at a large size may be an important element of managing evolving fisheries (Zhou et al. 2010, Hixon et al. 2014, Law and Plank 2018). Moreover, the regulatory framework that proved most effective in our simulations is widely applied in recreational, but not commercial fisheries, highlighting the importance of thinking outside of traditional management paradigms to incorporate approaches that have proven effective in the management of other fish and wildlife species. Indeed, although we did not consider such approaches here, adaptive harvest management (e.g., Nichols et al. 2007), which includes monitoring and management strategies for ecological (i.e., population size) and evolutionary (i.e., size-at-age distribution) outcomes of harvest may provide the greatest benefit to fisheries via eco-evolutionary modeling approaches such as we have outlined here.

Fish and wildlife harvest decisions are often based on immediately visible traits such as length, mass, sex, or secondary sexual characteristics. Harvest may at first glance seem to impose a straightforward selection pressure against traits such as weapons, ornamentation, or large body size; however, cues of desirable phenotypes are also tied to ontogeny (Festa-Bianchet et al. 2014, Festa-Bianchet and Arlinghaus 2021 [Chapter 12]). Harvest therefore affects not only the phenotypic cues that set the harvest decision point, but also patterns of growth and maturation as well as corresponding behaviors that influence survival (e.g., risk taking, Arlinghaus et al. 2017b) and reproduction (e.g., sneaker males, Foote et al. 1997). Our findings thus indicate that harvest can impose multiple dimensions of selection that relate to both the specificity of harvest (i.e., a phenotypic trait) and the overall risk of harvest mortality. Although such insights are hardly novel, they suggest that subtle changes to how we think about and thus manage fish, or for that matter other harvested populations, can have significant implications for the long-term prospects of important commercial and recreational fish and wildlife populations. When considering harvest regulations, we must realize that we are not simply managing population abundance, age, or size, but also population life history, which affects the dynamics of populations and the services they provide to humans.

It is important to note that our approach was intended to provide a simple life-history model that demonstrates the concept of life-history evolution in fisheries for this book chapter, not to forecast eco-evolutionary dynamics in a specific system. We consequently excluded significant factors that may impose costs on early and small maturation such as size-specific predation or resource competition. Future applications of eco-evolutionary modeling to freshwater systems will likely need to be system specific, using an understanding of fish life histories and angler behavior to forecast the effects of varying biotic, abiotic, and human social trends. One important caveat to our findings, for example, is that the results are dependent on the ecological processes and life history captured in the model. Fish growth is a complex process characterized by variation in the time, rate, and size at which different life-history events occur. We have deliberately simplified growth in this analysis to demonstrate the simplest possible example of life-history evolution. Future assessments of species that express alternative life-history strategies or occupy communities with alternative composition and structure are needed to assess the broad eco-evolutionary implications of protected slots or other management strategies. It would be highly insightful, for example, to examine the effects of different tradeoffs between growth and fecundity. Moreover, because the growth and ontogeny and even the reproductive behavior of many harvested fish also includes a significant degree of plasticity in response to rapidly changing environmental variables such as temperature and food availability (Ylikarjula et al. 1999, Chizinski et al. 2010b), advances in the modeling and management of fisheries harvest will be necessary to create predictably sustainable fisheries in a changing world. Additionally, we have modeled a population without gene flow from non-harvested populations. Previous modeling work has shown that gene flow from an unexploited population to an exploited population may dilute the evolutionary effects of harvest (Tenhumberg et al. 2004). It could therefore be highly informative to examine the effects of gene flow when seeking to predict the eco-evolutionary effects of harvest for specific systems, especially given the extensive history of fish stocking in many systems. To this end, we developed this model using a functional programming approach and have provided the modeling components as functions in the R programming language that may be accessed independently or together via a model wrapper function. In other words, all parameter values used in this analysis may be changed simply by providing different values to the model run function in the code (model functions and initial parameter values available at https://github.com/lsw5077/Harvest_Management) to create a readily modified life-history evolution model. We suggest that curious readers explore the effects of specifying different growth or harvest parameter values and encourage adventurous readers to try changing the life-history functions themselves to investigate the costs and benefits of alternative life-history strategies.

The social and ecological complexity inherent to all harvested fisheries, the very reason that they are difficult to understand and manage, also provides great opportunity for learning. By leveraging growing data and modeling efforts to understand the vast social and ecological diversity of fisheries, different underlying potential for evolution among species, and expert knowledge of systems distributed among researchers and managers, we may begin to understand how (and how much; Hutchings and Kuparinen 2020) evolution is likely to affect freshwater fisheries especially under projected changing environmental conditions. As we have demonstrated, eco-evolutionary modeling has great potential for showing how interactions between fish populations and human harvesters gives rise to emergent eco-evolutionary patterns.

To that end, greater data synthesis is needed among the rich and diverse datasets maintained by natural resources agencies. Monitoring datasets, especially those collected for freshwater fisheries, constitute a great resource for understanding and predicting the dynamics of harvested fisheries at varying spatial scales (Heino 2011). If the skillful and successful management of freshwater recreational fisheries, and indeed all fisheries, is to continue, researchers and managers must work together to understand how the variables managers can observe and influence—fish abundance, size distributions, and harvest behavior—respond to variations in regulations, angler behavior, and a changing environment (Fincel et al. 2015). Our ability to manage fish and wildlife populations for a rapidly changing world will likely depend on our ability to synthesize scientifically collected data with professional expertise and the traditional ecological knowledge of people that participate in subsistence, commercial, and recreational harvest. Modeling approaches and management solutions that leverage emerging science in cooperation with the institutional expertise and decision-making capacity of fisheries managers will be essential to crafting management decisions for increasingly dynamic fisheries.

ACKNOWLEDGMENTS

The authors thank the two reviewers for their extensive feedback that significantly improved this chapter. The authors additionally thank K. Coblentz for early feedback and the editors of this book for all their efforts.

14 Individual Heterogeneity in Annual Survival: Quantifying the "Doomed Surplus"

Todd W. Arnold

CONTENTS

INTRODUCTION

Paul Errington is widely credited with developing two important ideas explaining how harvest mortality could be compensated by other population processes so that wildlife populations could be sustainably harvested (Guthery 2012): (1) the *Principle of Inversity*, which posited greater reproductive success when spring breeding density was reduced (Errington 1945), and (2) the *Threshold of Security*, which hypothesized that large segments of the fall population comprised a "doomed seasonal surplus" that would not survive the winter, even in the absence of harvest (Errington and Hammerstrom 1936, Errington 1945). Although density dependence was a key component of both ideas (Conroy 2021 [Chapter 1]), Errington (1954) also articulated a role for among-individual heterogeneity in determining composition of the doomed surplus, based on his long-term studies of mink (*Neovison vison*) predation on muskrat (*Ondatra zibethicus*) populations. He observed that muskrat populations were composed of two distinct components: (1) healthy adults with regular home ranges that were rarely killed by mink, and (2) subordinate or diseased individuals, juveniles raised during years of surplus production, or otherwise healthy muskrats forced from their normal habitats by food shortage or drought that were highly vulnerable to mink predation. Errington (1954: 278) argued that these latter individuals were "doomed to die" even in the absence of mink predation, and this component of the population could therefore be taken (i.e., by mink or by muskrat trappers) with little consequence to next year's population.

Although most hypotheses about compensation of harvest mortality invoke density-dependent reductions in natural mortality or density-dependent increases in fecundity (e.g., Adaptive Harvest Management for North American waterfowl includes alternative models for both processes; U.S. Fish and Wildlife Service 2018a), variability in mortality risk among individuals (hereafter, individual heterogeneity) is another mechanism by which harvest mortality can be compensated at the population level (Lebreton 2005). However, nearly all assessments of individual heterogeneity to date have been based on verbal models or simple simulations (Lebreton 2005, Sedinger and Herzog 2012, Cooch et al. 2014, but see Lindberg et al. 2013).

My objectives in this chapter are threefold. I recast the concept of doomed surplus in terms of individual heterogeneity in survival, and explain how such heterogeneity can be relevant to harvest management. Then, I show how mixture models (Pledger and Schwarz 2002) can be used to estimate the proportion of pre-hunting season populations of game birds that demographically resemble a doomed surplus. Finally, I provide conceptual guidance on how knowledge of survival heterogeneity might be used in harvest management decisions, and point out remaining uncertainties that need to be addressed.

WHY HETEROGENEITY IN SURVIVAL IS IMPORTANT

Under the additive mortality hypothesis (Burnham and Anderson 1984), annual survival (S_a) is the product of natural survival in the absence of hunting (S_O) and hunting survival (S_h), which can be expressed as one minus harvest rate ($1 - h$), so: $S_a = S_O(1 - h)$. Factoring through, we get: $S_a = S_O - S_O h$, where S_O represents the y-intercept (i.e., survival in the absence of hunting), but also the slope indicating what proportion of a 0.01 increase in harvest mortality is translated into reduced annual survival. Importantly, although annual survival declines linearly with harvest rate under additive mortality, it does not decline with a slope of –1. Harvest has a weaker impact on total survival for species with fast life histories (low natural survival, high fecundity) than for species with slow life histories (high survival, low fecundity) because a greater proportion of harvest in the former case will fall on individuals that would have died anyway from natural causes (Sandercock et al. 2011).

Consider hypothetical populations of cranes, where $S_O \approx 0.95$, and coots, where $S_O \approx 0.50$ (Arnold et al. 2016*b*), but imagine that we adopt a naïve harvest framework that allows hunters to harvest “any 10 Gruiformes in aggregate.” The consequences of increased harvest for cranes are clearly much greater than for coots, because a 0.10 increase in harvest rate reduces annual survival by 0.095 for cranes, but only 0.050 for coots. But even beginning hunters can readily differentiate a crane from a coot, so harvest managers could readily accommodate among-species heterogeneity in survival by adopting more restrictive harvest regulations for cranes and more liberal harvest regulations for coots. If among-species heterogeneity in life-history traits is less pronounced, as for example among species of temperate-nesting ducks, then combined bag limits can represent a viable harvest strategy, but species with slower life histories will nevertheless be more at risk to over harvest (Johnson et al. 2019).

Age (juvenile vs. adult) and sex are often important sources of within-species heterogeneity in survival (Cooch et al. 2014, Arnold et al. 2016a), and represent characteristics that can be readily recognized when harvested animals are in hand. But hunters cannot necessarily discriminate age and sex before lethally harvesting animals, especially among small game species that exhibit rapid maturation with deterministic growth, with sexes that are at least superficially monomorphic. Typically, adult females have the greatest value to future population size, and juvenile males have the least value (Cooch et al. 2014). Relative contributions of each cohort to population size in year t + 1 can be readily assessed as the column sums from a population projection matrix, as in this matrix for lesser scaup (parameters from Arnold et al. 2016a, Koons et al. 2017):

$$\begin{bmatrix} 0.5F_J S_{JF} & 0.5F_A S_{AF} & 0 & 0 \\ S_{JF} & S_{AF} & 0 & 0 \\ 0.5F_J S_{JF} & 0.5F_A S_{AF} & 0 & 0 \\ 0 & 0 & S_{JM} & S_{AM} \end{bmatrix} = \begin{bmatrix} 0.204 & 0.575 & 0 & 0 \\ 0.416 & 0.602 & 0 & 0 \\ 0.240 & 0.575 & 0 & 0 \\ 0 & 0 & 0.402 & 0.689 \end{bmatrix}.$$

Here, the column sums total 0.860 for juvenile females, 1.752 for adult females, 0.402 for juvenile males, and 0.689 for adult males (columns 1–4, respectively). Hence, an adult female is worth ~2 juvenile females, ~2.5 adult males, and ~4.5 juvenile males. Even so, bag limits for scaup have historically disregarded age and sex because hunters cannot reliably distinguish juveniles from adults before harvest (and even determining sex and species can be difficult). If age-sex composition of the harvest varies dramatically among years, then total annual harvest rate could be an erroneous indicator of potential impact on the population. Nevertheless, any variation in age- or sex-specific harvest rates eventually gets estimated through the Parts Collection Survey (Raftovich and Wilkins 2013), and future harvest regulations could be fine-tuned based on average age and sex composition of the harvest (e.g., by reducing overall bag limits if adult females are most vulnerable to harvest, or increasing them if juvenile males are most vulnerable).

Individual heterogeneity in survival can also occur within a cohort within a species. For example, some adult male mallards (*Anas platyrhynchos*) might have a lower probability of surviving the coming year due to among-individual variation in physiological condition, genetic makeup, disease exposure, reproductive investment, habitat selection, or a myriad of other unmeasured and possibly unmeasurable attributes. The principles governing sustainable harvest strategies for populations with individual heterogeneity are identical to those for among-species or among-cohort variation—individuals with high intrinsic survival probability should be harvested more conservatively, whereas individuals with lower intrinsic survival probability could be harvested more liberally. But because the source of individual variation in survival cannot typically be identified before or after harvest (i.e., individual variation is an unobservable property of individuals within the population), harvest strategies recognizing individual heterogeneity must focus on the combined population response to harvest (Lebreton 2005, Cooch et al. 2014).

USING MIXTURE MODELS TO MEASURE HETEROGENEITY

The triple-*i* assumption of "*i*ndependence of fates and *i*dentity of rates among *i*ndividuals" posits no individual heterogeneity in survival or encounter probabilities. As such, the assumption was a cornerstone of early survival analyses because it allowed survival to be modeled as a binomial process (Brownie et al. 1985, Lebreton et al. 1992). Though long recognized as an important statistical assumption, identity of vital rates among individuals has never been accepted as a viable assumption among evolutionary ecologists (Cooch et al. 2002). Only in the last two decades have practical solutions to deal with individual heterogeneity in survival become widely available to biometricians (Pledger and Schwarz 2002, Gimenez and Choquet 2010, Kéry and Schaub 2012).

Earlier investigators demonstrated that individual heterogeneity in survival or encounter probability has minimal bias on estimates of average survival given modest levels of heterogeneity (Nichols et al. 1982, Pollock and Raveling 1982). But heterogeneity in survival provides an opportunity to increase harvest while reducing impact to future population size, provided that harvest can be focused on the most vulnerable component of the population (Lebreton 2005, Sedinger and Herzog 2012), so quantifying heterogeneity in survival is important.

Long before they were developed as formal estimating models, simulation models employing finite mixtures of survival (S, φ) or encounter probabilities (p, f, r) were used to assess the robustness of closed and open population mark-recapture and tag-recovery models to violations of the identity of fates assumption (Carothers 1973, Pollock and Raveling 1982). Later, formal mixture models were developed that allowed estimation of individual heterogeneity in survival probabilities (Pledger and Schwarz 2002, Pledger and Phillpot 2008). Finite mixture models only partially relax the identity of fates assumption by allowing the existence of two or more latent groups, each of which is assumed to have identical within-group survival or encounter probabilities. When Pledger and Schwarz (2002) first applied mixture models to the classic San Luis Valley mallard band-recovery data from Brownie et al. (1985: 28), they estimated that approximately 8% of 6875 adult males had estimated annual survival of near zero ($\hat{S}$ = 0.006). Hereafter, I will adopt a similar working definition for "doomed surplus" as the estimated proportion of the pre-hunting season population that has an estimated annual survival probability near zero.

TABLE 14.1
Band Recovery Data Sets[1] Used in an Analysis of Individual Heterogeneity in Adult Survival Among 12 Harvested Species of North American Birds[2]

Species-sex[3]	Years	Banded	Recovered	ΔAIC_c	π	SE(π)	S_2	SE (S_2)	r	SE(r)
Sage-grouse-F	21	1011	147	1.0	0.104	0.096	0.566	0.041	0.146	0.011
Sage-grouse-M	21	2102	295	1.9	0.031	0.112	0.403	0.036	0.141	0.008
Am. Coot	31	10,955	160	**–4.7**	0.243	0.082	0.592	0.039	0.015	0.001
Mourning dove	50	958,335	3392	2.0	0.000	0.000	0.635	0.005	0.0037	0.0001
Am. woodcock	21	2355	122	2.0	0.000	0.000	0.509	0.037	0.057	0.005
Blue-winged teal-M	51	278,033	7737	**–3.0**	0.025	0.011	0.629	0.005	0.028	0.003
Wood duck-M	51	36,643	4792	2.0	0.000	0.000	0.602	0.005	0.141	0.002
Mallard-F	49	44,025	4605	**–29.4**	0.091	0.016	0.599	0.007	0.107	0.001
Mallard-M	49	40,619	8047	**–67.6**	0.077	0.009	0.692	0.004	0.206	0.002
Am. black duck-F	16	6611	517	**–5.0**	0.139	0.048	0.589	0.024	0.086	0.004
Am. black duck-M	16	9400	1122	**–19.0**	0.120	0.025	0.692	0.014	0.136	0.004
Lesser scaup-F	62	8793	389	**–3.5**	0.123	0.050	0.622	0.021	0.045	0.002
Lesser scaup-M	62	39,617	1985	2.0	0.000	0.000	0.726	0.005	0.050	0.001
Redhead-F	50	21,053	1262	0.6	0.028	0.024	0.678	0.011	0.065	0.002
Canada goose	15	16,741	5262	**–39.9**	0.069	0.010	0.727	0.006	0.379	0.005
Snow goose	18	153,946	15,375	2.0	0.000	0.000	0.871	0.006	0.159	0.006

Notes

1 Data on sage-grouse from Zablan et al. 2003, American woodcock from Krementz et al. 2003, lesser scaup from Arnold et al. 2016a, and lesser snow goose from Alisauskas et al. 2011. Remaining data sets compiled from *Game Bird Data* (USGS Bird Banding Laboratory) for this analysis.

2 Model rankings (ΔAIC_c) are for a two-group Pledger mixture model positing a doomed population component with 0 survival versus a null model with no individual heterogeneity in survival, for 16 hunted populations of North American birds. Positive ΔAIC_c values of 2 indicate no support for heterogeneity, whereas large negative values (in bold) indicate strong support. Parameter estimates include π, the proportion of the marked population with 0 survival; S_2, survival of the remaining ($1-\pi$) of the population, and r, the dead encounter probability.

3 Am. = American, F = female, M = male. If sex not listed, both F and M included. Scientific names in order presented: *Centrocercus urophasianus, Fulica americana, Zenaida macroura, Scolopax minor, Spatula discors, Aix sponsa, Anas platyrhynchos, Anas rubripes, Aythya affinis, Aythya americana, Branta canadensis, Anser caerulescens.*

APPLICATION TO REAL-WORLD BAND-RECOVERY DATA

I compiled 16 band-recovery data sets from 12 North American bird species (see Table 14.1 for sources), including one upland game bird, three webless migratory game birds, and eight species of waterfowl (Table 14.1). Four species contributed separate data sets for each sex. For each data set, I first fit a null model where survival (S) and Seber reporting probabilities (r) were constant across all individuals and time intervals. Mixture models can readily accommodate temporal variation in both survival and encounter probabilities (Pledger and Schwarz 2002), but I kept model structure simple for these examples to facilitate interpretation of survival heterogeneity with respect to harvest management. Although it is reasonable to suppose that encounter probabilities might also exhibit individual heterogeneity, and that this heterogeneity might even be correlated with heterogeneity in survival (Nichols et al. 1982), heterogeneity in encounter probabilities cannot be estimated with dead-recovery models because individuals are only encountered once (White et al. 2013).

To estimate the proportion of the population with zero probability of survival (i.e., "the doomed surplus"), I fit a mixture model (Pledger and Schwarz 2002) that recognized two groups, where annual survival of the first group was fixed to $S1 = 0$. Remaining estimated parameters included π, the proportion of the banded population (if any) having zero probability of survival, $S2$, annual survival probability of the remaining $(1 - \pi)$ proportion of individuals (equivalent to S in the null model), and r, the probability that a banded bird would be harvested and reported to authorities. In comparison to the null model, the mixture model included one additional parameter (π), so in the absence of an estimable doomed surplus, expected Akaike Information Criterion (AIC_c) values corrected for small sample sizes for mixture models should be ~2 AIC_c units larger than the simple null model with no heterogeneity. All models were fit using modified dead-recovery models as implemented in program MARK (White and Burnham 1999), with probabilities modeled using logit links and parameters estimated using simulated annealing.

For eight out of 16 data sets, mixture models were supported by lower AIC_c and typically identified 3%–15% of the banded population as a doomed surplus (Table 14.1). Models with no individual heterogeneity best fit remaining data sets, with five data sets having point estimates of $\pi = 0.000$ and the remaining three data sets estimating that 3%–10% of the population had zero survival, albeit with the additional π parameter not supported by lower AIC_c (Table 14.1). Ability to detect a doomed surplus seemed unrelated to sample sizes, with strong heterogeneity detected in one of the smallest data sets (American coot; 160 recoveries), whereas no heterogeneity was detected for the largest data set (lesser snow goose; 15,375 recoveries). There were no obvious patterns with respect to detectable levels of heterogeneity and sex, taxa, or life histories among modeled species (Table 14.1).

IMPLICATIONS FOR HARVEST MANAGERS

Half of all data sets tested showed evidence of bimodal variation in survival, where a typically small component (3%–15%) of the population had near-zero probability of annual survival. These results have important implications for harvest management, because this component of the population could be readily harvested with minimal influence on future population size. However, because this component of the population is not recognizable to hunters or to managers, it can only be harvested by putting the high survivorship component of the population at risk. Most previous studies of compensation in harvest have looked for evidence that survival rates are independent of harvest rates, presumably because hunting mortality is negatively correlated with natural mortality due to density-dependent feedback processes (Burnham and Anderson 1984, Boyce et al. 1999, Sandercock et al. 2011). However, an alternative pathway for harvest compensation can occur if the harvested component of the population is naturally more vulnerable to both hunting and non-hunting mortality (Lebreton 2005, Cooch et al. 2014).

If we assume that the proportion of total mortality that is due to harvest is independent of survival, then population components with low intrinsic survival will comprise a larger proportion of the harvest. For example, if $\pi = 0.1$, $S1 = 0$, $S2 = 0.7$, and $r = 0.2$ and we assume that band reporting probability $(p) = 0.5$, then total kill rates (K) for each component will be:

$$K1 + K2 = \pi(1 - S1)r/p + (1 - \pi)(1 - S2)r/p = 0.040 + 0.108 = 0.148,$$

with 27% of the total harvest (0.040/0.148 = 0.27) coming from 10% of the population with $S1 = 0$. Under a hypothesis of additive harvest mortality within each population component (Cooch et al. 2014: Equation 17), average survival of a heterogeneous population can be estimated as:

$$\bar{S} = \pi S1_0(1 - K1) + (1 - \pi)S2_0(1 - K2).$$

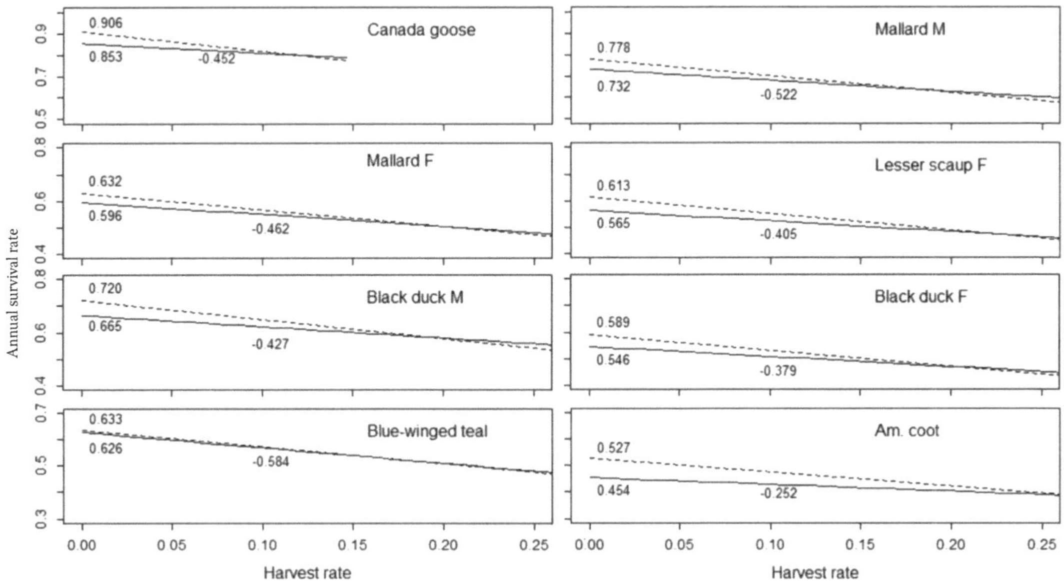

FIGURE 14.1 Annual survival in relation to harvest rate for the eight banded populations with evidence of heterogeneity in survival ($\Delta AIC_c < -2$). Dashed lines represent predicted survival under the additive mortality hypothesis in the absence of heterogeneity, whereas solid lines represent average predicted survival in the presence of heterogeneity. Values near the y-axis represent y-intercept values (i.e., S_0, survival in the absence of hunting), and for the null model the negative value is also the slope. For models including heterogeneity, the slope is plotted beneath the regression line, and it is often much shallower than under models that ignore heterogeneity. X-axis ranges are consistent (0–0.25) for all eight plots, whereas y-axis ranges are consistent within each row.

Two important differences are apparent when comparing mean survival in relation to harvest under fully additive models for models that assume heterogeneity versus identical models with no heterogeneity (Fig. 14.1). First, models that recognize heterogeneity have lower estimates of S_0 (survival in the absence of hunting). Second, models that recognize heterogeneity have a shallower slope for the relationship between annual survival and total harvest; they resemble models intermediate between completely additive and completely compensatory mortality. These effects were very weak for adult male blue-winged teal, where π and r were both < 0.03, but quite strong for Canada geese where π was 0.07 and r was 0.38 (Fig. 14.1).

There are at least three alternative sources of variation that could masquerade as a doomed surplus effect, as modeled herein. If average survival of the entire banded sample increased substantially from year 1 to year 2 after marking, it would be statistically indistinguishable from the doomed surplus model. Such a pattern would be expected if analyses were based on birds first banded as juveniles, where survival is inevitably lower during year 1 (Arnold et al. 2016a: table 2), but I used data sets from adults to specifically avoid this problem. McCrea et al. (2013) hypothesized that heterogeneity in survival within the San Luis Valley mallard data was due to ~30% of birds banded as adult males actually being incorrectly aged juvenile males, an error rate that seems highly unlikely for preseason bandings of mallards or other game birds (Schroeder and Robb 2005), but nevertheless demonstrates that alternative models can be equally supported by the data. Finally, if average encounter probabilities declined sufficiently from year 1 to year 2 after marking, it could also produce an identical likelihood for the data (see Brownie et al. 1985). It seems unlikely that simple temporal variation could contribute to such patterns in survival or recovery probabilities among years because all analyses involved long-term data sets (Table 14.1: 15–62 years), but if game birds are routinely banded in locations (e.g., major hunting hotspots) where they have greater probability of being harvested during the first year after banding than during subsequent years, such data would be indistinguishable from data generated under the doomed surplus model (see Brownie et al. 1985: 30–34).

Mixture models for dead-recovery data cannot identify heterogeneity in encounter probabilities, and although individual heterogeneity in recovery probability is unlikely to be confused with heterogeneity in survival (White et al. 2013), individual variation in recovery probability could be extremely important in determining whether heterogeneity in survival can be effectively exploited by harvest management. Analyses presented in this chapter assume equivalent encounter probabilities between groups with low versus high survival (i.e., a constant fraction of all mortality is due to hunting, regardless of overall mortality), but if individuals in the doomed surplus group die of natural causes before hunting seasons begin, or are otherwise unavailable for harvest, then harvest will fall disproportionately on the high survival component of the population, and no benefits of heterogeneity will be realized. This could occur, for example, if pre-season banding efforts on the breeding grounds targeted ducks in large molting concentrations that were more susceptible to avian cholera, that died after banding but before hunting seasons began. Conversely, if population components with lower survival are at even greater risk to hunting mortality, then compensation potential could be even greater. Although circumstantial evidence indicates that individuals in poor nutritional condition may be more vulnerable to both harvest and natural mortality (Hepp et al. 1986, Dufour et al. 1993, Fowler et al. 2020), such variation should be more clearly quantified through additional empirical studies before being used to justify more liberal harvest regulations. Incorporation of live-encounter data (Lindberg et al. 2013) or use of individual covariates recorded at the time of marking (Arnold and Howerter 2012, Gimenez et al. 2018) represent possible ways to formally measure such heterogeneity.

SUMMARY

Individual heterogeneity in survival and other vital rates is widespread among vertebrates (Gimenez et al. 2018: table 1), including hunted populations of North American birds (Table 14.1). Ignoring heterogeneity in survival can lead to two potential biases with respect to harvest management (Fig. 14.1): (1) overestimating survival in the absence of hunting (S_0), which overstates the impact of harvest under the additive mortality hypothesis, and (2) failing to take advantage of compensation that can occur under heterogeneity, even if harvest is fully additive within each population component. Nevertheless, taking advantage of survival heterogeneity by implementing more liberal harvest strategies relies on the untested assumption that population components with lower intrinsic survival probabilities will have similar or higher probabilities of being harvested by hunters. Testing this assumption, through use of relevant covariates measured at the time of marking, should be a priority for future banding studies.

ACKNOWLEDGMENTS

The author thanks the numerous banders and band-reporting hunters who contributed to the data sets used in analysis. G. White modified Program MARK to include Pledger mixture models as an option for dead recovery data, and he and E. Cooch provided constructive critiques of earlier analyses. K. Malone and N. Cole provided constructive reviews of this chapter.

Section IIB

Efficacies of Harvest Regulations

Duck hunters gather their large bags in a wetland in Minnesota, United States of America in ca.1899. Photo by T. W. Ingersoll, from the collections of the Library of Congress (cropped from original stereograph print).

15 Direct and Indirect Effects of Harvest Regulations on Wildlife Populations

Tyler M. Harms and Stephen J. Dinsmore

CONTENTS

INTRODUCTION

Harvest management in North America has a long and dynamic history that can be traced back to early efforts to prevent overharvest of certain game species (e.g., outlawing any take of certain species completely). Though early harvest management decisions were based largely on anecdotal observations and political motivations (Trefethen 1975), increased interest in sport hunting in the early 1900s required a more sophisticated and biological approach to wildlife management. Importantly, this approach incorporated information on population dynamics of game species as the guiding force of harvest management decisions (Connelly et al. 2005). Increased information on population dynamics of game species resulted in the development of several key wildlife management principles such as additive and compensatory mortality (Dasmann 1964, Anderson and Burnham 1976), maximum sustainable yield (Caughley 1985, Conroy 2021 [Chapter 1]), and harvestable surplus (Leopold 1933, Arnold 2021 [Chapter 14]), all of which are described in earlier chapters of this book. These principles continue to serve as the foundation of harvest management programs across North America.

Historically, harvest management programs included three basic components: (1) inventory of harvested populations, (2) identification of both population and harvest goals, and (3) development of regulations that meet those goals (Strickland et al. 1994). As a result, harvest management decisions were based primarily on data collected on harvested populations prior to the hunting seasons. In recent decades, however, managers have placed increased emphasis on measuring the impact of harvest on populations relative to natural population processes and other influences (e.g., human disturbance, environmental stochasticity). This process, combined with the three components mentioned above, forms the basis of adaptive harvest management (Nichols et al. 1995), an iterative, decision-theoretic approach that aims to simultaneously monitor population dynamics in response to different harvest management decisions (i.e., regulations packages) and reduce uncertainty surrounding population responses to those decisions in the face of other confounding factors (e.g., environmental stochasticity; Williams and Johnson 1995). To be successful, adaptive harvest management hinges on identification of management objectives, regulation alternatives to achieve each objective, and subsequent evaluation of population response to the selected regulation alternative each year (Williams and Johnson 1995, Runge 2021 [Chapter 7]). Thus, population monitoring is a critical component of establishing harvest regulations for hunted populations.

Harvest can result in demographic consequences to the harvested population. These consequences are often realized as changes in vital rates (e.g., survival, reproduction), and the ability to predict population response to these consequences is a key component of harvest management. Several studies have evaluated the influence of harvest on annual survival of various harvested populations such as elk (*Cervus Canadensis*; Biederbeck et al. 2001), brown bear (*Ursus arctos*; Gosselin et al. 2015), muskrat (*Ondatra zibethicus*; Clark 1987), willow ptarmigan (*Lagopus lagopus*; Sandercock et al. 2011), greater sage-grouse (*Centrocercus urophasianus*; Bloomberg 2015), northern bobwhite (*Colinus virginianus*; Williams et al. 2000, Rolland et al. 2010), and mourning dove (*Zenaida macroura*; Otis 2002). Among waterfowl, effects of harvest on annual survival have been investigated for several species in North America, including the snow goose (*Chen caerulescens*; Calvert and Gauthier 2005, Alisauskas et al. 2011), Canada goose (*Branta canadensis*; Rexstad 1992, Sheaffer et al. 2005, Iverson et al. 2014, Conover et al. 2015), mallard (*Anas platyrhynchos*; Burnham et al. 1984, Smith and Reynolds 1992, Sedinger and Rexstad 1994), American black duck (*Anas rubripes*; Francis et al. 1998, Conroy et al. 2002), mottled duck (*Anas fulgivula*; Moon et al. 2017), wood duck (*Aix sponsa*; Johnson et al. 1973), northern pintail (*Anas acuta*; Rice et al. 2010, Mattsson et al. 2012), redhead (*Aythya americana*; Geis and Crissey 1969, Péron et al. 2012), canvasback (*Aythya valsineria*; Geis and Crissey 1969), lesser scaup (*Aythya affinis*; Arnold et al. 2016a), ring-necked duck (*Aythya collaris*; Conroy and Eberhardt 1983), and various species elsewhere (Barker et al. 1991, Clausen et al. 2017). Monitoring demographics and population size of harvested populations is a common approach for measuring the impact of harvest and information from these efforts has been proven valuable for informing harvest management decisions.

Though the effects of harvest regulations on the demographics of harvested populations are both obvious and well documented, harvest regulations can also indirectly affect populations in a variety of ways. Research has demonstrated physiological changes in harvested populations in response to harvest regulations and pressure, such as higher levels of both stress hormones and reproductive steroids in a hunted population of gray wolf (*Canis lupus*) compared to a population with no hunting pressure (Bryan et al. 2015). Indirect effects of harvest pressure can also apply to non-target populations. For example, Pearse et al. (2012) documented decreased body condition of greater white-fronted goose (*Anser albifrons*) in areas subjected to spring hunting for snow geese, likely due to increased energy expenditure in response to hunting disturbance. Harvest can also affect demographic rates of non-target populations (e.g., Andreasen et al. 2018) and, although much less documented in the literature, can be considered when setting harvest regulations for target species provided some form of monitoring the non-target population occurs (e.g., Drewien et al. 1999). These and other studies increasingly highlight the importance of considering both direct and indirect effects on harvested wildlife populations.

CASE STUDY: WOOD DUCKS IN IOWA

The wood duck is one of the most important waterfowl game species in North America. Wood ducks are abundant in North American and populations have been stable to slightly increasing in the United States of America since 1966 (Sauer et al. 2013, Zimmerman et al. 2017). In 2018, a total of 881,195 wood ducks were harvested in the United States of America, making it one of the most harvested species in North America (Raftovich et al. 2019). A total of 407,754 wood ducks were harvested in the Mississippi Flyway, of which 16,944 were harvested in Iowa (Raftovich et al. 2019). Consistently high harvest rates and increasing population trends prompted the U.S. Fish and Wildlife Service to review the allowable harvest of wood ducks in the United States of America, which ultimately led to an increase in the daily bag limit for wood ducks from two to three birds in 2008 in Iowa and elsewhere. Additionally, in 2014 the U.S. Fish and Wildlife Service offered production states in the Mississippi Flyway the option to implement an early, special hunting season for teal species (mostly blue-winged teal [*Spatula discors*]). Iowa opted to participate in this special, early teal season in 2014 and maintained this season until 2018. The early teal season typically started on the first Saturday in September and ended on the third Sunday in September, except for 2016 when the season ended on the second Sunday in September rather than the third. There was concern among biologists and hunters that this early teal season would result in decreased survival of wood ducks due to incidental take because wood ducks breed in Iowa and are present in the state during the early teal season. Comparing demographic and harvest rates in years before and after harvest regulation changes is the most direct approach to investigate effects of such changes on target and non-target populations (Cooch et al. 2014). These two regulation changes offered a unique opportunity to evaluate the demographic effects of harvest regulation changes on wood ducks, both as a target and non-target species.

In this chapter, we present a case study comparing wood duck annual survival and harvest rates between years in which the daily bag limit for wood ducks was two birds (2000–2007), years in which the daily bag limit was three birds (2008–2016), and years in which a special, early teal season was implemented (2014–2016). Results from such an analysis are important to consider when making changes to harvest regulations to ensure sustainable populations of both target and non-target species.

METHODS

Each year, wood ducks are banded throughout eastern North America in both the Atlantic and Mississippi Flyways. For this analysis, we used only band-recovery data from Iowa for the years 2000 to 2016. This analysis could be replicated in states where similar harvest regulation changes occurred. We obtained band-recovery data from the U.S. Geological Survey Bird Banding Laboratory. Banding data included all birds banded in Iowa since 2000 and recovery data included all birds banded in Iowa that were recovered anywhere in the Mississippi Flyway since 2000. We selected this subset of banding data because we were most interested in examining impacts of the early teal season implemented in Iowa on wood ducks banded in Iowa. We included only birds banded in the months of June, July, and August in the analysis, which comprised >99% of all wood ducks banded in Iowa because we were most interested in recovery rates resulting from harvest. We included only those birds that were recovered via hunter harvest because we were most interested in evaluating harvest impacts on annual survival. Additionally, because hunter harvest comprised ~94% of all recoveries in the data set, including other recoveries (e.g., found dead) would likely not change our survival estimates nor adequately capture mortality from these other sources. In some states, waterfowl hunting seasons extend into the subsequent calendar year. Therefore, for any birds harvested in the months of January and February of the subsequent year, we set those birds as harvested in the previous year to represent all birds harvested during a single hunting season. These birds were mostly harvested in southern states in the Mississippi Flyway (e.g., Alabama, Arkansas, Louisiana, and Mississippi) where waterfowl seasons extend into the following calendar year (e.g., a bird harvested in January 2018 was recorded as harvested in the 2017 harvest year for our analysis). We removed all birds with unknown age or sex variables as well as birds recorded as *local* (age code 4), thus retaining only hatch-year and after-hatch-year male and female birds.

We estimated annual survival for wood ducks in Iowa using the Brownie dead-recovery model in Program MARK (Brownie et al. 1985, White and Burnham 1999). The Brownie parameterization estimates true annual survival (S; hereafter annual survival) in addition to an annual recovery rate (f), which is the product of the probability a bird is harvested, retrieved, and reported (Brownie et al. 1985). We assigned each bird in our data set to one of four age-sex groups: (1) hatch-year female, (2) hatch-year male, (3) after-hatch-year female, and (4) after-hatch-year male. Each year was considered an individual occasion in our encounter histories resulting in 18 total occasions. We fit a limited set of nine models that allowed us to compare annual survival and recovery estimates between years with different regulations (Table 15.1). We first fit a single model that estimated constant annual survival and recovery rates across all age-sex groups and years as a way to test our model with the data and used to assess model fit. Next, we fit a series of six models that evaluated the effects of the bag limit change and implementation of early teal season on annual survival and recovery rate both for each age-sex group and across all groups. One such model was structured to estimate both annual survival and recovery rates across all age-sex groups for years with and without the early teal season (i.e., S[teal] f[teal]), which allowed us to determine if this season influenced either parameter. Furthermore, comparing estimates of annual survival and recovery rates for different age-sex groups between models that did or did not include the effect of the early teal season (i.e., S[g*teal] f[g*teal] and S[g] f[g], respectively) allowed us to determine if the early teal season influenced annual survival or recovery rate for one or more age-sex groups. Additionally, we fit two models that estimated annual survival and recovery rates across all age-sex groups (no group-specific variation) and for each age-sex group (with group-specific variation) for three periods: (1) years during which the daily bag limit was two birds (2000–2007) with no early teal season, (2) years during which the wood duck daily bag limit was three birds with no early teal season (2008–2013), and (3) years during which the wood duck daily bag limit was three birds with the early teal season (2014–2016). We fit these models to estimate parameters for each age-sex group (i.e., *group*regulation periods*) and across all groups (i.e., regulation periods). These models removed the confounding effect of the early teal season implementation for years during which the daily bag limit was three birds. Lastly, we fit a model that allowed both annual survival and recovery rate to vary by age-sex group (i.e., group). For all models, we included the same effects on annual recovery and annual survival, again because we were most interested in how these parameters varied together in response to a single change in harvest regulations. We evaluated models using Akaike's Information Criterion (AIC; Akaike 1973a) adjusted for small sample sizes (AIC_c) and considered models with $\Delta AIC_c \leq 2$ to have strong support (Burnham and Anderson 2002).

TABLE 15.1

Model Selection Results From an Analysis of Wood Duck (*Aix Sponsa*) Annual Survival (S) and Recovery (f) Rates in Iowa 2000–2016 (g: Sex and Age Group Effect, Reg: Effect of Change in Daily Bag Limit from 2 to 3 Birds, Teal: Effect of the Early, Special Teal Season)[1]

Model	ΔAICc	*w*	*k*	Dev
S(g*regulation periods) f(g*regulation periods)	0.00	1.00	24	1211.66
S(g*teal) f(g*teal)	21.22	0.00	16	1248.89
S(g) f(g)	90.55	0.00	8	1334.25
S(g*reg) f(g*reg)	95.04	0.00	16	1322.72
S(regulation periods) f(regulation periods)	405.94	0.00	6	1653.62
S(teal) f(teal)	433.31	0.00	4	1685.00
S(.) f(.)	503.01	0.00	2	1758.70
S(reg) f(reg)	504.23	0.00	4	1755.91

Note

[1] Models are ranked by Akaike's Information Criterion, adjusted for small sample sizes (AICc), ΔAICc is the difference in AICc units from the top model (AICc = 68310.88), w is the model weight, k is the number of parameters in the model, and Dev is the model deviance.

RESULTS

A total of 47,737 wood ducks were banded in Iowa during 2000–2016 and fit our criteria for analysis. Of these birds, 15,750 were hatch-year females, 20,739 were hatch-year males, 3491 were after-hatch-year females, and 7614 were after-hatch-year males. We analyzed a total of 9289 recoveries during 2000–2016, which included 2705 hatch-year females, 4610 hatch-year males, 494 after-hatch-year females, and 1480 after-hatch-year males.

Mean annual survival and recovery rates (±SE) for wood ducks banded in Iowa were 0.49 (±0.004) and 0.20 (±0.002), respectively. Annual survival was greatest for after-hatch-year males (0.57 ± 0.1000) followed by hatch-year males (0.51 ± 0.006), after-hatch-year females (0.47 ± 0.020), and hatch-year females (0.40 ± 0.008). Recovery rate was greatest for hatch-year males (0.23 ± 0.003) followed by after-hatch-year males (0.20 ± 0.005), hatch-year females (0.18 ± 0.003), and after-hatch-year females (0.15 ± 0.006).

The model estimating survival and recovery rates for the four age and sex groups and three periods corresponding to regulation changes (*group*regulation periods*) was the best supported model and captured 100% of the model weight (Table 15.1). Annual survival was greatest during years with the three-bird bag limit and early teal season for all age-sex groups except after-hatch-year females (Fig. 15.1a). For hatch-year males, annual survival was greater during both years with the three-bird bag limit and no early teal season ($S = 0.53$, 95% CI = 0.51, 0.55) and years with the three-bird bag limit and early teal season ($S = 0.55$, 95% CI = 0.49, 0.61) than during years with the two-bird bag limit ($S = 0.49$, 95% = 0.47, 0.50; Fig. 15.1a). Annual survival did not differ across regulations changes for all other age-sex groups (Fig. 15.1a). Recovery rate was greater during years with the three-bird bag limit and no early teal season for both hatch-year females ($f = 0.19$, 95% CI = 0.18, 0.21) and hatch-year males ($f = 0.26$, 95% CI = 0.25, 0.27) than during years with the two-bird bag limit ($f = 0.18$, 95% CI = 0.17, 0.19 for hatch-year females and $f = 0.23$, 95% CI = 0.22, 0.24 for hatch-year males; Fig. 15.1b). However, recovery rate was lower during years with the three-bird bag limit and early teal season than during years with the two-bird bag limit for both hatch-year females ($f = 0.15$, 95% CI = 0.13, 0.16 compared to $f = 0.18$, 95% CI = 0.17, 0.19) and after-hatch-year females ($f = 0.09$, 95% CI = 0.06, 0.13 compared to $f = 0.15$, 95% CI = 0.14, 0.17; Fig. 15.1b). Recovery rate did not differ across regulations changes for after-hatch-year males (Fig. 15.1b).

DISCUSSION

With this analysis, we demonstrated the potential ramifications of harvest regulation changes on the wood duck, both as the intended and unintended target of such changes. Numerous studies have evaluated the demographic effects of harvest regulations changes on game birds and other harvested species. Our analysis, however, is one of few that has evaluated the demographic consequences of harvest regulation changes for a non-target species. Similar analyses could be conducted for other scenarios for which unintended consequences to a non-target species are possible, provided some form of monitoring data exist for the non-target species.

The overarching goal of most harvest management programs is to maximize harvest opportunity of the target species. Therefore, identifying the maximum harvest that can be sustained by a population with minimal consequences and subsequent evaluation of harvest impacts on the population are both critical components of successful harvest management programs. In this study, we demonstrated that increasing the daily wood duck bag limit had no negative effect on annual survival of wood ducks in Iowa; therefore, suggesting the population could sustain this increased harvest opportunity. Increasing the daily bag limit significantly increased recovery rates for hatch-year male wood ducks, but did not change recovery rates for all other age-sex groups. This result is contradictory to our prediction of increased recovery rates for all age-sex groups in response to increased daily bag limits, which has been demonstrated in another study for this species

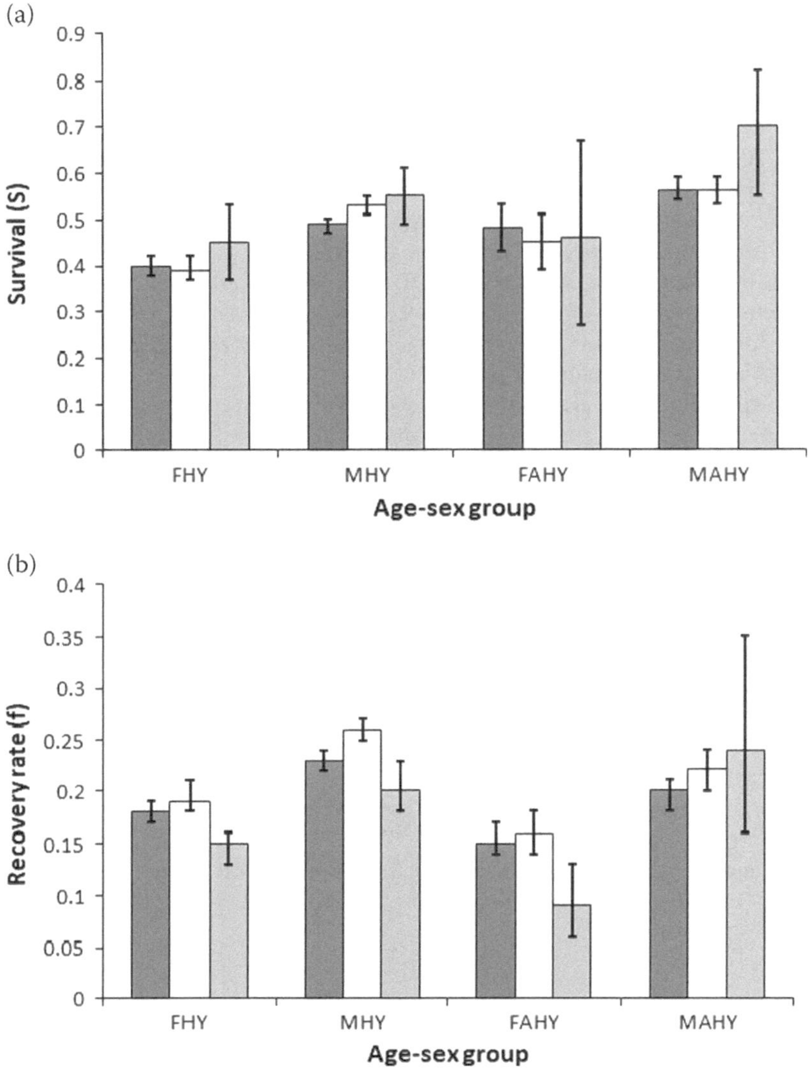

FIGURE 15.1 True annual survival (S; a) and recovery rate (f; b) estimates (95% confidence interval) of wood ducks (*Aix sponsa*) by age-sex groups for years with a two-bird bag limit and during which the special, early teal season was not implemented (2000–2007; dark gray), with a three-bag limit and no special, early teal season (2008–2013; white), and with a three-bird bag limit and the special, early teal season (2014–2016; light gray) in Iowa. *FHY* represents female hatch-year birds, *MHY* represents male hatch-year birds, *FAHY* represents female after-hatch-year birds, and *MAHY* represents male after-hatch-year birds.

(Balkcom et al. 2010) and for other waterfowl species (Allen et al. 1999, Wilkins 2007). However, harvest rates could be influenced by other factors in addition to harvest regulations changes such as species-specific biology and distribution and hunter behavior (Balkcom et al. 2010, Shirkey and Gates 2020), all of which may confound changes in harvest rates in response to increasing daily bag limits. Despite these confounding effects, evaluating the influence of harvest regulations changes on harvest rates provides valuable information to managers that leads to more informed regulatory decisions.

The implementation of the early teal season in Iowa generally resulted in an increase in annual survival and decrease in recovery rates of wood ducks. To our knowledge, no study has previously evaluated the demographic consequences of harvest regulations on non-target species, although we argue that these consequences are important to consider when making harvest management decisions. We posit two potential reasons for the increase in annual survival rates and decrease in recovery rates of wood ducks after the implementation of the early teal season in Iowa. First, because wood ducks breed in Iowa and are thus likely present on wetlands and other habitats targeted by teal hunters during this early season, we suggest this early hunting disturbance could cause local wood ducks to re-locate to refuges or other habitats not typically targeted by waterfowl hunters. To offer some validation to this hypothesis, we compared the proportion of wood ducks banded in Iowa and harvested in Iowa during years with the early teal season to those years with the two-bird bag limit and years with the three-bird bag limit and no early teal season, and ultimately found the proportion of Iowa-banded wood ducks harvested in Iowa during years with the early teal season (18% ± 2%) was slightly lower than the proportion during years with the two-bird bag limit (19% ± 1%) and years with the three-bird bag limit and no early teal season (22% ± 1%). This local redistribution of ducks in response to hunting pressure has been documented in other species (Cox and Afton 1997, Béchet et al. 2004a, Legagneux et al. 2009). Several refuges exist in Iowa to offer protection to waterfowl and other migratory birds during the hunting season, and these refuges often host large numbers of waterfowl throughout the hunting seasons (Iowa Department of Natural Resources, unpublished data). Hunting pressure has also been shown to cause regional movements of waterfowl, likely due to premature or hurried fall migration (Cox and Afton 2000). Wood ducks in Iowa typically commence fall migration in mid-September, so it's possible the hunting pressure in early September as a result of the teal season caused birds to prematurely move south. Further research on movements of local wood ducks in response to early hunting pressure would help shed light on this possibility.

A second possible driver of increased annual survival and decreased recovery rates of wood ducks as a result of the early teal season in Iowa is related to hunting season structure and hunter participation. The implementation of the early teal season in Iowa in 2014 resulted in a delay in the start of the regular duck season during which all ducks, including wood ducks, could be harvested. Prior to 2014, the regular duck season started on or around 15 September each year was open for approximately one week, and then closed again to be re-opened in early October. In 2014 and subsequent years, once the early teal season was implemented, the regular duck season started on or around 25 September. As mentioned above, wood ducks in Iowa typically begin migrating south in mid-September. Therefore, the delayed start of the regular duck season may have resulted in reduced harvest opportunity because many of the local wood ducks, which generally comprise a large portion of the daily bag early in the regular duck season, had already departed Iowa. This reduction in harvest opportunity is contradictory to the goals of most harvest management programs, and especially for waterfowl management programs. A consequence is that this leads to reduced hunter satisfaction because most waterfowl hunters highly value the opportunity to harvest birds (Schroeder et al. 2006, Brunke and Hunt 2008, Bradshaw et al. 2019, Schroeder et al. 2019b). In addition to reduced harvest opportunity, we also speculate that hunter participation overall may have decreased, or perhaps shifted, as a result of the early teal season. More specifically, we suspect that the early teal season resulted in fewer hunters in the field during the early portion (first two weeks) of the regular duck season (a period during which wood duck harvest opportunity is greatest in Iowa) because hunters already experienced an "opening day" (Connelly et al. 2005). Previous research has demonstrated that duck–hunter participation decreases drastically after the opening day of the season until waterfowl numbers increase again during peak migration (Iowa Department of Natural Resources, unpublished data) and that duck hunter numbers decrease with increasing season length (Haugen et al. 2015). Hunter participation and behavior directly impacts harvest (Haugen et al. 2015), and though challenging to predict, this result provides clear evidence for the need to consider how harvest regulations alter hunter behavior and what effects, if any, the resulting change in behavior has on harvest of both target and non-target species.

Our estimates of annual survival and recovery rates for wood ducks are the first, to our knowledge, for the Upper Midwest and are comparable to estimates from other regions of the United States of America. Annual survival for wood ducks ranged from 0.40 to 0.80 for adult and juvenile males and 0.30 to 0.85 for adult and juvenile females in eastern North America (Zimmerman et al. 2017), and annual survival for females averaged 0.63 in southeastern Missouri (Dugger et al. 1999). Harvest rates for wood ducks in eastern North America ranged from 0.045 to 0.093 in response to various season lengths and bag limits (Balkcom et al. 2010). Knowing these rates and how they change through time allows managers to evaluate and adapt management decisions in response to changes in rates, regardless if those changes are a result of previous management decisions or other factors (e.g., habitat variability). Furthermore, these rates can form the basis for more sophisticated models to estimate population trends for harvested species (e.g., Zimmerman et al. 2017) or to integrate impacts of harvest and habitat variability on demographics of harvested populations (e.g., Mattsson et al. 2012).

SUMMARY

With any decision in wildlife management, there is always a consequence, and this is especially true in harvest management. Such consequences must be carefully weighed against the intended benefit of a decision to ensure a sustainable system. This careful balance, along with the ever-present uncertainty associated with harvest management decisions and increased scrutiny of said decisions as public and hunter attitudes toward harvest management change, makes monitoring the effects of harvest on both target and non-target populations an increasingly critical component of harvest management programs (Cummings and Bernier 2021 [Chapter 10]). Our example of indirect effects on the harvest of wood ducks in Iowa should also serve as a catalyst to others to measure the consequences (direct and indirect) of future changes in harvest regulations using detailed demographic studies. Such an approach is consistent with a desire to adaptively manage wildlife populations.

ACKNOWLEDGMENTS

Thanks are due to the many wildlife professionals throughout the Mississippi Flyway who spent countless hours in the field capturing and banding wood ducks, and to the hunters who reported band recoveries. D. Bystrak with the U.S. Geological Survey Birding Banding Laboratory facilitated access to the data for this analysis. The thoughtful comments from two anonymous reviewers improved the quality of this chapter, and the authors are grateful to the editors for the opportunity to be a part of what is certain to be a great resource for fisheries and wildlife students and professionals.

16 What to Exploit When You're Exploiting

Angling Rates and Size Selection Responses to Changing Bag Limits

Zachary S. Feiner, Alexander W. Latzka, and Max H. Wolter

CONTENTS

INTRODUCTION

Human exploitation of wild populations is usually selective (Losey et al. 2008, O'Dea et al. 2014), removing individuals with certain physiological, behavioral, or physical traits (Hessenauer et al. 2015, Van de Walle et al. 2018). Importantly, harvest selection can outweigh natural or anthropogenic selection (Darimont et al. 2009, Fugère and Hendry 2018) and influence the phenotypic and genetic composition of populations (Edeline et al. 2009, Festa-Bianchet 2017, Festa-Bianchet and Arlinghaus 2021 [Chapter 12], Wszola and Fontaine 2021 [Chapter 13]). Selective harvest has implications not only for population performance, but also for stakeholder acceptability of exploited populations. For example, removal of large fish can reduce overall size structure, resulting in a less desirable fishery (Olin et al. 2017, Blackwell et al. 2019), but also selecting against fast growth or large size in the long term, resulting in reduced overall yield (Eikeset et al. 2013, Dunlop et al. 2015).

To counteract the effects of overexploitation and selective harvest, many recreationally harvested populations are regulated with size limits, bag limits, or both (Gangl 2021 [Chapter 24]). These regulations usually operate at the individual level, limiting the size and number of animals a single person can harvest. Regulations are often designed to both control overall harvest and protect or enhance size structure (Gwinn et al. 2015, Ayllón et al. 2019), as increased opportunities to harvest large animals is often a primary goal for stakeholders (Arterburn et al. 2002, Harper et al. 2012). Bag limits can also more equitably spread harvest among users (Radomski et al. 2001), as individual harvest rates tend to be highly skewed with a few individuals accounting for much of the harvest (Cook et al. 2001, McDougall and Amundson 2017).

Despite their widespread use, unpredictable relationships between regulatory actions and overall harvest or user behavior can result in surprising and unintended management outcomes (Fulton et al. 2011, Abbott and Haynie 2012, Gruntorad and Chizinski 2021 [Chapter 4]). For instance, angler effort that is unresponsive to reduced fish population size could increase the likelihood of overfishing and reduce the effectiveness of regulations (e.g., size limits) designed to reduce harvest (Allen et al. 2013). Strict regulation of a single species could also unpredictably shift effort and harvest onto other species, as was observed in the Alaskan charter boat fishery when increasingly strict regulations on Pacific halibut (*Hippoglossus stenolepis*) fisheries led to increased harvest of other species, including rockfish (*Sebastes* spp.) and Pacific cod (*Gadus microcephalus*, Beaudreau et al. 2018). Many regulations are enacted based on perceptions of resource quality and stakeholder acceptability, rather than ecological analysis, which is likely the underlying cause of many failed regulations (Radomski et al. 2001, Hiller et al. 2021b [Chapter 2], Kaemingk et al. 2021 [Chapter 3]). Moreover, regulating based on user acceptability can be ineffective because user acceptability tends to decline when regulations reduce yield, preventing implementation of policies sufficient to reduce harvest (Breisjøberget et al. 2017, Johnston et al. 2018). To avoid these problems, it is important to consider how user behavior will respond to new regulations, meaning the current paucity of information on these types of interactions represents a significant hurdle in fisheries and wildlife management (Fulton et al. 2011, Hunt et al. 2011, Gruntorad and Chizinski 2021 [Chapter 4]).

Landscapes with many recreational fisheries represent ideal systems for elucidating behavioral responses to regulation. Anglers make decisions about effort and harvest allocation within a mosaic of fish populations and regulations. Thus, pairing standardized, fishery-independent data with fishery-dependent data across a variety of systems can be a powerful tool for understanding variability in angler behavior. Here, we leverage extensive fisheries-dependent and fisheries-independent datasets (~600 lake-year observations) collected from Wisconsin panfish fisheries to examine how changes in bag limits affected (i) total effort, harvest, catch, and release rates; (ii) the distribution of harvest among anglers; and (iii) the importance of fish abundance, size structure, and regulation on the size selectivity of anglers. Our results provide information that managers can use to develop actionable and effective regulations while avoiding unintended changes in user behavior.

REGULATION EFFECTS IN WISCONSIN'S PANFISH FISHERY

Study System

Panfish, including bluegill (*Lepomis macrochirus*), pumpkinseed (*L. gibbosus*), black crappie (*Pomoxis nigromaculatus*), white crappie (*P. annularis*), and yellow perch (*Perca flavescens*), represent some of the most popular recreationally harvested fish species in Wisconsin (and North America), accounting for more than half of total effort and 90% of harvest in Wisconsin lakes (Wisconsin Department of Natural Resources, unpublished data). Wisconsin panfish have historically been liberally regulated—there was a 50-fish combined daily bag limit in place until 1997, when the statewide combined daily bag limit was reduced to 25 fish (Rypel et al. 2016). More recently, ten-fish combined daily bag limits have been instituted in a subset of lakes based on stakeholder interest and potential to protect size structure (Hansen and Wolter 2017).

Data Collection

Fisheries-independent data are collected annually on a rotating subset of Wisconsin lakes during spring (April–June, depending on species) each year by Wisconsin Department of Natural Resources staff.

Black crappie and yellow perch are collected using fyke nets when water temperatures reach 10°C–21°C and 4°C–21°C, respectively. Bluegill are collected using boom electrofishing when water temperatures are 10°C–21°C. All target species are counted and a subset are measured (total length) to the nearest 2.54 mm. Catch per unit effort was determined as the total number of fish caught per fyke net night or km of shoreline electrofished and was only included from lake-years when local biologists verified that sampling was reliable and sufficient effort was used (based on lake area, see Feiner et al. 2020*b*). As an indicator of population size structure, proportional stock density was calculated for each species in all lakes and years where at least 20 fish were measured as the number of preferred-size fish divided by the number of stock-size fish multiplied by 100, according to lengths suggested in Gabelhouse (1984a). We included sampling conducted in years from 1990 to 2018 in this analysis, after methods had been standardized. Lakes were sampled on an annual rotation, meaning some lakes were sampled multiple times during this period (mean sampling events = 1.7, median = 1, range = 1–14). For a more detailed description of field methods, see Feiner et al. (2020b).

Fisheries-dependent data are collected annually using standardized creel interviews performed on northern Wisconsin lakes by the Wisconsin Department of Natural Resources from May 1 to March 1 the following year, excluding November and April due to low effort and unsafe ice conditions. Creel clerks make daily counts of total anglers visiting the lake and randomly interview a subset of anglers, collecting information on effort (angler hours), targeted species, catch, harvest, and length of harvested fish. This information is used to estimate annual total angler effort, catch, catch rates (fish/hour), harvest, and harvest rates (fish/hour) for all targeted species on the lake for each year the lake has been surveyed (see Rasmussen et al. 1998 for detailed methodology). Lakes are selected for creel surveys on a stratified annual rotation so many lakes were surveyed multiple years (mean and median = 2 years, range = 1–8 years).

Regulation Effects on Effort, Catch, Harvest, and Release Rates

We assessed the effect of bag limits on effort, catch, harvest, and release rates using two methods. First, we estimated total directed effort (angling hours), total directed catch and harvest (number of fish), and release rates (1 – (total harvest/total catch)) by anglers targeting panfish for each lake and year in which a full-year creel survey was performed (May to February the following year) and at least one panfish species was targeted, caught, and harvested. Panfish regulations for each lake and year were retrieved from a Wisconsin Department of Natural Resources database; we chose to limit our analyses to lakes with 50-fish, 25-fish, and 10-fish bag limits (50-bag, 25-bag, and 10-bag) which provided sufficient data to compare angling metrics among regulation types (Table 16.1). We tested for differences in total effort, catch, harvest, and release rate among regulation types by using a linear mixed model with the respective fishing metric as a continuous response, regulation as a fixed factor, and random intercept terms for lake and year. These random effects were included to account for interannual fluctuations in angling metrics over time, as our dataset spanned nearly three decades, and for differences in fishing pressure among lakes (e.g., high vs. low-pressure systems). By accounting for spatial and temporal variation, these models allowed us to attribute some of the remaining variation in angling metrics to regulations. All regressions were performed with R packages "lme4" and "lmerTest" (Bates et al. 2015, Kuznetsova et al. 2016). Second, *post hoc* marginal means were used to assess the magnitude of differences between regulation types (R package "emmeans"; Lenth 2019).

Regulation Effects on Individual Harvest

Our second question was to determine whether reduced bag limits influenced individual angler harvest. Creel interviews were conducted with single anglers and with angling parties. Angling parties were interviewed in aggregate (i.e., a single person is chosen to speak for a fishing party and the number of anglers is recorded). Therefore, to estimate per-angler harvest for parties with multiple anglers, we randomly redistributed total party harvest among individual party members following the same per-angler harvest distribution observed in single anglers (we also performed the following analysis

TABLE 16.1
Sample sizes for numbers of lakes, lake-years, creel interviews, mean (SD) lake area (ha), and mean (SD) maximum depth (m) of lakes used to test for differences in total angling metrics among regulation types (Creels), and number of lakes, lake-years, mean (SD) targeted catch rate (fish/hour), and mean (SD) proportional stock density (PSD) for each species in lakes with both creel and standard survey information used to test for angler size-selectivity (Creels and Surveys)

		Creels			Creels and Surveys					
Species	Regulation	Lakes	Lake-Years	Interviews	Lakes	Lake-Years	Targeted catch rate	PSD	Lake area	Max depth
Panfish	10-bag	10	21	4620	–	–	–	–	946 (1617)	15.4 (6.7)
Panfish	25-bag	167	255	63,012	–	–	–	–	350 (389)	14.9 (8.5)
Panfish	50-bag	109	121	24,465	–	–	–	–	401 (371)	15.5 (7.8)
Bluegill	10-bag	10	20	2391	6	11	2.33 (1.29)	52.6 (15.1)	946 (1617)	15.4 (6.7)
Bluegill	25-bag	167	255	37,234	64	83	2.78 (1.34)	42.8 (22.3)	350 (389)	14.9 (8.5)
Bluegill	50-bag	108	119	14,557	8	8	2.66 (1.34)	43.1 (23.3)	392 (361)	15.5 (7.8)
Black crappie	10-bag	10	21	3182	6	10	0.84 (0.53)	55.8 (33.1)	946 (1617)	15.4 (6.7)
Black crappie	25-bag	166	247	28,964	60	69	1.04 (0.65)	65.3 (23.1)	350 (390)	14.9 (8.5)
Black crappie	50-bag	105	114	9819	15	16	0.60 (0.41)	56.4 (28.4)	404 (377)	15.5 (7.8)
Yellow perch	10-bag	10	21	1136	6	7	0.85 (0.76)	26.5 (19.8)	946 (1617)	15.4 (6.7)
Yellow perch	25-bag	163	251	26,380	50	57	1.09 (0.85)	28.7 (23.3)	356 (392)	14.9 (8.53)
Yellow perch	50-bag	108	118	8915	24	25	0.92 (0.67)	27.9 (19.1)	401 (373)	15.5 (7.8)

evenly distributing harvest among all anglers with no substantial effect on the results). To determine whether regulations influenced individual angler harvests, we compared the average number of fish harvested per angler among regulation types using a gamma regression with a log-link including random intercepts for lake and year and a fixed effect of regulation type. For this comparison, we included only anglers that harvested fish to remove potential catch-and-release anglers from the responses. Finally, we tested for differences in the proportion of anglers that reached the bag limit under each regulation type using a logistic mixed model with reaching the limit as the response (reached bag limit = 1, did not reach bag limit = 0), random intercepts for lake and year, and a fixed effect of regulation type.

Drivers of Size Selectivity by Anglers

The final question we sought to answer was how bag limits, fish population abundance, and fish population size structure interacted to shape angler size selectivity. This analysis was performed separately on bluegill, black crappie, and yellow perch, the three most popular panfish species in Wisconsin with the most data available for comparisons, using lake–year combinations where both standardized survey and creel survey data were available to describe the length distributions of the fish population and the recreational harvest, respectively (Table 16.1). We quantified the fish length preferences of anglers using Chesson's α, an index of selectivity commonly used in predator diet studies to understand which prey are utilized out of proportion with their occurrence in the environment (Confer and Moore 1987). This analysis assumed that the standardized fyke or electrofishing surveys adequately represented population size structure available to anglers. An initial analysis comparing size structure between electrofishing and fyke netting yielded similar results, giving us confidence that these gears were similarly characterizing size structure of our selected species. Moreover, these gears consistently captured fish smaller than those observed in the creel, suggesting that this method adequately characterized the availability of angling-susceptible fish in the population. Though the smallest individuals likely had low catchability in our sampling gears, which could inflate size selectivity should anglers harvest small fish, anglers were positively size selective across almost all lakes, suggesting our method was a useful way to quantify angler size selectivity.

For each lake-year that had paired observations of at least 20 fish lengths from standardized Wisconsin Department of Natural Resources lake and creel surveys, we binned both length distributions into 5-mm length bins and determined the proportional abundance within each bin. Chesson's α was then calculated for each size bin as:

$$\alpha = \left(\frac{r_i}{p_i}\right) / \sum_{i=1}^{n} \left(\frac{r_i}{p_i}\right) \tag{16.1}$$

where r_i and p_i are the proportional abundances of fish in size bin i in creel data and standardized sample data, respectively, and n is the number of size bins observed. Size bins exhibiting positive selection by anglers were identified where $\alpha > 1/n$. For each lake-year and species, we quantified angler size selectivity and accounted for variation in bin-specific harvest and selectivity by calculating the cumulative frequency of harvest across positively selected-for bins and determining the length corresponding to the 90th percentile of that cumulative harvest (i.e., the length greater than the lengths of 90% of fish in positively selected length bins).

We did not expect angler size selection to respond linearly or predictably to size structure, abundance, or regulation (see also Feiner et al. 2020*b*). Therefore, we used generalized additive mixed models (GAMMs) to flexibly determine the effect of these variables on the 90th percentile lengths of fish that were positively selected for by anglers (R package "gamm4," Wood and Scheipl 2017). The model included selected size as a continuous response, regulation as a fixed factor, random effects of lake and year, and cubic spline functions of population proportional stock density and catch per unit effort as continuous covariates.

TABLE 16.2
Mixed model results for tests of the effect of regulation (10-bag, 25-bag, or 50-bag limits) on lake-wide fisheries (total effort, catch, and harvest) and individual anglers (proportion reaching the bag limit, harvest/angler, and per-angler yield)[1]

Response (units)	Model	10-Bag	25-Bag	50-Bag	σ_{Lake}	σ_{Year}	$\sigma_{Residual}$
			Angler Responses				
Effort(hour/ha)	Gamma	38.5(26.2, 56.4)	41.1(34.9, 48.4)	36.7(30.0, 45.0)	0.895	0.169	0.503
Catch(fish/ha)	Gamma	55.5(33.5, 919.9)	55.5(44.7, 68.9)	37.7(28.7, 49.6)	1.19	0.233	0.66
Harvest(fish/ha)	Gamma	17.6(10.3, 29.9)	19.8(15.7, 24.9)	16.9(12.6, 22.7)	1.21	0.27	0.698
Release rate(proportion)	Linear	0.63(0.56, 0.71)	0.62(0.60, 0.65)	0.52(0.49, 0.55)	0.06	0.02	0.14
Limited out(proportion)	Logistic	0.09(0.06, 0.11)	0.008(0.006, 0.010)	0.001(0.001, 0.002)	1.20	0.46	--
Harvest/Angler(fish/angler)	Gamma	3.82(3.49, 4.19)	4.26(3.92, 4.63)	4.62(4.15, 5.13)	0.44	0.15	0.99
			Per-angler filet yield				
Bluegill yield (g)	Gamma	169(141, 202)	176(158, 197)	209(173, 252)	0.44	0.15	0.95
Black crappie yield (g)	Gamma	281(235, 336)	276(244, 311)	243(190, 310)	0.42	0.23	1.11
Yellow perch yield (g)	Gamma	85.4(65.8, 111)	91.8(76.1, 111)	148.9(100.6, 221)	0.73	0.35	1.13
Composite bag yield (g)	Gamma	267(229, 311)	270(239, 304)	242(200, 291)	0.39	0.22	0.98

Note
[1] The family of model used (linear, logistic, or Gamma with a log-link) was based on best describing the data. Least-squared means (95% confidence intervals) for each regulation type and standard deviations of random lake and year intercepts and residual variation are provided for each model.

Is Angler Size Selection Maintaining Filet Yields?

Differences in per-angler harvest and size selectivity (see Results) raised the question of whether behavioral responses of anglers to regulations were sufficiently compensatory to maintain per-angler yield. Lyons et al. (2017) previously defined species-specific equations to predict filet yield (g) from fish total length from analysis of filet weights produced by anglers fileting 321 bluegill, 139 black crappie, and 137 yellow perch across a range of sizes in Wisconsin. We used these relationships to examine how per-angler filet yield varied among regulation types. For each creel interview, we calculated mean per-angler harvest and mean harvested fish length, predicted filet weight (g) from mean harvested fish length (Lyons et al. 2017), and calculated filet yield (g) by multiplying by mean per-angler harvest. We then compared average per-angler yields across regulation types using a gamma mixed regression model with random effects for lake and year and fixed effects for bag limit as above. We performed this analysis on summed catches of bluegill, black crappie, and yellow perch to test how overall yields differed, and separately for each species to evaluate how yield varied for anglers targeting each species specifically, accounting for the fact that most anglers tended to target a particular species. All statistical analyses were performed in R (R Core Team 2017).

RESULTS

Regulation Effects on Effort, Catch, Harvest, and Release Rates

During the study period, 397 creel surveys were performed across 286 lakes, including more than 92,000 interviews of panfish anglers (Table 16.1). After accounting for variance across lakes and years, effort (angler hour/ha) was consistent across regulation types, averaging between 37 and 41 hours/ha across all three bag limits (Table 16.2; Fig. 16.1a). There were also minimal differences

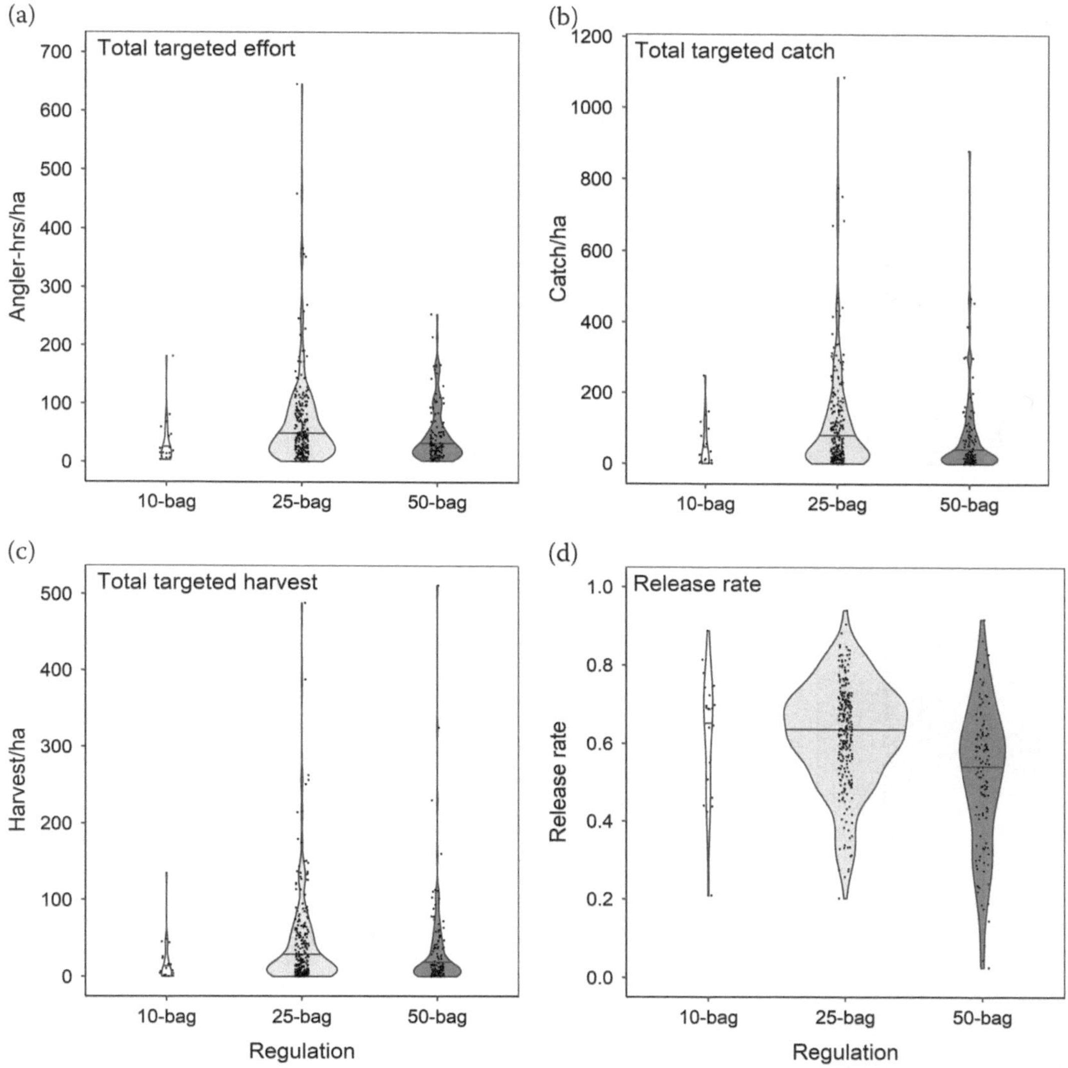

FIGURE 16.1 Violin plots comparing distributions of (a) total angler effort (angler hours/ha), (b) total catch (fish/ha), (c) total harvest (fish/ha), and (d) release rates (proportion of catch that was harvested) of panfish across Wisconsin lakes with either 10-, 25-, or 50-fish bag limits. Black points represent individual lake years, and line represents median. There were no clear differences in effort, catch, or harvest, but release rates were ~15% lower in 50-bag lakes compared to the other regulation types.

in total panfish harvest across regulation types, varying from 16.9 to 19.8 fish/ha (Table 16.2; Fig. 16.1c). In contrast, both total catch and release rates significantly differed among regulation types. Anglers caught substantially fewer fish/ha in 50-bag lakes than either 10- or 25-bag lakes (Table 16.2; Fig. 16.1b). Anglers in 50-bag lakes also released about 15% fewer of their caught fish than under either stricter bag limit (Table 16.2; Fig. 16.1d).

Regulation Effects on Individual Harvest

Regulations only affected individual harvest at the lowest bag limit (Table 16.2). Among anglers that harvested fish, those fishing in 10-bag lakes harvested about one fewer fish compared to

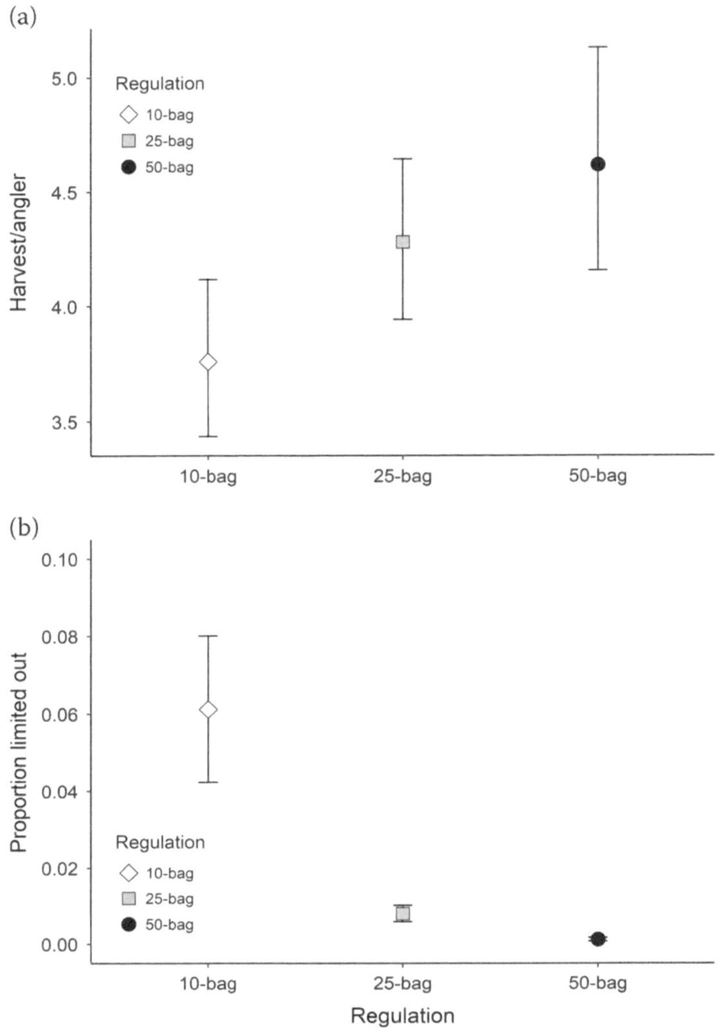

FIGURE 16.2 Model results comparing the effect of regulation on average per angler harvest (a) and probability of reaching the daily bag limit (b) for Wisconsin panfish anglers across regulation types. Points represent least-squared mean estimates accounting for lake and year effects with 95% confidence intervals. Anglers harvested fewer fish on average and limited out more often in 10-bag lakes than 25- or 50-bag lakes.

anglers fishing in 25- or 50-bag lakes (Fig. 16.2a). This pattern was also clear when comparing the proportion of anglers that limited out—essentially no anglers reached the bag limit in 50-bag or 25-bag lakes, whereas about 7% of anglers caught their limit in 10-bag lakes (Fig. 16.2b).

Drivers of Size Selectivity by Anglers

There were weak to moderate effects of regulation type, population size structure, and angler catch rate on angler-selected fish size (Table 16.2). The most consistent pattern across all three species was a weak trend for anglers to select for larger fish under stricter bag limits (Fig. 16.3a, d, g), although these effects were not statistically significant ($P > 0.2$; Table 16.3) and estimates of mean selected size were relatively uncertain with overlapping confidence intervals (Fig. 16.3a, d, g). Mean fish

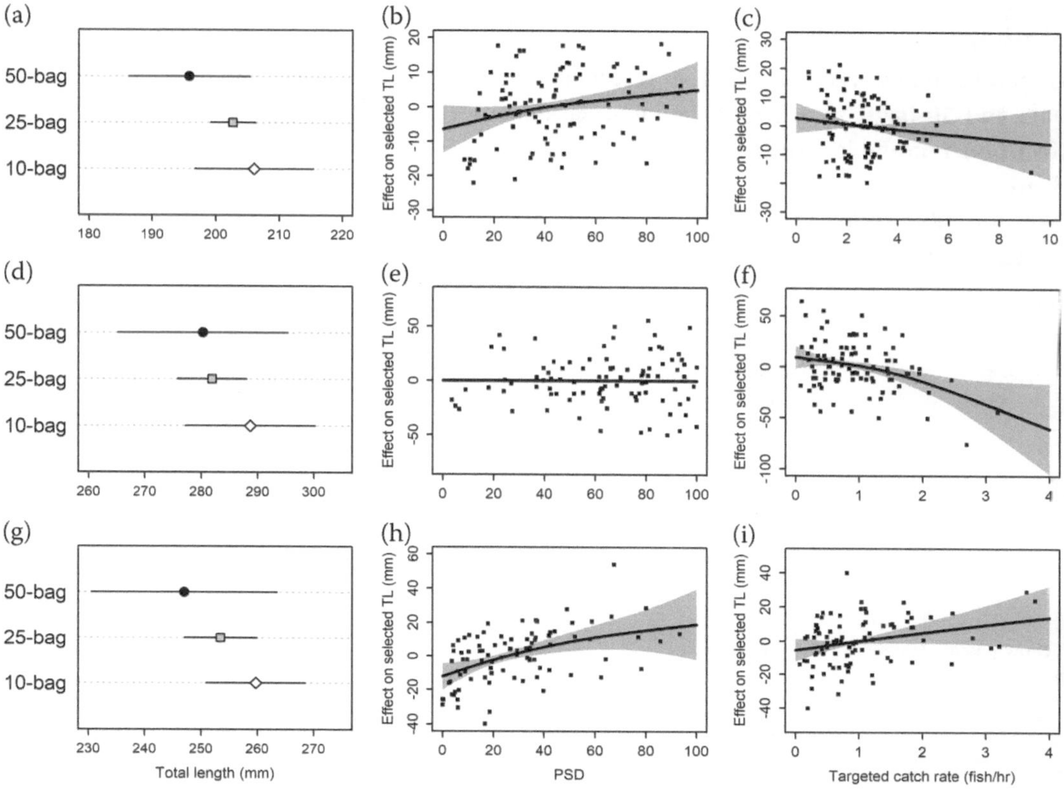

FIGURE 16.3 Results of generalized additive mixed models testing the effects of bag limit (first column), proportional stock density (PSD, middle column), and catch rate (right column) on the angler-selected length of bluegill (top row (a)-(c)), black crappie (center row (d)-(f)), and yellow perch (bottom row (g)-(i)) in Wisconsin lakes. Bag limit effects are presented as parametric least-squared mean estimates with 90% confidence intervals, and PSD and catch rate effects are shown with 95% confidence interval (shaded area) and lake-year residuals (points).

lengths were 10–15 mm longer for anglers when fishing 10-bag lakes as compared to 50-bag lakes, and 3–8 mm longer fish compared to 25-bag lakes. Size structure positively influenced size selectivity in both bluegill and yellow perch anglers (Fig. 16.3b, h). Neither relationship was strictly linear—rather, both were modestly curvilinear (effective degrees of freedom > 1; Table 16.3) with increases in selected size lessening with additional increases in proportional stock density. Lastly, angler catch rate negatively influenced selected size in black crappie anglers and slightly positively influenced selected size in yellow perch (Fig. 16.3f, i). We found no evidence to suggest that angler size selectivity had changed over time (no effect of year; Table 16.3).

Is Angler Size Selection Maintaining Filet Yields?

Increases in preferred harvestable size were slightly to completely compensatory, depending on the target species. In a combined panfish creel, this behavior was completely compensatory (Table 16.1; Fig. 16.4) with filet weight yields in ten-bag lakes matching or exceeding those in more liberally regulated lakes. However, the species-specific analysis showed that compensation varied substantially among species. Average black crappie yield was highest in 10-bag lakes, followed by 25- and 50-bag lakes. Average bluegill yields in 10-bag and 25-bag lakes averaged 81% and 84% of yields in 50-bag lakes. Average yellow perch yields in 10- and 25-bag lakes were

TABLE 16.3
Generalized additive mixed model results for tests of the effects of regulation (10-bag, 25-bag, and 50-bag), year, proportional stock density (PSD), and targeted angler catch rate (fish/hour) on the length of angler-selected fish across northern Wisconsin lakes, including degrees of freedom (df, for the fixed effect of regulation) or effective degrees of freedom (edf, for cubic spline fits of year, PSD, and catch rate), *F* values, *P* values, and adjusted R^2 for each species

Species	Parameter	df (edf)	*F*	*P*
Bluegill	Regulation	2	1.31	0.28
$R^2 = 0.06$	Year	0	0	1
	PSD	1.09	0.12	0.04
	Catch rate	0.63	0.07	0.15
Black crappie	Regulation	2	0.49	0.61
$R^2 = 0.24$	Year	0	0	1
	PSD	0.60	0.07	0.17
	Catch rate	1.60	0.19	0.02
Yellow perch	Regulation	2	1.29	0.28
$R^2 = 0.19$	Year	0	0	1
	PSD	1.44	0.168	0.02
	Catch rate	0	0	0.86

57% and 62% of yields observed in 50-bag lakes. To fully recoup yields, bluegill anglers would have to harvest one additional fish or only harvest fish 210 mm or larger (for reference, only 12% of harvested bluegills exceeded 210 mm in our creel dataset). Yellow perch anglers would be required to harvest one to two additional fish of average size or only harvest 271 mm or larger fish (only 6% of harvested yellow perch in the creel exceeded 271 mm).

DISCUSSION

Limits on individual harvest are commonplace in wildlife and fisheries management, implemented to prevent overexploitation of resources while allowing harvest opportunities. However, harvest limits can potentially be ineffective, caught between competing goals of protecting resources while maintaining publicly acceptable opportunities for harvest (Reed and Parsons 1999, Breisjøberget et al. 2017). The history of regulations through time and space in our study area allowed us to use a combination of fisheries-independent and dependent data to show that bag limits can reduce and redistribute individual harvest among users when limits are sufficiently low. However, we also show that they had little impact on overall effort or harvest, and we found modest evidence for angler behavioral shifts in size selection.

Our results suggest pathways toward more effective fisheries regulations and identify key behavioral responses by anglers that could lead to unexpected management outcomes. Reduced bag limits did not alter total overall effort or harvest in the Wisconsin panfish fishery. We reported a slight trend toward less harvest in 10-bag lakes compared to 50-bag lakes, but harvest estimates were highly variable across lakes. Angler motivation has been the subject of study for years and often shows that effort is allocated less by harvest opportunity than by aesthetic factors (Hunt et al.

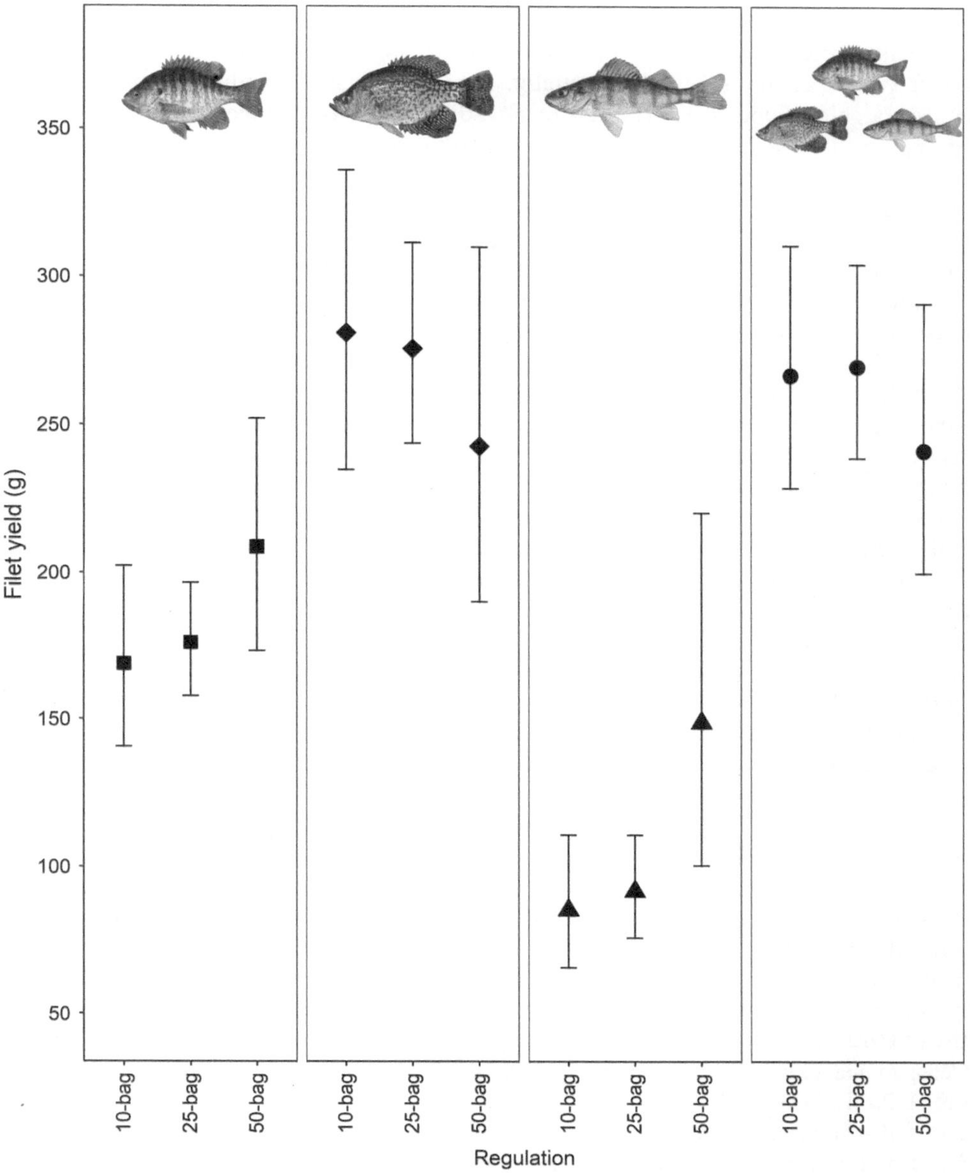

FIGURE 16.4 Predicted filet yield (g) across regulation types based on observations of mean per angler harvest, mean preferred fish size, and the relationship between total length and filet weight for (left to right) bluegill, black crappie, yellow perch, and a combined catch. Points are least-squared mean estimates accounting for lake and year effects, and error bars are 95% confidence intervals.

2019). This could mean that reduced regulations do not strongly influence the attractiveness of lakes to anglers unless changes are dramatic (e.g., Beard et al. 2003a). Thus, responses of angler effort across landscapes to changing regulations may be unpredictable and interact with other variables. For example, Carruthers et al. (2018) and Matsumura et al. (2019) found that a suite of variables, including variability in angler behavior and spatial patterns in fish and lake productivity altered how effort was distributed at local and regional scales.

The reason why total catch in our study was lessened in 50-bag lakes, even after accounting for lake identity and temporal variability in these fisheries, is unclear. The 50-fish bag limit only existed in Wisconsin until 1997—potentially, changes in fish communities over time toward higher centrarchid abundances (Hansen et al. 2015b, 2017a), changes in fishing techniques among anglers (i.e., technology, Feiner et al. 2020a), or shifts in angler motivations in more recent years may have led to higher catches in 25- and 10-bag lakes, which were instituted more recently. However, catch rates for the three main panfish species in Wisconsin have been stable or only moderately increasing over time (Feiner et al. 2020b) and we accounted for temporal variation using a mixed effects design which showed relatively less variation attributable to interannual differences as compared to inter-lake differences. This suggests that temporal associations between regulation changes and fishery characteristics are not solely responsible for the differences we observed. Moreover, our analyses show no differences in effort, catch, or harvest in the two regulation types still used today (10-bag and 25-bag) strengthening our contention that individual bag limits at these levels do little to regulate lake-wide harvest pressure. Our observations are supported by similar findings in terrestrial and aquatic systems. For instance, Eastern wild turkey (*Meleagris gallopavo silvestris*) harvest was not correlated with fall season length or season bag limits in Virginia even as season length was varied from two to eight weeks in length (Weaver and Mosby 1979), and season length had little effect on the number of days duck hunters harvested ducks and no effect on their species selectivity in the U.S. Central Flyway (Haugen et al. 2015). In fisheries, European eel (*Anguilla anguilla*) angler effort was non-responsive to a number of regulatory changes, including bag, rod, and seasonal limits, until regulations were severe (Beardmore et al. 2011a), whereas daily bag limits for rainbow trout (*Oncorhynchus mykiss*) failed to prevent fishery collapse when effort was uncontrolled in British Columbia, Canada (Post and Parkinson 2012). Weak responses to regulatory change, therefore, may be characteristic of both wildlife and fisheries management.

One mechanism behind the lack of overall harvest control appears to be that current bag limits are set far in exceedance of the ability of anglers to meet them. Virtually no anglers reached the 50- or 25-fish bag limits, and only 8% of anglers reached a 10-fish bag limit. This is a common occurrence, and criticism, of individual bag limits (Radomski et al. 2001). Studies in mallard (*Anas platyrhynchos*) and gray ducks (*A. superciliosa*) (Haugen et al. 2017, McDougall and Amundson 2017), willow (*Lagopus lagopus*) and rock ptarmigan (*L. muta*) (Breisjøberget et al. 2017), and black crappie and yellow perch (Isermann et al. 2007, Mosel et al. 2015), have found that the proportion of users that reach limits is usually low, and that bag limits generally need to be significantly reduced (often by 50%–90%) to reach objectives for harvest reductions. Therefore, setting bags without quantification of current harvest levels likely leads to ineffective practices that could result in overharvest when systems are not self-regulating (Hunt et al. 2011, Allen et al. 2013).

Harvest distributions in aquatic and terrestrial systems tend to be highly skewed, where a handful of highly motivated or skilled users achieve high harvests while most users harvest few individuals (Cook et al. 2001, Haugen et al. 2017, McDougall and Amundson 2017). Here, bag limits may have more equitably distributed harvest by increasing the proportion of anglers that reached their limit, thereby trimming the extreme rightward tail of the harvest distribution. We acknowledge, however, that without longitudinal angler data we cannot understand the true shape of the harvest distribution. Potentially, return trips by anglers could skew the harvest distribution at an annual scale, meaning understanding total individual harvest should be further investigated in systems that are similarly regulated.

Improved size structure is a consistent stakeholder request and management goal, and many regulations are designed to improve or maintain size structure in fish populations (e.g., Jacobson 2005, Rypel 2015). However, the few examinations of the relationship between population size structure and angler-acceptable sizes generally assume either static size preferences or a positive linear relationship (e.g., Mosel et al. 2015, Lyons et al. 2017). Importantly, our observations of modestly non-linear increases in size selection with increases in PSD may contradict this assumption. Non-linear responses to increasing fish size structure potentially indicates that anglers have a lower size limit beyond which fish become harvestable regardless of other considerations. This suggestion is supported by the observation of self-imposed minimum size limits in anglers targeting a number of species in Nebraska, United States of America, lakes (Chizinski et al. 2014a, Kaemingk et al. 2020). We should note that it is difficult to fully disentangle acceptable fish size (i.e., fish anglers are willing to keep) versus preferred size. Moreover, the trends we observed were moderate and only observed in two of three study species—black crappie anglers exhibited no change in size selectivity. Nevertheless, these patterns could have important insights for management. Non-linear relationships between fish size structure and angler size selection may result from anglers adjusting minimum acceptable sizes while readily harvesting preferred-size fish, resulting in diminishing returns for management to improve size structure Developing relationships between population size structure and preferred harvestable size across a broader suite of species (Chizinski et al. 2014a), in concert with angler outreach to understand the drivers of fish size preferences, could form the basis for realistic management goals able to balance the ecological potential of fish populations and stakeholder desires for yield (Cook et al. 2001, Arlinghaus 2006a).

In this study, anglers appeared to exhibit slightly compensatory behavior to maintain per-angler filet yields under stricter bag limits. Compensatory behavior is common for predators foraging in dynamic environments (Charnov 1976), and these tradeoffs have been increasingly noted among hunters and anglers. Lennox et al. (2016) observed anglers harvesting larger Atlantic salmon (*Salmo salar*) under more restrictive regulations in Norwegian rivers, whereas shorter seasons led to the harvest of more juvenile ducks in New Zealand (McDougall and Amundson 2017) and a higher probability for Norwegian hunters to shoot red deer (*Cervus elaphus*) when given the opportunity (Diekert et al. 2016). Hunters adapted strategies that maintained yields across wide variations of season length for multiple game species in Denmark (Sunde and Asferg 2014). Importantly, these types of behaviors may circumvent the goals of regulatory change because managers generally lack the ability to directly control harvest rate under many current management strategies (Conroy 2021 [Chapter 1]). Increased size selectivity in response to reduced bags could increase exploitation of valuable large, old individuals (Hixon et al. 2014), whereas seasonal restrictions may both concentrate harvest and alter selectivity patterns (Diekert 2012, Haugen et al. 2015, Diekert et al. 2016). Thus, consistent monitoring or adoption of adaptive management strategies may be critical to determining whether these dynamics are occurring at scales large enough to affect regulatory outcomes. Clearly, further unraveling angler harvest decisions and their impacts on population dynamics will be crucial in designing effective management strategies for exploited wild populations.

CONCLUSIONS

Our results offer several insights for the development of robust harvest regulations. To be effective, harvest limits must be set within the bounds of commonly observed harvest levels. Management can fail to achieve regulatory control when harvest limits are set outside these bounds, as when bag limits are increased, but stakeholders lack motivation to increase exploitation (Riley et al. 2003, Sass and Shaw 2020, Paukert et al. 2021 [Chapter 18], Sylvia et al. 2021 [Chapter 17]), or when bag limits are set at amounts within typical harvest levels (Beardmore et al. 2011a, Haugen et al. 2015), this study). Therefore, regulations must meet a difficult balance of being socially palatable while still having an ecological basis, which may require a strong public outreach program to gain public support (Conroy 2021 [Chapter 1], Fuller et al. 2021 [Chapter 8], Hiller et al. 2021a,b [Chapters 2, 23], Kaemingk et al. 2021 [Chapter 3]). In addition, the limited and unexpected effects of regulations on angler effort, harvest, and size selection in our study suggest the need to clearly understand relationships between user behavior, regulation, and exploited population dynamics to build sustainable management paradigms for exploited populations.

ACKNOWLEDGMENTS

The authors thank the many Wisconsin Department of Natural Resources biologists, technicians, and creel clerks that collected data for the duration of the study. Conversations with the Wisconsin Department of Natural Resources Panfish Team and G. Sass aided in the development of this project. A. Sylvia, L. Hunt, and L. Powell provided constructive reviews of earlier versions of this chapter. This project was funded by the U.S. Fish and Wildlife Service Federal Aid in Sportfish Restoration and the Wisconsin Department of Natural Resources Project F-95-P.

17 Shifting Angler Harvest Behaviors: A Case Study Using Largemouth Bass

Andrea Sylvia, Brandon Maahs, and Michael J. Weber

CONTENTS

INTRODUCTION

Understanding and regulating harvest rates within fisheries and wildlife populations is critical to effective management. Historically, high harvest rates resulted in the decline of many fish and wildlife populations. For instance, American bison (*Bison bison*) was nearly brought to extinction in the 1800s from unsustainable harvest rates (Soper 1941). Hunting by early settlers moving west decimated pronghorn antelope (*Antilocapra americana*) populations (Sexson and Choate 1981) and by 1900, white-tailed deer (*Odocoileus virginianus*) populations throughout North America were nearly extirpated (Woolf and Roseberry 1998). Similarly, freshwater fisheries managers in the 1920s recognized population declines in largemouth bass (*Micropterus salmoides*) due to increased recreational fishing and harvest rates (Long et al. 2015) and high exploitation of yellow perch (*Perca flavescens*) contributed to their population decline in Lake Michigan in the 1960s (Wilberg et al. 2005). In both wildlife and fish populations, implementing harvest regulations was necessary to protect targeted populations from overharvest (Conroy 2021 [Chapter 1]).

Freshwater fisheries serve an important role in understanding changes in harvest practices and management through time. Recreational angling continues to grow as a popular activity throughout the United States of America, with 35.8 million anglers spending 459 million days and US$46.1 billion fishing in 2016 (U.S. Fish and Wildlife Service and U.S. Census Bureau 2016). Among anglers, harvest has historically been an important component to recreational angling (Holland and Ditton 1992, Radomski 2003, Hunt et al. 2011, Embke et al. 2019). As harvest increases to unsustainable levels, targeted populations can experience growth overfishing as a result of unproportional harvest of larger, older individuals resulting in a truncated size structure (Gulland 1983, Beamish et al. 2006). Further, if harvest rates continue at a rate higher than the population's ability to replace itself, recruitment overfishing can occur (Allan et al. 2005, 2013). As a result, traditional management of fish populations relies on allowing anglers to remove a small enough portion of the population that fish can continue to both grow and replace themselves (Otis et al. 1998, Nobel and Jones 1999, Quinn 2001). Although overharvest can result in detrimental effects, some level of regulated harvest is often beneficial for maintaining healthy fisheries (Noble and Jones 1999, Radomski 2003, Isermann and Paukert 2010).

Though high harvest rates are still common for many species, stakeholder uses of natural resources have shifted in past decades for some fish and game populations. This paradigm shift has resulted in species that were historically targeted for harvest no longer representing a consumptive resource (Ankney 1996, Storm et al. 2007, Allen et al. 2008, Myers et al. 2008, Sass and Shaw 2020). Growing non-consumptive activities, such as wildlife watching and catch-and-release angling, may require a shift in management strategies given wildlife and fisheries management has traditionally been based on regulation of harvest. Harvest regulations continue to rely on the assumption that stakeholders will harvest animals; however, continued alteration in resource use may present new challenges for successful resource management (Ankney 1996, Nobel and Jones 1999, Storm et al. 2007, Koons et al. 2019). Therefore, understanding the importance of harvest versus other sources of mortality (catch-and-release mortality, tournament mortality, and natural mortality) for total mortality is central in designing management plans of fish and wildlife populations.

Largemouth bass is one of the most popular sportfish in the United States of America and represents a valuable case study in shifts in fishing practices. Historically, largemouth bass was harvested at high rates (e.g., 40%–69% exploitation in only four days of angling), resulting in truncated size structure and low abundance (Funk 1974, Holbrook 1975). During the 1970s, minimum-length and bag limits were set to protect fish from harvest to improve population structure (Larkin 1977, Isermann and Paukert 2010). In addition to more stringent regulations, both fisheries managers and angler groups began adopting catch-and-release angling practices to reduce high harvest rates (Barnhart 1989, Schramm and Gilliland 2015). By the 1990s, a rise in the popularity of competitive largemouth bass tournament angling coincided with a shift in fisheries from being predominantly harvest oriented to voluntary live-release (Quinn 1996, Noble 2002, Myers et al. 2008, Isermann et al. 2013, Long et al. 2015), with catch-and-release angling more than doubling in the United States of America by 1996 (Quinn 1996, Myers et al. 2011). Today, some waterbodies experience less than 3% harvest of bass populations (Cooke and Cowx 2004, Bartholomew and Bohnsack 2005, Allen et al. 2008, Sylvia 2019), while harvest may still represent an important source of mortality in other populations where harvest can approach 45% (Edwards et al. 2004, Waters et al. 2005, Kerns et al. 2016). Thus, understanding harvest rates relative to other mortality sources of fish populations is important to the future designation and implementation of management regulations.

We conducted the following case study to demonstrate changing patterns of harvest practices in a largemouth bass population in a Midwestern reservoir. A multistate live recapture Cormack-Jolly-Seber model in Program MARK using fishery-independent and dependent data was used to estimate and compare direct harvest mortality (defined here as fish intentionally kept for consumptive purposes) versus all other non-harvest mortality sources (defined here as catch-and-release mortality and natural mortality). Our results highlight common voluntary catch-and-release fishing practices in largemouth bass fisheries where goal-oriented harvest may no longer represent an important source of mortality for a recreationally important fish. We conclude by discussing the need to address such changes in future fisheries and wildlife management by developing creative new management options that no longer rely on harvest mortality for altering populations, but instead identify additional sources of morality in bass populations including catch-and-release mortality from recreational and tournament angling.

CASE STUDY METHODS

SAMPLING

Brushy Creek Lake is a 279-ha reservoir in Webster county, Iowa, United States of America. The lake has a mean depth of 8.9 m, a maximum depth of 22.9 m, and is densely covered in both emerged and submerged course woody structure along the perimeter of the lake (mean 2.000; SE = 0.004 trees/100 m^2). Emerged and submerged aquatic vegetation is also common throughout the lake. Brushy Creek is a popular destination for bass anglers with approximately 45 bass tournaments held between April and October (mean = 32.3; SE = 18.0 tournament angler hours/ha/year during 2015–2017) and open-water recreational angling pressure, excluding tournament events, of ~107 angler hours/ha during April–September 2001 (Miller and Herrig 2001).

Mark-recapture methods are widespread in fisheries literature as an effective tool to estimate survival and temporal trends in mortality. Tagging studies follow a basic format where some individuals in the population are captured, tagged, and released. Subsequent recaptures of tagged individuals are then used to estimate population-level survival and harvest (Seber 1982, Burnham et al. 1987). We distinguished recapture rates and fishing mortality sources of largemouth bass by using both an active high-reward tagging system and sampling fishery-independent and fishery-dependent sources (Kerns et al. 2016, Sylvia 2019). Though estimates of natural mortality rates are possible with traditional mark-recapture models, they are often difficult to estimate because natural deaths are rarely observed (Quinn and Deriso 1999), resulting in estimates that can be unstable (Hoenig et al. 1998). Assumptions underlying the estimation of natural mortality of fish populations rely strongly on reward-tag return rates (assuming 100%), mixing of newly tagged individuals, and similar survival of previously tagged fish (Pollock et al. 1991, Hoenig et al. 1998). To improve estimates of natural mortality and expand on fishing mortality estimates by increasing detection probabilities, we combined telemetry methods with a traditional mark-recapture model to improve detections of a subset of fish that were assumed to have similar survival to other fish in the population (Pine et al. 2003, Pollock et al. 2004, Bacheler et al. 2009, Hightower and Harris 2017).

Electrofishing (pulsed DC, 300 V and 8 amps) occurred during one period a month on Brushy Creek during the open-water season (April–November) in 2015, 2016, and 2017. Electrofishing periods lasted three to five consecutive days each month until the entire accessible shoreline had been sampled (mean 242; SE = 26 minutes/month). All bass ≥ 381 mm (statewide 15-in minimum length limit for bass in Iowa, United States of America, 2015–2017) were collected, weighed (g), measured (mm), and tagged on the top left jaw with a metal Monel butt-end band (selected due to their high retention for black bass; 0% tag loss after one year in smallmouth bass *Micropterus dolomieui*; MacCrimmon and Robbins 1979). To estimate reporting rates, 10% of bass in Brushy Creek were tagged with reward tags ($99) during each tagging event and the remainder of bass received non-reward tags.

All bass tournaments at Brushy Creek during the first week of the month from April 2015 to October 2017 were attended and censused to provide additional recapture information during our monthly sampling period (n = 40 tournaments; mean 26.7, SE = 4.8 anglers per tournament). Tournament events began in April and lasted until October of each year, with a minimum of zero tournaments per week and a maximum of three tournaments per week (two weekend tournaments and one evening weekday tournament). Tournament regulations included a 381-mm (15-in) minimum length limit and a three-fish/angler bag limit; all tournaments required release of bass back into the lake following weigh-in and anglers were penalized for mortalities prior to weigh-in. Following tournament weigh-in, all captured bass were placed in an insulated livewell with lake water and supplement oxygen. All bass were weighed (g), measured (mm), and evaluated for jaw tags.

Daytime electrofishing (pulsed DC, 300 V and 8 amps) was used in May and July 2015 to collect bass to be implanted with radio telemetry tags. Forty bass >700 g (<2% tag:body weight; Winter 1996) and >381 mm were collected across four sections (north, west, east, and south) of Brushy Creek and implanted with radio telemetry tags (Advanced Telemetry System [ATS], Isanti, Minnesota; F1835 14 g in water) in the intracoelomic cavity using established surgical methods (Adams et al. 2012). Transmitters operated on a frequency of 148.010–151.050 kHz and were programmed to be activated for 897 days with power on for 24 hours/day. Fish were held in a recovery tank with oxygen flow until considered recovered when ventilation and fining behavior had returned to normal, and then returned to initial capture location.

Tracking of radio tagged bass began one week following implantation of radio tags and occurred monthly thereafter. No mortalities occurred within one month of tagging, suggesting that subsequent mortalities were not a result of the capture and tagging process. Locations for radio-tagged bass were established from a boat using an ATS (Isanti, Minnesota) model R4000 receiver and a three-element folding Yagi antenna. Bass locations were determined once gain was the lowest achievable setting, and the fish was considered to be within a 2-m distance. Fish were assumed alive if movement occurred across consecutive months and dead when found at the same location for two consecutive months. Mortalities were assumed to have occurred during the period of first location.

Analysis

We analyzed bass encounter histories using maximum likelihood estimates for monthly survival (Φ; hearafter referred to as survival), detection probability (p), and transition probability (ψ) during 2015–2017 using a multistate live recapture Cormack-Jolly-Seber model in Program MARK (White and Burnham 1999). Multistate models use a finite number of states to estimate individual movement probabilities (Lebreton et al. 1992). Assumptions of the model include the transition of an individual from state A to state B requires the individual to survive state A then transition to state B at the last possible moment before the next period. Additionally, all individuals transition between states at the same time, individuals in state A at time i have equal movement probabilities, and movement or recapture probabilities for individuals in state r do not depend on the history of the individual. Basic notation of the estimation of survival, detection, and transition event follows probabilities associated with each capture occasion conditional on the fish's first release, where Φ_i^{AB} is the probability that a fish alive in state A at occasion i is still alive and in state B at occasion $i+1$, and p_i^A is the probability that a fish alive in state A at occasion i is recaptured at time i. For example, a recapture history of three occasions between two states A and B (AAB) would be modeled as

$$P(AAB) = \Phi^{AA} p^A \Psi^{AB} p^B$$

in the maximum likelihood function.

Survival was estimated for 25 periods adjusted to monthly survival estimates during the open-water season and a winter survival estimate consisting of five months and adjusted to a single monthly survival estimate for each of the five months. We used three separate groups to account for potential differences in detection probabilities: jaw tagged bass, reward jaw tagged bass, and telemetry tagged bass. We hypothesized that detection was much higher for telemetry bass through radio tracking than jaw tagged and reward jaw tagged bass through electrofishing and angler reports within the Brushy Creek lake state. Largemouth bass of each group could reside in one of two states: Brushy Creek (B) or harvested (H; Fig. 17.1). All largemouth bass were initially tagged through electrofishing and recaptures could occur at bass tournaments, electrofishing events, reports by recreational anglers (all tagged bass), or weekly radio-tracking events (telemetry tagged bass only). Transitions could occur from the Brushy Creek state to the harvested state (ψ B to H) or remain in the Brushy Creek state (ψ B to B). Fish could not move from the harvested state back into Brushy Creek; thus, survival in the harvested state was set to zero, the transition probability from the harvested state to the Brushy Creek state (ψ H to B) was set to zero, and probability of remaining in the harvest state (ψ H to H) was set to 1. Further, detection probabilities of telemetry bass in the harvest state were set constant at 1 because we knew the fate of all telemetry bass at the end of the study on 31 March 2018 (i.e., all instances of harvest of telemetry tagged bass were known). Alternatively, detection probabilities of recreationally harvested jaw tagged bass were

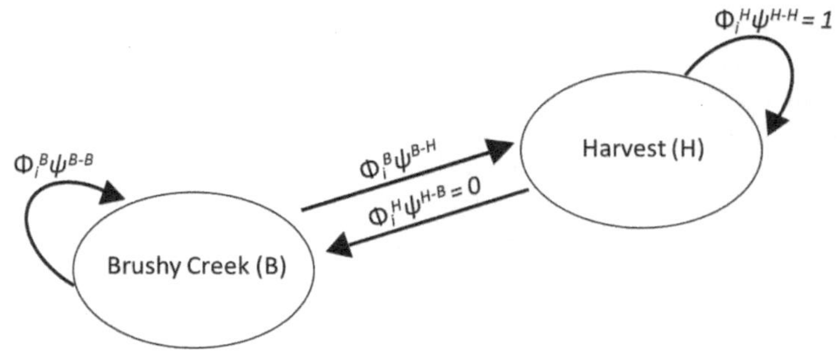

FIGURE 17.1 Conceptual diagram of multistate mark-recapture model design used to estimate natural and harvest mortality of tagged largemouth bass in Brushy Creek Lake, Iowa, United States of America from 13 April 2015 through 31 March 2018. Arrows represent transition probabilities (ψ), p represents detection probabilities within states, and Φ represents survival estimates of each state. $\psi^{B\text{-}B} = 1 - \psi^{B\text{-}H}$.

estimated and reflected the reporting rate of jaw tagged bass that were harvested in Brushy Creek across the study period.

Monthly capture histories were created for 1129 largemouth bass separated by tagging groups between 13 April 2015 and 31 March 2018 (Table 17.1) where an individual received a B if it was captured and released (jaw tags) or located alive (radio tags) in Brushy Creek, or an H if it was harvested. A time-varying covariate (i.e., a covariate that changed on each time interval) of water temperature was used in the analysis to describe potential variation in survival, detection probability, and transition probability. Mean monthly water temperature (°C) was sampled with continuous temperature loggers (Onset Corporation HOBO Pendant Temperature/Light 64 K Data Logger, 15-minute sampling intervals) from two locations within the lake at 0.1- and 4.6-m depth.

TABLE 17.1
Number of jaw tagged, reward tagged, and telemetry tagged largemouth bass that were harvested and recaptured or relocated during 2015, 2016, and 2017 in Brushy Creek Lake, Iowa, United States of America

				Number recaptured or relocated		
Tag type	**Year**	**Number tagged**	**Number reported harvested**	**2015**	**2016**	**2017**
Jaw tag	2015	455	5	60	62	7
	2016	245	7	–	18	22
	2017	258	1	–	–	32
	Total	958	13	60	80	61
Reward jaw tag	2015	36	0	0	0	1
	2016	2	0	–	0	0
	2017	93	0	–	–	1
	Total	131	0	0	0	2
Telemetry tag	2015[1]	40	0	41	160	46
	2016	0	1	–	–	–
	2017	0	1	–	–	–
	Total	40	2	41	160	46
Total bass		1,129	15	101	240	109

The most parsimonious model was selected through procedures based on Akaike's Information Criterion for small sample size (AIC_c), where lower AIC_c values and higher Akaike weights represent the most parsimonious model (Akaike 1973*b*). Models were developed to describe variation in survival, detection probability, and harvest probability. We evaluated differences in survival for largemouth bass in Brushy Creek across years (Φ [Year]), months (Φ [Month]), constant survival across periods (Φ [.]), effects of water temperature (Φ [Water temp]), and combinations of periods and water temperature. Models for detection probability evaluated differences across tag types as well as models that set jaw tag and reward jaw tags equal for the harvest state p [J = R,T]. Detection probability models for Brushy Creek included interactive group by time effects (p [g*t]), models that set jaw tag and reward jaw tags equal (p [J = R,T]), models with trends in time (p [T]), and models that included linear and quadratic water temperature (p [water temp]). Finally, we evaluated monthly (ψ [month]), yearly (ψ [Year]), and time-dependent (ψ [t]) transition probabilities to the harvest state, as well as additive and interactive patterns of month and year (ψ [year + month], ψ [year*month]).

Once all models were run, Markov chain Monte Carlo simulations were used to obtain better estimates on error for model parameters on the most supported model that were not initially estimated well using maximum likelihood estimates in program MARK. Uniform (flat) priors were specified for each parameter and original maximum likelihood parameter estimates from the top model were used as starting values. Five chains comprising 4000 tuning iterations, 1000 burn-in iterations, followed by 10,000 iterations were used in the final estimates. Model convergence was assessed using $\hat{R}$ statistics between duplicate chains (Gelman 1996) and evaluation of trace plots using the *coda* package in program (R Core Team 2018). Parameters and their standard errors were estimated by the mean and standard deviations from the Markov chain Monte Carlo iterations. All results are reported as mean parameter values, their standard deviations, and 95% credibility intervals from the simulations. Differences across groups and time were considered significant when 95% credibility intervals did not overlap zero.

RESULTS

Of the 1129 tagged bass ≥ 381 mm, 958 (85%) were tagged with jaw tags, 131 (12%) were tagged with reward jaw tags, and 40 (3%) were implanted with radio telemetry tags (Table 17.1). A total of 201 (21%) regular jaw tags and two (2%) reward jaw tags were recaptured, whereas telemetry tags were relocated 247 times across the study period (Table 17.1). A total of 15 (1%) tagged bass were reported harvested: 13 were jaw tags, two were telemetry tags, and none were reward jaw tags (Table 17.1). Of the harvested jaw tags, five (33%) were harvested in 2015, seven (47%) were harvested in 2016, and one (7%) was harvested in 2017 (Table 17.1). A telemetry-tagged largemouth bass was harvested in 2016 and another in 2017 (Table 17.1).

Convergence diagnostics of the final Markov chain Monte Carlo model indicated all trace plots showed low serial correlation and all $\hat{R}$ parameters were between 0.99 and 1.10. Of the 14 models evaluated describing detection, survival, and harvest probability, the most supported model ($\Delta AIC_c = 0.00$, $w_i = 0.31$) indicated that detection probability in Brushy Creek varied across months and years, was similar between jaw tags and reward jaw tags, and differed from telemetry tags. Detection probability of harvested bass was constant across time and between jaw tags and reward jaw tags, and differed from telemetry tags (Table 17.2). Harvest probabilities varied monthly but were similar across years. Survival of bass in Brushy Creek from mortality sources other than harvest did not vary across months or years, but was related to water temperatures.

The top model indicated that radio tags had higher detection probability compared to jaw tags in Brushy Creek (Fig. 17.2), but no difference in detection existed between non-reward and reward jaw-tag types. Average detection probability of radio tags in Brushy Creek was 0.59 (95% CI: 0.04 to 0.62) ranging from 0.09 (95% CI: 0.06 to 0.11) in September 2016 to 0.92 (95% CI: 0.94 to 0.95) in August 2015 (Fig. 17.2). Average detection of jaw tags in Brushy Creek was 0.02 (95% CI: 0.01 to 0.03), with

TABLE 17.2

Cormack-Jolly-Seber multistate models used to estimate largemouth bass recapture probability (p), transition probability (Ψ), and survival (Φ) in Brushy Creek Lake, Iowa, United States of America during 25 periods from 13 April 2015 through 31 March 2018[1]

Model	AICc	ΔAICc	w_i	K	Deviance
Φ (Water temp) *p* (Brushy [J = R, T * t], Harvest[J = R, T]) *ψ* (month)	4292.40	0.00	0.31	58	2333.16
Φ (Year + Water temp) *p* (Brushy[J = R, T * t], Harvest[J = R, T]) *ψ* (year + month)	4292.68	0.28	0.27	61	2326.97
Φ (Water temp) *p* (Brushy(J = R, T * t) Harvest(J = R, T)) *ψ* (year + month)	4292.98	0.59	0.23	59	2331.59
Φ (Year + month) *p* (Brushy[J = R, T * t] Harvest[J = R, T]) *ψ* (year + month)	4294.84	2.45	0.09	67	2316.12
Φ (Water temp) *p* (Brushy[J = R, T * t] Harvest [J = R = T]) *ψ* (year + month)	4296.75	4.36	0.04	58	2337.51
Φ (Water temp) *p* (Brushy[J = R, T * t] Harvest[J = T, R]) *ψ* (year + month)	4297.23	4.83	0.03	59	2335.83
Φ (Water temp) *p* (Brushy[J = R * Water temp, T * t] Harvest[J = R, T]) *ψ* (year + month)	4297.32	4.92	0.03	37	2382.67
Φ (.) p (Brushy[J = R, T * t] Harvest[J = R, T]) *ψ* (t)	4303.81	11.41	0.00	72	2314.17
Φ (Water temp) *p* (Brushy[J = R, T * t] Harvest[J = R, T]) *ψ* (year * month)	4306.00	13.61	0.00	73	2314.17
Φ (Water temp2) *p* (Brushy[J = R, T * t] Harvest[J = R, T]) *ψ* (t)	4308.20	15.80	0.00	74	2314.17
Φ (Year * month) *p* (Brushy[J = R, T * t] Harvest[J = R, T]) *ψ* (year + month)	4313.91	21.52	0.00	81	2304.44
Φ (t) *p* (Brushy[J = R, T * t] Harvest[J = R, T]) *ψ* (t)	4329.80	37.40	0.00	95	2289.01
Φ (t) *p* (Brushy(g*t) Harvest(J = R, T)) *ψ* (t)	4359.39	66.99	0.00	119	2263.56
Φ (t) *p* (Brushy[g*t] Harvest[J, R, T]) *ψ* (t)	4360.07	67.67	0.00	120	2261.91

Note

1 Parameters include K = number of parameters, Deviance = −2 x log-likelihood of the model less −2 x log-likelihood of the saturated models (same number of parameters and degrees of freedom), AIC_c = sample-sized corrected Akaike's Information Criterion, and w_i = Akaike weight. Φ is survival probability, *p* is recapture probability, *ψ* is transition probability. Water temperature (Water temp) was included as a continuous variable. Largemouth bass groups include Brushy Creek (Brushy), and harvested (Harvest) jaw tags (J), reward jaw tags (R), and telemetry tags (T). Models tested included time varying (t), constant (.), yearly (year), and monthly (month) variation.

the lowest detection probability in November 2016 at 0.004 (95% CI: 0.041 to 0.615) and the highest detection probability of 0.08 (95% CI: 0.04 to 0.12) in May 2015 (Fig. 17.2). Detection probability of harvested jaw tags was 0.21 (95% CI: 0.041 to 0.62), whereas detection probability of harvested radio tags was set constant at 1.

Survival of bass in Brushy Creek was negatively correlated with water temperature ($\beta = -0.02$, 95% CI: −0.05 to 0.02). Additional models indicated some support for annual variation in survival in Brushy Creek ($\Delta AIC_c = 0.28$, $w_i = 0.27$), resulting in survival probabilities across years ranging from 0.82 (95% CI: 0.78 to 0.99) to 0.98 (95% CI: 0.97 to 0.99). Non-harvest mortality was higher in June, July, and August compared to other months across all years (Fig. 17.3). Average non-harvest mortality of bass in Brushy Creek was 0.08 (95% CI: 0.04 to 0.12) and ranged between 0.03 (95% CI: 0.00 to 0.05) in winter 2017 and 0.14 (95% CI: 0.00 to 0.04) in July 2016. Conversely, monthly harvest probabilities were consistently low (average monthly harvest = 0.01; 95% CI: 0.01 to 0.03) and ranged from 0.01 (95% CI: 0.00 to 0.02) in October to 0.03 (95% CI: 0.02 to 0.04) in June (Fig. 17.3). Additional models receiving support indicated harvest mortality varied across years (model 2; $\Delta AIC_c = 0.28$, $w_i = 0.27$; Table 17.2), resulting in harvest rates ranging from 0.02 (95% CI: 0.00 to 0.09) to 0.15 (0.04 to 0.28). The third ranked model ($AIC_c = 0.59$, $w_i = 0.23$; Table 17.2) also indicated that harvest varied among months and years ranging from 0.02 (95% CI: 0.00 to 0.06) to 0.21 (95% CI: 0.06 to 0.37) but that survival in Brushy

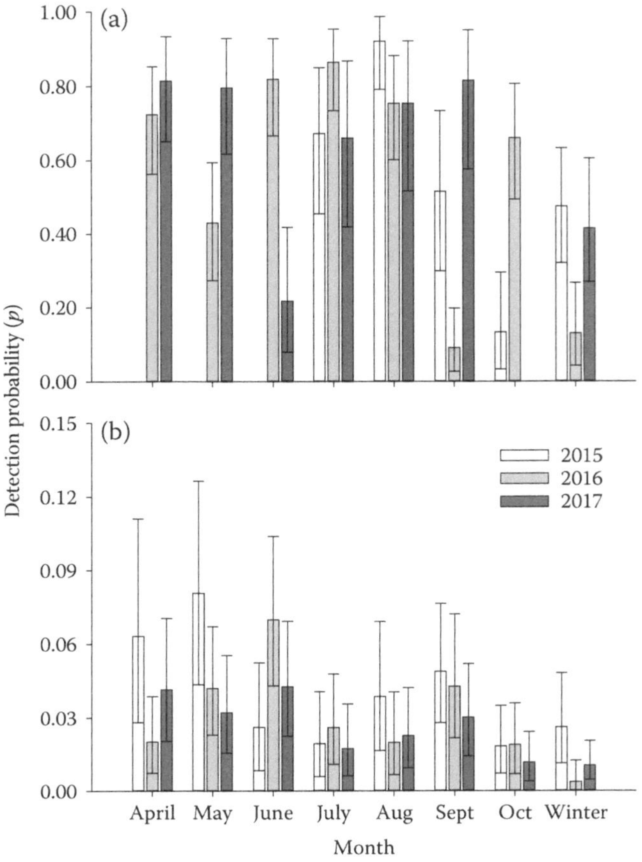

FIGURE 17.2 Estimated monthly and yearly detection probabilities of telemetry tagged (a) and jaw tagged (b) largemouth bass in Brushy Creek, Iowa, United States of America from 13 April 2015 through 31 March 2018. Error bars represent the 95% credibility intervals of the Markov chain Monte Carlo estimates.

Creek was similar among years. Harvest mortality comprised, on average, 0.15 (ranged from 0.07 in September 2017 to 0.25 in April 2016) of total monthly mortality (Fig. 17.3).

DISCUSSION

Results of our case study are indicative of shifting angler practices that are being observed in some locations and for some species where angler harvest no longer represents a major source of mortality. Despite Brushy Creek representing a very popular fishery (>107 angler hours/ha during April–September 2001; Miller and Herrig 2001), bass harvest in Brushy Creek Lake was low, accounting on average for <15% of total monthly mortality. Similarly, Miller and Herrig (2001) estimated that only 85 largemouth bass were harvested from Brushy Creek during 2001 (0.3 fish/ha) following >31,000 angler hours. Average bass harvest mortality across the United States of America has declined by about 50% since 1990 (Allen et al. 2008), resulting in bass harvest mortality as low as 0.07–0.11 in some states (Edwards et al. 2004, Driscoll et al. 2007, Kerns et al. 2016). However, bass harvest mortality is still highly variable, with recent exploitation estimates in other systems as high as 0.50–0.60 (Brown et al. 2015).

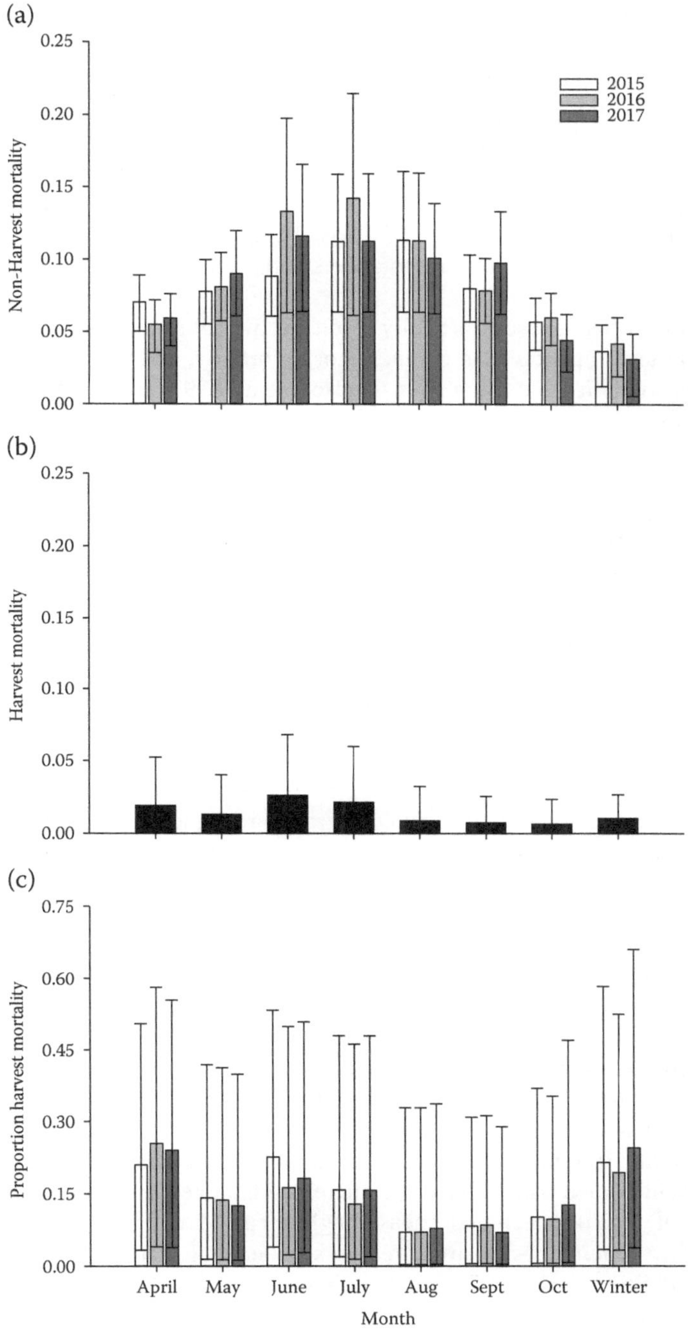

FIGURE 17.3 Monthly non-harvest mortality (a), harvest mortality (b), and proportion of harvest mortality (c) for largemouth bass in Brushy Creek Lake, Iowa, United States of America during 2015–2017. Error bars represent the 95% credibility intervals of the Markov chain Monte Carlo estimates. Note differences in y-axis scale among panels.

We determined that largemouth bass harvest was seasonal, increasing early in the summer and decreasing in late summer and winter. Seasonal fluctuations in harvest can be a result of changing angler and bass behaviors (Gabelhouse and Willis 1986, Fraser et al. 1993, Hanson et al. 2008). For example, warmer water temperatures may increase metabolism and foraging of bass, which can increase natural mortality (Waters et al. 2005, Holt and Jørgensen 2015, Weber et al. 2015). Warmer water temperatures may also make bass more accessible to anglers (Brett and Glass 1973, Fraser et al. 1993) and higher catch rates and angler participation during this time (Miller and Herrig 2001) may increase the probability of a bass being harvested. Temporal changes in non-harvest and harvest mortality are important to understand, as the timing of peak mortality can lead to divergent population effects.

Additive mortality, where harvest mortality results in overall reductions in survival, may result in larger population-level impacts than compensatory mortality, and is more common when both natural and harvest mortality occur together (Hudson et al. 1997, Boyce et al. 1999, Ratikainen et al. 2008). Alternatively, compensatory mortality, where increases in survival and reproduction in the population partially or fully offsets harvest mortality (Anderson and Burnham 1976, Allen et al. 1998), is more likely to occur if harvest precedes high natural mortality (Hudson et al. 1997, Boyce et al. 1999, Ratikainen et al. 2008). Compensatory mortality is also more common at low levels of harvest (Nichols et al. 1984, Weber et al. 2016). We observed highest non-harvest and harvest mortality of bass in Brushy Creek Lake occurred concurrently, suggesting additive mortality, but overall harvest mortality was low, suggesting compensatory mortality. We hypothesize that high levels of bass harvest mortality compared to low levels of non-harvest mortality were compensatory in Brushy Creek, but additional work is needed to elucidate these effects.

Instead of direct harvest for consumption, other sources of fishing mortality are likely now more important for catch-and-release fisheries. Angler activity plays an important role in variation in harvest among systems (Isermann et al. 2013). As seen in our case study, a considerable portion of angler effort was the result of tournament-angling events (~ 3 events/ week during the open-water season). Tournament anglers represent a rapidly growing segment of bass fisheries resulting in changing goals and attitudes toward fishing. Thus, shaping the perception of bass angling in some areas from one of traditionally consumptive-based goals (Holbrook 1975) to that of a specialized and even professional sport (Bernthal et al. 2015). Though only 15 bass were reported as harvested during the study period, 125 bass died as a result of initial tournament mortality (not including delayed mortality; Sylvia and Weber 2019), indicating other sources of angling mortality may be more important to consider when designing management objectives. Lethal effects of catch-and-release angling on bass have been well documented (Kwak and Henry 1995, Wilde 1998, Cooke et al. 2002) and represent patterns that are not unique to bass. Implications of catch-and-release mortality have been studied in marine and freshwater species, including muskellunge (*Esox masquinongy*; Landsman et al. 2011), bonefish (*Albula* sp.; Danylchuk et al. 2007), red snapper (*Lutjanus campechanus*; Rummer 2007), common snook (*Centropomus undecimalis*; Taylor et al. 2001), and even small-bodied fish targeted for microfishing, a newly popular form of recreational angling (Cooke et al. 2020). Catch-and-release angling can lead to lead to population-level impacts, such as altering population size structure (Hessenauer et al. 2018) and reducing reproduction success of fishes (Philipp et al. 1997), but is still widely accepted as a successful measure in sustaining fish populations (Siepker et al. 2007).

Though catch-and-release angling is highly accepted by many specialized North American anglers, concerns over ethical consideration of catch-and-release angling (Policansky 2002) and changes in composition of the angler population (i.e., increased immigration of individuals whose cultures traditionally use fish as a consumptive resource) may result in dramatic changes in angling practices in the future (Arlinghaus et al. 2007). Angler attitudes and behaviors toward harvesting vary across species (Chizinski et al. 2014*b*, Kaemingk et al. 2020), cultures (Lyman 2002, Wolfe 2006, Arlinghaus et al. 2007), demographics (Arlinghaus et al. 2007), and regions (Myers et al. 2008, Isermann et al. 2013, Miranda et al. 2017). Therefore, managers must be ready to address and incorporate angler practices, as well as identify factors associated with variation in exploitation across systems, into current management regulations.

Though reducing harvest can still be successful in managing traditionally highly exploited species such as muskellunge (Doss et al. 2019, Sass and Shaw 2020) and walleye (*Sander vitreus*; Stone and Lott 2002, Koupal et al. 2015, Sass and Shaw 2020), long-term reductions in harvest in some species may decrease the success of management goals (Bonds et al. 2008, Miranda et al. 2017, Sass and Shaw 2020). A growing body of research indicates that reductions in harvest by anglers have resulted in increases in relative abundance of fishes in many instances (Hansen et al. 2015b, Rypel et al. 2016, Sass and Shaw 2020). Presumably a desired thing, increased fish abundance as a result of catch-and-release practices can negatively impact fisheries management through negative intraspecific effects on growth and population size structure through density dependence (Hansen et al. 2015b, Sass et al. 2018, Sass and Shaw 2020), failure of length limits to restructure fisheries (Parks and Seidensticker 1998, Nobel and Jones 1999, Miranda et al. 2017), and the inability of harvest to control fish populations (Zipkin et al. 2009, Wenger et al. 2011, Weber et al. 2016, Sullivan et al. 2020).

Size-based harvest regulations are a common tool traditionally used in fisheries to direct angler harvest to certain subsets of a population, but may not always be effective (Noble and Jones 1999, Radomski 2003, Hessenauer et al. 2018). Selection of the appropriate regulation traditionally depended on fishery population dynamics (i.e., recruitment, growth, and mortality) and desired management outcomes (Isermann and Paukert 2010). Minimum-length-limit regulations are effective at protecting smaller individuals when abundance of these individuals is low, harvest of these fish is high, and growth rates are fast, allowing smaller fish to grow to larger sizes (Anderson 1980, Wilde 1997, Noble and Jones 1999, Isermann and Paukert 2010). For instance, minimum length limits were historically used to improve abundance and size structure of overharvested bass populations (Wilde 1997, Paragamian 1982, Eder 1984, Novinger 1984, Hoff 1995). Alternatively, slot limits are often used to protect intermediate and large fishes while encouraging the harvest of abundant, smaller individuals with the intent to improve growth rates and size structure by mitigating density-dependent processes (Anderson 1980, Isermann and Paukert 2010). When bass harvest still represents an important source of mortality, maximum size limits (that function similarly to slot limits in many instances) may result in improved bass size structure (Carlson and Isermann 2010, Dotson et al. 2013, Bonvechio et al. 2014) but are an unsuccessful tool for managing populations where voluntary catch and release is prevalent and anglers are unwilling to harvest smaller individuals (Nobel and Jones 1999, Parks and Seidensticker 1998, Bonds et al. 2008). Our results indicate that a 381-mm (15-in) length limit on harvested bass is likely ineffective for reducing harvest, as anglers self-regulate through catch-and-release practices. However, length regulations may limit the number and sizes of fish weighed in at competitive angling events, thereby reducing catch-and-release mortality of smaller fish. Thus, alternative and unique out-of-the-box approaches to managing catch-and-release dominated fisheries may be needed and more successful than traditional size limits for achieving management objectives.

In lieu of traditional size-based regulations, fisheries managers have responded to increases in catch-and-release fishing practices by altering harvest regulations (Wilde 1997, Sullivan et al. 2020), mechanical removals (Larson et al. 1986, Weidel et al. 2007), stocking additional prey (Wright and Kraft 2012, Maceina and Sammons 2015), mandatory harvest programs (Paul 2000, Paul et al. 2003, Johnson et al. 2009), introductory stockings of only female bass (Bonvechio and Rydell 2016, Maceina et al. 2016), and the use of rotenone (Coleman 2019) to control bass recruitment and subsequent density-dependent growth and size structure. However, these unconventional methods have been met with variable success. For example, many black bass are resilient to mechanical removal and require long-term removal practices to maintain low-level abundance in some waters (Weidel et al. 2007). Though increased harvest of smaller fish has been suggested to increase size structure of populations (Wilde 1997), reaching suitable harvest rates to improve the size structure of the fishery can be difficult to achieve (Langeland and Jonsson 1990,

Zipkin et al. 2009, Syslo et al. 2011, Sullivan et al. 2020). Even the removal of bag-limit regulations in brook trout (*Salvelinus fontinalis*) resulted in populations that could sustain harvest indefinitely (Paul 2000). Similarly, requirements to harvest fishes can be unsuccessful if catchability is lower and growth is faster than other species (Paul et al. 2003) or in some regions where sociological stigmas may inhibit necessary harvest to achieve management objectives (Rieman et al. 2006, Fausch 2008, Wenger et al. 2011, Gukian et al. 2018). Incentivizing harvest of fish, as is seen in some wildlife management programs (Witmer and DeCalesta 1991, Vercauteren et al. 2011, Pasko and Goldberg 2014), may be necessary to increase harvest to targeted levels. These approaches are generally limited in freshwater fisheries (but see northern pikeminow *Ptychocheilus oregonensis* in the Pacific Northwest; Beamesderfer 2000), yet have resulted in success in some marine species such as lionfish (*Pterois volitans*; Chapman et al. 2016, Wu 2017). Even so, the philosophy of catch and release has become so ingrained within the culture and mindset of many anglers that they may not view more liberalized regulations favorably, resulting in unintended consequences and lack of public support for management agencies (Bonds et al. 2008, Carey et al. 2011, Bass Anglers Sportsman Society 2016, 2018). Outreach targeting different angler groups regarding the importance of harvest for successful fisheries management will be vital to successful implementation of new and creative mechanisms (see Wilde et al. 2003) to manage fishes targeted for catch-and-release angling.

Changing patterns of harvest are not unique to fisheries. Urbanization, increased diversity in constituent goals, increased hunting complexity, and an increasing disconnect with nature have all led to reductions in harvest of wildlife populations (Dunfee et al. 2019). For example, expanding urban areas often overlap with deer habitat, reducing localized harvest in some instances (Hansen and Beringer 1997, Kilpatrick and LaBonte 2003). Complexities of urban deer hunting may also deter hunters, potentially resulting in less hunting pressure and lower harvest rates (Stradtmann and McAninch 1996, Kilpatrick and Walter 1999, Kilpatrick et al. 2002). In other instances, hunter selectivity (i.e., trophy deer hunting) may result in reductions in the number of deer being harvested (Roseberry and Klimstra 1974, Giles and Findlay 2004). Other examples of challenging harvest management scenarios include Canada goose (*Branta canadensis maxima*), greater snow goose (*Chen caerulescens atlantica*), and lesser snow goose (*C. c. caerulescen*) where abundant populations have resulted in habitat destruction or human conflict and liberalized regulations have failed to decrease populations (Ankney 1996, Koons et al. 2019, Paukert et al. 2021 [Chapter 18]). A lack of removal of the intended portion of managed populations may render management objectives in wildlife populations ineffective and lead to unintended consequences (Portatz et al. 2019), including increased human-wildlife conflict, depredation, and disease transmission (Messmer 2000, Portatz et al. 2019). Similar to fisheries, the development of non-traditional management techniques that no longer rely on hunter harvest for population control (e.g., sharpshooters, Doerr et al. 2001; commercial hunting, VerCauteren et al. 2011; aerial gunning, Davis et al. 2018) may also benefit wildlife populations in many instances.

The history of harvest in some fish and wildlife populations has seen an evolution from "tragedy of the commons" to "too much of a good thing" (Ankney 1996, Wenger et al. 2011, Koons et al. 2019, Sullivan et al. 2020). As such, management of fish and wildlife populations must shift in accordance with changing population demographics and stakeholder preferences. Natural resource managers should continue to focus on understanding primary mortality sources and targeting those sources to accomplish population objectives. For example, as documented here with largemouth bass, natural mortality is an important source of total mortality in many freshwater fisheries populations that is underevaluated and not well understood. Further, adaptive management as a tool to manage fish and wildlife populations, through traditionally integrating environmental conditions and management decisions (Williams 2011), may be valuable through the incorporation of social factors such as stakeholder practices and motivations (Fuller et al. 2021 [Chapter 8], Robinson et al. 2021 [Chapter 9], Runge 2021 [Chapter 7]). Integration of process uncertainty in angler and hunter habits in tandem with flexibility in management regimes will be necessary for future successful population management.

ACKNOWLEDGMENTS

The authors thank the technicians that assisted with data collection. Funding for this project was provided by the Iowa Department of Natural Resources through contract CRDFBGSCHO-0004, an Iowa State University Presidential Wildlife grant, and the Department of Natural Resource Ecology and Management at Iowa State University.

18 Harvest as a Tool to Manage Populations of Undesirable or Overabundant Fish and Wildlife

Craig P. Paukert, Elisabeth B. Webb, Drew N. Fowler, and Corbin D. Hilling

CONTENTS

HARVEST AS A MANAGEMENT TOOL

Harvest is a common management tool used for centuries to limit populations of game species (Caughley 1977, Redmond 1986). Managing populations using harvest regulations allow certain sizes, numbers, sex, and species to be harvested, and often include open or closed seasons. Regulated hunting opportunity and harvest are cornerstones of the North American Model of Wildlife Conservation, which developed gradually following unregulated harvest of wildlife populations that often were at risk of overharvest or extinction (Geist et al. 2001). Since then, many populations have recovered and expanded to the point where harvest regulations are now often used to limit or even reduce populations of some species. In general, harvest regulations have been well established as an effective way to control animal populations in many aquatic and terrestrial systems and are broadly accepted among the hunting and fishing public. For example, harvest regulations have been established or adapted to reduce or control populations of feral hogs (*Sus scrofa;* Hanson et al. 2009), white-tailed deer (*Odocoileus virginianus;* Simard et al. 2013) and cougar (*Puma concolor;* Cooley et al 2009), overabundant small black bass (*Micropertus spp.;* Isermann and Paukert 2010), northern pike (*Esox lucius;* Pierce 2010) or non-native species (Arlinghaus et al. 2016b).

Harvest has also been employed as a tool for controlling populations of invasive species. However, in many cases invasive species are so overabundant that a substantial commitment to harvest is necessary, which may exceed recreational harvest capacity and require commercial harvest or an active lethal control program by management agencies. Often removal of invasive species is challenging because the ultimate goal may be to eliminate the entire population, which may require impractical efforts. For example, controlling Asian carp in the Illinois River may require harvest rates of at least 70% (Tsehaye et al. 2013), whereas in the Great Smoky Mountains, an annual harvest rate of 40% would be necessary to decrease feral hog populations (Salinas et al. 2015). Many invasive species are known to negatively impact native species and ecosystems; thus eradication is an ideal outcome. However, there may be opportunity to use harvest to control populations of native (or non-native) species that have some value yet are still overabundant. In this chapter, we explore the process of using harvest to control overabundant populations that have some recreational value, provide two examples to control overabundant populations using harvest, and describe the challenges and effectiveness associated with these efforts and some of the unintended effects of using harvest to control populations.

Definition of Overabundant or Undesirable Populations

The term overabundance is somewhat subjective and context dependent. For the purposes of this chapter, we limit the focus to fish or wildlife species that have some recreational value, yet whose population size exceeds an acceptable level. Therefore, we generally exclude from the discussion species or populations that have strong universal support for removal (e.g., Asian carp in the Mississippi River Basin; brown tree snakes [*Boiga irregularis*] in Guam [Engeman et al. 2018]). We also focus on species that can be harvested recreationally and not primarily commercially harvested (although one example below does incorporate some elements of commercial harvest).

Factors that Influence Potential Harvest to Control Populations

Regardless of whether the population is a game species, invasive or introduced, the decision framework to use harvest to control the population is similar. Here, we outline a process that is based on structured decision making (Runge 2021 [Chapter 7]), and we emphasize specific steps of the process that are important to this unique situation. Questions to consider to determine if harvest management is an appropriate tool for controlling undesirable or overabundant populations (Fig. 18.1) are:

1. Is there sufficient evidence that population reduction is necessary?
2. Are there opportunities for hunters and anglers and are they willing to participate?
3. Are there sufficient (and accurate!) data to establish harvest goals to meet population objectives?
4. Is there a scientifically sound monitoring plan to evaluate the effectiveness of harvest in controlling the population?
5. What are potential unintended consequences of harvest?

Below we describe each of the criteria in more detail and explore two case studies to illustrate that these criteria are important and not always straightforward.

1. *Is there sufficient evidence that population reduction is necessary?*
   The first step is to determine if population abundance is an issue. Overabundance may take several forms from the mere presence of a species to actual abundance. For invasive species where a goal may be complete eradication, presence or occupancy may be more important than actual abundance. However, for species in which there is still an interest in maintaining a sustainable population, abundance estimates are useful to determine targets for any harvest management action. Ideally this would be a population estimate with high confidence in the number (i.e., narrow confidence interval). However, estimating actual abundance can often be challenging so estimates of relative abundance may be useful if they reflect actual abundance.

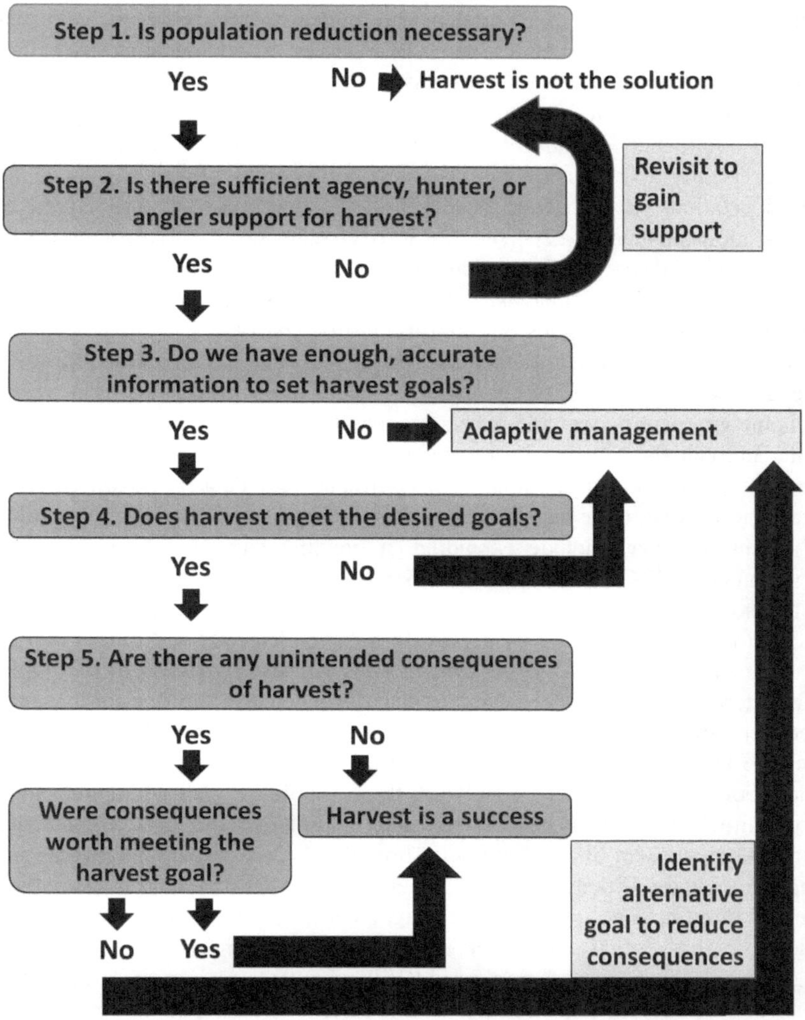

FIGURE 18.1 Components of a decision system to determine potential for harvest to be used to control undesirable or overabundant populations.

2. *Is there opportunity for hunters and anglers and are they willing to participate?* The second key step is to determine if there are sufficient hunters or anglers with the interest and opportunity to reduce the population through harvest. For commercial harvest, answering this question may involve developing a market for the species or population in conjunction with sufficient financial incentives to reduce the populations. However, recreational harvest may have less incentive or opportunity to participate because the purpose is typically for sport or food and not financial motivation. In addition, there may be a stigma associated with certain species (e.g., Asian carp; Morgan and Yun 2018) or the harvest of particular species (e.g., largemouth bass *Micropterus salmoides*; Miranda et al. 2017), thus creating resistance to harvest the population in sufficient numbers. Also, the commercialization of wildlife is contradictory to the North American Model of Wildlife Conservation, so developing a market is an outside-the-box option to be considered only after careful consideration of the implications of commercialization. Therefore, ensuring there is sufficient stakeholder willingness to harvest is a critical component of using recreational harvest as a management tool.

Understanding the motivations of hunters and anglers to participate in harvest is important when evaluating the potential effectiveness of a harvest management strategy (Gruntorad and Chizinski 2021 [Chapter 4]). The willingness to harvest may be linked with stakeholder trust in the conservation agency to make the correct decisions (Schroeder et al. 2017). Hunter satisfaction also influences retention and recruitment of hunters (Schummer et al. 2020). Though challenging to fully understand hunter satisfaction, it is often influenced by meeting a hunter's harvest expectations (Bradshaw et al. 2019). A key component to any harvest regulation is building relationships with the stakeholders through education and outreach so they understand the reasoning behind the decision to harvest (Sullivan et al. 2020). Therefore, managers may need to consider how particular harvest strategies or goals might be received by hunters if regulations increase expectations but are difficult to achieve.

Harvest strategies can often target certain size, age, or sex of an animal; thus, not only does there need to be a willingness to harvest, there needs to be willingness to harvest the targeted demographic groups. For example, when largemouth bass overpopulate in small lakes, harvest of smaller individuals is encouraged to reduce interspecific competition, but there is often an unwillingness to harvest these small fish (Isermann and Paukert 2010). In addition to the willingness to harvest, there also needs to be the opportunity to harvest. Opportunity can take several forms and may include extending or creating harvest seasons (e.g., Light Goose Conservation Order [LGCO]), or possible liberalized regulations that allow hunters and anglers to use technology or gears not typically allowed (e.g., elimination of shotgun shell limit in guns), or allowing opportunities of harvest in areas typically closed to hunting and fishing.

3. *Do we have sufficient and accurate data to set a harvest goal to meet the objective?* The third step of assessing the utility of harvest in controlling populations is to evaluate whether there are sufficient data to establish a harvest goal. As with any study, establishing an objective is critical to evaluating and selecting the best methods to enact management strategies. For an overabundant population there should be a clear targeted number or biomass of animals that needs to be harvested to achieve the desired management goal. Although this may seem like a simple question, the data to collect this information are often challenging to obtain. If there are not sufficient data to identify a harvest goal, there should at least be a commitment to an adaptive management framework where different management strategies (e.g., harvest) are employed and, if they do not meet the desired objective, other strategies are implemented and monitored (Williams 2011, Arlinghaus et al. 2016*b*). In some cases, accurate and precise estimates of the current animal population are not available, whereas in other cases lack of information on demographic parameters, such as age-specific mortality or reproductive rates can hinder efficacy of using harvest to control populations. Gathering or refining information on these parameters may be a necessary first step in determining whether harvest can be used to control populations and under which conditions it will be most effective. For instance, although feral hog survival was lower in areas of high-harvest intensity in Georgia, overall population growth rates did not differ between populations that experienced moderate- and high-harvest intensity, with increased immigration and reproduction likely compensating for higher mortality in the heavily harvested population (Hanson et al. 2009).
4. *Do we have a scientifically sound monitoring plan to evaluate the effectiveness of harvest?* The fourth step will provide a scientifically defensible plan to evaluate harvest. Monitoring is essential to assess the effectiveness of a harvest management action (Isermann and Paukert 2010, Cummings and Bernier 2021 [Chapter 10]), which often includes monitoring the population before and after harvest regulations are enacted. Agencies can use surveys of fish and wildlife to estimate population abundance before and after harvest regulations. However, this could also include harvest estimates by anglers or hunters to determine if changes in population abundance are related to increased harvest. Although hunter and angler surveys may be costly (Isermann and Paukert 2010), they are critical in evaluating the effects of

harvest. Such information can be used in an adaptive management process to evaluate the management action and adjust or modify strategies if harvest did not meet the desired goal (Williams 2011, Arlinghaus et al. 2016*b*). An adaptive harvest management process has been widely accepted in setting waterfowl harvest regulations (e.g., Williams and Johnson 1995), and has been used as a strategy to evaluate programs focused on reducing overabundant species, such as feral hogs (Engeman et al. 2007) and whitetail deer (deCalesta 2017). Using an adaptive management approach, or a structured process of simultaneously managing and learning about a natural resource, would allow agencies to incorporate uncertainty related to environmental variation, partial observability, partial controllability (such as harvest participation and harvest rates), and structural uncertainty (such as how a population will respond to increased harvest pressure) into a decision-making framework (Williams 2011, Runge 2021 [Chapter 7]). Specifically, once a harvest strategy has been developed and implemented, monitoring can help determine whether the population responds in the predicted way and, if not, then alternative management actions can be implemented.

5. *Are there unintended consequences of harvest?*
   The fifth step is consideration of unintended consequences of harvest. Unintended consequences of harvest can occur within the overabundant population, non-target species and ecosystems, the community of hunters and anglers, or the general public. Inability to correctly predict or control how harvest can influence a population can exacerbate the original issue or has potential to make the management action ineffective and could reduce public confidence in the ability of natural resource agencies to establish harvest regulations. Increased hunter or angler activity can result in indirect effects on non-target species (Harms and Dinsmore 2021 [Chapter 15]), including greater disturbance or redistribution to avoid hunters or anglers, as well as direct effects including incidental take (Dinges et al. 2015). Additionally, less restrictive harvest regulations enacted to increase harvest could alter hunter or angler perception of the resource and what is considered ethical under a *fair chase* approach (Posewitz 1994) to hunting or fishing.

CASE HISTORIES

We provide two examples evaluating use of harvest as a management tool for overabundant populations. Mid-continent light geese are native to the Mississippi and Central Flyways and have historically been a part of recreational waterfowl harvest, but their populations are considered overabundant. Individual blue catfish (*Ictalurus furcatus*) are reaching large sizes in the Chesapeake Bay watershed and are of interest to some anglers in the region. Below we present case studies for these two species on how harvest has been used to manage overabundant populations.

Controlling Overabundant Light Geese: Is the Plan Working?

A unique example where harvest has been used as a management tool in an attempt to control population size is the case of lesser snow goose (*Anser caerulescens caerulescens*) and Ross's goose (*Anser rossii*), collectively referred to as *light geese*. Both species are colonial nesters distributed throughout the Canadian arctic and subarctic tundra as well as Alaska. Historically, the midcontinent population of light geese (those breeding in the central Canadian arctic and migrating through the Central and Mississippi flyways) overwintered in coastal marshes, primarily throughout Louisiana and Texas (Batt 1997). As herbivores, light geese adapted to forage on rhizomes and tubers of aquatic vegetation, such as bulrushes (*Bolboschoenus* sp.) and the seeds of cordgrass (*Spartina* sp.), that provide carbohydrate-rich resources as well as trace minerals important for maintaining body condition throughout the winter (Alisauskas et al. 1988). However, agricultural advances in the late nineteenth century and early twentieth century began to change coastal and adjoining prairie landscapes. Early records in Louisiana suggested that overwintering light geese were constrained to the first 12 km of coastal habitat adjacent to the Gulf of Mexico (McIlhenny 1932). Yet by the late 1970s, light geese had moved inland by more than 100 km in Texas, capitalizing on expanded rice production (Hobaugh 1984). By the 1980s, cohorts of light geese were overwintering in agriculturally dominant landscapes in Kansas, Missouri, and Nebraska (Alisauskas 1998). Concurrently, light geese began to heavily use portions of the Mississippi Alluvial Valley in Arkansas where rice production became the predominate agriculture crop (Alisauskas et al. 2011).

Importantly, light geese population size increased in conjunction with winter habitat expansion leading to increased survival. High winter mortality associated with lower nutrition available in coastal marshes may have historically limited population size (Alisauskas 1998). With increased energetic agricultural food resources on the landscape, light geese survival, recruitment, and population size were all positioned to increase (Abraham et al. 2005). As light goose populations grew, the greatest densities and perception of overabundance occurred on the nesting grounds, especially in subarctic colonies in the Cape Churchill, Manitoba region (Batt 1997). With an estimated 5% annual increase in the midwinter count, high densities of nesting geese resulted in overgrazing in subarctic coastal marshes leading to increased soil salinities that led to landscape-scale habitat alteration (Batt 1997). Higher soil salinities reshaped plant community structure and succession and prohibited re-establishment of historic plant communities (Batt 1997). Areas that experienced heavy grazing by light geese transitioned from native grasses and sedges to denuded mudflats with saline algae crusts, prompting concerns that similar habitat *degradation* would spread throughout the arctic to other light goose colonies and result in negative consequences to other arctic wildlife species such as nesting ducks, shorebirds, and passerines (Batt 1997). In 1997, the Arctic Goose Joint Venture published a special report entitled *Arctic Ecosystems in Peril* (Batt 1997) that outlined perceived ecosystem alterations resulting from an overabundance of mid-continent arctic and subarctic breeding light geese. Therefore, abundance of light geese was considered problematic.

By the mid-1990s, midwinter counts used to index changes in light goose abundance on the wintering grounds increased by 300% from 0.8 million in 1969 to 2.7 million in 1994. Wintering abundance trends were monitored through the midwinter index, an aerial transect survey across wintering states conducted by the Mississippi and Central Flyway Councils. At the time, these surveys were thought to be the best available estimate of population size when paired with aerial photographs of nesting grounds intended to correct for presumed underestimation in the midwinter surveys. By 1996, the entire midcontinent population of lesser snow geese was estimated to be between 4.5 and 6.0 million birds, prompting agency and hunter support for additional harvest management strategies. Consequently, wildlife professionals began discussing management options available to sustainably control mid-continent light goose populations. Specifically, calls for increased hunter harvest opportunities were proposed as a practical way to slow population growth rate and decrease abundance by reducing adult survival rates. Proposals suggested that "*waterfowl hunters, once convinced of the gravity of the problem of goose overpopulations, could, if given appropriate opportunities, significantly reduce adult survival rates of geese at little or no cost to governments*" (Ankney 1996).

By 1997 the Arctic Goose Joint Venture recommended that "*responsible public agencies in Canada and the U.S. should implement proactive population reduction measures to reduce mid-continent [light] goose populations to a level of about 50% of current numbers by the year 2005*" (Batt 1997) and suggested there was sufficient (and accurate) information to set a harvest goal. These recommendations stemmed from staged-based matrix models that estimated population growth rates (λ) and projected abundances derived from long-term demographic monitoring efforts. Elasticity analyses evaluated the impact of alternative management actions on their proportional ability to decrease the population growth rate. Among the potential demographic rates evaluated, reducing adult survival rates was determined to be most effective in decreasing population growth rate. Given that annual changes in the midwinter indexes suggested a $\lambda = 1.05$ (5% annual growth), hypothetical simulations estimated the increase in mortality rates that would be required to decrease λ to 0.85 (15% annual decrease) until the total population size decreased by 50% (approximately 1.5 million individuals). These simulations assumed that increases in overall mortality would come from increased hunter harvest and model results concluded harvest rates would need to increase by a factor of 2.87 and 3.11 for three to seven years, respectively, to reduce the population to 1.5 million individuals. However, serious concerns were expressed that the necessary increase in harvest rate could not be achieved under the existing framework of harvest regulations. Of particular concern was the inability to increase available harvest days as a result of the Migratory Bird Treaty Act. Many of the states that harvested light geese were already using the maximum number of days (107) to harvest light geese under the Migratory Bird Treaty Act (Vrtiska 2021 [Chapter 20]), which prohibited harvest of migratory birds between March 10 and September 1. However, managers believed additional harvest opportunity existed in April and May.

Following recommendations from *Arctic Ecosystems in Peril*, the U.S. Fish and Wildlife Service provided a new regulatory framework and initiated the LGCO in February 1999 as a specific management action to address the problem of overabundant light geese. The LGCO was defined as a management action and not a new hunting season; thus, harvest could extend into May, beyond the hunting season lengths established by the Migratory Bird Treaty. Additionally, the LGCO allowed for special measures to facilitate increased harvest including electronic calls, unplugged shotguns, extended daily hunting hours, and removing daily bag limits. The conservation order was the first of its kind and a widespread communications campaign was initiated to rally hunters as agents of conservation to *save the tundra* by participating in the LGCO.

Despite support and participation by light goose hunters and an initial increase in total harvest during the first three years of the LGCO, light goose population size continued to increase. Importantly, long-term monitoring from band recoveries both prior to and after initiation of the LGCO has not shown a decrease in survival related to increases in total harvest (Alisauskas et al. 2011, Leafloor et al. 2012). In fact, more recent band-recovery analyses suggest that survival rates have continued to increase throughout the LGCO and substantially more adult and juvenile light geese have died annually of natural mortality than those by harvest mortality (Calvert et al. 2017). Therefore, using harvest to control light geese did not meet the management goals.

So why has the LGCO been an ineffective management action? Though *Arctic Ecosystems in Peril* was based on the best available data at the time and vetted by numerous arctic habitat and goose researchers, "*many of the harvest goals initially recommended for achieving a desired reduction in population size were based on incomplete or outdated knowledge about estimates of vital rates, harvest, and population size*" (Alisauskas et al. 2011). Inaccurate estimates of population size were likely detrimental in deriving harvest quotas that were severely underestimated. Original population models relied on ocular estimates of large discrete flocks of geese from the midwinter aerial survey; a subsequent, more statistically robust approach to estimate population size using band recoveries and harvest totals (Alisauskas et al. 2009) resulted in population estimates that were at least 2.5 times as large as midwinter counts (Alisauskas et al. 2011). Thus, from the start of the LGCO, expectations of required harvest totals were substantially below actual biological thresholds necessary to reduce survival rates. Although harvest rates initially increased with the beginning of the LGCO, they remained insufficient to affect survival rates and in fact harvest rates subsequently declined. As population size continued to expand, annual hunter harvest capacity became satiated and harvest rates declined following an inability to keep pace with increasing population size. In addition, the harvest of light geese using decoys is biased toward those individuals with poorer body condition which may have further contributed to the current compensatory nature of harvest by removing individuals who likely would have died of natural mortality later in the season (Fowler et al. 2020).

An important consideration when using harvest as a management tool to control population size is the acknowledgement of partial controllability (Williams and Johnson 1995). In the case of the LGCO, the decision to focus management on the harvest of adult geese resulted from an elasticity analysis that identified adult survival as the most influential parameter on population change. Though biologically and conceptually informative, elasticity analyses do not account for the political, social, or economic feasibilities of a management action (Alisauskas et al. 2011). In the case of the LGCO, decision makers may have overestimated the capacity of hunter engagement to effectively reduce survival rates given concurrent long-term declines in waterfowl hunters (Vrtiska et al. 2013).

Though well intentioned, decisions to implement the LGCO have resulted in unintended consequences. The LGCO was implemented with the expectation that it would be a discrete management action over three to seven years, after which the LGCO would be removed. However, the LGCO has been in place now for 22 seasons because populations did not respond in the expected way, and now the LGCO unintentionally functions as a spring hunting season for light geese. Given current population modeling, light geese could sustain greater rates of harvest and thus sustainably provide new hunting opportunities when hunter recruitment is in decline. Yet, this was not the intended purpose of the LGCO and the Migratory Bird Treaty Act would need to be revised or amended to continue spring harvest of light geese if the LGCO management action were ended. Further unintended consequences exist when considering economic growth generated in response to the LGCO. Professional guide services have capitalized on the existence of the LGCO as a framework to provide premier spring hunting opportunities after the end of regular waterfowl season. Guide services are large economies and have benefited from the continued messaging from non-profit hunting organizations (despite conflicting evidence) that hunters can make a difference on arctic ecosystems if they participate in the LGCO. Though empirical data are lacking, anecdotal evidence suggests persistence in this messaging appears to have de-valued light geese in the eyes of many hunters as *sky carp* and *tundra rats*, potentially compromising the integrity of the *public trust doctrine* in North American wildlife (Organ et al. 2014).

More recently, the midcontinent population of light geese has been declining (Calvert et al. 2017, Ray Alisauskas, Environment and Climate Change Canada, written communication, 15 October 2020) as annual productivity has continued long-term declines (Alisauskas 2002). This decline in productivity and subsequent population size is likely related to increasing strength of density dependence on terminal boreal and tundra staging areas before nesting (Ross et al. 2017) and reductions in gosling growth and survival, at least in parts of the midcontinent breeding range. If light goose populations continue the recent decline and begin to self-regulate, managers will need to re-assess the purpose of the LGCO and continue monitoring vital rates in response to spring harvest pressure.

The LGCO shows that overabundance was a problem (Fig. 18.1, step 1), and hunters were willing and had the opportunity to harvest, and there was (presumably) sufficient and accurate data to set harvest goals (Fig. 18.1, steps 2 and 3), which initiated a regulatory framework (the LGCO) that allowed harvest of these birds in spring. However, one challenge has been to develop accurate estimates of population vital rates that were used in population estimation. With continued monitoring and re-analysis, the population was estimated to be over twice as large as originally calculated, and thus control efforts based on the original analysis could not decrease the population growth rates and the management goal was not achieved (Fig. 18.1, step 4). Although there was opportunity and willingness of hunters to harvest light geese, the capacity of the hunters to harvest sufficient numbers of light geese (given the new population size estimates) was not sufficient, although spring hunting provided an economic boost to local hunting guides (Fig. 18.1, step 5). This case history highlights the importance of accurate population data and continued monitoring to control overabundant populations.

Blue Catfish in the Chesapeake Bay Watershed: Cause for Concern or an Economic Boost?

Blue catfish are native to watersheds emptying to the Gulf of Mexico from the Mobile Basin to Guatemala (Boschung and Mayden 2004), but were intentionally introduced to the Virginia portion of the Chesapeake Bay watershed from 1974 to 1985 in the James, Mattaponi, and Rappahannock Rivers (Jenkins and Burkhead 1994). During this period, the Chesapeake Bay was degraded due to habitat alteration, pollution, overfishing, hurricanes, and disease outbreaks (Houde 2006), leading fishery managers to seek new fishing opportunities for the public. The blue catfish has expanded its range to include most major tributaries of Chesapeake Bay in multiple management jurisdictions, including Maryland, Virginia, and Washington D.C. and is encroaching on Delaware and Pennsylvania. Much of the Bay is habitable by blue catfish during large freshwater outflow events, providing dispersal corridors to new systems (Nepal and Fabrizio 2019). Consequently, a fisheries management solution in a single jurisdiction has become a large-scale invasive-species problem.

Blue catfish has generated considerable concern for native species of the Chesapeake Bay because it has become incredibly abundant in some localities (e.g., James River System: 708 fish/ha; Bunch et al. 2018). High blue catfish densities coupled with omnivorous, opportunistic feeding strategies (Schmitt et al. 2019) have contributed to concerns for native aquatic species of cultural, ecological, and economic importance, some of which are declining. Blue catfish predation on species of management interest (e.g., *Alosa* species, American eel *Anguilla rostrata*, and blue crab *Callinectes sapidus*) was detected when and where prey species are prevalent (Schmitt et al. 2019). There is also evidence that blue catfish has contributed to declines in abundance of the native white catfish (*Ameiurus catus*) due to competitive interactions (Schloesser et al. 2011). In addition, commercial fishers targeting other species spend considerable time removing blue catfish from fishing gear (Invasive Catfish Task Force 2014). Recreational anglers specializing on largemouth bass (*Micropterus salmoides*) have reported blue catfish interference, reducing bass catch rates and angler satisfaction. Consequently, resource managers have expressed an interest in reducing blue catfish abundances as a mechanism to lessen impacts of blue catfish and support rehabilitation of Chesapeake Bay (Invasive Catfish Task Group 2014), and therefore abundance or even presence of blue catfish is considered a problem. Although monitoring and evaluations of blue catfish harvest are ongoing, the following provides an example of an overabundant fishery that is valued by some stakeholders (e.g., trophy anglers), making elimination of the species not universally accepted.

To reduce blue catfish abundance, there is a growing interest in increasing harvest via fisheries in the Chesapeake Bay region. Concerns for the ecosystem health of Chesapeake Bay and the desire to efficiently harvest blue catfish prompted Virginia Marine Resources Commission to approve a commercial electrofishing license, where licensees are permitted to harvest blue catfish using electrofishing equipment typically confined to research and monitoring. Although there was support from commercial fishers, seafood processors, and management agencies for increased harvest, federal regulations requiring catfish landings be inspected by the U.S. Department of Agriculture have impeded the expansion of commercial markets. The U.S. Department of Agriculture inspections create a burden on wild-caught seafood processors as fish arrivals can be unpredictable, making coordination with inspectors challenging and can require the processor to cover inspector overtime wages. Consequently, some processors have placed limits on the blue catfish they will accept or decline to process them altogether. Blue catfish is a low value product with an average value of Virginia commercial landings of US$0.75/kg (US$0.34/lb) during 2013–2018 (National Oceanic and Atmospheric Administration 2020*a*), which is a fraction of the value of other important fisheries in Virginia (e.g., Eastern Oyster $20.35/kg, or $9.23/lb, and striped bass $8.47/kg, or $3.84/lb; National Oceanic and Atmospheric Administration 2020*a*). Consequently, the limited market is likely partially responsible for underutilizing the resource, due to processing limitations and higher values for other species.

One substantial challenge to managing blue catfish is the scientific uncertainty in population dynamics, with a lack of objectives impeding development of specific harvest goals. There is regulatory and political will to harvest blue catfish, but managers have not determined how to evaluate management success, which is further confounded when stakeholder groups have different goals (e.g., trophy fisheries, harvest fishery, eradication of invasive species). Blue catfish abundances have reached a level where population sizes may need to be reduced as effectively and efficiently as possible if managers want to decrease impacts to the Chesapeake Bay ecosystem. Reducing large fish abundance may reduce predation on native species. However, this may also be detrimental to recreational fisheries and could expose consumers to higher contaminant levels (Luellen et al. 2018), which is an unintended consequence of blue catfish harvest.

Monitoring programs are in place to evaluate blue catfish responses to management intervention. In Virginia tributaries of the Chesapeake Bay, electrofishing surveys provide biological data and relative abundance information for catfishes, whereas a fish trawl provides information on the abundance of juvenile fishes in lower, more saline parts of the rivers (Tuckey and Fabrizio 2019). However, the two surveys do not overlap spatially, producing uncertainty in spatial dynamics. In addition, given the extensive spatial areas of these river systems, annual broad-scale electrofishing efforts are taxing on agency staff. Harvest is also monitored, but currently there are no surveys in place to estimate the size or age composition of the harvest. Improving monitoring of blue catfish would likely be beneficial to the development of robust population modeling efforts but may be challenging due to personnel and funding limitations.

Current sport fishing regulations are not a barrier to increasing harvest of blue catfish in Chesapeake Bay. However, some anglers are interested in maintaining a trophy fishery, so eliminating blue catfish in Chesapeake Bay is not universally accepted. The commercial and recreational fisheries in Virginia adhere to a maximum size limit (only one fish > 813 mm per person per day) in river systems that support trophy-angling opportunities (James and York rivers). Further, commercial electrofishing is restricted to fish > 635 mm as a compromise with recreational fishers. Beyond Virginia waters, the species is managed with no size or daily limits by agencies in Chesapeake Bay, allowing unlimited harvest. Calls for increased harvest have been met with opposition from many recreational anglers who fear increased harvest will decrease trophy-fish abundance, as harvest has increased concurrent with declines in growth rates and proportions of large fish (Hilling et al. 2018, in press). Growth rates have declined over time (Hilling et al. in press) with supporting evidence that growth rates are correlated with fish densities (Nepal and Fabrizio 2020). Consequently, conflicting interests from stakeholders have made managing blue catfish more challenging than simply implementing control measures.

Although commercial harvest of blue catfish appears to be the most logical solution to reducing abundance, blue catfish harvest could create unintended consequences for stakeholders and managers. Blue catfish in these systems can carry high contaminant loads and, although many contaminants levels are below U.S. Environmental Protection Agency safety thresholds, there is uncertainty surrounding how a combination of contaminants can influence health risks (Luellen et al. 2018). Consequently, increased harvest may expose consumers to greater health risks from environmental contaminants. In addition, the effects of low-frequency electrofishing on non-target fishes are currently unknown. Despite many scientists and commercial fishers reporting no by-catch beyond catfishes via low-frequency electrofishing, there have been claims of effects on non-target species in fishery-independent surveys (M.C. Groves, Maryland DNR, oral communication, Chesapeake Bay Program Invasive Catfish Workshop, VCU Rice Rivers Center, Charles City, VA, 29 January 2020). Increased fishing effort using traditional gears may increase the by-catch of depleted and declining species with alewife (*Alosa pseudoharengus*), American shad (*A. sapidissima*), blueback herring (*A. aestivalis*), Atlantic sturgeon (*Acipenser oxyrinchus*), and shortnose sturgeon (*Acipenser brevirostrum*) currently managed using harvest moratoria. Increased commercial fishing effort may result in increased by-catch and mortality events for these non-target anadromous species as they move upstream to spawn.

Finally, the most concerning potential consequence of increasing blue catfish harvest might be that increased harvest and demand for blue catfish could increase desires to sustainably manage the fishery. The wording of the regulation permitting commercial electrofishing states its purpose is to "*sustainably manage populations of nonnative catfish species*." Establishing value for non-native species may increase their spread to new systems via unsanctioned stockings. Consequently, increasing harvest could result in greater demand for the resource and an associated economic industry that facilitates blue catfish range expansion, transitioning management objectives from reducing abundances to maximizing fishery yields. Therefore, the future status of blue catfish in Chesapeake Bay remains to be seen, and conflicting objectives from multiple stakeholders is a challenge for the management of the species. An ongoing stock assessment will hopefully provide insights on population dynamics and possible responses to management. Objectives for fisheries management should be clearly defined, along with evaluation measures, before harvest is used as a tool to reduce abundances of unwanted species to avoid creating new problems.

The blue catfish example shows that identifying the problem was based on concerns from multiple fronts including impacts of native species and impacts to commercial and recreational fisheries (Fig. 18.1, step 1). Commercial fishers were supportive of increased harvest of blue catfish (Fig. 18.1, step 2), so increased harvest was considered a strategy by management agencies to control blue catfish. However, there was substantial uncertainty surrounding population vital rates and responses to harvest, which made setting harvest goals difficult (Fig. 18.1, step 3). Given the uncertainty of vital rates, it is difficult to determine if management actions are reaching the desired outcomes (Fig. 18.1, step 4). Nevertheless, population monitoring is continuing. Some stakeholder groups are supportive of maintaining blue catfish as a sport fishery, which has provided the unintended consequence of boosting the local economy (Fig. 18.1, step 5). Therefore, the primary challenge with Chesapeake Bay blue catfish is the lack of sufficient data to determine harvest goals, and stakeholders that have different goals for the fishery.

CONCLUSION

We provide a decision framework for applying harvest management strategies to control overabundant populations of fish and wildlife. Harvest can be a useful tool to control overabundant populations. In the Columbia and Snake Rivers in the Pacific Northwest, northern pikeminnow (*Ptychocheilus oregonensis*) is managed as a nuisance species due to concerns of predation on salmonids. Fishery managers have instituted several removal programs, including a sport-reward angling fishery to encourage harvest of large northern pikeminnow (Winther et al. 2020). Anglers harvested an average of 177,033 northern pikeminnow annually from 1991 to 2019, which has reduced northern pikeminnow predation on juvenile salmonids by 30% compared to pre-initiation of the Northern Pikeminnow Management Program (Winther et al. 2020). The greater snow goose (*Anser carulescens atlanticus*) shares a similar story of increasing abundance in the late twentieth century with the lesser snow goose (Gauthier et al. 2005). However, the population size of greater snow goose was much smaller (750,000–1,000,000) compared to the midcontinent population of lesser snow goose during the mid-1990s (Béchet et al. 2004*b*) and the spring harvest on greater snow goose had a greater impact on reducing population growth rate that led to at least a stabilization or decline in population size. Prior to spring harvest in 1999, adult harvest rate averaged 6%, but increased to 13% in the first four years of management actions, and adult survival declined from 83% to 73% (Lefebvre et al. 2017). This increased harvest activity also reduced breeding propensity and clutch size. The relative success of harvest management in reducing greater snow goose populations compared to that of lesser snow geese can largely be attributed to differences in the total harvest quota required to affect population growth. Harvest management actions had greater potential, because of smaller population size in greater snow goose, to effectively increase harvest rates despite finite numbers of hunters available to engage in harvest activity.

Although the application of harvest strategies for these populations is similar to conventional approaches to manage harvest for game or sport-fish populations, there are several important distinctions. Unlike popular fish and wildlife game species, there can be a social stigma associated with the overabundant species in this chapter. Outreach and education to inform hunters and anglers of the benefit of harvesting these species may be needed. In particular, the public may need assurance of the palatability of these species (Morgan and Yun 2018), which may have the stigma of *sky carp* or other unfavorable terms.

For our light goose and blue catfish examples, we can summarize how each example fits our five criteria above (Table 18.1). Both examples included a regulatory challenge to allow harvest: the Migratory Bird Treaty Act initially limited harvest efforts of light geese, and permitting of more efficient fishing gears encouraged blue catfish harvest. Both also had the unintended consequences of an increased economic boost from guide services. One notable difference in our two examples is the LGCO established harvest goals and calculated population estimates, whereas blue catfish management in Chesapeake Bay did not. Although initial light goose population estimates were biased low, agencies revisited those estimates and learned from their previous actions. Therefore, agencies and organizations may need to embrace that learning and uncertainty are part of the process of using harvest to control overabundant populations (i.e., adaptive resource management; Williams et al. 2011).

Explicit integration of adaptive resource management into the decision-making process involved in using harvest as a management tool to reduce overabundant populations may provide a more robust framework to evaluate the effectiveness of this practice. Although both the light goose and the blue catfish examples discussed within this chapter incorporated some aspects of adaptive resource management, the next steps in the decision framework would likely be clearer and more justifiable if all components outlined in the structured decision making and adaptive resource management process had been deliberately integrated (Williams 2011, Runge 2021 [Chapter 7]). By acknowledging and incorporating uncertainty in how ecological systems function and respond

TABLE 18.1

Criteria used to determine if harvest can be used to control overabundant light geese in the United States of America and blue catfish in the Chesapeake Bay watershed

Issue and Step	Light Geese	Blue Catfish
Step 1. Can we estimate abundance?	Yes but difficult	In progress, but difficult
Step 2. Is there agency, hunter, or angler support and opportunity to harvest?	Yes	Mixed-some anglers want to protect large fish
Step 3. Do we have reliable estimates of harvest that are needed to control the population?	Some	No
Step 4. Is there a sound plan to evaluate the effectiveness of harvest to meet objectives?	Yes	To be determined
Step 5. Are there unintended consequences of harvest?	Non-target harvest Devaluation of the resource Disturbance to non-target wildlife Economic increase in local guides	Increased value, possible spread Human contaminant exposure Non-target mortality (by-catch) Economic increase in local guides

to management practices from the initial consideration of using harvest as a management tool to control overabundant populations, natural resource agencies can specifically articulate objectives and formulate management alternatives based on multiple, contrasting hypotheses. Specifically, steps 3–5 of the approach we outline for using harvest to control populations (Fig. 18.1) depend on the incorporation of adaptive management, thereby allowing for the iterative process of learning and adapting harvest strategies and approaches despite structural and environmental uncertainty.

ACKNOWLEDGMENTS

The authors thank M. Groves and R. Alisauskas for helpful comments that improved an earlier version of this manuscript. Support for C. Hilling was provided through a Virginia Sea Grant Graduate Research Fellowship. Support for D. Fowler was provided through Nebraska Game and Parks Commission. The Missouri Cooperative Fish and Wildlife Research Unit is jointly sponsored by the Missouri Department of Conservation, the University of Missouri, the U.S. Geological Survey, the U.S. Fish and Wildlife Service, and the Wildlife Management Institute.

19 The Efficacy of Antler Harvest Regulations in Meeting Management Objectives

Daniel L. Morina, Christopher P. Hansen, Lonnie P. Hansen, and Joshua J. Millspaugh

CONTENTS

INTRODUCTION

Harvest regulations are the primary lever used by state wildlife agencies to manage game populations. For ungulate populations, antler harvest regulations (AHRs) are commonly used to affect harvest of segments of the male population. When AHRs are coupled with a quota system (i.e., limiting the number of hunters), managers can control numbers of wildlife harvested along with the structure of the harvest. Most white-tailed deer (*Odocoileus virginianus*), mule deer (*Odocoileus hemionus*), elk (*Cervus canadensis*), and moose (*Alces alces*) AHRs have been point restrictions (i.e., branched antler or minimum number of points to be eligible for harvest) designed to protect young males (Bender and Miller 1999, Biederbeck et al. 2001, Bender et al. 2002, Demarais et al. 2005, Child et al. 2010, Kuzyk et al. 2011). However, there is wide variation in AHR application, including minimum antler spread or antler points, main beam length, and various permutations (Figs. 19.1–19.3). Each AHR is tailored to the specific population attributes (i.e., antler size), management objectives, and the ease with which regulations can be implemented and enforced.

Implementation of AHRs have had varied underlying motivations in North America. Early experimentation with AHR began for elk and mule deer in western states (Boyd and Lipscomb 1976, Bender and Miller 1999) and moose in Canada and Alaska (Schwartz et al. 1992, Child et al. 2010). Decreases in post-hunting season juvenile:adult female ratios in elk and mule deer in the later part of the twentieth century, often associated with population declines, generated concerns among agency biologists and hunters (Bender and Miller 1999, Bender et al. 2002). Harvest-driven, female-skewed, population sex ratios, with many more females than males, often coincided with this apparent poor productivity (Bender et al. 2002, Bishop et al. 2005). Some evidence suggested low male:female ratios during the breeding season could cause extended and late breeding seasons and a lower proportion of females being bred (Follis 1972, Gruver et al. 1984, Noyes et al. 1996). These potential outcomes could increase predation on juveniles and prevent adequate body development of late-born young prior to critical periods, thus potentially explaining the declining recruitment (Mysterud et al. 2002, Milner et al. 2007). This early hypothesis became dogma among hunters and agency staff, leading to the conclusion that a simple solution to the productivity problem was to adjust harvests for more balanced population sex ratios that were less female biased (White et al. 2001). AHRs on moose in Alaska and Canada had similar objectives in that unlimited male moose hunting resulted in female-skewed population sex ratios with associated biological concerns (Schwartz et al. 1992, Laurian et al. 2000, Child et al. 2010).

AHRs in white-tailed deer in the eastern United States of America have a somewhat more recent history and different objectives than AHR on ungulates in the west and Canada. As in elk, mule deer, and moose populations, conservative male-biased harvests in the mid-twentieth century designed to promote population recovery in eastern white-tailed deer populations had produced female-skewed population sex ratios and young-dominated male age structure (Demarais et al. 2000, Miller et al. 2003, Adams and Hamilton 2011, Demarais and Strickland 2011). In most productive white-tailed deer ranges, populations were growing rapidly and generating increasing concern about problems associated with high deer densities. As a result, controlling growing deer populations in the east was a greater concern for most agencies than any possible effects of female-skewed population sex ratios on productivity (Diefenbach et al. 2019, 2021 [Chapter 22]).

Quality deer management, a change of management philosophy in which male harvest is regulated and doe harvest increased, arose in response to high white-tailed deer densities and young age structure of males (Marchington and Miller 2007). Quality deer management originated in southeastern states late in the twentieth century and spread to eastern and midwestern states shortly thereafter (Adams and Hamilton 2011). The quality deer management concept was promoted by many agency biologists, especially in southeastern states, to improve population management while addressing hunter interests in seeing and harvesting male deer in older age classes. Quality deer management's efforts to change hunter philosophy have been important; however, voluntary implementation of the concept on a large geographic scale is not effective so

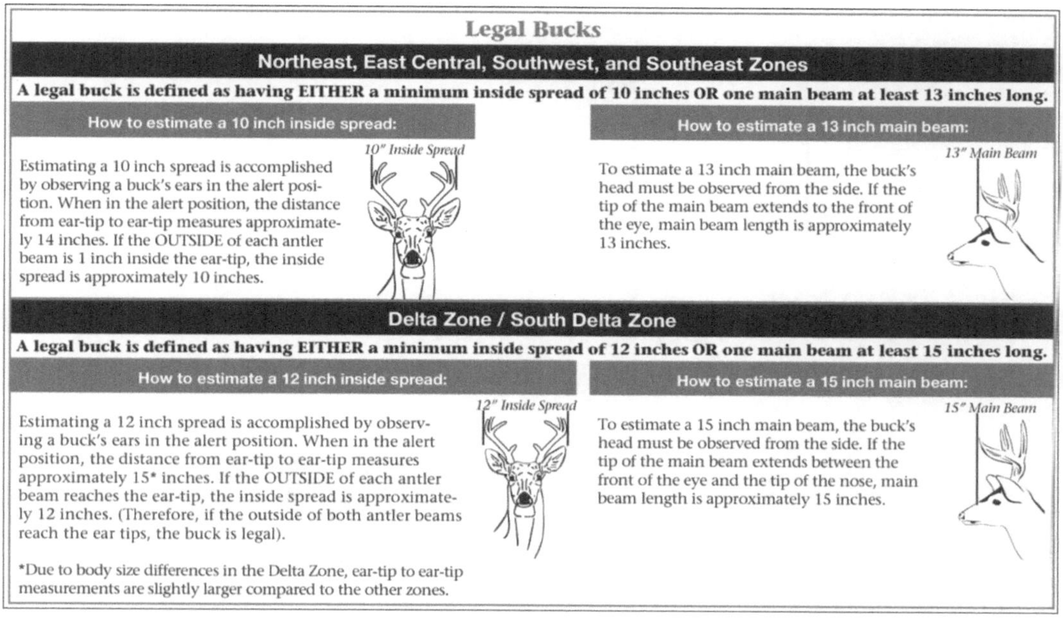

FIGURE 19.1 Example of main beam length and inside spread combination antler harvest regulations used by Mississippi. Image courtesy of the Mississippi Department of Fisheries, Wildlife, and Parks.

FIGURE 19.2 Example of an antler point restriction for elk. Males must have four points or more on one antler or a brow tine of at least 13 cm (5 in) long in several game management areas in Colorado. Image courtesy of Colorado Parks and Wildlife.

state-implemented AHR resulted. AHRs implemented throughout the eastern United States of America have been diverse, including number of points, antler spread, antler beam length, and combinations of these. The diversity of AHRs is primarily due to concerns about potential genetic high-grading effects of antler point restrictions (Strickland et al. 2001, Webb et al. 2012). Generally, multiple antler criteria were considered the best method of differentiating age classes (Marchington and Miller 2007). Compliance with these often complex and judgment-based regulations has been difficult for the average hunter and challenging for agency staff to enforce and is perhaps best applied voluntarily on private land and not for large-scale regulations. Regardless of possible biological concerns, antler point restrictions are simple for hunters to follow and have been the most common AHR recently used. Most AHRs now are socially motivated, primarily designed to produce more adult males for hunters to see and harvest (Adams and Hamilton 2011). However, more productive areas in many states have liberalized antlerless harvest regulations in concert with AHR to facilitate achievement of population goals (Hansen et al. 2017*b*, Wallingford et al. 2017).

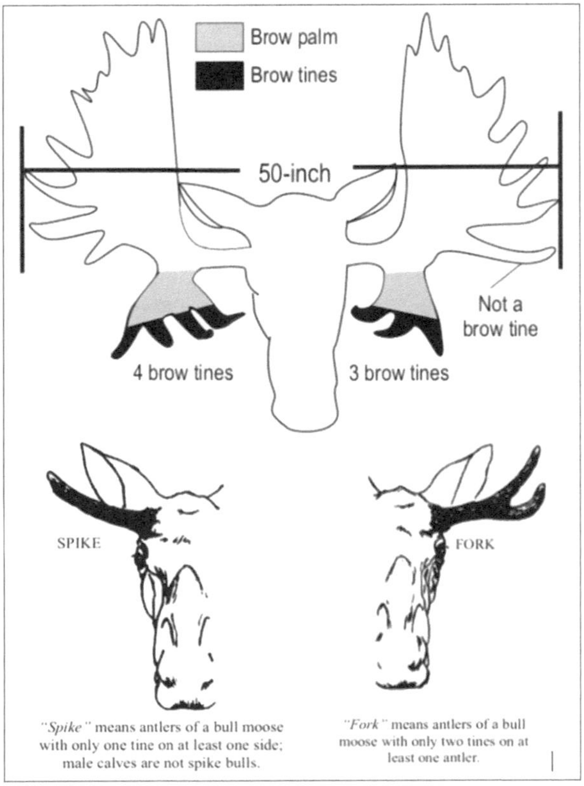

FIGURE 19.3 Example of a moose with a spike or fork antler, with antlers greater than or equal to 127 cm (50 in) in spread, or with three times on at least one brow palm as required under antler restrictions in various hunting units in Alaska. Image courtesy of the Alaska Department of Fish and Game.

Despite widespread application, perceived and real efficacies of AHRs are variable across taxa and region. AHRs that have met objectives in some regions and species do not work for others. Productivity of the population, cooperation from hunters, management legacies, and factors affecting harvest vulnerability, among others, all affect the efficacy of AHRs. Despite their prevalence in harvest management, there is no synthetic review of the efficacy of AHRs. Such a review is needed to provide a comprehensive understanding of AHRs and their efficacy in meeting management objectives. Much of the literature on AHRs exists in agency reports and is not readily available. Thus, our objective was to compile data on AHR use and their results across North America to examine which systems have met management goals by region and species. We believe it most effective to summarize the existing information using a case-study approach that allows us to provide detailed examples in hopes of providing the necessary context to understand application of AHRs in various situations. We have included all North American examples of AHR evaluation we could find in the published literature as well as a few well-designed studies in agency reports. We report successful and unsuccessful results as well as unintended consequences. We identified five main objectives ungulate managers have typically implemented using AHRs: (1) increasing number of older males, (2) increasing male sex ratios, (3) influencing population size, (4) increasing average antler size, and (5) increasing hunter satisfaction. Thus, we structured our review to address these objectives. Our synthesis of the various AHRs will help managers considering their use to meet specific management goals according to their species and location.

APPLICATION OF AHRS

Increasing Number of Older Males and Balancing the Population Sex Ratio

Increasing the number of mature males within a given ungulate population or within harvest has been a main focus of implementing AHRs. Managers have focused on increasing the number of older males for two reasons. First, recreational hunters have become more selective about animals they harvest, electing to allow younger males with smaller average antler sizes to pass while waiting for opportunities to harvest older males. The hunting public has become more vocal about their preference for older males and their desire for state wildlife agencies to manage for older males in populations (Adams and Hamilton 2011). The second justification relates to breeding efficiency resulting in more synchronous birth dates, higher pregnancy rates, and higher recruitment owed to a larger proportion of older males in the population (Geist 1991, Squibb et al. 1991, Noyes et al. 1996). Some managers focus their efforts on simply increasing the male proportion of the population, to better balance female-skewed population sex ratios. Many agencies set a minimum male:female ratio as their management objective, especially in areas where harvest is directed toward the male segment of the population and female harvest is limited (Schwartz et al. 1992).

The following examples evaluated instances where AHRs were implemented to increase the number of older males within a population or to balance the female-skewed population sex ratio. We note that the definition of an "older" male differed among examples with some including all males 2.5 years or older and others establishing higher benchmarks (>2.5 years of age). Our assessment of success in each example was based on the specific age objective identified by managers.

Colorado Elk

In 1971, the Colorado Wildlife Commission implemented antler harvest regulations on male elk harvest with the objective of protecting yearling males. The first AHR implemented for the 1971 hunting season restricted harvest to males with at least one antler ≥15 cm (6 in) in length with ≥2 points. The 1972 hunting season AHR restricted harvest to males with ≥4 points on one antler (Fig. 19.2). Boyd and Lipscomb (1976) compared pre-AHR data to the 1971 and 1972 hunting season data from the White River elk herd using documented male elk numbers after each hunting season recorded in ratios of branch-antlered males and yearling males per 100 females and the number of males harvested in each age class per season.

From post-hunt male estimates, branch-antlered males decreased from four to two per 100 females, and yearling males increased from five to 13 per 100 females after one season under the ≥2-point AHR. Branched-antlered males and yearling males each comprised 49% of the total harvest with the remaining 3% comprised male calves. Total male harvest declined from 759 the previous year's average to 328. After one season of the ≥4-point AHR, the ratio for branch-antlered males went from two to four per 100 females, whereas yearling males increased from 13 to 19 per 100 females. Branched-antlered males and yearling males comprised 79% and 10% of the total harvest, respectively. The total harvest was estimated at 500 males, which was higher than the previous year but still lower than the three-year average prior to AHR implementation.

Both the ≥2 point and ≥4-point AHRs protected yearling males from harvest, allowing them to survive to at least 2.5 years of age, which met the primary objective of the agency. However, the ≥2-point AHR resulted in a decrease of branch-antlered males and both AHRs decreased the total male harvest. The authors also reported an increase in abandoned illegal harvests, with record high of 32 yearling and two mature bulls reported as abandoned in the study area. The previous highs were in 1968 and 1969, when 13 bulls were reported abandoned each year. Inference is limited because each AHR was only implemented for one season.

Washington Elk 1

In 1989, The Washington Department of Fish and Wildlife implemented a unique variant of a traditional AHR in three of their elk herd units. Their proximate objective was to increase mature male elk numbers without reducing hunting opportunity, with the ultimate objective of increasing breeding efficiency measured in the number of calves per 100 cows. The AHRs allowed open-entry harvest, harvest by any hunter with a valid statewide tag, of one-point males and limited harvest of branched males (≥2 points) by permit (spike-only branched male by permit). Bender et al. (2002) evaluated the effectiveness of this AHR strategy by determining herd composition by helicopter counts and analyzing trends in population total male:female ratios, branched male:female ratios, and yearling male:female ratios and their relation to calf production in these three herds. They compared mean ratios from three years prior and six years after AHR implementation. Male:female ratios were used as a proxy for the number of males in the population because total male population was not estimated.

The average total male and branched male ratios for the three elk units increased from five to ten and one to six males per 100 females after the implementation of the spike-only branched-male-by-permit AHRs, respectively. Average yearling male ratios remained unchanged at approximately four males per 100 females. However, average calf ratios decreased from 38 to 29 calves per 100 females after spike-only branched-male-by-permit AHRs. The study reported either no relationship or negative relationships between both total male:female ratios and branched male:female ratios in relation to calf:female ratios.

Washington's spike-only branched-male-by-permit AHR met their proximate objective by increasing mature male elk numbers, based on male:female ratios as proxies, in all three herd units. They also increased the total number of males in these units. However, increased breeding efficiency was not achieved. Factors other than male ratios were likely driving calf production in these elk units such as female nutritional condition (Bender et al. 2002).

Washington Elk 2

The Washington Department of Fish and Wildlife managed their elk populations with the objective of maintaining existing population levels with a minimum post-hunting season ratio of 12–15 males per 100 females in their open-entry game management units and ≥20 males per 100 females in limited-entry units. They used four different AHRs to accomplish their objectives. In 1989, open-entry units were managed using either *any-male* harvest or AHRs limiting male harvest to males with ≥3 points on one antler. Limited-entry units were managed using any-male harvest with either moderate limited entry (overall male mortality rate <50%) or light limited entry (overall male mortality rate <40%) regulations since 1983. Bender and Miller (1999) compared herd composition and mortality data of the four management strategies to evaluate their relative efficacy at achieving Washington's objectives among all strategies.

Male:female ratios in any-male units (22:100) and ≥3-point units (27:100) were similar. Moderate limited entry units (42:100) and light limited entry units (53:100) had much higher ratios. Male mortality rates for any-male, ≥3-point, moderate limited entry units, and light limited entry units were 0.70, 0.54, 0.41, and 0.34, respectively. The proportions of mature males were greater in moderate limited entry (22%) and light limited entry (26%) units than in any-male (10%) and ≥3-point (10%) units. Calf production did not differ based on harvest strategy.

All four of the harvest strategies were successful in achieving the sex ratio objectives of the Washington Department of Fish and Wildlife. In fact, the any-male and ≥3-point units met the minimum objective of limited entry units. However, the moderate and light limited entry units produced much less female-skewed population sex ratios and higher proportions of mature males in the populations. The ≥3-point units did not produce a higher percentage of mature males than the any-male units. If increasing the number of mature males becomes an objective in Washington, moderate and light limited entry harvest strategies were the only proven regulations to achieve that goal.

Oregon Elk

From 1979–1998, the Oregon Department of Fish and Wildlife applied an AHR in a single wildlife management unit with the objective of increasing older-aged males (≥4.5 years old) in its population. The AHR restricted harvest to males with ≥3 points on one antler in the Saddle Mountain wildlife management unit, whereas any male was legal for harvest in all other units. Biederbeck et al. (2001) evaluated the efficacy of this AHR by comparing herd monitoring and harvest data from the Saddle Mountain wildlife management unit to data from two units (Wilson and Trask) where any male elk could be legally harvested from 1994 through 1998. Specifically, they compared mortality rate of males by age 4.5 years, average age at harvest, and proportion of males available for harvest under both regulations.

Male survival by age 4.5 years was low with 7%, 4%, and 3% of 4.5-year-old males surviving in Saddle Mountain, Wilson, and Trask wildlife management units, respectively. The average age at harvest was 2.50 years for Saddle Mountain compared to 1.58 and 1.71 years for Wilson and Trask, respectively. An average of 74% of all males in Saddle Mountain were available for harvest each year, whereas 67% and 58% of the males in Wilson and Trask were available, respectively. The study also reported average male ratios during the study period of 20.2, 5.8, and 6.8 males per 100 females in the Saddle Mountain, Wilson, and Trask wildlife management units, respectively.

Oregon's ≥3-point AHR in Saddle Mountain wildlife management unit was unsuccessful in achieving the objective of increasing older-aged males in the population. The mortality rates of males by age 4.5 years were similar among all wildlife management units. The ≥3-point AHR delayed the harvest of 1.5-year-old males by one year and produced less female-skewed population sex ratios as well as greater proportions of males available for harvest, though these were not explicit objectives of the regulation. Limiting hunter numbers may be a better alternative to AHRs in these units, because populations are low, hunting pressure is high, and the units are easily accessible due to their close proximity to city centers.

Kenai Peninsula, Alaska Moose

In 1987, the Alaska Board of Game implemented AHRs for moose harvest in two open-entry game management units in response to low male:female ratios. The main objective was to improve female-skewed population sex ratios of moose, and secondary objectives included increasing the number of "prime-aged" males (≥4.5 years old as best interpreted from the article), increasing opportunities to view male moose, and maintaining hunting opportunity. The AHRs required male moose to be either yearlings (spike or forked antlers) or males with ≥127 cm (50 in) in antler spread or ≥3 tines on one brow palm to be legal for harvest (Fig. 19.3). Schwartz et al. (1992) compared population and harvest statistics from five years prior to AHRs to similar data from the first five years of AHRs to evaluate the efficacy of the new regulations.

Under the new AHRs, the male:female ratio of moose increased from 16 males per 100 females to 25 males per 100 females. Before AHRs, male harvest comprised 46% yearling, 38% 2.5- to 3.5-year-old, 11% 4.5- to 5.5-year-old, and 5% ≥6.5-year-old moose. After AHRs, those proportions changed to 64% yearling, 17% 2.5- to 3.5-year-old, 12% 4.5- to 5.5-year-old, and 7% ≥6.5-year-old moose. The average annual male harvest declined from 636 to 443 moose and the number of hunters declined from 3602 to 2605 under the new AHRs.

Alaska's AHRs successfully improved population sex ratios of moose on the Kenai Peninsula making them more balanced. Although they did not directly document increases in "prime males" in the population, harvest pressure shifted from 2.5- to 3.5-year-olds to yearlings most likely leading to increases of 2.5- to 3.5-year-old moose. Total moose harvest declined and coincided with a decline in number of hunters. Some hunters indicated they were not willing to participate in restricted male moose hunts, whereas others were uncomfortable identifying legal males. Finally, male moose viewing opportunities increased resulting in increased public support for the AHRs.

Interior Alaska Moose

From 1996 to 1999, high harvest rates of male moose in an open-entry game management unit in interior Alaska resulted in a decline in post-hunt male:female population sex ratios below the management objective of 30 males per 100 females. The Alaska Department of Fish and Game responded by shortening the 2000–2001 hunting season to reduce male harvest, but failed to increase male:female ratios. In 2002, they decided to implement AHRs to improve the balance of sex ratios. The AHRs restricted male harvest to yearlings (spike or forked antlers), males with ≥127 cm (50 in) in antler spread, or ≥3 brow tines on one antler. Young and Boertje (2008) evaluated the efficacy of these AHRs at meeting objective and reported other metrics of interest to managers.

After two years of AHRs, male:female population sex ratios increased from 26 males per 100 females to 32 males per 100 females. Comparing data from two years before to two years after AHRs, average male harvest decreased from 25% to 12% of males available for harvest in the given year. Comparing the same two-year periods, the number of hunters declined 24% (1568 to 1187) and hunter success rate declined from 34% to 29%. The mean proportion of 76- to 102-cm (30- to 40-in) males declined from 30% to 6% of total male harvest and the mean proportion of ≥127-cm (50-in) males increased from 26% to 49% of the total male harvest.

Alaska's moose AHRs allowed them to exceed their management objective of 30 males per 100 females in their game management unit after two years of use. The AHRs also resulted in a greater proportion of larger antlered males harvested. As a result of AHR success, Alaska increased the hunting season length by five days in 2004 and started to issue a limited number of any-male permits in 2006.

Mississippi White-tailed Deer

The AHR implemented statewide by the Mississippi Department of Wildlife, Fisheries, and Parks in 1995 prohibited harvest of male white-tailed deer with less than four total antler points. Though not explicitly stated, we assumed one of their objectives was to increase the proportion of males harvested older than 1.5 years old and therefore based success on this objective. The secondary objective of increasing antler size of harvested males is also addressed here (and below). Demarais et al. (2005) collected harvest data from 21 Wildlife Management Areas across Mississippi four years before and five years after the AHR was initiated. They compared the pre- and post-AHR data to evaluate its effects on age at harvest, harvest rate, and antler size of males harvested.

Yearling males comprised 58% of the total harvest before the ≥4-point AHR was installed and dropped to 16% under the new rules. The proportions of 2.5-year-old males and ≥3-year-old males in the total harvest increased from 25% and 17%, respectively, to 42% each. However, the number of 2.5- and ≥3-year-old males harvested remained relatively constant with only a slight increase in ≥3-year-old male harvest after AHR implementation. The total male harvest decreased from 3.8 to 2.2 males harvested per 5 km^2. The antler sizes of harvested males generally declined during the ≥4-point AHR period.

Mississippi was successful at increasing the proportion of ≥2.5-year-old males in the harvest. Although the proportion of ≥2.5-year-old males harvested increased, the total number remained the same due to an overall decrease in total harvest resulting from lower yearling harvest. The ≥4-point AHR also resulted in decreasing antler sizes of males harvested. Yearling males that reach the minimum requirements of this AHR for harvest may have a higher potential for antler growth at older ages and removing them early may negatively impact average antler sizes in older age classes (Strickland et al. 2001, Demarais and Strickland 2017). This effect is known as high grading and may or may not result from AHR use depending on regional variation in population demographics, hunter culture, and harvest intensity.

Pennsylvania White-tailed Deer

In 2002, the Pennsylvania Game Commission adopted an AHR with the objective to increase the proportion of older males in their deer population by protecting over half of the yearling population from harvest. The Pennsylvania Game Commission established a ≥4-point per side AHR in ten western management units and a ≥3-point per side AHR in remaining units. The difference in AHR used was determined by average antler characteristics of yearling males harvested in each management unit the previous year. To aid in hunter compliance of the new regulations, the Pennsylvania Game Commission established policies for handling violations of the AHRs. Hunters who mistakenly killed antlered deer who did not meet AHRs and voluntarily reported their violation paid a US$25 fee and lost their harvest, but were given a replacement male tag for that season, whereas those who failed to report their violation faced increased penalties and did not receive a new tag. Wallingford et al. (2017) evaluated the effects of the new AHRs. They compared pre- and post-AHR harvest data and age structure estimates tracking changes for yearling (1.5 years old) and adult (≥2.5 years old) males.

After AHRs were implemented, yearling harvest decreased from 82% to 52% and adult harvest increased from 18% to 47% of the total male harvest. The new AHRs protected 69% of the yearling population from harvest as opposed to 20% under the previous regulations. The estimated proportion of adult males in the population increased from 20% to 50% after AHR implementation, whereas yearling proportions decreased from 80% to 50%. Overall, total male harvest slightly decreased after AHRs.

The AHRs protected more than half of the yearling population from harvest, successfully meeting their objective of the new regulations. They also achieved a 29% increase in the proportion of adult males harvested and a 30% increase in adult males in the population. The total number of males harvested was decreased partly due to the new regulations, but also due to an already decreasing deer population documented the year prior to AHRs by the authors of the study. The decline in male harvest may also have been affected by hunters choosing to purchase antlerless tags, because the Pennsylvania Game Commission increased its allocation of antlerless tags and increased the season length for antlerless harvest.

Missouri White-tailed Deer

In 2004, the Missouri Department of Conservation implemented an AHR in 29 of its counties with the objectives of increasing harvest of adult (≥2.5 years old) male white-tailed deer and improving hunter and landowner satisfaction with their deer management. They also had an additional objective of increasing antlerless harvest to reduce the overall deer population (addressed below). The chosen AHR was a minimum of four antler points on ≥1 side for a male to be legal for harvest. Hansen et al. (2017) compared harvest data from the 29 AHR counties to data from adjacent non-AHR counties to quantify the effects of this AHR on harvest composition. Expected harvest for each age class in AHR counties was calculated assuming the proportions would be the same as the adjacent non-AHR counties if the AHR had no effect. They evaluated counties in northern and central Missouri separately, because they believed antler characteristics, land use, and deer demographics differed between these regions enough to influence the response to the AHR.

After the first year of AHRs, total male harvest in both northern and central AHR counties was 35%–36% lower. Male harvest did recover somewhat but was still 17%–18% lower than expected by the fourth year of AHRs. Yearling male harvest decreased by 49%–56% in northern counties and 44%–63% in central counties during the study period. Harvest of 2.5-, 3.5-, and ≥4.5-year-old males was greater in both northern and central counties after four years with AHRs.

Missouri's ≥4-point AHR was successful in achieving their objective of increasing adult male harvest. Yearling harvest was also significantly decreased under the AHR. However, the total antlered-male harvest declined, possibly due to the decrease in yearling harvest or hunters becoming more cautious when making harvest decisions in an effort to avoid harvesting sub-legal males (Hansen et al. 2017*b*).

Influencing Population Size

In areas where populations are above management objectives, AHRs may be counterproductive, unless agencies increase antlerless harvest to reduce the population growth rate. AHRs that require hunters to harvest an antlerless animal before they can legally harvest an antlered animal aim to accomplish such a shift in harvest. These types of regulations are commonly referred to as *earn-a-buck* programs. The following examples evaluated instances where AHRs were implemented to influence population size, either to increase, decrease, or maintain total numbers. Our assessment of success in each example was based on changes in population size of an intended trajectory.

Wisconsin White-tailed Deer

Since 1996, the Wisconsin Department of Natural Resources has selectively implemented two different season frameworks in addition to a standard season with the objective of increasing antlerless harvest to reduce overabundant deer populations. The first season type added a four-day antlerless firearm season in December (4DAY). The second season type included the same four-day antlerless firearm season in December with an additional four-day antlerless firearm season in October (8DAY). Wisconsin managers also implemented occasional concurrent earn-a-buck regulations to these two season types to further increase antlerless harvest. Antlerless harvest was unlimited during the entire period these regulations were applied. Van Deelen et al. (2010) quantified changes in antlerless and antlered harvests relative to these regulations from 1996 to 2008 to evaluate their efficacy in achieving increased antlerless harvest.

The additional antlerless harvest opportunity of the 4DAY season type was associated with an average increase of 1.10 deer per km^2 in antlerless harvest. Average antlerless harvest increased 1.32 deer per km^2 under the 8DAY season type. Average antlerless harvest increased 2.03 deer per km^2 with the addition of the earn-a-buck regulations to both 4DAY and 8DAY season types. Average antlered harvest decreased 0.02, 0.09, and 0.60 deer per km^2 under the 4DAY, 8DAY, and 4DAY+8DAY+earn-a-buck season types, respectively.

Wisconsin's objective of increasing antlerless harvest was achieved under all three regulation types, but the inclusion of earn-a-buck regulations had the greatest effect. The additional earn-a-buck regulations resulted in a slight decrease in antlered harvest. The authors suggest this decrease may be a source of hunter dissatisfaction with earn-a-buck, among several other sources (Holsman et al. 2010).

New York White-tailed Deer

In 1999, Cornell University implemented earn-a-buck regulations at a university-owned research forest with the objective of increasing antlerless harvest, thereby reducing herd density. They believed that reducing deer numbers would decrease deer-vehicle accidents and damage to agriculture, sawtimber regeneration, research plots, and flora. Their earn-a-buck regulations required hunters to document the harvest of two female deer before they could harvest one male deer. Boulanger et al. (2012*b*) quantified female harvest before and after earn-a-buck regulations to evaluate their effectiveness at increasing female harvest.

Average annual antlerless harvest increased from 13.5 deer per year before earn-a-buck to 34.0 deer per year four years after earn-a-buck. Average antlerless harvest the last four years of data collection (9–12 years after earn-a-buck) decreased to 15.8 deer per year. Average annual deer harvest increased from 30.9 deer per year before earn-a-buck to 61.3 deer per year four years after earn-a-buck but then decreased to 44.8 deer per year by the last four years of the study. Hunter effort was relatively consistent before and after earn-a-buck.

Cornell's earn-a-buck regulations allowed them to successfully reach their objective of increasing antlerless harvest within the first four years of their implementation, but the effectiveness may have decreased after that period. The decline in antlerless harvest after the four-year period may have been a result of greater effort required to harvest a deer due to the smaller population. Total deer harvest responded similarly. After implementation of earn-a-buck, there was an increase in regeneration of desirable sawtimber species in the research forest (Boulanger et al. 2012*b*).

Missouri White-tailed Deer

We previously discussed the Missouri Department of Conservation AHR in 29 counties. In addition to increasing the harvest of adult males in all counties, they also wanted to increase antlerless harvest in its northern counties where deer populations were above management objectives. Antlerless harvest opportunity was unlimited in most AHR counties throughout the duration of the open deer hunting season. Overall, they predicted the AHR would shift harvest pressure from yearling males to females in northern counties where deer densities were higher, thus increasing female harvest. Results were reported for northern AHR counties and central AHR counties as comparisons to expected results based on adjacent non-AHR counties in each region.

Yearling male harvest decreased by 49%–56% in northern counties and 44%–63% in central counties during the study period. Harvest of 2.5-, 3.5-, and ≥4.5-year-old males increased in both northern and central counties after four years with AHRs (Hansen et al. 2017*b*).

Missouri's use of AHRs was unsuccessful in achieving their objective of increasing female harvest in northern counties. After the implementation of AHRs, female harvest in the northern counties was 1%–5% lower than expected and was 12%–18% greater than expected in the central counties. Even though yearling harvest decreased, results suggested that harvest pressure was not diverted to females in high-density areas, but may have shifted to older-aged males instead. In contrast, female harvest was increased in central counties where population management was not intended. Harvest pressure may have shifted slightly from yearling males to females in central counties, as there were fewer legal males available for harvest under the ≥4-point AHR (Hansen et al. 2017*b*).

Increasing Average Antler Size

Many managers use AHRs for the objective of increasing the number of mature males within their populations. Managers may expect average antler sizes to respond similarly to changes in age structure of their population in response to AHRs because antler size typically increases with age (Strickland and Demarais 2000). However, this outcome is not necessarily the case. Objectives related to antler size should specify if their intent is to increase average antler size within age class rather than a general increase in antler size in the total male population. The following examples evaluated instances where AHRs were implemented to increase antler sizes and provide examples of how the results differ between all males and within age class.

Florida White-tailed Deer

In 1994, the Florida Game and Fresh Water Fish Commission implemented AHRs in northwest Florida with the objective of increasing average antler sizes of harvested males. Before the change, males were required to have ≥3 cm (1 in) of visible antler above the hairline to be legal for harvest. The new AHR required males to have one antler ≥13 cm (5 in) in length to be legal. Shea and Vanderhoof (1999) compared harvest data collected before and after the new regulations (1985–1997) on 13 wildlife management areas in northwest Florida to evaluate the effects of the length restriction on average antler size of the male harvest.

Average antler beam length, circumference, and number of points increased from 20.1 to 23.5 cm, 5.7 to 6.2 cm, and 4.0 to 4.4, respectively, for deer harvested under the new antler restrictions. The mean age of males harvested increased from 2.24 to 2.41 years old. Average antler beam length, circumference, and number of points of 2.5-year-old males harvested decreased from 27.1 to 26.2 cm, 6.5 to 6.3 cm, and 4.9 to 4.5, respectively.

The ≥13-cm (5-in) length antler regulation did meet the objective that Florida had intended by increasing the mean antler size of the total harvest. However, it also caused a decrease in average antler size of harvested males within the 2.5-year-old age class. Yearling males that produce comparatively larger antlers relative to other 1.5-year-olds will most likely produce comparatively larger antlers relative to their age class as they get older (Demarais and Strickland 2011). So, the decrease in antler size documented in the 2.5-year-old age class may have been a result of high-grading yearling males with larger antlers meeting the ≥13-cm (5-in) length minimum for harvest. Strickland et al. (2001) demonstrated that high grading may only be apparent at high levels of harvest intensity based on the results of their simulation model. Managers implementing antler restrictions to increase antler size should ensure they protect as much of the yearling age class as possible and consider harvest intensity in their location.

Mississippi White-tailed Deer

In a previous example, we detailed Mississippi's experience in using a ≥4-point AHR to meet their management objective of increasing the proportion of adult males in their harvest. Strickland et al. (2001) also evaluated Mississippi's AHR to determine its effects on average antler size of males harvested within each age class. They compared pre and post-AHR harvest data for three regions including the Delta, Upper Coastal Plain, and Lower Coastal Plain soil regions, given antler sizes varied by soil region across Mississippi. They quantified changes in age-specific harvest proportions and mean antler size within age-specific harvest.

Comparing data from four years before to three years after the ≥4-point AHR implementation, average antler size declined for the 2.5- and 3.5-year-old age classes in the Delta region, whereas no significant change was detected in either the Upper Coastal Plain or Lower Coastal Plain regions (Table 19.1). Demarais et al. (2005) continued to study the effects of the AHR on age-specific antler size through 2001 and reported that antler sizes decreased for at least one of the two age classes (2.5 and ≥3.5 years) across all regions. In those cases, antler sizes decreased 13–23 cm (5–9 in) for 2.5-year-old males and 25–43 cm (10–17 in) for ≥3.5-year-old males from pre to post AHR. Demarais et al. (2005) suggested this lag in antler degradation in the Upper Coastal Plain and Lower Coastal Plain regions could be due to limited nutrition and later fawning dates in those regions, resulting in yearlings not phenotypically displaying their genetic potential, which protected them from harvest. They also suggested that variability in harvest rates of vulnerable males

TABLE 19.1

Age-Specific Mean Boone and Crockett Scores of Male White-Tailed Deer Harvested *Pre* (1991–1994) and *Post* (1996–1998) Implementation of a Four-Point[1], Selective-Harvest Regulation From Three Soil Resource Regions in Mississippi

		Pre Regulation			Post Regulation			
Region[2]	Age Class	n[3]	$\bar{x}$	SE	n	$\bar{x}$	SE	p-value
Delta[4]	2.5	105	86.7	1.7	44	77.5	3.1	0.005
	3.5	25	113.1	3.6	5	94.0	8.0	0.041
UCP[5]	2.5	40	65.3	3.0	36	58.4	3.4	0.129
	3.5	19	92.4	5.1	16	83.6	5.7	0.255
LCP[6]	2.5	87	70.0	2.2	103	67.6	2.1	0.439
	3.5	69	86.7	3.4	33	87.9	4.1	0.828

Notes

1 Males must have ≥4 antler points to be legal for harvest.
2 Represents Delta, Upper Coastal Plain (UCP), and Lower Coastal Plain (LCP) soil resource regions as defined by Pettry (1977).
3 Sample size (n), mean ($\bar{x}$), and standard error (SE) presented for each group.
4 Data collected from Sunflower Wildlife Management Area.
5 Data collected from Choctaw Wildlife Management Area.
6 Data collected from Caney Creek (#3), Chickasawhay (Jones County), and Old River Wildlife Management Areas.

across regions could influence the rate at which the effects of the AHR are realized (e.g., Strickland et al. 2001). Given these findings, Strickland et al. (2001) and Demarais et al. (2005) advise against using AHRs to increase age-specific antler sizes.

Florida White-tailed Deer 2

Shea and Vanderhoof (1999) reported that the 13-cm (5-in) AHR implemented on white-tailed deer in Florida resulted in a reduction in antler size (see above); thus, soon after, some wildlife management areas in Florida adopted new, more restrictive AHRs, requiring males to have ≥2 points on one side. In 2004, even more restrictive AHRs, requiring males to have ≥3 points on one side, were implemented across all wildlife management areas with the objective of protecting more 1.5-old-year males and increasing age-specific antler size of harvested males. Cohen et al. (2016) used deer harvest data from 23 wildlife management areas across Florida from 1999 through the 2012–2013 hunting season to identify whether implementation of the new three-point AHR reversed the degradation of antler size. They estimated gross Boone and Crockett score of each male using measurements of basal circumference, main beam length, inside spread, and number of points for each antler, and then compared antler sizes across treatment periods (pre and post three-point AHR), original AHR (13-cm [5-in] and two-point AHR), and region.

Cohen et al. (2016) reported that implementation of the three-point AHR had different effects on antler size, depending on the original AHR that was in place. Gross Boone and Crockett scores of harvested males increased an average of 9 cm (3.52 in) after implementation of the three-point AHR in areas that switched from a 13-cm (5-in) AHR, whereas there was no difference in antler size in areas that switched from a two-point AHR. Regardless of the modest average increases in overall antler size, there were no large differences in age-specific antler sizes. Thus, even though the estimated proportion of protected yearlings changed from 44% under the 13-cm (5-in) AHR to 83% under the three-point AHR, the additional restrictions may have been insufficient to increase age-specific antler size. The change from a 13-cm (5-in) AHR to the three-point AHR also resulted in a decrease in harvest per unit effort, suggesting it took hunters longer to harvest a deer under the more restrictive AHR. Thus, Cohen et al. (2016) recommend that managers carefully consider the tradeoffs associated with restrictive AHRs.

Georgia White-tailed Deer

In Georgia, AHRs were implemented that restricted male white-tailed deer harvest to either (1) males with a main antler beam ≥ 41-cm long or an outside spread ≥ 38 cm; (2) males with ≥ 4 antler points on one side (including the brow tine); or (3) any male with antlers visible above the hairline (i.e., no AHR), depending on location. Gulsby et al. (2019) used hunter harvest data from wildlife management areas in each of the three AHR categories to identify whether AHRs affected antler size of harvested males from 2003 to 2013. Antler sizes were calculated by summing measurements of the basal diameter, main beam length, outside spread, and number of points.

Gulsby et al. (2019) reported no difference in antler sizes of males harvested on wildlife management areas with the four-point AHR, compared to areas with no AHR, but reported that 2.5- and 3.5-old-year males had larger antlers, on average, in areas with the beam/spread AHR. These results differed from studies that suggested AHRs could result in smaller antler sizes, due to limited protection of larger individuals in the 1.5- or 2.5-old-year age cohort (e.g., Strickland et al. 2001, Demarais et al. 2005). Gulsby et al. (2019) suggested the AHRs in Georgia, which protected >95% of 1.5-old-year males, were restrictive enough to prevent *high grading*, as identified in other studies. However, they caution that increased restrictions, and complexity of restrictions, could result in lower hunter satisfaction, due to lowered opportunity to harvest adult males (e.g., 77% of 2.5-old-year males were protected from harvest under the beam/spread restrictions). Thus, a balance in the restrictiveness of AHRs that meets both biological and social objectives should be considered (Fuller et al. 2021 [Chapter 8], Kaemingk et al. 2021 [Chapter 3], Robinson et al. 2021 [Chapter 9]).

Increasing Hunter Satisfaction

Resource agencies are mandated to manage publicly owned wildlife resources on behalf of their beneficiaries under the Public Trust Doctrine (Batcheller et al. 2010), which requires management strategies that are biologically sound, but also address the diversity of concerns of constituents (Decker et al. 2016). Hunter participation and satisfaction is a substantial concern for state agencies because a large portion of state funding originates from hunting- and fishing-related activities (Voyles and Chase 2017, Graham et al. 2021 [Chapter 6]), either through license fees or excise taxes on firearms, ammunition, and fishing gear (i.e., Pittman-Robertson and Dingell-Johnson Acts; U.S. Fish and Wildlife Service 2015*c*). Thus, another objective of AHRs is to increase hunter satisfaction and participation. Identifying the factors affecting hunter satisfaction and participation is also an important consideration (e.g., Barro and Manfredo 1996, Gruntorad and Chizinski 2021 [Chapter 4]) so future regulations may be more effectively implemented.

Though many state agencies conduct hunter surveys evaluating satisfaction with hunting experiences or regulations, there is relatively limited peer-reviewed literature on the social consequences of AHRs, particularly in the western contiguous United States of America. It is important to identify hunter perceptions of AHRs before implementation, and this topic has received much attention (e.g., Bull and Peyton 2001, Kandoth et al. 2010, Cornicelli et al. 2011, Harper et al. 2012, Siemer et al. 2015, Robinson et al. 2016*a*).We limited our review to peer-reviewed literature and published reports that described hunter satisfaction during or after implementation of an AHR to determine whether social objectives were achieved.

Alaska Moose

We previously discussed the Alaska Board of Game AHR on the Kanai Peninsula. In addition to increasing the proportion of males in the adult sex ratio, they were also interested in hunter participation, attitude, and ethics in relation to the AHRs. Schwartz et al. (1992) used moose harvest tickets, collected from all moose hunters, to acquire information related to hunter participation and success, and performed informal surveys at check stations and public meetings to evaluate attitudes toward the AHRs.

Schwartz et al. (1992) reported a 25% decrease in the number of hunters after implementation of the selective harvest system, largely related to unwillingness to participate in the restrictive hunt and concerns about ability to identify a legal male. However, surveys at check stations suggested most hunters were supportive of the AHR. Reported illegal harvest of male moose increased under the AHRs, averaging 7% of the legal harvest, mostly due to misclassification of sublegal males as meeting the AHR criteria.

Building on results from Schwartz et al. (1992), Alaska Department of Fish and Game initiated a study to formally evaluate the influence of the AHR on hunter satisfaction. Fulton and Hundertmark (2004) surveyed a random sample of hunters in the AHR game management areas using a five-point scale to identify support and beliefs about the AHRs, and satisfaction with their most recent moose hunt in an AHR area. Respondents were also asked a series of questions identifying moose hunting-related experiences that affected hunting satisfaction, respondent beliefs related to positive or negative outcomes of the AHR, and the extent to which respondents agreed that the AHRs led to the positive or negative outcomes.

Fifty eight percent of moose hunters in the AHRs either strongly (28%) or somewhat (30%) supported the system, whereas 29% were either strongly (11%) or somewhat (18%) opposed, and 14% were neutral. Sixty five percent of respondents believed the AHR was good for moose populations, whereas 46% believed the AHR was good for moose hunters. Hunting success (moose harvest) explained 17% of the variation in overall hunting satisfaction, and support for the AHR increased the amount of variation in satisfaction by 14%. Evaluation of the beliefs related to the effect of AHRs on moose and moose hunters explained 71% of the variation in level of support for the AHR. Results suggested a majority of hunters supported the AHRs, with those supporting the

AHRs believing it was good for moose populations and hunters. Those opposed to the AHRs cited limited opportunity for acquiring meat and difficulty in identifying legal males as factors discouraging them from hunting.

Minnesota White-tailed Deer

In 2005, the Minnesota Department of Natural Resources introduced an AHR on white-tailed deer in Itasca State Park, which limited adult hunters to harvest antlerless deer or males with three points on at least one side. Schroeder et al. (2014) evaluated the influence of the number of deer seen by hunters, the type of deer pursued, agency trust, hunter experience, and length of regulation on hunter support for the AHR. All hunters who expressed intent to hunt Itasca State Park were surveyed after each nine-day deer season from 2005 to 2009. The effect of the hypothesized factors influencing support for the AHR was evaluated.

Hunter support for the AHR was relatively neutral and consistent across years, averaging 2.88 on a scale of 1–5 (1 = strongly opposing, 3 = neutral, 5 = strongly supporting the AHR). Hunters were slightly dissatisfied with the number and quality of deer seen, ranging from 1.92 to 3.22, and this satisfaction decreased over time. There was slight trust in Minnesota Department of Natural Resources, ranging from 3.07 to 3.59, but this trust also decreased over time. Most (~75%) hunters shot the first legal deer they saw, 17%–23% targeted big males, 2%–5% targeted any male, and 2% targeted antlerless deer. The type of deer pursued and agency trust were the factors most related to support for the AHR; those targeting big males and who had trust in Minnesota Department of Natural Resources had the most support. Hunters who saw more deer and who had fewer years of hunting experience were also more supportive of the AHR. The influence of most factors on support of the AHR stayed relatively constant across years, although the relationship between the type of deer pursued and support for AHR decreased over time. The decrease in satisfaction with deer seen over time was likely related to the reduction in deer densities in the park, suggesting that seeing and harvesting deer is paramount for many hunters.

Pennsylvania White-tailed Deer

Wallingford et al. (2017) evaluated hunter acceptance of the Pennsylvania white-tailed deer AHR in 2002 (see above) by monitoring hunter opinions related to the AHR during pre and post deer season periods from 2002 through 2004. A random sample of hunting license buyers was selected for both pre- and post-season surveys, and individuals who were surveyed each year were included in a longitudinal panel analysis. Survey questions measured hunter support for the AHR and influence of AHR on hunting experience; satisfaction with deer seen or harvested and quality; and perceptions about deer survival and breeding activity on a five-point scale. For the panel analysis, the change in responses to survey questions was calculated to identify whether attitudes changed over time.

Proportion of hunters who either agreed or strongly agreed with survey questions related to support for the AHR ranged from 53% to 73% in the three-point AHR and from 44% to 61% in the four-point AHR, and support for the AHR was always at least two times greater than opposition. Overall, the proportion of hunters who supported a statewide AHR varied between 61% and 70%. Support for the AHR did not change much over time, with 48% of hunters unchanged, 23% more supportive, and 29% less supportive. Hunters opposed to the AHR were most concerned with illegal harvest and reduced antlered deer harvest. Approximately 33% of respondents believed that hunters would not retrieve males if they were sublegal. Though the AHR received steady support throughout the study, support for the Pennsylvania Game Commission deer management program declined, with 41% rating it lower and only 21% rating it higher.

High support for the AHR in Pennsylvania suggested the regulation was a social success. Decline in support for deer management in the state was believed to be related to declines in deer densities and not solely due to the AHR, given AHR support was steady across years. Support for the AHR was expected to increase with time, as biological changes (e.g., higher proportion of older males) became more visible to hunters. However, this increase in support was not observed, likely because hunters formed initial opinions about the AHR and those opinions were not readily changed. Wallingford et al. (2017) believed obtaining public support for the AHR before implementation facilitated its social success, given initial opinions did not change substantially throughout the study.

Missouri White-tailed Deer

Hansen et al. (2018) initiated a study to evaluate hunter perceptions and satisfaction of a white-tailed deer AHR instituted in Missouri in 2004 (see above; Hansen et al. 2017*b*). During 2001, 2004–2008, and 2011–2013, mail-back questionnaires were sent to firearms deer hunting permit holders to identify hunter characteristics, perceptions of deer populations, attitudes toward deer management, and satisfaction with their hunt. Respondents were grouped into pre-AHR (2001), experimental (AHR initiated; 2004–2008), and long-term AHR (2011–2013). Further, respondents were separated into control counties with no AHR, treatment counties with AHR, and by geographic region (central or northern counties). The long-term effects of the AHR were assessed by establishing trends in hunter satisfaction and support for the AHR between the experimental and long-term periods.

Over 70% of respondents across counties supported the AHR, and this support remained throughout the study. Respondents generally perceived there to be more adult males in AHR counties, compared to control counties, but support for the AHR or hunter satisfaction did not change as a result. Respondents who thought deer management was good or excellent, who targeted antlered males, and who hunted counties with greater male harvest were more likely to support the AHR. Respondents with more experience, hunted both public and private land, and who reported harvest of an antlerless deer were less likely to support the AHR.

The AHR was considered a social success, given high hunter support for the program. Support for the AHR was consistent over time, possibly because strong initial opinions about the AHR were difficult to change, even with increased experience. Hunters who targeted antlered males were more supportive of the AHR, likely due to the perceived increase in adult males, whereas hunters targeting any deer were less supportive, possibly because of a reduction in opportunity, particularly on public lands. Declines in hunter satisfaction with the hunting season and deer management in some counties were likely related to lower deer densities and lower deer vulnerability to harvest, rather than the AHR.

New York White-tailed Deer

In 2005, the New York State Department of Environmental Conservation initiated an AHR that limited harvest of antlered white-tailed deer males to those with three points or greater on one side in eight wildlife management units in New York. This program was continued and expanded to four more wildlife management units in 2006. Hunters were surveyed in wildlife management units with an AHR each season from 2005 through 2007 and following the 2010 season to evaluate hunter experiences and attitudes about the AHR. Questions included in surveys were related to hunter satisfaction with hunting experiences (particularly for antlered males), change in hunting satisfaction since the AHR began, whether expectations were met, attitude about the AHR program, and whether the AHR program should continue. A series of reports were published with survey results, the results of the two most recent (2007 and 2010) we describe here (Enck and Brown 2008*b*, Enck and Decker 2011).

More than half of hunters from all AHR wildlife management units were satisfied with their overall hunting experience in 2007, whereas approximately one-third were dissatisfied. Approximately one-third of hunters believed their hunting satisfaction had increased since the beginning of the AHR, whereas one-third believed their hunting satisfaction had decreased. Slightly less than half of hunters were satisfied with their antlered-male hunting experience in 2007, whereas 37%–44% were dissatisfied. A majority of hunters reported that the number of legal males seen, male:female sex ratios, legal:sublegal male ratios, and freedom to shoot any male were "too low to be satisfied." Between 30% and 46% of hunters in 2007 said they were more supportive of the AHR since inception, whereas 14%–25% were less supportive. Between 60% and 77% of hunters believed that the AHR should be continued, whereas only 14%–29% believed that the AHR should be discontinued. The AHR had no influence on hunting participation for >70% of local hunters and >60% of non-local hunters in 2010. Similar to 2007, 62%–80% of hunters considered the AHR either very or moderately acceptable in 2010.

Despite some unmet expectations regarding the number of legal males seen and male:female ratios, hunters were supportive of the AHR, suggesting it was a success from a social standpoint. Hunters may have been positive of the AHR, regardless expectations not being met, because they believed expectations would be met over time. Success may also be attributed to the willingness of the agency to consider and incorporate public comment and interests into regulations before implementation.

British Columbia Mule Deer

Prior to 1992, an 80- to 85-day any-male mule deer season was held in southern British Columbia. Due to concerns related to low male:female ratios, a 23- to 40-day AHR season was implemented from November through December in 1992–1997. The AHR required antlered males to have at least four points on one side, excluding the brown tine. In 1998, the AHR was extended to also include September in order to further reduce male harvest. Each year, a questionnaire was sent to a random set of hunters to determine hunter success, sex-age composition, hunter numbers, and hunter days. Kuzyk et al. (2011) evaluated these survey data to determine how the AHR affected hunter opportunity, compared to an any-male season.

Male harvest was greatest before implementation of the AHR, and AHR seasons accounted for only 31%–37% of the male harvest, whereas any-male seasons accounted for 63%–69% of the harvest. Antler point restriction seasons appeared to shift timing of hunting pressure and success, as harvest of males doubled in October during the 1998–2009 period (AHR in September and November through December) relative to the 1987–1991 period (no AHR) and the 1992–1997 period (AHR only in November–December). Hunter numbers and hunter days stayed relatively consistent during 1987–2009, suggesting implementation of the AHR did not affect hunter participation. The maintenance of hunter participation was likely related to the combination of AHR and any-male seasons, which provided hunters ample opportunity to harvest any male they wanted, while also meeting agency objectives to reduce male harvest.

CONCLUSIONS

Efficacy in Affecting Male Age Structure and Population Abundance

Results of AHRs have varied according to the objective they aimed to meet. In most cases, AHRs allowed more males to reach 2.5 years old, but increased harvest pressure on older age classes, and almost always resulted in a decrease in total male harvest. As a result, many western agency biologists concluded that AHRs were either unsuccessful or only temporarily successful at producing the desired effects on male age and abundance (Bender and Miller 1999, Zornes et al. 2012), although exceptions exist (Schwartz et al. 1992, Kuzyk et al. 2011). In some situations, limited entry or quota systems for harvesting males have replaced AHRs or have been simultaneously implemented to improve the number of adult males in the population.

Some agency biologists have resisted AHR because of potential negative genetic effects, some failures to achieve desired outcomes (Demarais et al. 2005), or simply because they deviate from traditional management strategies and philosophies. Nevertheless, AHRs have been successfully employed in the east to increase the number of adult males in the harvest without reducing hunting opportunity, although some reduction of total antlered male harvest has occurred (Hansen et al.

2017*b*, Wallingford et al. 2017). Success without loss of hunting opportunity may be attributed to ungulate demographies. Liberal harvest opportunities for antlerless ungulates in the more productive populations may reduce the high harvest pressure on legal males experienced in less productive areas. AHR failures to produce desired biological or social effects may be a result of implementation in less productive areas where ungulate populations are low.

Social Acceptance

Overall, AHRs have been largely successful from a social standpoint; however, there are several important factors that affect whether hunters support or disapprove of these regulations. The opportunity to see and harvest game as well as the type of game (sex and age) pursued influences hunter satisfaction (e.g., Duda et al. 1996, Heberlein and Kuentzel 2002) and were also common factors affecting support for AHRs and resource agencies (Fulton and Hundertmark 2004, Enck and Brown 2008*b*, Enck and Decker 2011, Kuzyk et al. 2011, Schroeder et al. 2014, Wallingford et al. 2017, Hansen et al. 2018). Support for AHRs will likely be high when they are perceived to increase opportunity, but support will decrease when opportunities are limited. For example, hunters pursuing adult males may be more likely to support an AHR that is intended to increase the number of adult males in the population. However, hunters interested in harvesting the first deer they see may feel a loss of opportunity to harvest an animal if AHRs are in place. Regulations that provide opportunity to a diversity of constituents, either through strategically timed AHRs (e.g., Kuzyk et al. 2011) or through liberal antlerless seasons (where possible) may be effective ways to address biological objectives while maintaining broad public support.

Support for AHRs did not change greatly over time, even when support for management agencies declined due to a reduction in deer densities (Wallingford et al. 2017, Hansen et al. 2018). Further, hunters with more experience were less likely to support antler point restrictions, likely because they had pre-conceived expectations about harvesting deer and were therefore less tolerant to change (Schroeder et al. 2014, Hansen et al. 2018). Attitudes are formed by a variety of evaluative beliefs and can be difficult to change (Ajzen and Fishbein 1980). Addressing beliefs that influence attitudes toward AHRs may be important, given perceptions affect participation (Barro and Manfredo 1996). Many of the studies evaluating hunter satisfaction with AHRs (Enck and Decker 2011, Schroeder et al. 2014, Wallingford et al. 2017, Hansen et al. 2018) suggested obtaining public input before implementation was instrumental in gaining support for the AHR. Public participation in the planning process may be crucial in establishing positive beliefs and attitudes about the regulation and trust in the agency, which may lead to increased support for AHRs and other management actions (Schroeder et al. 2014).

One of the primary concerns with AHRs is that the difficulty in identifying legal individuals will result in an increase in illegal take and wanton waste of meat (Carpenter and Gill 1987, Schwartz et al. 1992, Schroeder et al. 2014, Wallingford et al. 2017). Though this is often a concern among hunters, evidence supporting a significant increase in illegal take of animals under AHRs is somewhat mixed (Wallingford et al. 2017, Schwartz et al. 1992, Erickson et al. 2003, Boyd and Lipscomb 1976). There will undoubtedly be animals mistakenly harvested when antler restrictions are established. Reduced penalties for hunters who report mistakenly harvested animals (e.g., Wallingford et al. 2017) may help reduce the number of animals that are wasted. Further, establishing AHRs that are clear and easy to identify in the field could reduce the number of mistakenly harvested animals and ameliorate the concerns of hunters.

Recommendations and Future Direction

AHRs have been implemented under a highly diverse set of human and ungulate demographic and geographic criteria, hunting traditions, and agency objectives and philosophies. It is not surprising that perceived and real success and failure of AHRs have varied considerably under this complex set of conditions. Unfortunately, it is difficult to sort out consistencies in successful and failed AHRs because of these complexities and often the lack of thorough evaluations of these diverse programs. Some general observations:

1. AHRs are popular with the hunting public because of the logical perception that reducing harvest pressure on young animals will produce more males in older age classes without a loss of hunting opportunity. This perception has driven recent implementation of AHRs, often through the political process. Agencies should be proactive and objective in developing simple, innovative, and enforceable regulations that address their unique set of ungulate population management issues, satisfy hunter interests, and provide adequate hunting opportunity.
2. In some cases, AHRs can be successful with little loss of hunting opportunity where robust ungulate populations allow liberal harvests of females that may deflect potential increased harvest pressure on adult males. These opportunities are most prevalent for white-tailed deer in the more productive parts of the eastern United States of America. However, liberal antlerless harvest can decrease overall population size, which may be undesireable to some hunters (Wallingford et al. 2017). Managers must take the regional differences in hunter cultures and expectations into consideration when planning AHR implementation.
3. Where ungulate productivity and populations are low and female harvests tightly controlled, as in much of the west and north, AHRs protecting young males will usually produce more two-year-old males but increased harvest pressure on legal males may reduce the number of older-aged males. Limited entry permits on males may be necessary if goals include balanced male age structures.
4. Hunting regulations designed to achieve ungulate population objectives that satisfy constituent interests often reduce hunting or harvest opportunity. Agencies should monitor and be sensitive to constituent satisfaction and participation.
5. Some reduction of antlered harvest will occur under AHRs because a male generally must survive an additional year before becoming legal to harvest. The amount of natural mortality and illegal harvest may vary geographically. Acceptable levels of illegal harvest may reflect agency and staff philosophy as much as demographic consequences.
6. AHRs that protect smaller antlered males and allow harvest of larger antler males may result in high grading of males with larger antlers within their age class when harvest pressure is high, such as southeastern states that have long seasons and liberal bag limits. Managers should consider this as well as regional variation in antler development when developing AHRs with the objective of increasing antler size.
7. Traditions and general resistance to change in hunters (and agency staff) will hamper implementation of innovative harvest regulations required to achieve population objectives. Data-driven justification for these regulations and public involvement in their development will be necessary for their successful implementation.
8. Resource agencies should also consider other biological ramifications associated with AHR management strategies before implementation. For example, chronic wasting disease (CWD), a prion disease that has been detected in free-ranging and captive cervids in 26 states in the United States of America, is of great concern to resource agencies. Belsare and Stewart (2020) estimated the probability of CWD outbreak in white-tailed deer increased from 37% under current harvest practices in Montcalm county, Michigan to 45% under an AHR that reduced yearling harvest rate from 47% to 32%. Thus, the diverse benefits and consequences associated with AHRs should be addressed and discussed among agency staff and stakeholders to facilitate sound management decisions.
9. AHRs have been widely implemented yet rigorous evaluation of their effectiveness is limited. More evaluation is needed to learn whether agency objectives are being met. Our review indicates that AHRs differ in their effectiveness based on species, region, societal

acceptance, and agency objectives. Although before and after monitoring of AHRs and pairing AHR regions with non-AHR regions is difficult, such designs allow for strong inference. At a minimum, we encourage agencies to monitor herd composition, abundance, and other important demographics along with hunter perceptions over time.

10. AHRs are likely to be most effective in regions with extremely high harvest rates. Allowing more young (e.g., 1.5-year-olds) individuals to survive will result in a more readily observable change in antler structure of the population. For species such as white-tailed deer, this expected outcome is more probable because the change in antler size between ages 1.5 and 2.5 years can be considerable. Other antlered species might not see such differences which is why white-tailed deer are an ideal candidate for AHRs. However, there might not be much observed difference unless harvest rates are high, which might explain why some western state agencies have perceived AHRs to fail.
11. One way to evaluate expectations of AHRs is through use of simple demographic models. For example, Raedeke et al. (2002) demonstrated how accounting-type models could be used to assess the effects of AHRs on elk populations in British Columbia. Those simulations demonstrated how three-point and six-point antler regulations influence important demographic features of the population. Such work could also demonstrate the important role that harvest rate plays in affecting the efficacy of AHRs and allow agencies to test hypotheses about how their AHRs might affect population features.

ACKNOWLEDGMENTS

The authors thank S. Demarais, D. Diefenbach, K. Pope, L. Powell, and B. Strickland for comments that improved the clarity and quality of this chapter. The authors gratefully acknowledge support from the Wildlife Biology Program and Boone and Crockett Wildlife Conservation Program at the University of Montana, Robert M. Lee Endowment Fund, and the Jack and Bonnie Russell Endowment.

Section III

Management Alternatives

Artist's Impression: Managing natural resources, as well as the wants and needs of stakeholders, is a balancing act performed by wildlife and fisheries managers. The task, as well as the responsibilities, can seem overwhelming at times. There is no perfect management scenario, as all plans have priorities, tradeoffs, and risks.

–Jennifer Clausen

Section IIIA

Harvest Regulation Paradigms for Wildlife and Fisheries

An engraving by Theodor de Bry published in *Americae XI* in 1618 showing early perceptions of abundance of game and fish in Virginia in North America with anglers and musket- and falcon-bearing hunters. From the collections of the Library of Congress (LC-USZ62–49747, book page is cropped to the borders of the original illustration).

20 Harvest Management of Migratory Game Birds

Mark P. Vrtiska

CONTENTS

INTRODUCTION

Migratory game birds are most likely the group of wildlife with the highest take from recreational harvests in North America, if not the world. Indeed, the mourning dove (*Zenaida macroura*), with an annual harvest of 10–15 million birds (Seamans 2019*b*), may be the most recreationally harvested animal on the planet. Harvest contributes to management and conservation of migratory game birds and their habitats directly through hunters' purchase of licenses and stamps, as well as excise taxes on arms and ammunition. For example, in the United States of America, sales of the federal Migratory Bird Hunting and Conservation Stamp (commonly known as the *duck stamp*) alone have generated close to US$1 billion used to conserve about six million acres of waterfowl habitat (U.S. Fish and Wildlife Service 2018*c*). Mourning dove hunting has generated approximately $8 million annually to state agencies through the Pittman-Robertson Act from only the number of shotgun shells purchased (Baskett and Sayre 1993).

Aside from the economic benefits, there are cultural traditions and other important benefits (Seng et al. 2001, Williamson et al. 2001, Peterson et al. 2011) associated with migratory game bird hunting.

Some hunters, especially those in Alaska, harvest migratory game birds for subsistence purposes. Similar to recreational harvest, there also are cultural traditions associated with subsistence harvest (Bromley 1996, Benson and Decker 2001). Although recreational hunting accounts for the majority of harvest (e.g., Padding et al. 2006), subsistence hunting by Indigenous groups, permitted to harvest outside of standard dates for hunting seasons, may have an impact on some populations given the timing of the harvest (i.e., breeding season). However, current subsistence harvest levels appear to be sustainable and more formal inclusion and processes have been developed to account for this type of harvest (Bromley 1996, Baldassaree and Bolen 2006, Natcher et al. 2011).

Market hunting was an important source of harvest pressure until the early twentieth century. In North America, large numbers of migratory game birds, primarily waterfowl, were harvested using a variety of methods and sold commercially (Baldassarre and Bolen 2006, Sawyer 2012). Although market hunting is no longer legal in North America and many other countries, it still occurs in some countries (Balmaki and Barati 2006, Kanstrup 2006). Illegal harvest of migratory game birds does occur (e.g., Gray and Kaminski 1994), but there have been few attempts at measuring its extent and impact. Mourning doves have been of particular concern (Braun et al. 1993).

Given the importance of migratory game bird harvest to humans and obligation for responsible management, it is critical that managers attempt to understand the factors that influence populations. Although factors such as habitat and weather may exert greater influence over populations than harvest, a major focus of migratory game bird management and research has been related to recreational harvest. The importance of harvest management is probably driven mostly by the *controllability* by managers, because establishing or changing harvest regulations is easier than affecting landscape change or other factors that affect populations. Managers seek to maintain harvest levels compatible with desired migratory game bird abundance and distribution while also providing hunting opportunity to meet hunter desires and maintain support for conservation and management efforts (Runge 2021 [Chapter 7]). Meeting these dual goals requires an understanding of the various regulations that affect harvest, but also how those regulations may affect hunters.

Migratory game birds are different from other game species because they may cross multiple national and state boundaries during their annual cycle. Although the types of regulations used to manage resident game birds such as grouse may be similar to those for management of migratory game birds (Dahlgren et al. 2021 [Chapter 21]), the structure and processes for setting population or harvest objectives are unique and provide the context for management actions. In this chapter, I review the authority, structure, processes, and actions (regulations and monitoring programs) used to manage recreational harvest of migratory game birds in North America, and primarily the United States of America. The relation of hunter opinions, behaviors, and activities are also included given the aforementioned dual goals, as well as recent concern regarding declining hunter numbers (e.g., Schulz et al. 2003, Vrtiska et al. 2013). Finally, I provide thoughts on future management and research needs of harvest management for migratory game birds.

Regulatory Authority and Structure

Given the scale of movements exhibited by migratory game birds, numerous government agencies, at multiple levels, often have an interest in the population objectives and harvest management of migratory game birds. These dynamics have resulted in the development of relatively complex infrastructures and processes that enable participation in harvest management by concerned entities. The multi-agency structure for management was also a response to the decimation and extinction of some migratory game bird populations during the unregulated recreational and market hunting experienced in North America in the early twentieth century (Baldassarre and Bolen 2006, Sawyer 2012).

In the United States of America, the initial attempt at establishing a higher level of authority than states began with the Migratory Bird Act of 1913 (Weeks-McLean Act). Unique within the U.S. framework for public trust (Hiller et al. 2021*b* [Chapter 2]), the Weeks-McLean Act placed migratory game birds under the jurisdiction of the federal government including hunting seasons that were to be restricted by opening and closing dates. Although the act was a major step forward in protecting migratory birds, it did not protect migratory birds outside of the borders of the United States of America. It was not until the signing of the Convention for the Protection of Migratory Birds (Convention; Migratory Bird Treaty) by representatives from the United States of America and Great Britain (for Canada) in 1916 that federal authority and responsibility for migratory birds were established. Protection of migratory birds in North America was afforded when a similar treaty was signed between the United States of America and Mexico in 1936 (Baldassarre and Bolen 2006). Other treaties with Japan and Russia and with subsequent updates have provided for the cooperative management of migratory birds at international scales.

Final authority and responsibility for the Convention was realized in both the United States of America and Canada by the passage of Migratory Bird Treaty Acts in both countries, which gave the U.S. Secretary of the Interior and the Canadian Minister of Environment and Climate Change the responsibility for implementing the Acts. The U.S. Fish and Wildlife Service within the Department of the Interior is the agency under which migratory game bird seasons are authorized and permitted in the United States of America. Likewise, the Canadian Wildlife Service is the agency responsible for migratory game bird seasons in Canada.

Although primary authority rests with federal agencies, migratory game bird management also requires state and provincial involvement (Vrtiska and Oldenburger 2018). Initial banding efforts (Lincoln 1935) guided the development of general waterfowl migration pathways commonly called flyways that led to the to the division of the United States of America into four administrative Flyway Councils (i.e., Atlantic, Mississippi, Central, Pacific; Fig. 20.1) for establishing annual hunting migratory game bird regulations (Hawkins et al. 1984). Band recovery information (i.e., recovery derivation and distribution) also led to the division of the United States of America into three management units (Eastern, Central, and Western) for mourning doves, and two (Eastern and Central, aligning with the Atlantic and Mississippi flyways, respectively) for American woodcock (*Scolopax minor*). Each Flyway Council has a waterfowl (ducks, coots, geese, and swans) and a *webless* (doves, rails, snipe [*Gallinago delicata*], and sandhill crane [*Grus canadensis*]) technical committee. A subcommittee under the *waterfowl* technical committee is responsible for webless species in the Pacific Flyway (Vrtiska and Oldenburger 2018). The technical committees review, discuss, and make recommendations regarding harvest management. These recommendations are forwarded to the Councils who then make decisions about recommendations that are forwarded to federal authorities. There have been the formation of other issue-specific or technical groups such as the Harvest Management Working Group, National Dove Task Force, Human Dimensions Working Group, and the National Science Support Team that can provide deeper examination of management issues than the technical committees. However, these groups do not have any regulatory authority.

The final layer in regulatory authority for migratory game bird hunting lies with state or provincial fish and wildlife agencies or departments. In the United States of America, most state agencies are governed by a board of commissioners that make final decisions regarding hunting regulations (Hiller et al. 2021*b* [Chapter 2]). A board does not govern provincial agencies; decisions may be vetted through advisory boards to gather input, but decisions are made through the agency and by the appropriate ministers.

REGULATORY PROCESSES

Annual meetings of the Flyway Councils and associated technical groups provide an effective forum for federal, state, and provincial representatives to engage and interact with each other in discussing and coordinating migratory game bird conservation and management activities. Representatives from other governmental (e.g., U.S. Geological Survey) and non-governmental agencies or organizations (e.g., Ducks Unlimited, Inc.) may attend to provide input or insights on various issues. The flyway

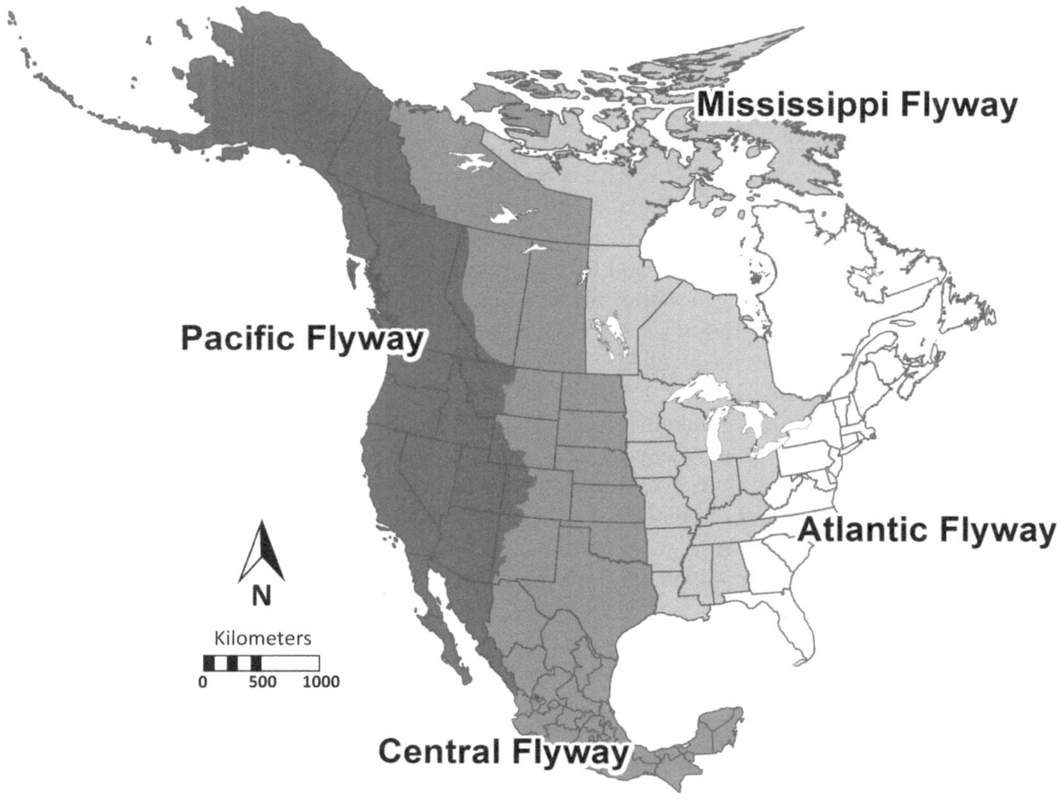

FIGURE 20.1 The four administrative flyways in North America used to make regulatory decisions on migratory game birds (U.S. Fish and Wildlife Service, public domain image).

system also acts as a means to provide public input on the federal regulations. Although the state and provincial agencies provide individual input, the flyways allow state and provincial agencies to discuss regulations and provide their input to the federal authorities via a unified voice. Such unified input typically carries more weight with the federal governments than individual agency comments.

Migratory game birds are unique, relative to resident game, with regards to their hunting seasons in one additional, important characteristic. An annual, federal process is required to *open* migratory game bird hunting seasons, as all seasons are closed until opened by the federal process. In contrast, most resident game hunting seasons are open until regulations establish closed periods. Thus, some state process and regulations governing resident game populations do not need to be opened or approved on an annual basis whereas an annual process is a requirement for migratory game birds.

Changes to migratory game bird regulations or management actions may come "top-down" or "bottom-up." For example, a top–down change may be a result from a change in the population status of a species or a population (e.g., Pacific and mid-continent populations of greater white-fronted geese *Anser albifrons*, Atlantic and Pacific populations of brant *Branta bernicla*) that may compel a harvest reduction commensurate the current population status. Such a case would cause the U.S. Fish and Wildlife Service or Canadian Wildlife Service to dictate some level of harvest reduction by working within the Flyway Council framework to determine the best action or method to accomplish the harvest reduction. Conversely, hunters of a particular states or provinces may express a desire in regulations to accommodate some indicated preference (e.g., increase in bag limit) or open a new season on a migratory game bird. Hunters express that preference to state or provincial wildlife agencies who then bring the issue to the Flyway Councils for discussion.

Either way, the status quo or proposed changes are formulated into hunting season recommendations and submitted by each Flyway Council to the U.S. Fish and Wildlife Service for consideration on an annual basis. The Service Regulations Committee (commonly referred to as *the SRC*), the decision-making body for the U.S. Fish and Wildlife Service comprised upper-level administrators, considers the Council's recommendations along with input from their staff and other stakeholders in setting annual regulations. The SRC then makes their decisions on those recommendations and forwards them to the Director of the U.S. Fish and Wildlife Service and Department of Interior for final approval. They then are published through the Federal Register process (https://www.federalregister.gov/) for public comment. It should be noted that prior to the recent approval of the *Final Supplemental Environmental Impact Statement* on issuance of annual regulations permitting the hunting of migratory birds (U.S. Fish and Wildlife Service 2013), regulations were set based on that year's breeding population survey and other information. More recently, migratory game bird seasons have been established a year in advance to breeding population survey and other information. For example, the 2020–2021 hunting seasons were established during fall 2019.

The current process is different in Canada, where regulation changes occur every two years and has three steps. The Canadian Wildlife Service first reviews migratory game bird population status information and publishes a report (Canadian Wildlife Service Waterfowl Committee 2020*a*). Then in consultation with the provinces, territories, and other stakeholders, regulation proposals including changes are developed for each province or territory. The proposals may start following the previous regulation period and may be included in annual reports (Canadian Wildlife Service Waterfowl Committee 2020*b*). Final regulatory proposals include proposed changes, their rationale, and predicted effects, which are then published in a final report (Canadian Wildlife Service Waterfowl Committee 2018). For example, the formal process for the 2020–2021 and 2021–2022 hunting seasons began in September 2019 with review of population status and development of regulation proposals in November 2019. The population status report was published in January 2019 to support the development of the regulation proposals report. Further public input and other consultation occurred, and the regulations were finalized in June 2020.

Typically, prior to setting regulations, states gather input on stakeholder preferences for regulatory options via surveys, public meetings, and other means (Hiller et al. 2021*b* [Chapter 2]). As mentioned previously, provincial (and some state) governments may gather input from advisory boards that comprised interested stakeholder groups. States then develop regulation or hunting season proposals based on that input as well as other biological, social, or political information. Although varied, state processes generally include notices or decrees notifying the public about hunting seasons (Hiller et al. 2021*b* [Chapter 7]). The public can provide input through their commissioners or other representatives, written (including email) comments, or attend public commission meetings and provide oral input. Regulations and hunting seasons are then finalized and the U.S. Fish and Wildlife Service is formally notified of hunting season selections. Final regulations enacted by states may be the same or more restrictive, but not more liberal, than federal regulations.

HARVEST MANAGEMENT PROCESSES

Although federal, state, and provincial government agencies responsible for migratory game birds have a variety of interests, the formal, cooperative establishment of objectives allows partners to formulate and select regulations and other actions. Primary emphasis is generally placed on maintaining populations of migratory game birds at appropriate levels, dictated by treaty obligations and codified through federal documents outlining overall management objectives for populations and recreational harvest (U.S. Fish and Wildlife Service 1975, 1988, 2013). Migratory game bird populations in relation to recreational harvest is incorporated in those documents, but they do not provide or define overall management objectives for hunter numbers or hunter satisfaction.

Management of some populations of migratory game birds use prescribed strategies. However, the *Final Supplemental Environmental Impact Statement* on migratory bird hunting directed that harvest management policies for migratory game birds would use an *informed management* approach (U.S. Fish and Wildlife Service 2013). Informed management is an iterative process that comprises explicit management objectives, a (limited) set of management actions, monitoring programs, and a set of models that predict or describe system dynamics. All of these elements are discussed, examined, and approved by all stakeholders before implementation, which allows for transparency in decision making. However, some elements, such as objective setting, are difficult because typically there are multiple objectives that may conflict or occur at different scales (Humburg et al. 2000, Johnson et al. 2015, Runge 2021 [Chapter 7]).

Formal adoption of the informed management approach arose from the implementation of *adaptive harvest management* (often referred to as *AHM*) for duck harvest management in the United States of America in 1995. Adaptive harvest management was an extension of adaptive management (Walters 1986) adopted by the U.S. Fish and Wildlife Service and Flyway Councils to make regulatory decisions more transparent and learn about population responses to those decisions (Johnson et al. 1993, Williams et al. 1996). Adaptive harvest management is an informed management approach that recognizes sources of uncertainty in the management process including: (1) *environmental variation*, which is the temporal variation in weather and large-scale habitat conditions that affect waterfowl population dynamics; (2) *partial controllability*, recognizing that management actions may not produce precisely predictable results, partially due to variation in hunter participation and behavior; (3) *partial observability*, which results from incomplete or imprecise monitoring programs (e.g., population surveys, banding) that do not produce accurate measures of the variables of interest (e.g., population size, harvest rate) by the management actions; and finally, (4) *structural uncertainty*, which describes the limited knowledge about the processes affecting the dynamics of the populations. The latter uncertainty also includes lack of knowledge of the relationships between management actions and objectives and populations as well as incorporating different beliefs (e.g., compensatory vs. additive mortality) among managers about management actions on populations (Nichols et al. 1995, Johnson et al. 1997). Adaptive harvest management allows for this last source of uncertainty to be formally addressed and reduced.

Adaptive harvest management for ducks has been examined extensively, and the process is used as the premier example of adaptive management as well as harvest management (e.g., Johnson et al. 1993, Williams et al. 1996, Johnson et al. 1997, Humburg et al. 2000, Johnson and Case 2000, Nichols 2000, Nichols et al. 2007, Johnson et al. 2015, U.S. Fish and Wildlife Service 2019*a*). Since implementation in 1995, the use of adaptive harvest management and several other species-specific harvest management strategies have been developed and implemented (e.g., Boomer et al. 2007, U.S. Fish and Wildlife Service 2007, 2014). These advances served as the new paradigm for harvest management in the last two decades (Conroy 2021 [Chapter 1]). The application of these duck harvest management strategies also presents managers of other species groups with examples of processes and methods to establish objectives, transparency in decision making, and connecting information from monitoring programs.

The relationship between informed and adaptive harvest management approaches is that the set of management actions that can be taken may be comprised multiple components of regulatory alternatives. For example, framework dates, season length, daily bag limit, and shooting hours are combined as a set of regulatory alternatives called *package*s in adaptive harvest management (U.S. Fish and Wildlife Service 2019*a*). Thus, independent changes for season length or bag limits are not possible in regard to annual duck regulations. Establishment of regulatory packages causes some management flexibility to be lost, because changes in packages must be discussed and agreed upon by all stakeholders. However, combining these regulatory alternatives allows evaluation of their cumulative impact in regard to the other elements of adaptive management (e.g., objectives, population models) and allows for more focused discussion on those elements. Thus, managers spend less time on discussions or disagreements about appropriate season length and bag limits.

TABLE 20.1
Current regulatory alternatives (packages) for the four flyways in the United States of America based on adaptive harvest management approaches used in each flyway (U.S. Fish and Wildlife Service 2019*a*)[1]

	Flyway			
Regulation and season	**Atlantic**	**Mississippi**	**Central**	**Pacific**
Shooting hours	One-half hour before sunrise to sunset			
Opening date				
Restrictive	October 1	Saturday nearest October 1		
Moderate	October 1	Saturday nearest September 24		
Liberal	October 1	Saturday nearest September 24		
Closing date				
Restrictive	January 31			
Moderate				
Liberal				
Season length				
Restrictive	30	30	39	60
Moderate	45	45	60	86
Liberal	60	60	74	107
Daily bag limit				
Restrictive	3/—	3/2/1	3/3/1	4/3/1
Moderate	6/–/–	6/4/1	6/5/1	7/5/2
Liberal	6/–/–	6/4/1	6/5/2	7/7/2

Note

[1] Daily bag limit numbers in format *X/X/X* refer to total daily bag/drake mallard total/hen mallard total. In the Atlantic Flyway, the mallard daily bag limit is not prescribed by the regulatory alternatives under their multi-stock adaptive harvest management protocol.

Flyway management plans for some species or populations (e.g., mid-continent sandhill crane) may not use an informed approach, but these flyway-specific plans do provide guidelines for changes in management actions in response to population status (e.g., Mississippi Flyway Council 1996, Central, Mississippi, and Pacific Flyway Councils 2015, Central Flyway Webless Migratory Game Bird Technical Committee 2018). When management plans include prescriptions of regulatory alternatives (e.g., season lengths and bag limits, similar to packages found in the adaptive harvest management process; Table 20.1) to be considered, there typically are a limited set of alternatives to be taken.

REGULATORY ALTERNATIVES

Managers may employ a number of regulatory actions that are designed to meet migratory game bird population objectives, recognizing that substantial changes are typically needed to produce measurable changes in harvest (Johnson and Moore 1996). Additionally, those regulatory actions need to meet hunter or other user group's objectives, while also avoiding an objectionable perception that might create opposition. In the same manner, regulatory actions must avoid being too cumbersome or complex to induce unintentional violations. More recently, there has been an increased emphasis on how regulations affect hunter participation with attempts at formulating regulations that increase hunter participation or decrease loss of existing hunters.

Managers use two types of regulations to meet the above stated goals: basic and annual regulations (Reeves 1993, Baldassarre and Bolen 2006). The United States of America and Canada (and Mexico) have *basic regulations* that primarily deal with hunting methods and are not subject to annual consideration (e.g., see the U.S. Code of Federal Regulations, Part 20, Subpart C). Basic regulations include acceptable and unacceptable methods of take (e.g., types of weapons, baiting, shot types, use of live decoys), and importing and exporting migratory game birds (Reeves 1993). The two basic regulations most pertinent to migratory game birds are baiting and use of lead shot. There has been a long history of trying to define baiting and what is legal or illegal for both dove and waterfowl hunting (e.g., Reeves 1993). The banning of lead shot in the 1990s in the United States of America and Canada for waterfowl hunting was controversial, despite evidence that lead poisoning caused substantial and widespread mortality (Bellrose 1959, Sanderson and Bellrose 1986, Havera et al. 1992). Although still legal in most instances, the use of lead shot for other migratory game birds, such as doves, has also become a recent topic of discussion (e.g., Schulz et al. 2002, 2006). Although basic regulations protect migratory game bird populations, this type of regulation is not the set of "tools" by which managers use to meet migratory game bird population and hunter objectives.

Annual regulations are the primary set of actions managers use to achieve harvest management objectives. Annual regulations include framework dates, season length, daily bag limits, and shooting hours. The process for change in annual regulations is different from basic regulations, and the regulatory and harvest management processes described above typically deal with annual regulations. These annual regulations are briefly described below including statements regarding their effectiveness in managing migratory game bird populations and hunter acceptance of those regulations. A number of other sources also describe and provide histories of various annual regulations (e.g., Martin and Carney 1977, Blohm 1989, Baldassarre and Bolen 2006, U.S. Fish and Wildlife Service 2013).

Framework Dates

The Migratory Bird Treaty Act specifies that hunting seasons must take place between September 1 and March 10 to avoid breeding seasons. However, for most migratory game birds, the opening and closing season dates (i.e., framework dates) established by the U.S. Fish and Wildlife Service and the flyways typically have been more restrictive. Framework dates for duck hunting have been manipulated in recent years with changes in 2002 and 2019. The extension of earlier framework dates for doves seems to be a constant source of debate among hunters as well. There may be some impact of changing framework dates on population levels (Lercel et al. 1999, Johnson and Dubvosky 2001) as well as the spatial distribution of harvests (Haugen et al. 2014). To date, framework date changes have been more associated with hunter satisfaction or participation. Over 70% (35 of 48) of states in the continental United States of America selected the earliest possible opening date, latest possible closing date, or both, for the 2019–2020 duck hunting seasons, suggesting hunter support for recent extensions to framework dates.

Season Length

The Migratory Bird Treaty Act specifies that hunting seasons, including falconry, can be a maximum of 107 days within the established framework dates (see above). Season length has perhaps the most important influence on annual harvests and is of particular importance to hunters. Numerous studies have shown that season length was positively related to annual estimates of harvests and harvest rates (e.g., Martin and Carney 1977, Hindman et al. 1998, Rexstad 1992,

Afton and Anderson 2001, Heusmann and McDonald 2002). However, relatively large changes in season length may be necessary to produce significant changes in harvest or harvest rate (Johnson and Moore 1996, Johnson et al. 1997) because so many other factors influence harvest (e.g., weather, habitat conditions, and hunter activity). Additionally, proper evaluation of the effect of season length on harvest and harvest rate has been difficult because changes in daily bag limit or other regulations have typically accompanied changes in season length (Miner and Bart 1989, Nichols and Johnson 1989, Reeves 1993).

Although most hunters only hunt a few days during the season (Dubovsky 2019*b*), longer seasons allow individuals more opportunities to harvest migratory game birds (but see Haugen et al. 2015). Surveys of waterfowl hunters all indicated season length is of primary importance (Ringelman 1997, National Flyway Council and Wildlife Management Institute 2006, Slagle and Dietsch 2018*a,b,c,d*). Little information is available on season length and hunter preferences for other migratory game birds. However, other migratory game birds migrate early enough that season length is not as important a factor as daily bag limit.

Other than increasing or decreasing overall season length, the only other modification of season length was for a shorter season on a particular species within the regular duck season. This was used in the United States of America and was called the "season within a season" (SWAS) approach (Gammonley et al. 2010). This approach was used initially for canvasback (*Aythya valisineria*) and northern pintail (*Anas acuta*) in the early 2000s, where the harvest strategies indicated some harvest was allowable, but the general season length would result in harvests that exceeded objectives for these species (Gammonley et al. 2010). Flyway Councils requested and were granted shorter seasons for canvasback and northern pintail within the longer, overall duck season (e.g., 39 days within a 74-day season) instead of having a closed season (i.e., no harvest allowed) for these species. Although the SWAS approach was used for other duck harvest strategies, it has more or less been dismissed as it added regulation complexity, but still put hunters at risk of unintentional violations as part of the season was still closed on certain species (Gammonley et al. 2010). Even so, this approach seemed to acceptable by most (62%) duck hunters as an alternative to having shorter seasons for all ducks (National Flyway Council and Wildlife Management Institute 2006).

Daily Bag Limit

Daily bag limits are one of the more frequent regulatory alternatives used by managers to influence harvest. The daily bag limit is the number of migratory game birds that can be harvested by an individual in a day. An important distinction needs to be made regarding daily bag limits between migratory game birds and resident game birds in that daily bag limits for migratory game birds apply across jurisdictions. For example, an individual hunting in Nebraska can shoot 15 mourning doves (migratory species), but cannot legally go to Kansas and shoot 15 more in the same day (but could shoot eight in Nebraska and seven in Kansas). However, the same individual could shoot the state limit of three ring-necked pheasants (resident species; *Phasianus colchicus*) in Nebraska, and then drive to Kansas and shoot a limit of four there.

The utility of daily bag limits in affecting harvest is somewhat mixed, probably due to the species or population and their life-history traits in regard to mortality (more additive or compensatory). Additionally, daily bag limits typically have been manipulated along with season length; therefore, confounding evaluations (Nichols and Johnson 1989, Péron et al. 2012) or changes made are not sufficient enough to detect changes (e.g., Lyons et al. 2020).

Nonetheless, changes in daily bag limit are easy for hunters to understand and the sociological implications of daily bag limits may be more important than biological ones. Although the importance to hunters and their preferences for bag limits appear to be variable (e.g., Slagle and Dietsch 2018*a,b,c,d*), evidence is emerging that at least among waterfowl hunters, their satisfaction is linked to their harvest (Slagle and Dietsch 2018*b*, Gruntorad et al. 2020). Interestingly, if asked about the importance of harvest or achieving a full daily bag limit, it is generally ranked low (Schroeder et al. 2006,

2019*b*, Gruntorad et al. 2020). Additionally, information from waterfowl hunters indicates that satisfaction is linked to their expectations (Brunke and Hunt 2008, Bradshaw et al. 2019). Given that few duck hunters harvest a full daily bag limit on a particular hunt (Haugen et al. 2017), setting bag limit regulations in relation to expected harvest levels may be an important consideration for managers.

Managers have altered daily bag limits in an attempt to reduce regulatory complexity while protecting some stocks of migratory game birds and maintaining opportunity for others. One such alteration is the use of an aggregate bag, where only a certain number of either species or even sex-specific limits can be taken within an overall daily bag limit. For example, the daily bag limit for doves (white-winged [*Zenaida asiatica*], mourning, and white-tipped [*Leptotila verreauxi*]) in Texas in 1994 was 10, but a legal bag could only include two mourning and two white-tipped doves (George et al. 1994). Rails (*Rallus* and *Porzana* spp.) are still managed using aggregate daily bags in the United States of America.

A more recent attempt using the aggregate bag limit concept was the *Hunter's Choice* experiment conducted in the Central Flyway (Gammonley et al. 2010). Hunter's Choice was an attempt to direct harvest pressure at some duck stocks and away from others. Hunter's Choice used a one-bird aggregate daily bag limit within the overall daily bag limit—as an example, a daily bag limit is five ducks, with only one total duck from any of the four components of the following group: female mallard, mottled duck (*A. fulvigula*), northern pintail, and canvasback. Thus, the stocks needing protection were included in the one-bird aggregate bag and buffered by the overall harvest of each individual stock as well as by all other ducks in the overall daily bag limit, in addition to lowering the overall bag limit (from six to five). Hunter's Choice retained a common season length for all ducks and allowed at least one duck in the daily bag that the hunter may not accurately identify before harvesting, thereby avoiding some unintentional take. The name of this system was derived from the decision that would occur if a duck in the one-bird aggregate bag was harvested. At that point in the hunt, it was the "hunter's choice" to continue hunting and risk harvesting another duck in the one-bird aggregate (Gammonley et al. 2010).

Unfortunately, Hunter's Choice did not meet expectations in sufficiently reducing harvest of stocks needing harvest protection (Gammonley et al. 2010). One main factor was that stocks needing protection were not adequately buffered given their co-occurrence in the daily bag with other species (Haugen et al. 2015). For example, northern pintail, canvasback, and female mallard are typically not harvested within the same daily bag in the Central Flyway (Haugen et al. 2015).

Another attempt in modifying or altering conventional bag limits was the *Point System* approach (Smith and Dubovsky 1998). This approach also had the objective of directing harvest away from those stocks needing protection and maintaining or affording more harvest opportunity on others. The Point System assigned different point values to each species (and sexes for mallard) with higher points assigned to those needing more protection and lower values to abundant or less harvested species. Thus, for example, canvasback = 100 points, hen mallard = 90 points, male mallard = 20 points, and northern pintail = 10 points, and the daily bag limit was reached when a hunter met or exceeded 100 points for the day. Thus, the approach was that hunters would harvest species and sexes with lower point totals to increase their overall daily bag and avoid those with higher point totals.

Although the Point System was used for decades, it was eliminated by the U.S. Fish and Wildlife Service as an option for duck season frameworks in 1994. The main reasons the Point System was discontinued were uncertainty about its effectiveness and enforcement concerns, particularly the "re-ordering" of ducks (i.e., claiming a 100-point bird was harvested last, but was actually harvested first, thus allowing the hunter to continue hunting) in a hunter's daily harvest (Smith and Dubovsky 1998).

The most recent alteration to daily bag limits is the *two-tier system*, with initial evaluation efforts beginning in the 2021–2022 hunting season in Nebraska and South Dakota. The objective of this system is unique in that the changes in daily bag limits are not primarily directed at protecting or affording more hunting opportunity at duck stocks, but rather increasing waterfowl hunter participation. In this system, duck hunters have the option of choosing one of two daily bag limits: one has the current, full daily (i.e., 6/day) bag limit with all species- and sex-specific regulations, or the other that has a reduced bag limit (i.e., 3/day) and no species- or sex-specific regulations (i.e., can shoot any three ducks). Thus, the two-tier system is proposed as a solution to a constraint to participation by individuals with little or no experience duck hunting who may have difficulty identifying duck species (Hinrichs 2019).

Bonus Birds

The *bonus bird* regulation is a unique combination of the concepts of daily bag limits and special seasons. Bonus birds have been allowed either on some lightly harvested species of duck or in lieu of adding another season (U.S. Fish and Wildlife Service 1988). Bonus birds were used primarily for scaup (*A. marila* and *affinis*) and blue-winged teal (*Anas discors*), and bonus bird options were discontinued during the restrictive seasons in the late 1980s and early 1990s. However, with recent expanded teal hunting opportunity (Teal Harvest Potential Working Group 2013), bonus birds were again offered for blue-winged teal. Two additional blue-winged teal are permitted during the first part of the regular duck season in northern production states, in lieu of providing an additional, special September teal season.

Zones and Split Seasons

Zones are geographic areas within a state separated by distinct boundaries that have different hunting season dates (primarily, but other hunting regulations may differ). Hunting seasons also may be *split*, where the total allotted season length is divided into two or three segments (e.g., an open period followed by a closed period and then an open period again). Zones and splits are used to spread out harvest opportunity based on type or availability of habitat or typography, align hunting seasons with temporal peaks in migratory game bird abundance, and allow for multiple *opening days* (initial season opening dates, which typically have higher participation rates) to increase hunting success, and hunter preferences for hunting seasons. Zones and splits have been combined for ducks (i.e., the number of zones a state may want then determines the number of splits they are allowed), but may not be combined for other migratory game birds.

Although they complicate regulations, zones and splits appear to be popular among hunters as they spread out or increase hunting opportunity. Further evidence of support is that the majority of states in the continental United States of America use zones or splits for migratory game birds. Zones and splits increase the exposure of hunting on migratory game birds. However, the effects of increased hunting opportunity on migratory game bird populations from use of zones and splits remain largely unknown due to the different scales of monitoring programs measuring population demographics and the scales at which zones and splits are applied (Nichols and Johnson 1989).

Shooting Hours

Shooting hours are those hours in which a game species can be legally harvested. In the United States of America, the shooting hours for migratory game birds are one half-hour before sunrise to sunset, whereas in Canada, the closing time is one half-hour after sunset (with some state and provincial differences). Shooting hour restrictions are typically used to reduce harvest (Martin and Carney 1977, Kirby et al. 1983, Blohm 1989), provide temporal refuge, improve or maintain hunt quality, or help facilitate hunter identification of waterfowl (Baldassarre and Bolen 2006).

Probably the most widely known application of shooting hours has been the U.S. Fish and Wildlife Service restriction of shooting hours to sunrise to sunset for most waterfowl seasons in 1988 to reduce harvest (Blohm 1989, Baldassarre and Bolen 2006). Waterfowl shooting hours restrictions were also in place for the opening day of duck seasons during the 1940s–1960s (Martin and Carney 1977). Generally, use of further restricted shooting hours (e.g., shooting hours end at noon) has been limited or implemented at small scales (e.g., public wildlife management areas).

Hunter acceptance of restricted shooting hours is mixed, depending on the application and reason behind the restriction (Pierce et al. 1996, Schroeder et al. 2006, Dinges 2013). Managers encountered strong opposition to shooting hour restrictions in 1988 for waterfowl hunting. Opposition may occur when hunters perceive regulation changes to result in an apparent loss in hunting opportunity, whereas support would follow from a perceived increase in hunt quality or opportunity.

Season or Area Closures

Regulated, recreational harvest seldom has been a primary factor contributing to a species' or population's decline in numbers (Leopold 1939). However, when species or population levels have declined due to other factors, the typical, initial response by federal or state authorities is to close hunting seasons to help reverse the decline. The most recent hunting season closures occurred in 2002 and 2008 for canvasback (with the exception in the Central Flyway in 2008 for the aforementioned Hunter's Choice experiment). A more dramatic closure for Canada goose (*Branta canadensis*) in the Atlantic Flyway was enacted from 1995 through 1999 (Hindman et al. 2004). The effectiveness of season closures may depend on the factors associated with the species or population decline and the life history of the species or population in question. Hindman et al. (2004) suggested that the hunting season closure helped facilitate the recovery of the Atlantic population of Canada goose.

Area closures are used to a lesser extent than season closures and prohibit harvest of a species or group of species at a smaller spatial scale than closures aimed at the continental or flyway level. Managers may use such closures when hunters may encounter a majority of the species or population at a specific, local area and a reduction in harvest is necessary. Typically, the size of the area closures may be slightly larger than an individual refuge or wildlife management area, but less than a zone, and no other restrictions are in place for other species or populations. The recovery of Aleutian cackling goose (*B. hutchinsii*) contained area closures to hunting as part of their recovery (Springer and Lowe 1998).

Closed seasons are not popular among managers or hunters. Managers grapple with the philosophical issue regarding reductions in harvest, which is often the first action taken for species with declining populations even when the decline is not a result of recreational harvest. Thus, these actions may send a message to stakeholders that recreational harvest has a greater impact on a species or population than it does in reality. Further, the unpopular nature of closed seasons among hunters reverberates up to managers. Unintended violations of restrictions may occur when hunters encounter multiple species of migratory game birds, including those for which the season has been closed. Greater understanding and acceptance of season or area closures by hunters appears to depend on how much other hunting opportunity for other species is maintained (National Flyway Council and Wildlife Management Institute 2006).

Permits and Quotas

Permit or quota systems result in the harvest of a limited number of individuals from a population. Permits or quotas are imposed to allow some harvest opportunity, but require a more intensive system of limited take from a population for either biological or sociological reasons. Increased emphasis is placed on the monitoring program (e.g., aerial surveys) used to track population trends because of the regulated harvest and need for specific numbers of permits available or quota to be harvested. Population trends are then used to determine the number of permits available or the quota to be harvested. Permit or quota hunts are currently used on tundra swan (*Cygnus columbianus*) and sandhill crane populations, but also have been used for some Canada goose populations (U.S. Fish and Wildlife Service 2013).

The allocation of permits or quotas can be controversial (e.g., Miller 1998), and these systems require a higher degree of cooperation and agreement among management agencies involved in the harvest than other regulatory alternatives. A good example of the background, application, and potential for controversy of a permit or quota system can be found in the management of the Mississippi Valley population of Canada goose in the Mississippi Flyway (Miller 1998). For hunter acceptance, permit or quota systems require fair and equitable allocation (e.g., Glass and More 1992). Otherwise, permit or quota hunts tend to be of high quality (e.g., limited number of other hunters) or provide opportunity for a "trophy" species.

Special Seasons

Special seasons have been implemented for migratory game birds with the primary objective of providing additional hunting opportunity on populations capable of sustaining additional harvest (Ladd et al. 1989). Special seasons are typically of shorter duration from regular hunting seasons. The best example of additional hunting and harvest opportunity may be the September teal season in the Atlantic, Mississippi, and Central Flyways, directed primarily at early migrating blue-winged teal. However, special seasons can be more local or regional, such as those for white-tipped dove hunting seasons in Texas (Waggerman et al. 1994).

Special seasons also have been implemented to help alleviate nuisance problems, primarily resident Canada goose breeding in the conterminous United States of America. These seasons are used to reduce local populations causing nuisance or depredation problems while avoiding negative impacts to migrant populations (U.S. Fish and Wildlife Service 2005). Although these goose seasons increase hunting opportunity and may help alleviate nuisance problems, their effectiveness in reducing populations is questionable (Heusmann 1999*a*, Groepper et al. 2012, Iverson et al. 2014, Paukert et al. 2021 [Chapter 18], but see Lyons et al. 2020).

Finally, there are special seasons that have been implemented where the primary objective is to provide separate hunting opportunity for a set of hunters rather than focus on additional harvest. The best example are youth waterfowl seasons that have been in place in the United States of America since 1996. The primary objective of this special season is to introduce and recruit new hunters into waterfowl hunting. More recently, special seasons for veteran's and active military personnel were granted through the John D. Dingell, Jr. Conservation, Management, and Recreation Act in 2019. Both these seasons are limited to a few days. Although the modest amount of data indicates that youth waterfowl seasons do not contribute much to the overall harvest, participation and harvest during veteran's and active military seasons may exceed participation in youth seasons and has not yet been evaluated. The effectiveness of youth waterfowl seasons in recruiting hunters has not been assessed.

Duck Hunting Management Units

As the name implies, these are areas or regions providing additional hunting opportunity for ducks, primarily mallard, in the Pacific (Columbia Basin) and Central Flyways (High Plains). Duck management units are designated using banding, and other biological data indicating more hunting opportunity (i.e., longer seasons) for mallard within the units are justified (Ladd et al. 1989). Impacts to other species are relatively low and as noted above with season length, additional hunting opportunity through longer seasons is preferred by hunters.

Conservation Orders

The Migratory Bird Convention Act in Canada and the Migratory Bird Treaty Act in the United States of America authorized the use of conservation orders to address nuisance, depredation, or other negative issues or impacts with migratory game birds. Conservation orders are implemented when other management actions, particularly recreational hunting seasons, are not effective in addressing the problem. To be clear, conservation orders are not hunting seasons and are not implemented to increase or provide additional hunting opportunity. However, techniques used during conservation orders may be the same ones used to regulate recreational hunting.

A conservation order can take place outside the Migratory Bird Treaty Act framework dates for which regular hunting seasons are allowed (i.e., before September 1 and after March 10), remove bag and possession limits, and permit other methods of take that are prohibited during regular hunting seasons (e.g., shooting hours, use of unplugged shotguns).

Recent depredation issues with resident Canada goose and overabundance issues associated with light geese and degradation of subarctic habitats have resulted in use of conservation orders (U.S. Fish and Wildlife Service 2005, 2007). However, the use of conservation orders in addressing problems with light geese have had mixed results (Reed and Calvert 2007, Alisauskas et al. 2011, Calvert et al. 2017, Paukert et al. 2021 [Chapter 18]), and no formal evaluation has been conducted on resident Canada goose. However, implementation of conservation orders may have had other impacts. For example, managers found that it was more difficult to find support for more intensive management actions (e.g., direct control on breeding grounds) among those participating in the conservation order for light geese (Dinges et al. 2014).

State and Provincial Regulations

Within the constraints of the frameworks outlined in the Migratory Bird Treaty Act, and federal and flyway regulations, states and provinces may enact their own additional regulations to allocate harvest or hunting opportunity as long as they are not currently covered by higher, authoritative-level regulations. Probably the most common example across states and provinces is the use of spatial and temporal refuges or sanctuaries (i.e., derivatives of closed areas and shooting hours). Use of spatial and temporal refuges is primarily for maintaining abundance of migratory birds in an area or for protecting a particular species or population. Changes in refuges can be controversial, as closing or opening an area will affect some segment of hunters by altering their hunting opportunity. Similar to permits and quotas, states also may use a draw or lottery system on public areas to limit the number of hunters using the property each day. Restricting the number of hunters may address crowding problems, but also may provide a high-quality hunt experience. States or provinces also may place limitations or restrictions on hunting equipment and tactics, which can be on either local or statewide scales. Examples include limiting the number of shotgun shells allowed by an individual onto a public area or the type of decoy allowed (e.g., spinning-wing decoys).

MONITORING PROGRAMS

Recreational hunting is allowed by federal mandates provided it is commensurate with current population status. Therefore, estimating or indexing migratory game bird populations is necessary, and it may be necessary to obtain information regarding harvest or hunting participation. A tradeoff to the benefits of using an informed management approach is that such processes may require intensive and extensive monitoring systems to track measurable attributes of the management objectives. Monitoring wildlife populations as well as hunter-related objectives may require large investments of resources. These requirements have been the primary limiting factor in successful implementation of an informed management approach beyond migratory game birds. The current status of migratory game bird populations is a requirement of federal mandates, and thus extensive population and harvest monitoring programs have been developed for this group of game species in North America (Martin et al. 1979, Hawkins et al. 1984, Blohm 1989).

Population Surveys

The best example of a long-term, wildlife population survey is the annual May Waterfowl Breeding Population and Habitat Survey conducted since 1955 (Reynolds 1987, Smith 1995). This survey covers an extensive portion of northern breeding grounds primarily for ducks and has been continuous through 2020. Habitat conditions (i.e., number of ponds) are also recorded on this survey. Both duck population counts (number of mallard) and the number of ponds are used in the adaptive harvest management process and other stock-specific harvest strategies for determining hunting season recommendations (U.S. Fish and Wildlife Service 2019*a*). Concurrent with that survey, other states may conduct their own aerial transect surveys with similar methodology to assess duck and other waterfowl populations that also are incorporated into the regulatory process (U.S. Fish and Wildlife Service 2019*a*).

Another important, long-term survey used to index or count populations is the Mid-Winter Waterfowl Survey (Martin et al. 1979). Conducted in early January, this survey provides abundance, distribution, and trend information for waterfowl along major wintering areas. For some species or populations, the Mid-Winter Waterfowl Survey may be the only operational survey to assess populations and has been used to formulate hunting regulations for tundra swan and some populations of brant and Canada goose. However, management decisions based on data from the Mid-Winter Waterfowl Survey have tapered in recent years as breeding surveys or other data streams are being used in its place. Additionally, some states are discontinuing the survey because of varying methodology from state to state, few attempts at estimating detection rates, and safety concerns from use of aircraft (Smith et al. 1989, Heusmann 1999*b*, Fleming et al. 2016). Additionally, important wintering areas in Mexico have not been surveyed since 2006 due to security and logistical challenges. The Mid-Winter Waterfowl Survey still provides, in most cases, the only information on relative abundance and distribution of wintering waterfowl in states where it is still conducted. Thus, these data have been used in regional habitat conservation planning (Fleming et al. 2016).

Other species or populations may also be indexed by separate, specifically designed aerial surveys that are then used to formulate hunting regulations. Special aerial surveys are conducted for the various populations of sandhill crane (Dubovsky 2019*a*), midcontinent population of greater white-fronted goose (Central, Mississippi, and Pacific Flyway Councils 2015), and other goose populations (Trost et al. 1990). The timing (i.e., breeding or migration), methodology, and degree of statistical rigor vary among these surveys.

Other migratory game bird populations are indexed via other methodology. For a number of years, managers assessed mourning dove populations using call count surveys (Dolton 1993, Seamans 2019*b*). However, managers stopped using data from the call count survey and replaced it with banding data to make management decisions in 2013. Banding data provide an index to the population as well as other population parameters of interest such as harvest and survival rates (Otis 2006). The population status of American woodcock is annually assessed with a singing ground survey (Seamans and Rau 2019) and a Mineral Site Survey is used for the Pacific Coast population of band-tailed pigeon (*Patagioenas fasciata*; Casazza et al. 2005, Seamans 2019*a*).

One major gap in migratory game bird population surveys is that currently there are no systematic monitoring programs for rails, snipes, or coots, and for other populations of the band-tailed pigeon. Although coots are counted in the Waterfowl Breeding Population and Habitat Survey, their entire range is not surveyed, and detection of coots from air may be difficult (Alisauskas and Arnold 1994). The Breeding Bird Survey (Robbins et al. 1986, Sauer et al. 2017*b*) or other local surveys may be used to provide an index of these populations.

Banding Programs

Arguably, the most important monitoring program for migratory game birds is the use of banding data. Banding involves the capture and marking of individuals with leg bands. Recovery information is obtained when a hunter harvests a banded bird and reports it to a central database, either the U.S. Bird Banding Laboratory or the Canadian Bird Banding Office. Banding data were first used to delineate general migration pathways (Lincoln 1935) and now have become formally incorporated into harvest management decisions (Otis 2006, U.S. Fish and Wildlife Service 2019*a*). Compared to some other techniques, banding is relatively inexpensive, especially given the amount of population information that is gleaned from banding returns. Banding data allow managers to estimate survival and harvest rates, recovery or harvest distributions, and harvest chronology (e.g., Powell et al. 2004, Seamans and Braun 2016, Bartzen and Dufour 2017). More recently, banding data have been used to estimate abundance of some migratory game bird populations using Lincoln estimation procedures (e.g., Alisauskas et al. 2009, 2014), which holds promise for other species where population surveys would be difficult.

Reliance on banding data does require that annual, operational banding programs be maintained. Such an effort is a large responsibility for the U.S. Bird Banding Laboratory and Canadian Bird Banding Office, which control, distribute, and maintain bands and associated banding and recovery databases. Additionally, biologists must conduct periodic analyses to determine *reporting rates*, the probability that on retrieval of a banded bird, the hunter reports the band to a federal bird banding office (Royle and Garrettson 2005, Zimmerman et al. 2009, Boomer et al. 2013). Reward band studies are typically used for this analysis, in which a subset of banded birds is marked with an additional aluminum band that notifies the finder that they are entitled to a monetary reward after reporting the band (Williams et al. 2002). Sufficiently high reward amounts (e.g., $100) have been shown to result in close to 100% reporting rates, which allows investigators to estimate reporting rates by comparing the proportion of reported reward bands to standard bands (Nichols 1991, Royle and Garrettson 2005).

Harvest and Hunter Surveys

Managers may also obtain annual harvest estimates for migratory game birds as a source of information to use to make decisions or assess impacts of recreational harvest on populations. Annual harvest estimates have been conducted longer and in a more standardized fashion for waterfowl than for other migratory game birds. Annual waterfowl harvest estimates in the United States of America and Canada are obtained by using an overall sampling framework in which hunters are asked about their harvest and hunting activity via a mail survey. Additionally, a sample of hunters from that survey are also asked to participate in a survey where they mail in wings of ducks and tail feathers of geese from each bird harvested during the hunting season. These parts are identified to species, sex, and age by federal, state, and private biologists during annual "wing bees" conducted after each hunting season. The combination of these surveys results in detailed estimates of the species, sex, and age composition of the annual waterfowl harvest. Similar mourning dove and woodcock "wing bees" are also conducted to obtain age and sex information.

Each state conducted its own harvest survey for other migratory game birds such as doves, rails, snipe, and woodcock until 1999. The sampling frames and methodology were different for most states, so the derivation of reliable annual harvest estimates for these species was difficult. Additionally, federal harvest surveys (i.e., Mail Survey Questionnaire) only sampled hunters that purchased a federal duck stamp. A federal duck stamp is not required to hunt species such as doves, rails, and woodcock; thus, only estimates of duck hunters hunting these other species were obtained. An effort to standardize sampling frames and methodology was initiated through the Harvest Information Program ([known as *HIP*; Elden et al. 2002). Each hunter is required to register for the program in each state they hunt migratory game birds, and registered hunters are then sampled for their harvest and activity. Although there are problems with the Harvest Information Program (e.g., vendors registering hunters that do not hunt migratory game birds; see Ver Steeg and Elden 2002), the program has provided more reliable harvest estimates for non-waterfowl species. These harvest surveys also provide estimates of the number of active waterfowl hunters and measures of their activity, such as number of days hunted per season and average harvest per hunter (Raftovich et al. 2019, Gendron and Smith 2019). Finally, annual totals of the number of federal duck stamps sold can also provide an index to trends in hunter participation (Vrtiska et al. 2013) as well as individual states through their license or permit databases (see Graham et al. 2021 [Chapter 6]).

Managers have recently prioritized obtaining hunter activity, expenditures, and attitudes and preferences toward migratory game bird management and regulations at regional or national scales (e.g., Ringelman 1997, National Flyway Council and Wildlife Management Institute 2006, Slagle and Dietsch 2018*a,b,c,d*). Although information for these surveys has not yet been formally incorporated into management decisions, managers have recognized the need to consider this aspect when formulating or implementing harvest regulations. The recent increased emphasis on hunter recruitment, retention, and reactivation (known as *R3*) efforts throughout the United States of America has further underscored the need for such information (Council to Advance Hunting and the Shooting Sports 2016).

INFORMATION NEEDS AND THE FUTURE

Managers have still much to learn, for some species or populations of migratory game birds, about basic life history, abundance, and demographics, outside the context of harvest management. Despite much research, questions and debate remain about the true nature or impact of recreational harvest on migratory game bird populations. Adaptive management provides the framework to examine the relationship between harvest and populations, but questions remain regarding density dependence and compensatory thresholds of harvest mortality for different migratory game bird populations (Conroy 2021 [Chapter 1]). Combining harvest, biological, and ecological information in a more comprehensive management that combines all approaches (e.g., Osnas et al. 2014) may be the future model for migratory game birds.

The backbone of migratory game bird management in North America is the unique suite of monitoring programs. The potential loss or erosion of monitoring programs could cause a negative feedback loop that leads to the loss or reduction of recreational hunting opportunity. Such losses of opportunity, in turn, have potential to lead to additional declines in hunter participation, and subsequently losses to habitat conservation and management programs. The cumulative effect, of course, is the potential for population declines (Vrtiska et al. 2013). Consequently, administrators must be informed and reminded of the value of maintaining monitoring programs relative to the whole enterprise of migratory game bird conservation and management. Demands on budgets are omnipresent, and in the future managers must explore and develop new technologies and methods that make monitoring programs more efficient. Additional monitoring programs are needed for some species or populations of migratory game birds that are under-represented or for which we have information gaps.

Managers also need to develop and refine hunter-related objectives that include understanding factors that influence hunter recruitment, retention, and reactivation (Graham et al. 2021 [Chapter 6], Gruntorad and Chizinski 2021 [Chapter 4]), understanding influences on hunter's support for conservation, and evaluating management actions that are intended to affect hunter participation. Additionally, further efforts are needed to formally model and incorporate hunter participation into harvest management decisions (Runge 2021 [Chapter 7]). Agencies also will need to try new strategies to increase participation (e.g., two-tier duck regulations) and increase flexibility in regard to regulations. Without more concerted effort toward hunter recruitment, retention, and reactivation, the future of recreational harvest may deviate from the North American Model of Wildlife Conservation (e.g., Organ et al. 2013) and irrevocably change to whom and how harvest management is applied with regard to migratory game birds.

The collaborative flyway process is essential to address the aforementioned needs. The flyway system provides the structure for those most knowledgeable, passionate about, and directly involved in or responsible for migratory game bird management to come together in the decision-making process. The coordinated effort will continue to foster cooperation. Perhaps more importantly, the flyway system has proven essential to bring resources to support the work, as evidenced by goose monitoring programs and the process used to identify priority information needs for webless migratory game birds.

In conclusion, the future of migratory game bird harvest management will be challenging, because of the future status of habitat and population levels of migratory game birds. Climate change, continued conversion and degradation of habitats to meet human demands, and invasive species are just some of the factors that will affect migratory game bird populations. Without abundant populations, harvest management will need to reduce uncertainties (Conroy 2021 [Chapter 1], Dahlgren et al. 2021 [Chapter 21]) to justify recreational harvest. It is clear that the future of migratory game bird management will require biologists, managers, and other conservationists to engage a dynamic base of public stakeholders to change behaviors and actions that affect all wildlife.

ACKNOWLEDGMENTS

Development and organization of this chapter was assisted by J. Gammonley and S. Boomer. Readability and clarity was significantly improved with reviews by G. Balkcom and T. Cooper. Reviews by B. Haw, V. Griess, L. Nelson, and N. Ott were extremely helpful in catching errors, and the help of editors K. Pope and L. Powell were greatly appreciated. Finally, K. Vrtiska had the patience to put up with the sometimes stress-filled individual writing this chapter.

21 Upland Game Bird Harvest Management

David K. Dahlgren, Erik J. Blomberg, Christian A. Hagen, and R. Dwayne Elmore

CONTENTS

INTRODUCTION

North American upland game birds consist of 24 species with extant wild-propagated populations. Twenty two of these species currently have regulated hunting seasons managed primarily by state wildlife agencies. However, herein we will focus our attention on Galliformes of the continental United States of America that are hunted: grouse ($n = 10$), quail ($n = 6$), partridge ($n = 3$), ring-necked pheasant (*Phasianus colchicus*), and plain chachalaca (*Ortalis vetula*). All are native to North America, except partridges and pheasants. Although American woodcock (*Scolopax minor*) and various species of dove (Columbidae family) may generally be considered upland game bird species, harvest is federally regulated as migratory species (see Vrtiska 2021 [Chapter 20]), and we will only address resident game birds. Wild turkey (*Meleagris gallopavo*) have typically been managed differently than all other upland game birds and we will not consider them here, nor will we address harvest of upland game birds that are artificially stocked.

We have a long history and deep-rooted culture of upland game bird hunting in the United States of America. Each year more than two million hunters seek upland game birds and spend over US$1 billion on hunting-related expenses (U.S. Fish and Wildlife Service and U.S. Census Bureau 2016). Therefore, the best available harvest management strategies are warranted for the conservation and long-term sustainability of these species and future hunting opportunities. Considering the recent declines in hunter numbers in the United States of America (U.S. Fish and Wildlife Service and U.S. Census Bureau 2006, 2016) and the increased scrutiny by the general public concerning consumptive activities such as hunting (Hiller et al. 2021*a,b* [Chapter 2, 23]), there will likely be an increased demand for justification and transparency in harvest management decisions. Society's tolerance for hunting, including the future of upland game bird harvest, will depend on having defensible harvest management strategies. State wildlife agencies and hunters should strive to adapt and provide the resources necessary for the progression of science-based harvest management for upland game birds.

For most upland game bird populations in the United States of America, the underlying goal of harvest management is to ensure sustainability of populations while providing hunting opportunities to the public. Early scientific literature on wildlife harvest management in the mid-twentieth century reported high levels of compensatory harvest mortality for upland game birds (Errington 1945, Campbell et al. 1973, Campbell 1985). The concept that upland game bird harvest was generally compensatory was taught broadly in wildlife education and many of these students became professionals that eventually helped set upland game bird harvest regulations (Leopold 1933). This and the fact that some wildlife species were increasing in abundance led to a general liberalization of harvest regulations for upland game birds throughout much of the United States of America in the 1960s and 1970s (Connelly et al. 2005). However, recent studies have suggested that some of those early findings were likely biased or overgeneralizations (Hudson and Dobson 2001, Kokko 2001, Williams et al. 2004). For example, biologists often did not account for the mechanisms of emigration and immigration, which may have led to incomplete conclusions about population changes due to harvest mortality (Williams et al. 2004). Relatedly, metapopulation theory has rarely been incorporated into harvest theory for upland game birds.

The characteristics of a species' life cycle has important implications for potential harvest impacts, and therefore regulations. Upland game birds vary in life-history strategies (Johnsgard 1973, 1983). For example, quail and partridge exhibit shorter generation times than most grouse species. Short-lived species tend to have lower adult survival and higher reproductive output (e.g., larger clutch sizes), whereas the opposite tends to be true for longer-lived species. Two useful concepts of potential population-level harvest impacts include additive and compensatory harvest mortality. Additive harvest mortality occurs when individuals are harvested and their mortality reduces annual survival in the population, such that fewer animals remain relative to what would be expected without harvest (Conroy 2021 [Chapter 1]). Compensatory harvest mortality depends on a density-dependent feedback in which the removal of individuals through harvest results in more per-capita resources available to the remaining non-harvested individuals, increasing their subsequent survival and resulting in no net loss to the population. Though all upland game birds have potential for both compensatory and additive harvest mortality, longer-lived species inherently have a lower threshold of harvest where compensation can operate compared to shorter-lived species (Cooch et al. 2014). Harvest rarely operates at either fully compensatory or fully additive, but this gradient between compensatory and additive harvest mortality provides a model for us to understand population dynamics as they relate to harvest.

Harvest management strategies should ideally reflect the life-history characteristics of a given species. However, the general liberalization of upland game bird regulations set by state wildlife agencies in the first half of the twentieth century has primarily been a result of trial and error, tradition (i.e., maintaining the previous year's regulations), and a conservative approach to changing regulations (i.e., reduction in season length and bag limits; Powell et al. 2011). One recent species-specific exception would be harvest of greater sage-grouse (*Centrocercus urophasianus*) for which most states with current hunting seasons have implemented some form of *adaptive harvest management* that has allowed harvest regulations to be adapted in a formal process that includes monitoring data (see detailed explanation below).

In this chapter, we will address (1) the sportspersons who harvest upland game birds, (2) the current process for setting harvest regulations, (3) the tools used by managers for harvest management, (4) evaluations of harvest management strategies, and (5) how to improve upland game bird harvest management in the future.

SPORTSPERSONS

Wildlife managers spend a considerable amount of time focused on managing and working with people. The human dimensions of wildlife management broadly includes conservation education, landowner technical assistance, public relations, and policy guidance. Considerations for human dimensions of harvest management include hunter expectations, food and subsistence needs, access issues, demand, and ethical considerations (Graham et al. 2021 [Chapter 6], Gruntorad and Chizinski 2021 [Chapter 4]). As wildlife harvest inherently involves people, managers must understand this stakeholder group.

Survey data exist that allow us to look at trends related to human dimensions of harvest management over time. One of the most important data sets regarding hunter demographics and participation is the *National Survey of Fishing, Hunting, and Wildlife-Associated Recreation*, which is conducted every five years. In 2016, there were approximately 11.5 million hunters in the United States of America and about 3.5 million of those hunted small game (upland game birds, rabbit and hare, and squirrel). This represents a 27% decline since 2006, the largest decline among hunter groups during that period. During the same period, the decline across all hunters was only 8% (U.S. Fish and Wildlife Service and U.S. Census Bureau 2006, 2016). In 1991, there were an estimated 7.6 million small-game hunters resulting in nearly a 50% decline in the number of small-game hunters in the last 25 years (Fig. 21.1; U.S. Fish and Wildlife Service and U.S. Census Bureau 1991). We see similar rates of decline for pheasant and quail hunters over the last 25 years, but a much higher rate of decline in grouse hunters. Collectively, quail, pheasant, and grouse hunters hunted 16 million days in 2016 compared to 77 million days in 1991. Not only did the number of upland-bird hunters decline by half between 1991 and 2016, but those that hunted in 2016 spent only half as many days afield as hunters in 1991.

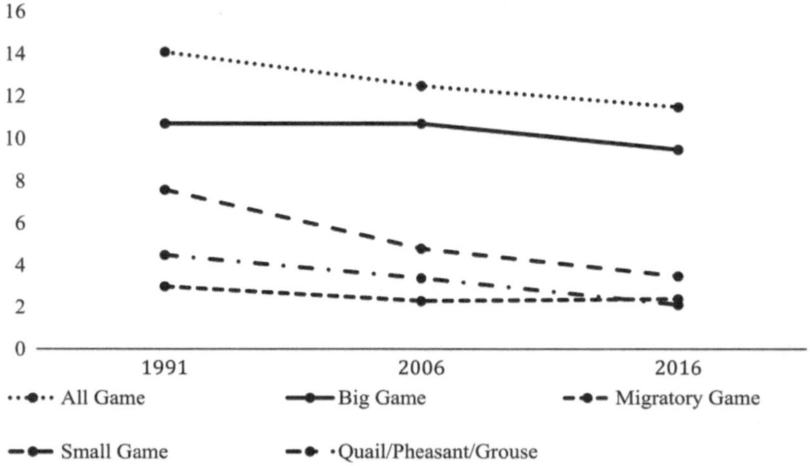

FIGURE 21.1 Estimated number of hunters (in millions) for various hunter groups as reported by the U.S. Fish and Wildlife Service's National Survey of Fishing, Hunting, and Wildlife-Associated Recreation for 1991, 2006, and 2016.

Despite declines in participation and effort, small-game hunters still provide a significant economic benefit. Small-game hunters spent an estimated $1.7 billion in 2016, which is approximately 15% of the total hunting expenditure that year (U.S. Fish and Wildlife Service and U.S. Census Bureau 2016). Private land supports a large proportion of small-game hunter effort as only 24% of small-game hunters hunted public land nationally. Therefore, opportunities to expand small-game hunting (and management) on public land or public access to private land programs may be warranted in many states. However, opportunities will vary based on regional differences in the availability of public versus private lands and the willingness of private landowners to participate in public access programs. Even in states that have areas available for public hunting, these lands are often poorly distributed, which can be a barrier for small-game hunters (Wszola et al. 2020*b*). Small-game hunters were evenly distributed between urban (45%) and rural (55%) residents. Similar to big-game hunters, small-game hunters were likely to be older (i.e., ≥ 55 years old), male (90%), and white (97%). They also tended to be more affluent and had slightly higher education levels than other hunters (U.S. Fish and Wildlife Service and U.S. Census Bureau 2016).

In summary, upland game bird hunters are rapidly declining, hunting fewer days, aging, more educated, and lack underrepresented social and economic groups. Generally, they are more affluent and spend significant monetary resources on hunting and incidental expenses, and this trend is especially pronounced for hunter groups for some species in certain regions such as leased hunting on private lands for northern bobwhite (*Colinus virginianus*) in Florida and Georgia. There are several reasons that likely contribute to these demographic trends. For many upland game bird hunters, success and satisfaction is often higher using bird dogs. Bird dogs require special training to be effective, which takes time and resources. Additionally, access to large areas of land is needed for most upland game bird hunting and lack of access can be a barrier. Many upland game bird populations have experienced long-term declines; northern bobwhite, greater sage-grouse, gray partridge (*Perdix perdix*), and ring-necked pheasant are far less abundant than they were a few decades ago (Sauer et al. 2017*a*). These population declines, combined with changes in land access (e.g., increases in leased or posted land, or greater distance to available hunting land), has hindered hunter opportunity and their justification for continued participation. For some upland game bird hunters, shifting to other forms of recreation is likely and this trend may be expected to continue unless upland bird populations and access are substantially increased. Despite these trends, we believe that upland game bird harvest will continue for the foreseeable future because of its long-standing tradition and a seeming increase in the dedication of upland hunters who continue to participate.

STRATEGIES FOR HARVEST MANAGEMENT

Upland game birds are considered non-migratory (i.e., resident species), state-managed species or fall under tribal jurisdiction. As such, they are not subject to federal management, including regulations of the Migratory Bird Treaty Act of 1918, and only fall under federal oversight if listed under the Endangered Species Act of 1973. Harvest of upland game birds has been managed solely by state or tribal wildlife agencies, which have historically focused on limiting overexploitation. Regulations have been largely relegated to season length and daily bag and possession limits. Although little has changed for upland game bird harvest management over the last several decades, some species (e.g., greater sage-grouse) are now managed more adaptively via permit or tag and quota systems.

States and tribes vary in their approach to setting harvest regulations. Many agencies have transitioned away from setting annual regulations for upland game birds to establishing a multi-year (i.e., three to five years) framework. Such an approach enables managers to provide hunters with consistent expectations for hunting opportunities, but may limit adaptability. Most state wildlife agencies have a public input process with oversight from a board of commissioners, usually governor appointed (Hiller et al. 2021*b* [Chapter 2]). As a group, board members ultimately set harvest regulations, usually based on recommendations from agency staff and public input. Staff evaluate reasonable, public-suggested changes and provide analysis as to potential outcomes of such changes. Notably, approval of the framework by the board does not preclude annual changes to regulations, especially if there is a stochastic event necessitating imminent actions to alleviate the potential for unacceptable population reduction resulting from harvest.

Statewide regulations are common for upland game birds, although regulations may vary spatially within a state for a given species or set of species. Sometimes these delineations of management subunits are driven by climatic, physical geography, or animal density continua (or combinations thereof), but generally align with administrative units (e.g., county, parish, or region) that facilitate monitoring of harvest and hunter participation. States may use the same boundaries established for big game management units, or managers may use counties or groups of counties to define a spatially meaningful unit. The biological relevance of these units for management is largely untested.

Wildlife agencies, in a few situations, may manage harvest of upland game birds at fine spatial scales using a process resembling adaptive harvest management in which previous harvest and population status and trend guide annual adjustments of permits and quota. For example, greater sage-grouse harvest in Oregon and Utah uses lek counts (i.e., breeding period monitoring), production information, and hunter effort and success to establish annual permit or tag numbers for each management unit.

TOOLS

Here, we summarize common tools used by state wildlife agencies to manage upland game bird harvest, and we present our list in declining order of relative frequency of use. States usually administer particular tools as regulations with implementation at a statewide scale. However, some regulations are specified within geographic subareas such as wildlife management districts. Thus, managers may elect to implement variants of any particular tool (e.g., alternative bag limits or season lengths) to reflect differences in population densities or variable management goals. We provide an example (see Fig. 21.2) of one state's regulatory structure for the 2020–2021 hunting season, which illustrates how combinations of these regulatory tools are implemented in practice.

Bag Limits

A *bag limit* is the number of individuals of a species that a sportsperson hunting upland game birds is allowed to harvest in a specified period. To our knowledge, every present-day harvest management system for wild-propagated upland game birds in the United States of America includes some form of a bag limit, although in some cases seasonal limits or quotas are used rather than a daily bag limit.

Managers may consider several variants of a bag limit. In some cases, states have implemented an aggregate bag limit for multiple species. For example, in many western states, species of forest grouse (e.g., spruce grouse *Falcipennis canadensis*, dusky grouse *Dendragapus obscurus*, sooty grouse *Dendragapus fuliginosus*, ruffed grouse *Bonasa umbellus*) and quail (northern bobwhite, California quail *Callipepla californica*, mountain quail *Oreortyx pictus*, Gambel's quail *Callipepla gambelii*, scaled quail *Callipepla squamata*) have had aggregate bag limits. Aggregate bags may specify that a single species only comprise a subset of the daily bag. For example, in Washington State hunters are allowed to harvest four forest grouse daily, but no more than three of any one species (Fig. 21.2).

Another variant on the concept of bag limits is that of sex-specific limits. The sex of some species of upland birds can be readily identified based on plumage and other physical characteristics. These range from cases of extreme sexual dimorphism such as ring-necked pheasant to subtle differences as in ruffed grouse and sharp-tailed grouse (Johnsgard 1983). Some partridges and species of quail cannot reliably be sexed without dissection. Currently, sex-specific bag limits

SPECIES	AREA	SEASON DATES	DAILY BAG LIMIT	POSSESSION LIMIT
Forest Grouse (Blue*, Ruffed, and Spruce) *Includes Sooty & Dusky	Statewide	Sept. 1 - Dec. 31	4 of any species, to include not more than 3 of each species	12 of any species, to include not more than 9 of any one species
Sage and Sharp-tailed Grouse, Ptarmigan	**Closed Statewide**			
Pheasant *At the Samish release site pheasants will only be released during the youth and senior seasons. Please see the WDFW website (***https://wdfw.wa.gov/hunting/locations/upland-bird***) for alternative sites.	Western Washington	Sept. 19 & 20 (***Youth Only***)	2 either sex	4 either sex
		Sept. 21-25 (***Hunters 65 Years or Older, Hunters with Disabilities***)	2 either sex	10 either sex
	Western Washington ***Regular Season***	8:00 a.m. to 4 p.m. Sept. 26 - Nov. 30	2 either sex	15 either sex
	Western Washington ***Extended Season (no pheasants released)***	8:00 a.m. to 4 p.m. Dec. 1-15 ONLY at Belfair, Fort Lewis, Kosmos, Lincoln Creek, Scatter Creek, Skookumchuck, & Whid- bey Island (except Bayview) release sites	2 either sex	15 either sex
	Eastern Washington	Sept. 19 & 20 (***Youth Only***)	3 cocks only	6 cocks only
		Sept. 21-25 (***Hunters 65 Years or Older, Hunters with Disabilities***)	3 cocks only	15 cocks only
	Eastern Washington ***Regular Season***	Oct. 24 - Jan. 18	3 cocks only	15 cocks only
California (Valley) Quail and Northern Bobwhite	Western Washington	Sept. 26 - Nov. 30	10 mixed bag	30 mixed bag
	Eastern Washington	Sept. 26 & 27 (***Youth Only***)	10 mixed bag	20 mixed bag
	Eastern Washington ***Regular Season***	Oct. 3 - Jan. 18	10 mixed bag	30 mixed bag
Quail (Mountain)	Western Washington	Sept. 26 - Nov. 30	2	4
	Eastern Washington	**Closed throughout Eastern Washington**		
Partridge (Chukar & Gray)	Eastern Washington	Sept. 26 & 27 (***Youth Only***)	6 chukar & 6 gray	12 chukar & 12 gray
		Oct. 3 - Jan. 18	6 chukar & 6 gray	18 chukar & 18 gray

FIGURE 21.2 An example of current, species-specific upland game bird harvest regulations for the state of Washington during the 2020–2021 season (from Washington Department of Fish and Wildlife 2020*b*). Area indicates portion of state, daily bag limit is maximum number of birds for harvest in one day, possession limit is maximum number of harvested birds possessed by a hunter at any time, and cocks refers to male birds.

have only been applied to ring-necked pheasant. However, sex-specific limits on other species, especially polygynous species, have potential to be used where increasing male harvest would likely have minimal effects on population growth.

Managers may also use dynamic bag limits, where the number of birds allowed to be harvested on a given day changes throughout the hunting season. For example, more restrictive bag limits may be set during the early portion of the season (e.g., opening weekend or the first week) when hunter numbers, and thus harvest pressure, are greatest. Alternatively, bag limits could be made more restrictive during later portions of the season when harvest is more likely to be additive to natural mortality. Such dynamic bag limits have not been common, but could be useful to maintain or increase opportunity while limiting the total realized harvest. The average daily bag of upland game bird hunters would need to be considered in relation to the total realized harvest. Most states report that hunters have average daily bags of far less than bag limits allow, so altering bag limits may or may not translate to substantial changes in the total harvest (Vrtiska 2021 [Chapter 20]).

Season Length and Timing

The use of limits on season length, which restricts the number of days per year that sportspersons are allowed to harvest a particular species, has been nearly ubiquitous across harvest management regulations. Season dates determine the timing of harvest and can have implications for the relative vulnerability of animals to harvest, hunter success, and the effect of harvest on survival to breeding season, as each of these factors may be related to environmental variation (e.g., snowfall) occurring throughout the season. Season lengths can vary widely from as short as a few days to nearly half the year. Of the tools available to managers, season length may have the greatest influence on hunter opportunity. Restricting a daily bag limit may alter the number of birds a hunter can harvest, but a shortening of the season length can reduce the total amount of time available to the hunter.

Managers may also choose to alter the length of time during a particular day when birds may be taken, otherwise known as *shooting hours*. Nearly all management systems restrict harvest to daylight hours, often sunrise to sunset (± 30 minutes), although in some cases certain periods of the day are also restricted. For example, the state of South Dakota has traditionally restricted pheasant hunting to after 1000 or 1200 hours, depending on the time of season. Several heavily hunted wildlife management areas in Oklahoma restrict hunting until 1630 hours for quail and ring-necked pheasant. Decisions to limit shooting hours may be to reduce harvest potential during times of day when birds may be more vulnerable, or may be for social reasons, such as to encourage hunters to frequent local businesses before or after their hunts.

Possession Limits

A possession limit restricts the total number of animals that may be possessed by one person at any given time, for example the number of harvested birds that a hunter may carry in their vehicle. Under most regulations for possession limits, once a hunter consumes the harvested birds they are allowed to harvest more birds. The definition of what constitutes possession varies. Possession can be taken to mean whole birds, such as hunters would carry during a multi-day trip afield, or can be interpreted as the total number of unconsumed birds they possess (including those left at home in their freezer). Restricting possession to two or three times the daily bag limit is most common, but not all states explicitly regulate possession.

Quotas, Seasonal Limits, and Permits

This type of regulation reflects limitation for the total number of animals an individual sportsperson can take within a particular season, in contrast to the use of simple possession limits without quotas or seasonal limits. Managers use quotas, seasonal limits, and permits as more restrictive tools than bag and possession limits. A quota applied to the entire season is the same as a seasonal limit, but quotas may be applied to a specific time interval within a season. Permits may reflect *individual bird* permits (e.g., one permit allows a hunter to harvest one bird) with the option to issue multiple permits per hunter. However, permits may also be used in combination with bag limits or quotas to restrict the number of *hunters* allowed to pursue a given species. Managers distinguish between these tools to achieve different management goals: a bird-based permit places a direct upper cap on the number of birds that may be harvested, whereas a hunter-based permit system restricts opportunity, usually with an indirect effect on the total numbers harvested. Across upland game bird species, quotas, seasonal limits, and permit systems have not been applied as widely as more generic bag limits, but they have been used regularly for specific taxa and circumstances. For example, a number of state agencies use permit- or quota-based systems to manage harvest of greater sage-grouse.

Special Hunts

These represent dedicated periods where only a distinct subset of hunters are allowed to pursue and kill upland birds. Examples can include youth or first-time hunters, veterans, or persons with disabilities. Although managers may be typically motivated by a goal of increasing opportunity to under-represented groups, these hunts most often occur before the start of general hunting seasons. As such, special hunts are both a tool for managers to implement and a source of harvest mortality and harvest opportunity to be considered when setting harvest management goals and implementing tools for the general hunting seasons.

Method of Take

The majority of hunters pursue upland birds with a shotgun and birdshot and in some cases harvest is explicitly restricted to this method of take. However, there are cases where upland game birds are allowed to be taken by any firearm either by explicit regulation or omission of restriction on type of firearm. In addition, some states allow harvest by archery or falconry. Unlike big-game hunters, upland-bird hunters are normally expected to follow the same set of regulations within a regular season regardless of method of take. The one clear exception is falconry, where state agencies often restrict bag limits, but allow longer season lengths relative to other harvest methods.

EVALUATION AND MANAGING UNCERTAINTY

SOURCES OF UNCERTAINTY

Williams (1996, 1997) broadly classified uncertainties related to harvest management into four general groupings that include environmental variation, structural uncertainty, partial observability, and partial controllability. *Adaptive harvest management* is an iterative process with linkages between decision making, population monitoring, and data analysis that attempts to explicitly account for each level of uncertainty for harvest management decision making (Conroy 2021 [Chapter 1], Runge 2021 [Chapter 7]). Such an approach, regardless of level of sophistication, can be useful and an improvement to current harvest strategies. Among upland game bird harvest systems in the United States of America, even relatively well-developed approaches, such as those for greater sage-grouse, rely on a large number of assumptions and do not fully address all elements of uncertainty. A new paradigm for upland game bird harvest management should embrace the challenge to both account for uncertainty in the decision process and reduce uncertainty when possible.

Environmental Variation

The stochastic nature of environmental conditions and the response of population dynamics to the stochasticity may reduce the ability to predict future populations with certainty. Upland game bird populations have often been used as model organisms to understand the principles that govern population dynamics (e.g., Martinez-Padilla et al. 2014). As such, this form of uncertainty has probably received the greatest attention from past research, albeit not always in a harvest management context. Nevertheless, for the vast majority of upland game birds we still lack fundamental understanding of extrinsic and intrinsic processes affecting population growth, such as the role of weather and density dependence, respectively.

When developing harvest management strategies, managers can address uncertainty associated with environmental variation through use of alternative mechanistic models that relate measures of population dynamics (e.g., time series of counts) to explanatory predictors such as weather and habitat. Models fit to time series data can then be used to detect the presence of intrinsic properties and their interaction with extrinsic factors. In cases where multiple competing factors may be responsible for driving population dynamics, these can be posed as competing hypotheses represented by alternative model forms, with greater weight assigned to the model that better predicts subsequent year's count data.

Structural Uncertainty

Managers often have incomplete knowledge of the degree to which harvest produces additive mortality, which would lower the subsequent breeding population relative to what would be expected absent harvest (Williams 1996, 1997). Structural uncertainty in general has received some research attention, but even for some of the best-studied upland game birds there is considerable debate concerning compensatory versus additive harvest effects. The most robust studies have either made use of experimental designs to evaluate responses in survival under differing harvest strategies (Williams et al. 2004, Devers et al. 2007, Sandercock et al. 2011) or have used time series of mark-recapture data to explore co-variance between annual survival and harvest rates (e.g., Sedinger et al. 2010, Péron 2013). Unfortunately, these sound study designs have been relatively rare for upland game birds. Other studies have correlated spatiotemporal variation in harvest regulations with population indices (e.g., Connelly et al. 2003), yet this approach may have limited inference because of confounding potential due to variation between harvest management decisions and other factors affecting population dynamics (Sedinger and Rotella 2005). Another method used to assess evidence for additivity can be found in applications of adaptive harvest management where alternative forms of population models that assume either additive or compensatory harvest are fit to data, and their predictive ability tested as harvest regulations are implemented (Powell et al. 2011). Through time, greater cumulative weight for one model form over the other may be taken as evidence for presence of additive or compensatory harvest in governing the population dynamics. However, this approach may be particularly sensitive to other sources of uncertainty in the model structure, and should probably be approached with caution (Conn and Kendall 2004, Sedinger and Herzog 2012).

Partial Observability

All monitoring designs have some degree of imperfect ability to sample upland bird populations, thus limiting the ability of harvest management to respond to changing populations. Partial observability can be overcome to a great extent through careful design of population surveys (e.g., randomized or stratified random sampling) and use of robust analysis techniques (e.g., incorporation of detection probability). An extreme form of partial observability occurs when information on population status is totally absent, either because it is ignored during the decision process or is not collected and available for use. In a review of North American upland game bird monitoring programs conducted by Sands and Pope (2010), just over half of managers (56%) reported using population monitoring data to inform their harvest management, suggesting that in nearly half of management systems for upland game birds, there is a complete lack of observability. Such a situation, of course, extends this issue beyond simple uncertainties derived from sampling design.

Another source of partial observability occurs because of time lags between the collection of data that informs management decisions and the timing of harvest under those decisions. Rules and regulations are normally set well in advance of the onset of hunting seasons; thus, some degree of time lag is probably unavoidable, but as the gap between information gathering and management implementation widens, so too does the uncertainty associated with partial observability. For example, many agencies conduct spring breeding surveys for various upland game birds to assess population trends, and managers could use these survey data to inform fall harvest regulations. However, spring surveys may not necessarily predict summer brood production, and thus changes in the numbers of birds reflected in spring counts may not be representative of the fall population available for harvest. Information on summer productivity can be used to adjust spring estimates to fall abundance and help mitigate time lag issues. However, harvest regulations for upland game birds are not typically changed on an annual basis. In fact, mean review time for regulations within state agencies in North America in 2004 was 2.4 years (Sands and Pope 2010), so significant time lags likely exist for upland game bird harvest management decisions.

The primary solution to address partial observability is to collect more and better data on game bird populations that follow best practices for survey design (e.g., sampling stratification) and analysis (e.g., detection probability or observer bias). Ideally, these would include annual information on abundance and population trend, along with information on reproductive output if significant time lags exist between population surveys and the onset of the hunting season. Conceptually, abundance and reproductive surveys could be completed concurrently, but such approaches are uncommon in the United States of America. Another useful tool is that of modeling and simulation, which can be used to highlight population parameters to which harvest decisions are highly sensitive and identify areas where more effective data collection is necessary.

Partial Controllability

Managers face incomplete control of harvest rate, also referred to as *implementation uncertainty* (Moa et al. 2017). That is, harvest management includes ambiguity of whether changes in regulations actually change the level of harvest. Managers do not affect upland bird abundance directly, but rather attempt to change hunter behavior to affect harvest (Gruntorad and Chizinski 2021 [Chapter 4]). For example, in response to population declines a manager may elect to reduce the daily bag limit. However, if only a small subset of hunters ever successfully fill their daily bag (likely, especially under high bag limits; Guthery et al. 2004) then this regulatory change may have little effect on the total harvest. A better alternative strategy for reducing harvest in that scenario may be a change to season structure (e.g., shortening the total length of season). Implementation uncertainty is noteworthy because it is directly related to actions under managers' control, but it has probably received the least attention from upland game bird researchers compared to the other three types of uncertainty (Moa et al. 2017). To our knowledge, for most upland game birds in the United States of America we lack answers to fundamental questions such as whether bag limits or season lengths have a greater effect on total harvest, whereas for other taxa, such as waterfowl, many of these questions have been addressed (e.g., Balkcom et al. 2010, Haugen et al. 2015).

Decreasing this form of uncertainty is perhaps most within the scope of managers to address. Basic data collection involves estimating how many birds are harvested within a particular management region, which can be accomplished through a variety of methods including check stations, mandatory harvest reporting, or post-hunt surveys. The more challenging aspect involves conducting harvest management under an experimental design to explicitly test the outcomes of the management action. For example, consider our earlier scenario of changing bag limits. The management region of interest could be subdivided into two or more experimental units, where the bag limit is reduced in some units (i.e., treatment) and left unchanged in others (i.e., control). Additionally, managers should explore the magnitude of change (i.e., reduction in harvest relative to control) to assess whether the change met management goals. These data would provide the manager the information needed to either lower the bag limits at a larger scale or (if the results are inconclusive) consider another management alternative such as shortening the hunting season. We note the dynamic of partial controllability may be exacerbated in the future by low overall game bird hunter numbers and their changing demographics. Managers may have to contend with shifting hunter success rates and resulting changes in the efficacy of particular harvest regulations.

Currently, information needs with the highest priority for evaluation of upland game bird harvest include assessment of the uncertainty associated with environmental variation, structural uncertainty, partial observability, and partial controllability. Unfortunately, to date relatively few resources have been dedicated to filling these information gaps, perhaps related to decision makers' past assumptions of exclusive compensatory harvest mortality for most upland game birds. Clearly, a paradigm shift in harvest management strategies for upland game birds is needed.

Adaptive Harvest Management for Upland Game Birds

Current harvest management for most upland game birds does not use an adaptive harvest management approach, but such a process could ultimately result in improved science-based harvest strategies (Williams and Johnson 1995). We suggest that adaptive harvest management can be implemented as either *active and complete* or *reactive and incomplete* (Fig. 21.3). *Active adaptive harvest management* uses population modeling to assess harvest impacts when adjusting future harvest regulations. *Reactive adaptive harvest management* forgoes harvest impact assessment and adjusts future harvest regulations based on population changes only. Whether active or reactive, adaptive harvest management provides opportunity to defensibly adjust harvest regulations based on environmental stochasticity, changes in harvest, and fluctuations in population size (see Hauser and Possingham 2008, Allen et al. 2011).

Not all upland game bird populations will require full implementation of adaptive harvest management, but there are opportunities for improvement in many cases. Contemporary harvest of greater sage-grouse in some western states provides one of the few examples of adaptive harvest management for a species of upland game bird. Increasing conservation concerns at the turn of the

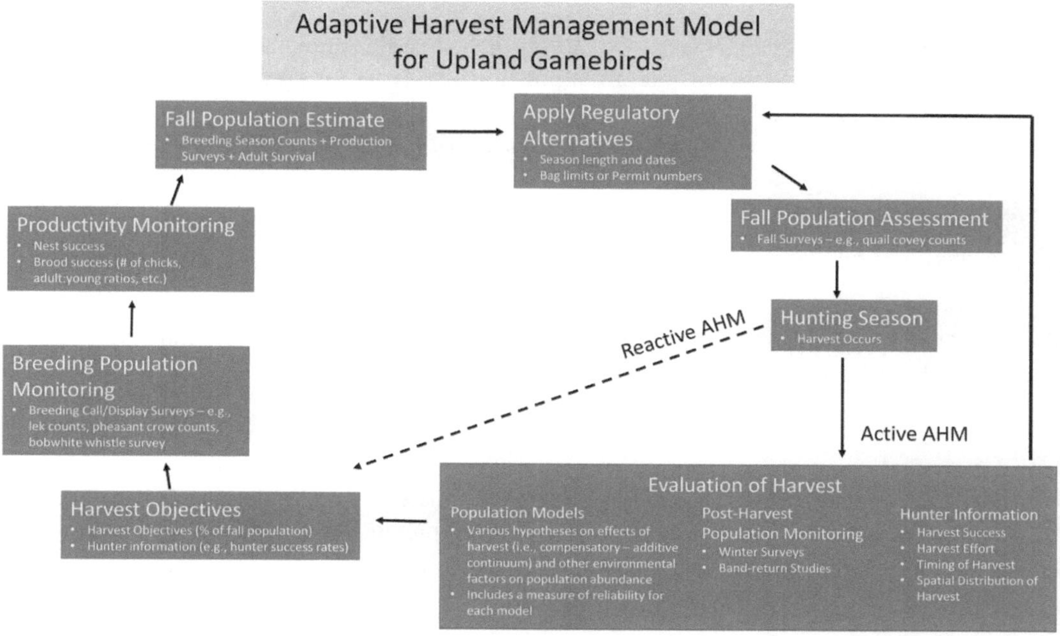

FIGURE 21.3 A proposed adaptive harvest management model for upland game birds, with active and reactive pathways denoted. Moving toward this type of model would aid managers in making harvest decisions that are defensible and more likely to ensure population persistence.

twenty-first century provided the impetus for states to adopt adaptive harvest management strategies. Managers developed the approach to minimize the risk of harming populations, while maintaining hunting opportunity for this iconic species. Below, we explore sage-grouse harvest management in Oregon as an example of a reactive adaptive harvest management system.

BOX 21.1 OREGON GREATER SAGE-GROUSE HARVEST STRATEGY: ADAPTIVE HARVEST MANAGEMENT

Here, we provide an example of an adaptive harvest management strategy for greater sage-grouse harvest in Oregon, and we define quantitative relationships to help the reader understand how each biological metric can be used to inform harvest management decisions (Equations (21.1)–(21.6)).

Early September hunting seasons are established to ensure wing collection (prepaid envelopes) contains adequate molt patterns to determine population composition. Tags are offered through a lottery system with the goal of allocating enough tags to collect ~100 wings per hunt unit. Tag numbers are adjusted annually to ensure that <5% of the fall population is at risk of harvest. Fall populations are estimated (Equation (21.3)) using spring male lek count data, five previous years of wing data (male:female sex ratio), and current year summer production (juvenile:adult female) surveys. Post-hunt surveys estimate past harvest success and provide foundation for future projections.

$$\mathbf{Tags}_i = maxH_i \div (HPR_i \times HSR_i) \tag{21.1}$$

where i represents hunt unit, H is harvest, HPR is hunter participation rate from previous five years and HSR is hunter success rate from previous five years for unit i, and

$$maxH_i = N_{fall} \times 0.05 \tag{21.2}$$

where N_{fall} is estimated fall abundance of sage-grouse for unit i, and

$$N_{fall} = M_i + F_i + J_i \tag{21.3}$$

where M_i is fall male abundance for unit i, estimated from spring male lek counts ($minM_i$) adjusted downwards (10%) to account for losses over summer,

$$M_i = \ min\ M_i \times 0.90 \tag{21.4}$$

and, where F_i is fall female abundance for unit i, estimated by multiplying spring male lek counts ($minM_i$) by male:female sex-ratio ($M{:}F$) from previous five years of wing data for unit i,

$$F_i = M_i \times M{:}\ F_i \tag{21.5}$$

and, where J_i is fall juvenile abundance for unit i, estimated by multiplying female:juvenile sex-ratios estimated from the current year's production surveys,

$$J_i = F_i \times J{:}\ F_i \tag{21.6}$$

Although the hunting season for sage-grouse in Oregon is stringently regulated by the Oregon Department of Fish and Wildlife (ODFW), sage-grouse hunting remains popular and permits are allocated through a lottery system (see *Oregon Greater Sage-Grouse Harvest Strategy*).

Oregon's reactive adaptive harvest management model integrates annual and longer-term data for adaptation of harvest regulations to current year population size, but it lacks an explicit feedback loop found in active adaptive harvest management processes to assess harvest impacts. Thus, the current Oregon model is incapable of assessing impacts, if any, from harvest on fall population size, nor can managers assess how changes in tag numbers, season length, or other regulations may affect abundance the following year. This example represents one the most progressive harvest strategies for an upland game bird, but a few additional steps toward active adaptive harvest management would provide invaluable information toward reducing many of the forms of uncertainty. Similar to waterfowl adaptive harvest management (Conroy 2021 [Chapter 1], Vrtiska 2021 [Chapter 20]), population models based on historic or published data (e.g., survival, reproduction, lek counts) could be developed with assumptions of varying degrees of compensatory or additive mortality and best model fit used to assess annual harvest effects (e.g., Powell et al. 2011). A simpler approach might include using collected data to build annual population simulations that include varying harvest rates to assess the level at which fall harvest rate would begin to impact the breeding population. We encourage managers and researchers to work together to implement more adaptive harvest management strategies in the future for upland game birds.

FUTURE

Most upland game bird species are currently managed with relatively liberal bag limits and season lengths in the United States of America (Table 21.1). Justifications for such liberal harvest regulations vary, but generally rely on an underlying, and often unacknowledged, assumption of compensatory harvest mortality for upland game birds or that much of the population is not hunted (inaccessible on private land or remote access on public land) and serves as a source. For many species, there is simply insufficient data to support this assumption, even for species where large amounts of data exist (e.g., northern bobwhite). Obviously, harvest of any wildlife species becomes additive mortality at some point as harvest rate increases (Conroy 2021 [Chapter 1]). The most important questions for guiding harvest strategies may be (1) is harvest likely to approach a level at which a population would be negatively impacted, (2) what factors are more likely to lead to additive mortality in harvest, and (3) what is the optimum level of harvest that can sustain a population?

As an example, we will consider the northern bobwhite, which are faster-lived species relative to other game birds. Correspondingly, most states have a liberal harvest strategy, with an eight- to ten-bird daily limit and season length > 90 days per year being common (Table 21.1). The seasons

TABLE 21.1
Summary of Major Harvest Regulations for Species[1] of Upland Game Birds in the Contiguous United States of America, Based on Information Obtained from State-Specific, Online Regulation Information for the 2019–2020 Hunting Season: Number of States with Hunting Season and Mean (minimum and Maximum) Season Length (Days), Bag Limit (Birds), and Possession Limit (Birds)

		Season Length			Bag Limit			Possession Limit		
Species	Number of States	Min	Mean	Max	Min	Mean	Max	Min	Mean	Max
Chukar	15	73	114.6	164	2	5.3	8	10	17.8	32
Ruffed grouse	30	31	108.1	163	1	3.2	5	4	9.1	16
California quail	7	60	112.6	153	5	10.0	15	15	30.0	45
Gambel's quail	7	51	95.7	129	5	10.7	15	15	30.3	45
Scaled quail	5	81	99.8	129	8	11.8	15	20	31.3	45
Greater sage-grouse	7	2	9.2	30	1	1.8	2	2	3.2	4
Northern bobwhite	38	14	84.8	151	2	7.2	15	6	17.8	45
Montezuma quail	2	65	78.5	92	5	6.5	8	10	17.0	24
Sooty grouse	4	31	106.8	153	2	2.8	3	6	9.0	12
Dusky grouse	9	51	105.6	155	3	3.2	4	6	10.0	12
Spruce grouse	4	109	127.3	155	3	3.8	5	10	11.5	12
White-tailed ptarmigan	3	9	42.3	71	2	3.0	4	2	8.0	12
Mountain quail	4	35	104.3	153	2	6.6	10	2	18.9	30
Gray partridge	17	68	109.4	145	2	5.0	8	9	16.9	32
Ring-neck pheasant	38	4	71.9	141	2	2.4	4	2	8.1	30
Greater prairie-chicken	5	10	94.3	122	2	2.5	3	2	6.8	15
Sharp-tailed grouse	10	15	78.7	123	2	2.8	4	2	7.9	16

Note

1 Scientific names in order of presentation: *Alectoris chukar*, *Bonasa umbellus*, *Callipepla californica*, *Callipepla gambelii*, *Callipepla squamata*, *Centrocercus urophasiaus*, *Colinus virginianus*, *Cyrtonyx montezumae*, *Dendragopus fuliginosis*, *Dendragopus obscurus*, *Lagopus leucura*, *Oreortyx pictus*, *Perdix perdix*, *Phasianus colchicus*, *Tympanuchus cupido*, *Tympanuchus phasianellus*.

typically start in early November and end in mid-February. The current approach to harvest management of bobwhite includes important underlying assumptions: (1) that harvest throughout the season is largely compensatory or inconsequential if additive due to overall low harvest rates, (2) uneven distribution (temporally or spatially) of harvest rates, (3) highly connected populations, and (4) high reproductive potential. However, data indicate that harvest rates can certainly result in additive mortality for bobwhite (Roseberry 1979, Guthery et al. 2000, Williams et al. 2004). Not only is total harvest relevant, but timing of harvest requires consideration (Kokko 2001, Kaemingk et al. 2021 [Chapter 3]).

Examining data from a wildlife management area in Oklahoma that receives regular quail hunting pressure illustrates that as the season progresses the probability of additive harvest mortality likely increases (Fig. 21.4). Bobwhite alive at the beginning of the season have a relatively low probability of becoming potential breeders the following spring (Fig. 21.4); therefore, the probability that harvest mortality will be compensated by density-dependent dynamics is greater earlier in the season. In contrast, birds that survive into February (late in the hunting season) have a high probability of surviving to the breeding season, which decreases the likelihood that density-dependent dynamics will compensate (and thus, increases the potential for additive mortality) for harvest mortality later in the season. These dynamics are partially due to there being less time for natural mortality to occur, but the relationship is not linear. In fact, data from Oklahoma suggest strong inflection points in January (mid-season) and late February (Fig. 21.4).

One might conclude from our example (Fig. 21.4) that ending the hunting season in January would allow more birds to survive to breeding season. However, the mortality used in this example includes harvest mortality and shortening the season could simply shift the inflection point earlier in the season if hunting constitutes a large percentage of total mortality. To make an informed decision regarding hunting season end date, we need information about hunter effort and temporal and spatial distribution of harvest throughout the season (Kaemingk et al. 2021 [Chapter 3]).

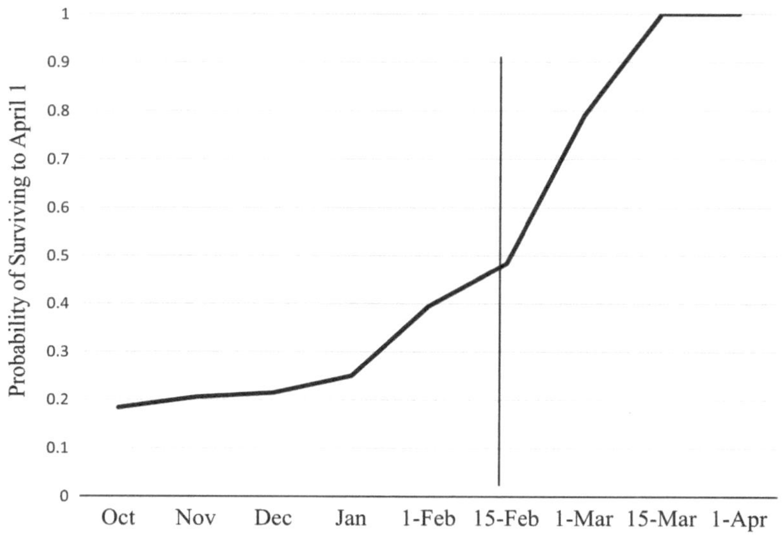

FIGURE 21.4 Probability of northern bobwhite (*Colinus virginianus*) alive at a given time surviving to the beginning of the subsequent breeding season (April 1). Crude survival determined from telemetry data from a Wildlife Management Area in western Oklahoma during 2013–2014. This illustrates that harvest occurring early in the hunting season (e.g., November–December) has much less potential impact to the subsequent breeding stock than harvest in late season (January–February).

Unfortunately, that type of data is rarely collected for upland game birds, even on heavily hunted wildlife management areas.

Another consideration is that immigration from surrounding areas could lead to dubious conclusions regarding harvest impacts as coveys coalesce throughout the season (Williams et al. 2004). Managers may be tempted to consider high harvest irrelevant due to the potential for immigration to compensate. However, compensation due to immigration relies on several assumptions that require careful examination: (1) habitat availability and quality is uniform across the landscape, (2) harvest rate is variable across the landscape and lower off the management area, (3) no barriers to immigration or emigration exist, and (4) dispersal is immediate and complete prior to the next hunting season. We suggest that managers carefully evaluate these assumptions to help provide justification for their harvest management approach. In states with more discrete habitat patches (higher levels of habitat fragmentation) and lower abundance of game bird, these assumptions become even more problematic (Kaemingk et al. 2021 [Chapter 3]).

Habitat for many species of upland game birds may be appropriately viewed as islands in a fragmented landscape as in metapopulation theory (MacArthur and Wilson 1967). In addition to the isolated nature of habitat, hunter harvest can be highly concentrated on specific areas (e.g., public access) increasing risk of local declines (Small et al. 1991, Kaemingk et al. 2021 [Chapter 3]). This spatial mismatch suggests that identification of population source-sink dynamics, knowledge of dispersal potential, harvest levels, and seasonal harvest distribution data are needed. Accordingly, functional population units need to be defined with estimates of abundance, as well as assessment of the effect of abundance on probability of dispersal. Predicting dispersal probability requires an understanding of barriers (e.g., maximum distances and landscape barriers), as well as filters to dispersal (i.e., landscape conditions that lower the probability of dispersal). Often, minimum patch size relative to occupancy probability is not known (see Crosby et al. 2013). There is currently a paucity of this type of data for most harvested upland game bird species, and contemporary harvest management often operates without considering these issues.

In certain situations, current harvest strategies are likely inconsequential, though they still may not be defensible over time. For many upland game birds, high population abundance and large connected intact habitat were generally the norm when current regulations were first developed. However, the human footprint has increased commensurate with habitat fragmentation and hunter access to private land is more restricted. In many situations, this combination has resulted in hunter pressure concentrated on fewer areas potentially reducing their capacity to sustain high levels of harvest, although this dynamic may be masked if not carefully monitored (Williams et al. 2004). Relevant examples for concern include northern bobwhite and ruffed grouse in eastern United States of America. Although habitat conditions and populations of upland game birds have largely declined over the past few decades, in most cases, harvest regulations have remain unchanged.

We are not necessarily suggesting that every harvest unit, management area, county, or parish requires unique regulations. The move from statewide regulations, when needed, will be a challenging step. However, our review suggests that, at a minimum, monitoring and assessment are needed to make informed defensible regulatory decisions, even if regulations do not ultimately change. In many cases, these data are completely lacking, and uncertainty may result from combined dynamics of aggregate bag limits, limited hunter effort and distribution, limited harvest data, and poor population data for game birds. We suggest a shift toward using quantitative and technical tools that can evaluate harvest management strategies and provide rigorous science-based regulatory decisions. To that end, we offer practical approaches to achieving this standard.

Paradigm Shift

The first step in improving upland game bird harvest management would be to assess whether an adaptive harvest management framework is needed for a particular species, population, or hunt area. Regulations and the decision process should be streamlined to achieve the objective.

Changes to the current decision-making system may not be needed for species with relatively high abundance and large connected habitats or where harvest mortality is unlikely to have any measurable impact on long-term population growth (e.g., male ring-necked pheasant harvest). However, in the majority of cases of contemporary harvest management we simply do not know the impact of harvest, and the broad assumption that harvest mortality is usually compensatory for upland game birds is no longer tenable. Perhaps it is not necessary to focus on the question of additive versus compensatory mortality. Populations may be sustainable even with additive harvest mortality, and we suggest it is more important to understand how various harvest levels—based on particular harvest regulations—affect long-term population dynamics. Moreover, agencies have not tended to prioritize the resources needed to assess harvest impacts on upland game birds for this purpose. Consequently, most management processes do not currently include adequate information to make informed decisions about upland game bird harvest. In many cases, managers do conduct breeding season surveys, but there is often no feedback loop to adjust harvest regulations (Sands and Pope 2010) or assess harvest impacts. For some states, there is simply little to no population monitoring for upland game birds, other than post-season hunter surveys. More resources and critical evaluations of current regulations are needed within upland game bird harvest management if we are to move into the future with defensible and justified approaches.

If managers decide adaptive harvest management is warranted after an initial assessment, then information tradeoffs need to be weighed against available resources. We recognize the decision process includes constraints and the reality of limited resources (Runge 2021 [Chapter 7]), and not all management strategies can be optimized. However, we highly encourage progression from our current approach to more data-driven, defendable harvest management of upland game birds, even if assumptions are required to do so. Various intensity levels of adaptive harvest management could be implemented (Fig. 21.3), some with only a minimum amount of new data or resources required. For example, species that commonly have breeding population monitoring (e.g., ruffed grouse, northern bobwhite, species of prairie grouse) are prime candidates for developing an adaptive harvest management strategy. Breeding population data could be extrapolated, using published information and educated assumptions, to fall population estimates. Then post-hunt surveys, and wing collections with a few adjustments that provide more detailed information, could provide a feedback loop and begin the process of assessing the effects of harvest (i.e., active adaptive harvest management; Fig. 21.3). At minimum, breeding population and productivity monitoring, or even informed assumptions of productivity, could be used to adjust regulations for the following fall hunting season (i.e., reactive adaptive harvest management). We encourage managers to explicitly consider sources of uncertainty in this approach, and where appropriate, conduct targeted studies to reduce those areas of uncertainty.

In addition to gathering more precise and accurate population information, more data about upland game bird hunters need to be collected and made available to those assessing and monitoring management strategies. Significant benefit could come from collecting data concerning (1) the timing and intensity of hunter effort as the season progresses, (2) the spatial distribution of harvest rate across season, (3) changes in hunter success through a season, (4) bias in hunter harvest such as sex or age of the birds, (5) temporal and spatial information on hunter effort focused on public access areas, (6) the temporal nature of harvest rate on public areas, (7) hunter motivations, and (8) hunter activity patterns (e.g., die-hard vs. casual hunters). Certainly, this list is not comprehensive, but managers could use information about hunters to adjust upland game bird harvest regulations and decrease the risk of unacceptable levels of harvest. Human dimensions research can also support managers as they seek to avoid unnecessary regulation that reduces hunter opportunity without achieving management outcomes.

Setting harvest regulations within an experimental design such that valid comparisons can be made has been rare, but could be highly beneficial. For example, nearby hunt units could have varying regulations (e.g., season dates, season length, area closures, bag limits, controlling hunter effort), and comparisons could be made through post-hunt surveys of hunters and populations. At first, it may be advisable to isolate differences in regulations (i.e., adjusting the bag limit, but not the other regulations) until a broader understanding is achieved. Such an approach has been evaluated with other species including mourning dove (*Zenaida macroura*; Bonnot et al. 2011).

Other issues regarding regulations need to be addressed to support future assessment of harvest management strategies. For example, aggregated bag limits of multiple species, such as western forest grouse, may become problematic when attempting to assess potential impacts of harvest on a specific species or population, especially when species (e.g., ruffed grouse and dusky grouse) vary considerably in their life-history strategies. Although concerns over identification of forest grouse species exist, with increased education and higher expectations forest grouse hunters would likely readily adapt to non-aggregated bag limits. For example, waterfowl hunters in every state must identify duck species in flight even in low-light conditions up to 30 minutes before sunrise. However, currently in most western states forest grouse hunters are not expected to know the difference between aggregated species, even after a forest grouse has been bagged. Relatedly, there is evidence that female grouse, especially successful brood females, have a higher probability of being harvested during early fall hunting seasons prior to brood breakup in late September or early October (Ellison 1991). Yet to date we do not have enough data to make an informed decision on the potential impacts of moving start dates to later in the fall.

We recognize hunters may be sensitive to regulation changes that address the issues we have identified. Communication between stakeholders and managers could provide opportunities for improved education to maintain the privilege and right of upland game bird hunting in the future. Adaptive harvest management for waterfowl began in 1995. Although challenges emerged at first, waterfowl hunters have accepted adaptive harvest management as the new norm, and the detailed regulations have not necessarily discouraged hunter participation nor the recruitment of new hunters. Many adult-onset hunters are seeking opportunities to hunt small game, which has traditionally been an important gateway for new hunters. We encourage special hunts and programs that focus on recruitment, retention, and reactivation via upland game bird hunting. Additionally, a large portion of upland game bird hunters also hunts waterfowl and has already been exposed to adaptive harvest management at some level. We realize that significant data streams, coordinated at the federal level, exist for all species of waterfowl (e.g., population counts, banding data, and harvest surveys; Vrtiska 2021 [Chapter 20]) that simply do not exist for nearly all upland game birds. The lack of data may produce challenges in presenting justifications for future regulatory changes if they are not supported by the best available science.

Another significant challenge to adopting a more formal decision-making framework for harvest of upland game birds is the unintentional devaluation of upland game resources due to past management paradigms. Historic research on upland game harvest has often supported the notion of high levels of compensatory harvest mortality, and many wildlife managers (including some of the authors) were taught, and subsequently adopted, the idea that upland game populations cannot be harmed by harvest as long as regulations were reasonable. However, such an approach may have produced an unintended consequence of general complacency regarding upland game harvest, resulting in less critical evaluation of harvest management systems, and even allocation of resources toward management of upland game populations, compared to other game species. Given past paradigms and the needs of state wildlife agencies to prioritize limited resources, we understand the low attention often paid to upland game bird harvest. However, we have now seen multiple prominent large-scale declines of upland game birds, and we have and will likely continue to experience persistent drivers of global change that affect our upland game bird populations. We should not assume the same harvest management practices that have been implemented for decades can continue in such a dynamic system absent critical evaluation. We ask agencies to consider fully the uncertainties in their harvest of upland game birds, and to seek clarity in those uncertainties by collecting data to test critical assumptions of current harvest management programs. We predict that in some cases justifications could be made to support continuing the status quo, whereas for other cases, evidence to support new directions will be found. Regardless, the issue of harvest management for upland game birds needs a closer evaluation and justification, and in some cases harvest regulations will need to be modified.

We have worked closely with upland game programs in state wildlife agencies, and we recognize the difficulties in shifting a paradigm for upland game bird harvest management. We join others who have been pushing for this shift, mostly unsuccessfully, for many years. We believe the socio-political process toward a new paradigm for harvest management of upland game birds will be among the most significant challenges for the twenty-first century. The vast majority of state wildlife agencies make policy changes, including harvest regulations, through a politically appointed board or committee, which most often includes a public input process. Change is always difficult and often met with resistance by the public, especially in cases where opportunities may become more limited. Such challenges can be overcome by using data, modeling scenarios, and transparency and honesty with the public. As upland game birds continue to be harvested in the future, we foresee increasing societal pressures concerning the intentional killing of animals (Hiller et al. 2021*a,b* [Chapters 2, 23]). As wildlife professionals, if we can adopt the best available science and use greater accountability and transparency, then we believe that the public will find value in and support the continued hunting of upland game birds.

SUMMARY

Harvest of upland game birds in the United States of America has been a longstanding tradition providing significant economic benefit and recreational opportunities for millions of hunters. Although hunter numbers have been declining, we foresee upland game bird hunting continuing into the foreseeable future and harvest of upland game birds will primarily be regulated by state and tribal wildlife agencies.

Over time, state wildlife agencies have largely managed upland game bird harvest through trial and error with an overriding conservative approach when making regulation changes. As wildlife professionals and hunters, we have assumed that this approach has worked without deleterious harvest impacts on populations. However, limited data have been collected, and the overall lack of experimental manipulation of harvest regulations to test for harvest effects results in a lack of clarity for what effects, if any, harvest has on most upland game bird populations. Moreover, this approach does not attempt to address the uncertainty inherent with management of harvest. When we consider habitat and population declines during the last few decades for most upland game bird species in the United States of America, our current approach becomes tenuous. We suggest a paradigm shift is needed if upland game bird harvest is going to continue with harvest strategies that support sustainable populations. Science-based and defensible harvest strategies are needed if the majority of our society is going to continue to support hunting these species.

Managers can start by considering ways to use the data and information they already have to assess harvest impacts. State wildlife agencies may then consider prioritizing information gaps for upland game bird populations they manage and dedicate resources to improve our understanding of harvest impacts for the highest priority situations. We recognize that not all circumstances call for an adaptive harvest management strategy, but more formal decision processes would start to address the uncertainty associated with harvest. If managers, supported by upland game bird hunters and organizations, begin to take steps toward an adaptive harvest management approach, we believe regulations will be more rigorous and defensible and that harvest will be less likely to negatively affect game bird populations.

ACKNOWLEDGMENTS

This chapter was improved by two peer-reviewers, S. Gangl and M. Schroeder. Recognition goes to the academic institutions and endowments that support the authors, namely Utah State University and the Berryman Institute, University of Maine, Oregon State University, and Oklahoma State University and the Bollenbach Chair in Wildlife Management. Past mentors and fellow scientists (and more than a few good bird dogs!) influenced the authors and the ideas presented herein, such as C. Braun, W. Burger, J. Connelly, F. Guthery, R. Robel, J. Sedinger, M. Schroeder, and E. Thacker. Lastly, acknowledgment goes to the yeoman efforts of both editors, especially L. Powell's keen interest in the topic covered by this chapter.

22 The Future of Managing Ungulate Species: White-tailed Deer as a Case Study

Duane R. Diefenbach, W. Matthew Knox, and Christopher S. Rosenberry

CONTENTS

INTRODUCTION

Ungulate populations provide benefits and accrue costs for people, economies, and the environment. In the United States of America, viewing of large mammals accounts for 50% of wildlife-watching trips away from home and is participated in by 11.8 million people, of which 75% are not hunters or anglers (U.S. Fish and Wildlife Service and U.S. Census Bureau 2016). Almost US$76 billion was expended on all types of wildlife watching in 2016 in the United States of America, and the non-consumptive enjoyment of wildlife dwarfed the $14.8 billion spent related to the consumptive use of big game species (U.S. Fish and Wildlife Service and U.S. Census Bureau 2016). Consumptive use of big game species funds most wildlife conservation. Nearly eight of ten hunters hunt deer (Fuller 2016) and hunting-license sales provide more than a third of wildlife agency funding (Association of Fish and Wildlife Agencies and Arizona Game and Fish 2017).

In addition to economic benefits, recreational hunting serves as the primary method of regulating deer populations to achieve management goals. As a polygamous species, white-tailed deer (*Odocoileus virginianus*) populations can readily sustain harvests of males. As a result, most wildlife agencies allow each licensed hunter the opportunity to harvest at least one antlered deer. On the other hand, female survival is the driving factor influencing population growth rate (Gaillard et al. 1998) and hunting is generally regulated through some type of permitting system to control the number of females harvested. By adjusting the number of females taken by hunters, deer populations can be increased, stabilized, or decreased to achieve management goals. Historically, hunters and administrators of wildlife agencies have embraced conservative harvest regulations to ensure sustained or increased deer populations (Nugent and Mawhinney 1987, Decker and Connelly 1989, Fuller and Gill 2001, Frye 2006).

In North America, wildlife is held in the public trust and hunting is regulated by government agencies rather than by landowners. This approach to wildlife management has been termed the North American Model for Wildlife Conservation, in which hunting and hunters are considered the foundation for wildlife conservation and provide the bulk of conservation funding (Heffelfinger et al. 2013). Elsewhere in the world there are different approaches to wildlife management where private landowners have greater control over the harvest of big game (Gill 1990). In some countries, landowners have historically had significant control over management decisions for ungulates under both public and private approaches because the right to shoot deer usually cannot be separated from land ownership.

Both the North American Model for Wildlife Conservation and landowner models of ungulate management have shortcomings identified as resulting in increased human-wildlife conflicts (Milner et al. 2006, Lindqvist et al. 2014, Peterson and Nelson 2017). In the United States of America and Canada, it has been difficult for wildlife agencies to manage deer to balance competing objectives because hunters are the primary constituency for ungulate management in North America and fund most wildlife conservation (Leopold et al. 1947, Diefenbach and Palmer 1997). Similarly, in Europe hunting organizations are primarily responsible for setting wildlife population goals, which may conflict with other activities such as agriculture or forestry. Conflicts that arise with abundant ungulate populations include deer–vehicle collisions that cause human fatalities and injuries and economic losses (Seiler 2004, Bissonette et al. 2008), agricultural damage (Bleier et al. 2012), and disease transmission (Martin et al. 2011). Abundant ungulates also have environmental effects that may conflict with human goals and objectives. Environmental effects of ungulate herbivory have been identified all over the world with many different ungulate species, including moose (*Alces alces*) in Scandinavia (Hörnberg 2001), red deer (*Cervus elaphus*) in northern Europe (Lilleeng et al. 2016), sika deer (*Cervus nippon*) in Japan (Tamura and Nakajima 2017), white-tailed deer in North America (Tilghman 1989), roe deer (*Capreolus capreolus*) and fallow deer (*Dama dama*) in Europe (Apollonio et al. 2010), and multiple ungulate species introduced to New Zealand (Cruz et al. 2017).

Minimizing conflicts with humans may require population goals that are below ecological carrying capacity (Zoë et al. 2010). Similarly, ensuring a sustainable population requires keeping ungulate populations below ecological carrying capacity, but large enough to ensure long-term viability of the population. Maximizing recreational opportunity, either for consumptive or non-consumptive purposes, can be in direct conflict with human-conflict and ecological objectives. In general, ungulate populations have increased worldwide as suggested by the number of species and locations throughout the world where ungulate herbivory is affecting ecosystem processes.

The future challenge to managing ungulate populations to meet objectives is likely to become more difficult as participation in recreational hunting declines and ungulate populations become more abundant. We use the white-tailed deer in North America as a case study to illustrate the management challenges facing decision makers. First, we show that declining participation combined with an older age structure will likely lead to dramatic declines in hunting participation in the coming decades. Second, we argue that traditional regulation changes intended to increase hunter efficiency may be ineffective given demographic factors involved. Third, we identify potential strategies that could be considered, which may help wildlife managers meet management objectives.

POPULATION DEMOGRAPHICS OF HUNTERS

Compared to the general population, demographics and residency of big-game hunters have changed little over the past 30 years. According to the National Hunting, Fishing, and Wildlife-Associated Recreation surveys of 1991 and 2016, big-game hunters have remained ≥90% male and 97% white; whereas the U.S. population is 48% male and non-whites have increased from 15% to 22% of the population (U.S. Fish and Wildlife Service and U.S. Census Bureau 1991, 2016). Similarly, the percentage of big-game hunters who live in urban areas has not changed (44%–45%), but the percentage of the general population living in urban areas has increased from 73% to 82%. Fewer than a quarter of hunters lived in an area of >1 million people (21%–23%) between 1991 and 2016, yet in the general population the percentage living in large metropolitan areas has increased from 43% to 57%. Lack of change in characteristics of big-game hunters, relative to the general population, indicates hunting does not seem to appeal to the portion of the U.S. population that is growing most rapidly.

Despite efforts to increase hunter numbers, age structure of big-game hunters indicates the decline in numbers will continue in the future. Big-game hunters are aging faster than the general population. In the general U.S. population, there has been an increasingly older age structure (Fig. 22.1). In 1991, 41% of the U.S. population was ≥45 years old, which had increased to 52% by 2016. However, hunters ≥45 years old increased from 28% to 60% during the same period. In Pennsylvania, the average age of deer hunters increased from 40 years in 1991 to 51 years in 2016 (Pennsylvania Game Commission, unpublished data; Table 22.1). Moreover, it appears a significant proportion of hunters drop out of the sport after age 65 years (Fig. 22.1 [bottom]). Deer hunters are not being replaced as they age out of the hunting population. Also, regardless of age, hunter numbers have been declining for the past several decades. According to the National Hunting, Fishing, and Wildlife-Associated Recreation survey of 2016, the number of hunters decreased 16% from 2011 to 2016, which included a 20% decline in big-game hunters. The 2016 level of hunting was at the lowest level in the past 25 years (U.S. Fish and Wildlife Service and U.S. Census Bureau 2016).

To illustrate the potential decline in deer hunters that may occur in the United States of America, we used demographic data from hunters in Pennsylvania (2009–2017) and Virginia (2008–2018) to predict the number of hunters 10 and 20 years from now. We used age-structure data of general hunting-license buyers in Pennsylvania (80% of license buyers hunted deer in 2018) and licensed deer hunters in Virginia (hunters who purchased a deer license) to predict population growth and age distribution of hunters. In Pennsylvania, the number of 12-year-olds purchasing hunting licenses declined approximately 408 licenses per year during 2009–2017. In Virginia during 2007–2018, 12-year-olds purchasing deer licenses declined 7% per year. In addition, we calculated the average proportional change in license buyers between age i to age $i + 1$ ($i = 12 - 89$). We used the number of 12-year-olds purchasing a hunting license (recruitment) and

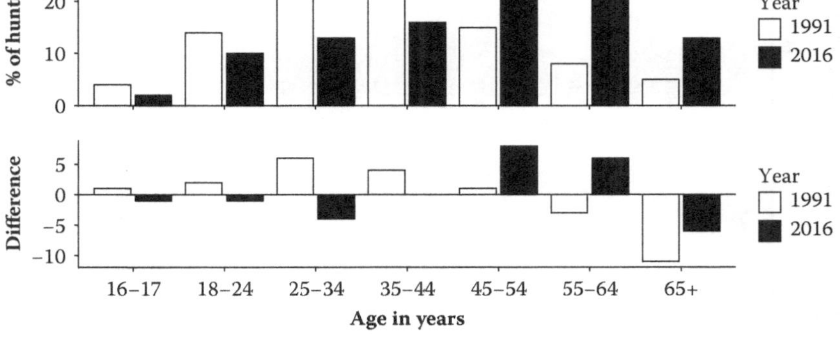

FIGURE 22.1 Age structure of hunters according to the National Survey of Hunting, Fishing, and Wildlife-Associated Recreation, 1991 and 2016 (U.S. Fish and Wildlife Service and U.S. Census Bureau 1991, 2016), (top) general U.S. population, (middle) big-game hunters, and (bottom) difference between big-game hunters and general population. Big-game hunters are older than the general population and aging out of the hunting population faster.

TABLE 22.1
White-Tailed Deer Harvest Statistics, Hunter Characteristics, and Estimated Deer Population Size by State and Province in the United States of America and Canada, 2018 or 2019 Hunting Seasons

	Harvest[1]				Hunters[2]					
State or Province	Antlered	Antlerless	% Antlerless	Density km^{-2} (mi^{-2})	Number	Density km^{-2} (mi^{-2})	Trend	Age	Success	Pre-Season Population
Alabama[3]	79,084	123,956	61	0.7 (1.7)	191,054	1.6 (4.1)	–2	50		1,250,000
Arkansas[3]	97,607	112,458	54	1 (2.5)	239,629	2.4 (6.2)	–15		60	1,000,000
Connecticut[3]	5747	5411	48	0.6 (1.5)	25,573	2.6 (6.8)	–17	50	25	110,000
Delaware[3]	4861	12,108	71	1.2 (3.1)	15,638	3.8 (9.8)	4	52	51	46,000
Florida[3]	48,250	26,724	36	0.7 (1.8)	91,432	1.3 (3.3)	–17	52		
Georgia[3]	81,323	157,921	66	0.8 (2.1)	242,057	2.4 (6.3)	0	46	56	1,000,000
Illinois[3]	71,186	81,988	54	1 (2.6)	228,329	3.3 (8.5)	–17	43	44	
Indiana[3]	51,646	63,236	55	1.7 (4.3)	212,719	1.3 (3.4)	–11	41	36	
Iowa[3]	42,073	51,519	55	0.3 (0.8)	237,235	1.7 (4.3)	–30	41	28	450,000
Kansas	41,056	38,902	49	0.2 (0.5)	106,896	0.5 (1.3)	–1	43	53	690,000
Kentucky[3]	70,362	78,023	53	0.7 (1.8)	262,887	2.6 (6.8)	–2	50	37	908,291
Louisiana[3]	62,816	57,284	48	0.9 (2.4)	136,000	2 (5.1)	–20	47	46	500,000
Maine[3]	20,093	8230	29	0.3 (0.7)	209,000	2.8 (7.2)	7	50	15	310,000
Maryland[3]	29,233	46,777	62	1.3 (3.3)	55,000	2.4 (6.3)	–5	44	56	240,000
Massachusetts[3]	7764	6156	44	0.8 (2.1)	50,000	4.1 (10.7)	0	51	20	100,000
Michigan[3]	211,754	148,912	41	2.3 (6)	554,331	6.1 (15.8)	–34	44	48	
Minnesota[3]	101,910	85,677	46	0.5 (1.3)	464,086	2.3 (5.9)	–5	42	36	
Mississippi[3]	90,697	106,200	54	0.8 (2.1)	137,983	1.2 (3.2)	–18		63	1,475,000
Missouri[3]	134,092	151,781	53	0.8 (2.1)	483,745	2.9 (7.6)	–5	40	43	1,400,000
Nebraska[3]	29,899	19,191	39	1.4 (3.7)	84,804	4.1 (10.6)	–3	41	47	300,000
New Hampshire[3]	7870	4436	36	0.4 (1)	55,853	2.7 (7)	–1	52	17	100,000
New Jersey[3]	19,240	26,412	58	1.4 (3.7)	88,025	6.4 (16.7)	–11	50	21	133,500
New York[3]	120,403	103,787	46	1 (2.6)	545,536	4.5 (11.6)	–5	49	30	1,200,000
North Carolina[3]	82,724	79,217	49	1 (2.6)	229,711	2.5 (6.4)	–4	53	47	1,000,000

(Continued)

TABLE 22.1 (Continued)

State or Province	Harvest[1]				Hunters[2]					Pre-Season Population
	Antlered	Antlerless	% Antlerless	Density km^{-2} (mi^{-2})	Number	Density km^{-2} (mi^{-2})	Trend	Age	Success	
North Dakota[3]	22,660	14,120	38	0.2 (0.4)	86,000	0.5 (1.4)				
Ohio[3]	77,027	107,441	58	2 (5.2)	287,875	7.5 (19.3)	–20	40	34	
Oklahoma[3]	69,927	39,333	36	0.7 (1.9)	184,032	1.9 (4.9)	4	48		750,000
Pennsylvania	163,240	226,191	58	1.4 (3.6)	660,000	5.7 (14.8)	–10	51	32	
Rhode Island[3]	1072	1213	53	0.5 (1.2)	4834	2.1 (5.4)	–9	48	30	
South Carolina[3]	100,201	94,785	49	1.8 (4.6)	145,535	2.5 (6.6)	–1	45	67	730,000
South Dakota[3]	25,389	19,079	43	0.1 (0.3)	69,252	0.3 (0.9)	–15	42	43	350,000
Tennessee[3]	71,884	63,288	47	1.1 (2.8)	547,635	8.2 (21.3)				
Texas[3]	508,155	375,408	42	1.1 (2.9)	808,464	0 ()	11	42	63	5,585,497
Vermont[3]	10,058	6492	39	0.5 (1.2)	77,289	3.4 (8.9)	–25	46	20	140,000
Virginia[3]	99,994	106,985	52	1 (2.6)	185,000	1.9 (4.9)	–24	50	60	1,030,000
West Virginia[3]	56,189	43,248	43	0.9 (2.4)	183,000	3.1 (8)	–17	45	50	562,000
Wisconsin[3]	137,877	152,335	52	1.4 (3.7)	610,146	6.3 (16.2)	–5	43	34	1,780,000
New Brunswick[3]	6025	1278	17	0.1 (0.2)	42,483	0.7 (1.7)	–11	53	17	80,700
Nova Scotia[3]	7797	2458	24	0.2 (0.5)	46,990	0.8 (2.2)	9	62	21	44,743
Ontario[3]	34,898	25,014	42	0.1 (0.2)	187,954	0.3 (0.9)	2		31	
Quebec[3]	26,091	21,509	45	0.3 (0.7)	130,513	1.3 (3.4)	–16	50	34	

Notes

1 Antlered = number of antlered deer harvested; Antlerless = number of antlerless deer harvested; % Antlerless = percent of harvested deer comprised antlerless deer; Density = number of antlered deer killed per square km (square mile) of estimated deer habitat.

2 Number = number of hunters; Density = number of hunters per square mile of estimated deer habitat; Trend = ten-year change in number of hunters; Age = mean age of hunters; Success = percent of hunters who harvested at least 1 deer.

3 Hunter numbers are number of deer hunters, all other states are total number of hunters.

the proportional change in number of hunters by age to project the number of hunters through 2030 and 2040. We predicted the number of hunters, 12–90 years old, in Pennsylvania would decline by 17% by 2030 (compared to 2020; decline from 807,227 to 670,302 hunters) and by 37% by 2040 (to 512,263 hunters; Fig. 22.2). In Virginia, we predicted the number of deer hunters would decline by 32% by 2030 (compared to 2020; decline from 173,058 to 118,027 deer hunters) and by 57% by 2040 (to 75,178 deer hunters; Fig. 22.2).

Although much of the decline in hunters was driven by the decline in number of new hunters recruited each year, aging of older hunters out of the population was a significant factor (Fig. 22.2). For example, Pennsylvania would have to recruit approximately 450 additional 12-year-olds every year for the next 20 years and retain hunters and increase reactivation of older hunters by 1%, to result in a no net loss of hunters by 2040. Similarly, Virginia would have to increase recruitment of 12-year-olds >6% each year compared to the previous year through 2040 and retain hunters and increase reactivation of older hunters by 2%, to result in no net loss of hunters. Aging and numerical decline of big-game hunters will create real challenges for resource managers and recreational hunting as a management method in the near future.

GENERAL FRAMEWORK FOR MANAGING DEER

Management of white-tailed deer in North America is a top–down process by which wildlife agencies develop a management plan with input from the public, agency staff monitor the deer

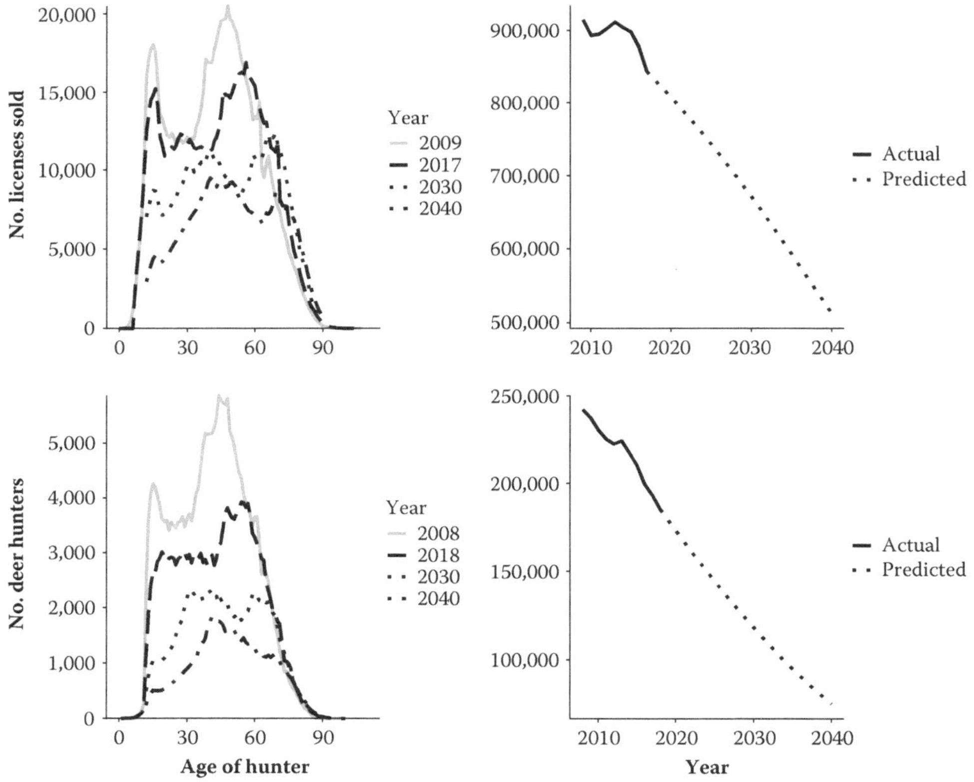

FIGURE 22.2 Age distribution and number of actual (pre-2019) and predicted (2030 and 2040) deer hunters in Virginia, United States of America (bottom) and hunting-license buyers in Pennsylvania, United States of America (top).

population, and then agency staff make recommendations to decision makers regarding hunting regulations to meet established goals. There are four components to any management program. First, application of management actions and deer harvest and population monitoring generally occurs within defined *management units*. Management units can be based on political boundaries or possess ecological and social characteristics that are as homogeneous as possible within physical boundaries such as roads and rivers (Karns et al. 2016, Swihart et al. 2020). Effective management units should be sized to meet desired precision of a monitoring program given cost and logistical constraints. For management and monitoring purposes, smaller political units are not necessarily more effective than larger ecological and social units (Rosenberry and Diefenbach 2019). Even so, many agencies use management units that are based on political boundaries rather than ecological-based units (Table 22.2).

Second, *management goals and objectives* must be developed for each management unit that represent public values within the context of the wildlife agency's mission and legal authority. To identify the goals and objectives, public engagement often takes the form of citizen advisory committees, public meetings or open houses, and public surveys (e.g., Stout et al. 1996, Fleegle et al. 2013). Management objectives for each goal must be defined such that they can be quantitatively evaluated by a monitoring program (Artelle et al. 2018). The importance of scientifically rigorous monitoring programs has been highlighted in recent years with lawsuits and legislatively sponsored reviews and audits of agency deer management programs (Millspaugh et al. 2007, Wildlife Management Institute 2010, Office of Legislative Auditor 2016).

Third, within each management unit data are collected to *monitor* either deer abundance directly or indicators that can provide trends in deer population characteristics (Table 22.1). Most states monitor harvest statistics (e.g., buck kill per unit area; Table 22.1) and some use sex-age-kill models or accounting-type population models (Diefenbach and Shea 2011). Some states use annual surveys of direct abundance (Kaminski et al. 2019). In addition to deer abundance, other data may be collected to assess how well objectives are being achieved, such as stakeholder opinions (Curtis and Hauber 1997, Gruntorad and Chizinski 2021 [Chapter 4]) and habitat conditions (Rosenberry et al. 2009).

Fourth, a *strategy* to attain population management goal needs to be identified. For white-tailed deer, changes in abundance are accomplished primarily through manipulation of the harvest of female deer. Manipulation of the male population can affect changes in social behavior and sex-age structure of the population (e.g., Wallingford et al. 2017, Morina et al. 2021 [Chapter 19]), but has little effect on population trends. Strategies to achieve management goals may be applied at the management unit scale (e.g., allocation of antlerless licenses) or at smaller scales to address local deer–human conflicts, such as excessive crop damage or deer–vehicle collisions.

These four components of a management program are repeating-loop processes (Runge 2021 [Chapter 7]) that occur at different temporal scales. On an annual basis, the result of harvest strategies with regard to management objectives are evaluated based on monitoring data (3 and 4 above). New recommendations for harvest strategies are then developed for the next year. However, at longer time intervals agencies may revise management goals and objectives based on stakeholder input (2 above). Although deer management programs vary by agency, the challenge remains the same: achieving deer management goals in a manner that balances the values of stakeholders in a transparent manner that is defendable.

TRADITIONAL HARVEST MANAGEMENT STRATEGIES

White-tailed deer populations increased throughout the twentieth century in North America, but with these increasing populations hunter participation also increased. Consequently, a successful strategy for managing deer populations was to regulate hunting-season length (number of days of opportunity to harvest a deer) and the number of deer each hunter could harvest each year (bag limit). In most situations, a longer season and a smaller bag limit could achieve the same level of harvest as a shorter season and a larger bag limit.

TABLE 22.2
Special Seasons, Bag Limits, Season Length, and Management Characteristics by State and Province in the United States of America and Canada, 2019

State or Province	Special Seasons/ Regulations[1]	Season Bag Limit		Firearms Season (days)	Plan[2]	Unit Type[3]	% Management Units Relative to Population Goal			Method[4]
		Antlered	Antlerless				At	Below	Above	
Alabama	P,D	3	1/day	82	N	A				D
Arkansas	P,D,U	2	6	62	Y	E	70	25	5	D
Connecticut	D,U,EAB	7	6	36	Y	E	62	15	23	Q
Delaware	D	2	∞	44	Y	E	83	0	17	D
Florida	P,D	3–5[5]	≤2[5]	129	Y	E	58	42	0	D,Q
Georgia	P,D,U,EAB	2	10	85	Y	E	58	42	0	D,Q
Illinois	D,U	2	∞	10–17	N	A	75	5	20	Q
Indiana	D,U	1–2		16	Y	A				Q
Iowa	D,U	2	0–∞	42	Y	A	56	25	19	Q
Kansas	D,U	1	6	12	N	E	50	33	17	D
Kentucky	D	1	1–∞	16	N	A	39	18	43	Q
Louisiana	P,D	3	4	84–93	Y	E	40	20	40	D
Maine	D,U	1	0–∞	25	Y	E	21	59	21	Q
Maryland	D,U	2 or 3	3, 35, or ∞	18	Y	E,A	50	0	50	D
Massachusetts	D,U	2	1–∞	12	Y	E	73	0	27	Q
Michigan	P,D,U	2	≤10	16	Y	E,A	5	5	90	Q
Minnesota	D,U	1	0–∞	9–23	Y	E	36	18	46	Q
Mississippi	P,D	3	5	77	N	A				D
Missouri	P,D	2	3–∞	30	Y	A	88	11	1	D,Q
Nebraska	D,U,EAB	2	0–∞	9	N	A	80	15	5	Q
New Hampshire	P,D	3	2	26	Y	E	75	0	25	D
New Jersey	P,D,U,EAB	6	1–∞	19–65	N	A	17	17	65	Q
New York	P,D,U	2	0–∞	23–44	Y	E				Q
North Carolina	P,D,U	2	4–∞	18–80		E,A	77	13	10	D
North Dakota				16.5			38	75	25	
Ohio	D,U	1	5–6	9	Y	A				
Oklahoma	P,D,U	2	6	16	Y	E,A				D,Q
Pennsylvania	P,D	1	2–∞	13	Y	E,A	56	0	44	Q
Rhode Island	D	1–2	2–3 or ∞	17	N	E,A	50	0	50	Q
South Carolina	P,D,U	5+	3–∞	83–140	N		25	25	50	Q
South Dakota	D,U	-[6]	-[6]	25	Y	A	34	64	2	Q
Tennessee	P,D,U,EAB	2	1–330	49–63	Y	A				Q
Texas	P,D	1–3	2–5	79–93	N	A				D
Vermont	D	1	≤4	16	Y	E	52	0	48	Q
Virginia	P,D,U,EAB	2–3	3–∞	14–49	Y	A	41	10	46	D

(Continued)

TABLE 22.2 (Continued)

State or Province	Special Seasons/ Regulations[1]	Season Bag Limit: Antlered	Season Bag Limit: Antlerless	Firearms Season (days)	Plan[2]	Unit Type[3]	% Management Units Relative to Population Goal: At	Below	Above	Method[4]
West Virginia	D,U,EAB	≤ 3	≤8	27	Y	A	63	37	0	D,Q
Wisconsin	P,D,U	2	0–∞	9	N	E,A				Q
New Brunswick	P,U	1 deer/year		28	Y	E,A	11	82	7	Q
Nova Scotia	P,U			39	N	E				Q
Ontario	D,U	1 deer/year		5–93	Y	E,A				Q
Quebec		2 deer/year		9–16	Y	E	16	42	42	Q

Notes

1 Increase antlerless harvest to address: P = landowner concerns, D = deer damage, and U = urban deer problems. EAB = earn-a-buck regulation requiring harvest of an antlerless deer before an antlered (or second antlered) deer can be harvested.

2 Does the agency have a deer management plan: Y = yes; N = no.

3 Type of management units: A = administrative; E = ecological.

4 Whether antlerless harvest is regulated through number of days of hunting (D) or quotas (Q).

5 Total harvest cannot exceed five deer of which ≤2 are antlerless.

6 Archery license allows either sex deer statewide; otherwise, licenses issued via hunter application and random drawing by management unit.

In states like Pennsylvania, which had >1 million deer hunters in the late 1980s, changes in season length or antlerless harvests were sufficient tools to maintain stable deer populations and provide sustained harvests (Diefenbach and Palmer 1997). In the early 1900s, deer hunting in Pennsylvania consisted of antlered-only harvests with sporadic antlerless hunting seasons, and it was not until 1953 that antlerless seasons were held annually (Kosack 1995). Even with declining hunter numbers, in 2018 Pennsylvania had 660,000 deer hunters (6 hunters/km^2; 15 hunters/mi^2; Table 22.1) and limited hunters to one antlered deer per year and two antlerless deer for most management units. Changes in season length and antlerless license allocations in the early twenty-first century allowed Pennsylvania to reduce the statewide deer population by 23% in three years (Wallingford et al. 2017). The population reduction was achieved by shifting from a 12-day antlered deer and three-day antlerless deer seasons to a single 12-day concurrent antlered and antlerless deer season and increased allocation of antlerless licenses. Except in more urban areas surrounding large cities, hunting was sufficient to control deer populations in Pennsylvania (Table 22.2).

In states with fewer licensed deer hunters and large urban areas, such as Virginia (2 hunters/km^2; 5 deer hunters/mi^2; Table 22.1), changes solely to season length and bag limits may be insufficient to control deer populations. Northern Virginia (Arlington, Fairfax, Loudoun, and Prince William counties; 3372 km^2, 1301 mi^2) is highly urbanized (666 people/km^2, 1724 people/mi^2) and was experiencing increasing deer populations by the 1990s. To reduce the deer population, the Virginia Department of Wildlife Resources implemented a series of regulatory changes over >20 years (Table 22.3). Changes to hunting regulations were intended to increase hunter access to public green space, motivate hunters to harvest antlerless deer, and provide more opportunities to harvest deer through longer seasons and larger bag limits.

Perhaps the most effective tool to increase the antlerless harvest in northern Virginia was implementation in 2008 of an earn-a-buck regulation, in which a second antlered deer could not be harvested until an antlerless deer was harvested (Table 22.3). Traditionally, earn-a-buck has been

TABLE 22.3

Regulatory and Policy Actions in Northern Virginia (Arlington, Fairfax, Loudoun, and Prince William Counties) and Resulting Deer Harvest during 1991–2019[1]

Year	Action	Harvest Antlered	Antlerless	Antlerless: Antlered Ratio
1991	Bonus deer permits[2] created and limited to one per hunter per year	4071[3]	3417	
1995	Extended deer seasons on public lands in Fairfax County	3553	4952	1.4 : 1
1998	Fairfax County government implements deer management program on public lands; bonus deer permits are antlerless only and unlimited	3439	4159	1.2 : 1
2002	Urban archery season implemented in Fairfax County	4310	5340	1.2 : 1
2004	The number of antlerless-only tags on basic deer license is doubled from 2 to 4	4216	5358	1.3 : 1
2006	Establish a late four-week antlerless-only firearm deer season (January)	3859	5808	1.5 : 1
2008	Late antlerless-only firearm season extended through March (three months); require harvest of an antlerless deer before a second antlered deer can be harvested	3739	6834	1.8 : 1
2009	Number of antlerless-only deer tags on bonus permits increased from 2 to 6	3561	6837	1.9 : 1
2011	Establish unlimited daily/season antlerless deer bag limit	3600	7079	2.0 : 1
2013	Establish September antlerless-only firearms deer season; require harvest of two antlerless deer before 2nd antlered deer allowed	3543	8477	2.4 : 1
2019	Same regulations 2014–2019	2825	5453	1.9 : 1

Notes

1 Deer hunters are restricted to three antlered bucks per year during October–December archery, muzzleloader, and firearm deer seasons.

2 Two deer tags; one either sex and one antlerless only but antlerless only since 1998.

3 Includes male fawns.

used to increase antlerless harvest by requiring hunters to harvest an antlerless deer before harvesting an antlered deer. Although an effective regulation to increase antlerless harvest, such implementation of the earn-a-buck regulation is disliked by hunters (Van Deelen et al. 2010). However, implementation of the earn-a-buck regulation only after the first antlered deer is harvested is more acceptable and has allowed Virginia to increase the percentage of antlerless deer in the harvest (Tables 22.1 and 22.3) and accomplish removal of 12 deer/km^2 (30 deer/mi^2) in 2018 on 30 square miles of public lands in Fairfax County.

A STRATEGY TO MAINTAIN EFFICIENCY OF HUNTER HARVEST

Simply extending season lengths and bag limits will fail to control deer when there are too few hunters willing to harvest a sufficient number of antlerless deer. Also, the effectiveness of increasing seasons and bag limits is reduced when recreational hunting cannot occur either because restricted access to private land or safety issues require less effective sporting arms (e.g., bows or crossbows rather than firearms), and is somewhat similar to concerns about anglers regulating fish population in catch-and-release fisheries (see Sylvia et al. 2021 [Chapter 17]). The regulatory changes implemented in northern Virginia have been effective where hunting opportunity has been maximized (Table 22.3) such that they currently have the longest hunting season (eight months) and most liberal bag limits (unlimited for antlerless deer daily and season) in North America for white-tailed deer.

In coming decades, more wildlife agencies may reach the limits of recreational hunting to control deer populations because of the declining trend in hunter numbers (Winkler and Warnke 2013; Figs. 22.1 and 22.2). Hunters generally are not motivated by ecological concerns or deer–human conflicts because other priorities inform their motivation for hunting (Diefenbach et al. 1997, Holsman 2000). Consequently, most hunters desire to harvest only one to two deer/year (Brown et al. 2000), and some are interested in primarily harvesting antlered deer (Bhandari et al. 2006).

Season lengths and bag limits are a top–down management technique that fits with the North American Model for Wildlife Conservation, but regulations and policy can be changed to also support a bottom–up approach to deer management. For example, providing private landowners and government land management agencies with methods to increase deer harvest can address localized problems. Many states have implemented Deer Management Assistance Programs that provide landowners with property-specific means to increase antlerless harvests (Table 22.2). Similar programs have been established to address crop damage or urban deer problems (Table 22.2). The success of bottom–up approaches to deer management depends on cooperation from private landowners, as well as agency resources to administer and promote such programs, but the motivations for landowners to allow hunting are complicated. Fee-based hunting may work in some situations (Guynn and Schmidt 1984), but landowner attitudes toward hunting and property rights may be impediments (Wright et al. 1988, Raedeke et al. 1996).

Incentivizing antlerless harvest beyond traditional reasons of recreation and sustenance may be necessary. The ideal method would encourage antlerless harvest within traditional hunting seasons and methods while minimizing agency costs and upholding the value of white-tailed deer as a native wildlife species. For example, Virginia has employed two strategies that have increased antlerless harvest via hunting. First, venison donation programs created by charitable non-profit organizations allow hunters to donate harvested deer, with the deer processing cost paid by the non-profit organization, to support community assistance programs. About 25% of successful Virginia deer hunters harvest ≥3 deer and venison donation programs distribute about 300,000 pounds of venison annually, which represents about 7500 deer or 7% of the annual antlerless harvest (Table 22.1). Second, Virginia has found that an effective way to encourage hunters to harvest more antlerless deer may be to implement an earn-a-buck regulation, such as implemented in Virginia, in which antlerless deer must be harvested before a second antlered deer may be harvested. Biologically, this type of regulation is compatible with white-tailed deer's polygamous breeding system. Socially, such a regulation may encourage hunters who primarily hunt for antlered deer to harvest antlerless deer (e.g., Bhandari et al. 2006, Stedman et al. 2008). Although applied in a limited area, this type of regulation in Virginia was successful in increasing the percentage of females in the harvest (Table 22.3).

THE FUTURE OF DEER MANAGEMENT

In locations where deer populations do not exceed management objectives, hunting likely will remain the primary method of deer management because it is cost-effective. Hunting provides recreational opportunity for sportspersons and can regulate deer populations with limited or no cost to society. Even if hunter numbers decline, in northern deer populations hunting likely will continue to be effective in controlling deer numbers because antlerless harvest is prohibited or limited where winter mortality can limit deer population growth (e.g., DelGiudice et al. 2002), although climate change could influence population dynamics and distribution of deer (Dawe et al. 2014).

However, in areas with limited mortality factors other than hunting, declining hunter participation and harvest will create significant challenges. Such challenges have existed in urban and suburban areas for decades where the effectiveness of hunting has been limited because of safety and access to land (e.g., Weckel et al. 2011, Williams et al. 2013).

We envision in the future that multiple methods will be required to control deer populations that likely will require an adaptive management approach (Nielsen et al. 1997). Methods to control deer populations, other than hunting, will incur costs to landowners and government agencies and acceptable methods will depend on resident attitudes toward lethal and nonlethal control measures and costs (Kilpatrick et al. 2007). Culling deer either through sharpshooting or capture and euthanasia can be effective, and the results are comparable to hunting (Etter et al. 2000, DeNicola et al. 2008). Culling can target deer in locations where they are causing problems and the population reduction is immediate. However, culling is expensive ($119–310/deer; Etter et al. 2000).

Another alternative to hunting is fertility control where populations are regulated by reducing fecundity. In principle, use of immunocontraceptive vaccines can reduce deer populations (Rutberg and Naugle 2008), but for rapid population reduction some lethal control methods would be required because adult female white-tailed deer can reproduce for >10 years. Boulanger et al. (2012*a*) applied sterilization of females and estimated that ≥80% of deer would need to be treated at a cost of approximately $1000 per surgery. Boulanger and Curtis (2016) evaluated the efficacy of a sterilization program and concluded that as a stand-alone method it was ineffective at reducing abundance in an open deer population.

Commercializing harvest of deer has been suggested as an approach to address deer population control, although it would have to be carefully integrated with the principles of the North American Model for Wildlife Conservation (VerCauteren et al. 2011, Hygnstrom et al. 2014). Mawson et al. (2016) provided an example where commercial harvest accomplished population reduction goals for western grey kangaroos (*Macropus fuliginosus*) and was acceptable to the public at relatively low cost. Commercial harvest of deer has potential benefits, but also potential conflicts under the North American Model for Wildlife Conservation, for wildlife management. Besides potentially reducing deer abundance, commercialization of deer could provide a natural source of protein and benefit the economy (VerCauteren et al. 2011). However, implementation would require changes to laws and regulations and would create challenges for law enforcement and meat processors. Also, recreational harvest and commercial harvest would be competing for the same resource and may create conflicts. Regulation of commercial and recreational harvest of fisheries might provide insights into how to implement commercial harvest of deer (e.g., Sutinen and Johnston 2003).

Some combination of hunting, culling, fertility control, and commercialization of wildlife has the potential to address future challenges to deer management. However, disease issues could reduce the likelihood of successful application of these tools. For example, chronic wasting disease (CWD) is an example of what happens when hunting alone cannot achieve management objectives and is a problem that may result in further erosion of the effectiveness of hunting. In managing deer to address CWD, reducing deer abundance is often the primary objective. When hunting cannot reduce deer abundance, agencies have used sharpshooting (Manjerovic et al. 2014, Mysterud et al. 2019). Thus, management actions related to CWD provide a look to the future of how hunting may be supplemented with alternative methods to achieve management objectives.

The problem with diseases is that they can also reduce the effectiveness of some potential management actions. For example, the presence of CWD could reduce hunter participation and recruitment (Needham et al. 2007). In Wisconsin, hunters rejected agency goals to manage CWD for a number of reasons and harmed efforts to control the impact of the disease (Holsman et al. 2010). Wisconsin hunters rejected the deer density goal, did not support extended hunting opportunities that conflicted with hunting traditions and consumption norms (i.e., number of deer killed per hunter), and questioned severity of the effect of CWD on deer populations in part due to lack of agency credibility. Holsman et al. (2010) questioned whether recreational hunting could achieve disease management objectives of the scale attempted in Wisconsin. Furthermore, disease could nullify the commercialization of wildlife if there is no economic value to be derived from deer carcasses (e.g., sale of meat for human consumption).

Recreational deer hunting will continue to have a role in deer management, but it may transition from the primary method to one of many alternatives. The magnitude of hunting's contribution to the wildlife manager's toolbox will likely be determined by future hunter participation, as well as hunter and general public acceptance of changes in regulations or management actions. We note that societal interest in how domestic animals are raised and processed has increased interest in wild game as a source of protein and this interest has, in part, increased hunting participation by non-traditional hunters, especially women (Pollan 2006). In addition, many state agencies are engaged in efforts to recruit and retain hunters (Price Tack et al. 2018). However, we identified declining trends in hunter numbers as an important factor that will determine the future and importance of deer hunting to achieve social, ecological, and disease-related management objectives.

ACKNOWLEDGMENTS

The authors thank the many agency biologists who provided the information for Tables 22.1 and 22.2. W. Knox was funded in part by the U.S. Fish and Wildlife Service's Wildlife Restoration Grant Program, Grant Award: VA F18AF00664.

23 Harvest Management of Furbearers

Tim L. Hiller, Roger D. Applegate, and Michael S. O'Brien

CONTENTS

INTRODUCTION

Harvest management is a cornerstone of wildlife management in North America, particularly for state and provincial fish and wildlife agencies. Yet, there seems to be a paucity of comprehensive works dedicated to practical guidance decisions for harvest management. The management of furbearers is no exception. To be most effective in dynamic systems, harvest management should be considered a learning process (see Conroy 2021 [Chapter 1], Runge 2021 [Chapter 7]), because decisions must be made and revised as conditions change. The complexity of these ecological systems, coupled with uncertainty in outcomes, may result in conservative decisions related to harvest. The challenges of data collection for a diverse set of species may make harvest management of furbearers especially prone to conservative decisions.

Furbearers may be defined pragmatically using a management-based approach as, "...the group of mammalian species either currently or historically harvested primarily for their pelts" (Hiller et al. 2018: 117). Such an approach is not ecologically based, and our definition is not all inclusive in many states, provinces, and territories for what we generally perceive to be furbearing species. For example, some jurisdictions use legal definitions of furbearer, unprotected mammal (or unprotected wildlife), predator (or predatory animal), or others for which the non-furbearer classifications typically denote fewer restrictions (e.g., no closed season) associated with harvest. These classifications may be in place to allow for real or perceived damage to life or property attributed to these species to be more readily addressed. Such legal definitions are sometimes defined in state, provincial, or territorial statutes to which jurisdiction agencies must adhere when developing and implementing regulations. For our purposes, we will focus on the broad array of mammalian species that are harvested for their pelts, which excludes other species that may be utilized for pelts or skins, but typically are classified differently (e.g., big game) by jurisdictional wildlife agencies (e.g., American alligator [*Alligator mississippiensis*], American black bear [*Ursus americanus*], mountain lion [*Puma concolor*], pinnipeds, and white-tailed deer [*Odocoileus virginianus*]).

Based on our approach, there are at least 27 furbearing species in North America, which includes species within the orders Carnivora, Didelphimorphia, and Rodentia (Hiller et al. 2018). The most popular (and widespread) species targeted by U.S. trappers include northern raccoon (*Procyon lotor*), coyote (*Canis latrans*), muskrat (*Ondatra zibethicus*), North American beaver (*Castor canadensis*), red fox (*Vulpes vulpes*), and bobcat (*Lynx rufus*; Responsive Management 2015). In Canada, the list is similar, with the addition of American marten (*Martes americana*), North American river otter (*Lontra canadensis*), fisher (*Pekania pennanti*), and Canada lynx (*Lynx canadensis*). Furbearers include a diverse group of taxa with substantially different life histories. Therefore, managers often use some level of species-specific (or at least within-group) harvest management.

Our focus is on harvest management of furbearers, although some of the information contained in this chapter may be applicable to harvest management in general. Furbearer management presents some aspects that are unique, particularly at the global scale. For example, the pelts of furbearers have economic value, and therefore there is some level of commercial use that should be integrated into some decisions (e.g., potential for revising regulations and harvest based on fluctuating conditions of global markets). Also, there are international agreements in place such that capture devices are tested based on several criteria, including animal welfare, and use of approved (or certified) devices is either mandated nationally via coordinated and consistent provincial–territorial regulations (Canada) or may be voluntarily integrated into regulations at the individual state level (United States of America). Further, this chapter is focused primarily on state, provincial, territorial, and tribal-First Nations management authority for harvest of furbearers in North America. However, certain federal regulations may affect harvest management of furbearers and therefore are also included. Finally, we include harvest management as it relates to avocational or vocational harvest, but limited discussion about topics specifically associated with damage management. Damage management, which ideally addresses specific species in specific locations that cause issues (e.g., livestock depredation), is an important component of furbearer management programs, but the general goal of harvest management of furbearers is to ensure that long-term harvest remains sustainable.

The fur trade in North America became expansive and is credited for financing continental exploration, influencing political boundaries, transforming cultures and traditions of native peoples, and engaging trade relations on and beyond the continent as never before (Ray 1987). However, the early era of the fur trade was also unregulated, and led to substantial population declines and local extirpations (e.g., North American beaver, sea otter [*Enhydra lutris*]) and unfortunately, extinction for a very few furbearing species (e.g., sea mink [*Neovison macrodon*]; Manville 1966, Ray 1987). Beginning in 1907, the U.S. Department of Agriculture, Bureau of Biological Survey, started predator control programs that focused on addressing livestock depredations and predation on big game species in western states and territories (Young and Goldman 1944: 381–385). These control measures were responsible for widespread population declines, particularly of large carnivores (e.g., gray wolf [*Canis lupus*]) in that region (Robinson 1953). Recent population recovery efforts for several of these species of large carnivores have been successful throughout western North America, although this sometimes presents a challenging management scenario in terms of available habitat, coexistence with humans, and other factors (see Peek et al. 2012).

Given that furbearers are harvested primarily for their pelts, but may also provide food and other benefits, their importance to the global economy can be substantial, even despite fluctuating markets. At the global scale, the fur trade (wild and farmed) results in >US$4 billion in raw pelts and about $30 billion in retail sales (Hansen 2017, Fur Commission USA 2020). Canada, Russia, and the United States of America are countries that lead origination of raw pelts, whereas China, Italy, Russia, and the United States of America lead retail sales of fur products (Fur Information Council of America 2020). The human societal value that accrues from the management (through harvest and sustainable use) of certain furbearer species (e.g., beavers, coyotes, gray wolves, raccoons) is less tangible and more difficult to quantify, but nonetheless significant. Such a value is created when harvest reduces what would otherwise be substantial risk, conflict, and damage to agriculture (crops and livestock), infrastructure (roads and other property), and human health (International Association of Fish and Wildlife Agencies 2005).

Finally, to assist with interpretation of discussions that follow, we clarify the difference between animal rights and animal welfare using descriptions that seem to be acknowledged by those that adhere to either set of principles. *Animal rights* is the concept that animals have the same, or similar, rights as humans, and animals may not be used by humans for any purpose, including food, clothing, entertainment, or experimentation. Conversely, *animal welfare* is the concept that such use of animals by humans is acceptable if humane standards are followed (see Truth About Fur 2017, People for the Ethical Treatment of Animals 2020).

FUR HARVESTERS: WHO THEY ARE AND WHAT THEY DO

People who trap (*trappers*, those who use trapping devices to harvest animals) or hunt (those who use primarily firearms, or to a lesser extent, archery equipment) furbearers may collectively be called fur harvesters. Trapping furbearers involves a wide range of capture devices (e.g., bodygrip traps, foothold traps, cable restraints, snares) and deployment techniques (*sets*; see Krause 2007), which are highly regulated by jurisdiction agencies. Several traps and sets have been developed to be species specific or to target a narrow group of furbearing species. Hunting furbearers involves two primary methods: predator calling (using mouth-operated or electronic sounds to attract predatory species within shooting range) and use of trained dogs (primarily to track, pursue, and tree a particular species). The use of trained dogs is popular for species such as bobcats, coyotes, and raccoons. State, provincial, territorial, and tribal regulations vary substantially for trapping and hunting of furbearers, and should be consulted for details and periodic revisions.

The mean annual income for fur harvesters from trapping is <$1,000, with 80% of U.S. trappers indicating that trapping does not provide an important source of income (Association of Fish and Wildlife Agencies 2015*b*). Rather, motivation of fur harvesters often includes factors such as interaction with nature, self-sufficiency or subsistence, and a rural lifestyle (Daigle et al. 1998, Zwick et al. 2006, Dorendorf et al. 2016). Thus, in the United States of America, fur harvesting is similar to other consumptive activities such as fishing and hunting, in that financial incentive is less of a factor that motivates individuals (Gruntorad and Chizinski 2021 [Chapter 4]). However, the harvest of furbearers and subsequent sale of pelts is an important source of income for many trappers in Alaska, the western United States of America, and many areas of rural Canada, particularly for Indigenous and other communities in northern and other more remote areas (Responsive Management 2015). Although there is some annual variation, up to 50,000 trappers may be engaged in commercial fur harvesting in Canada in any given year (Fur Institute of Canada, unpublished data). The contribution to family and community incomes varies regionally and annually, but tends to be most important in rural and northern regions. Income from fur harvesting is often an additive component to other seasonal natural-resource-based income sources, such as timber harvesting and commercial fishing. Regardless, the close connection to the outdoors and nature interaction associated with successful trappers may explain why trappers generally are considered to be highly knowledgeable about nature (Kellert 1980), which allows them to often use their expertise to serve as highly effective citizen scientists or otherwise assist with conservation and management efforts (e.g., Webb and Anderson 2016).

Management of the human component of harvest often includes interactions between agencies and stakeholder groups. Wildlife agencies and trapping and hunting organizations strive for and make substantial efforts to maintain collaborative and cooperative working relationships. Such an approach proactively avoids extreme negative relationships and interactions that might affect harvest management decisions. Each U.S. state and Canadian province and territory typically has at least one state- or provincial-level trapping organization and perhaps one or more within-state or province chapter or local council organization. In addition, there is one national organization (Canadian National Trappers Alliance) in Canada and two national organizations (Fur Takers of America, National Trappers Association) in the United States of America. About 32% of U.S. trappers are members of ≥1 trapping organization (Responsive Management 2015). Each organization may play a different role in harvest management decisions, and the relationship between a management agency and a given organization may range from little interaction to extremes of positive or negative, depending on the jurisdiction. Positive interactions may include close coordination for any management or conservation decisions designed to benefit conservation and sustainable use of wildlife, including an organization suggesting or supporting informed and science-based decisions on regulatory changes (including more restrictive, when warranted). Trapping organizations also facilitate their members becoming directly involved with capture of furbearing species for data collection, reintroductions, or research. Additionally, members of these organizations are often involved in other conservation activities, such as focused captures of common furbearing species that may be negatively impacting endangered species or that may require capture to reach other goals for population management.

The most recent estimate of >176,500 trappers in the United States of America during 2015 was an increase of 24% compared to the estimated number during 2004. Over half of trappers in the United States of America are located in the Midwest region (Responsive Management 2015; Fig. 23.1). Other survey results included mean number of days trapped/trapper was 36.7 for the 2014–2015 season, although 10% of trappers surveyed did not trap during that season; the mean number of traps set was 27.6/day. In Canada, there were approximately 41,000 licensed trappers during the 2015–2016 harvest season (Fur Institute of Canada, unpublished data). Overall estimates are in the range of 50,000 active trappers for Canada after accounting for additional Indigenous community trappers in jurisdictions where Indigenous trappers may not be required to be licensed.

TRAP TESTING AND ITS INFLUENCE IN NORTH AMERICA

In North America, formal testing of capture devices, including to quantify and improve efficiency and animal welfare, has been conducted for many decades (Novak 1987*a*, Barrett et al. 1988, Boggess et al. 1990, Jotham and Phillips 1994). During the 1990s, the European Union passed and

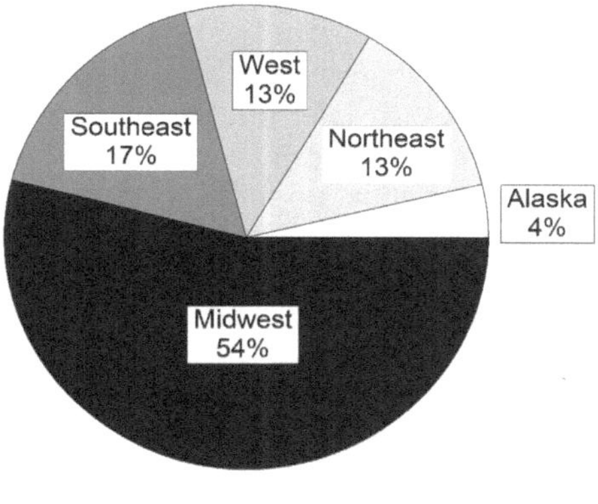

FIGURE 23.1 Distribution of trappers in the United States of America during 2015, based on an estimated total of 176,573 trappers (Responsive Management 2015).

implemented the Wild Fur Regulation EEC 3254/91, which would have eliminated the importation of furs and fur-related goods to Europe from Canada, the United States of America, and other countries if certain stipulations were not met (see Hamilton et al. 1998, Harrop 1998, and White et al. 2021 for more information). This included that countries allowing the use of foothold traps must show that the specific devices allowed met internationally agreed-upon humane trapping standards (European Commission 1991, Hamilton et al. 1998, Andelt et al. 1999).

In Canada, the Agreement on International Humane Trapping Standards was negotiated by the Government of Canada, and with the support of the Canadian provinces and territories (with whom the authority for management of most non-migratory wildlife and, in particular, most furbearing mammal species resides), was ratified in 1999 (Fur Institute of Canada 2019). Beginning with the 2007–2008 fur harvest season, the provinces and territories have modified harvest regulations to require trappers to use capture devices certified through the Agreement on International Humane Trapping Standards (which currently includes >200 different types of traps) for the 12 furbearing species listed within the Agreement (European Commission 1998*a*, Fur Institute of Canada 2019). Similarly, in the United States of America, international treaties and trade agreements are negotiated at the federal level, and the management authority for wildlife resides primarily with states and tribes. The realities of history and the relationships of the states to the federal government made this process of concluding an agreement much less straightforward in the United States of America. Ultimately, the United States of America and the European Union signed a non-binding bilateral understanding called an Agreed Minute (European Commission 1998*b*), which referenced the standards described in the Agreement on International Humane Trapping Standards and outlined a commitment by the United States of America to evaluate trap performance and advance the use of improved traps through development of Best Management Practices for Trapping for 23 furbearing species (Association of Fish and Wildlife Agencies 2019*a*, White et al. 2021). The Best Management Practices for Trapping program serves to provide information on program-approved capture devices to state fish and wildlife agencies, trappers, researchers, and others. The program's resources and information may be used to support revisions of regulations, but the program has no regulatory authority.

The results of trap testing in Canada have led to many mandated regulatory changes to require, at the species level, the use of traps certified as meeting the Agreement on International Humane Trapping Standards. In the United States of America, trap testing helps guide management decisions, including selection of devices for capture during research projects, avocational trapping, and damage management. For example, the Agreement on International Humane Trapping Standards Certified Trap List (Canada) and the Best Management Practices for Trapping (United States of America) can be relied upon by Animal Care and Use Committees for approval of capture devices for research purposes. Results have also either confirmed selection of or provided guidance for appropriate styles of trap jaw (e.g., offset, laminated, double) of the numerous models of foothold traps based on one or more target species. Interestingly, the innovations of trappers have led to numerous improvements in capture devices and techniques, with trap testing more often serving the role of determining what works best as opposed to directing the actual innovations.

HARVEST REGULATIONS

Some regulations associated with harvest management are common to harvest management of other species, whereas some regulations are quite unique for furbearers. We describe the former as general regulations, which include licensing, season timing and length, hunting, harvest limits, and harvest data reporting. We describe the latter as trapping-specific regulations, which include trap and equipment restrictions and specifications, trapping sets, trap-check intervals, registered traplines, and management of incidental take. The enactment of a particular regulation may be related to ensure sustainable harvest for some species. However, the abundance of many furbearing species is such that sustainable harvest is not the primary concern. For example, some regulations are designed to minimize or prevent incidental capture of non-target species (e.g., ungulates, raptors, threatened or endangered), to address potential concerns about animal welfare (e.g., trap-check intervals), to prescribe appropriate, tested traps or trapping-device categories for various species, or to manage the distribution of trappers for social reasons (e.g., registered traplines). Finally, some regulations (e.g., season timing) may be purposefully consistent for multiple furbearing species because trappers often target multiple species on their traplines.

General Regulations

Licensing, Permits, and Tags

The Convention on International Trade in Endangered Species of Wild Fauna and Flora (CITES) entered into force in 1975, with the goal of protecting species from unsustainable trade; currently, CITES has a membership of 183 parties, including Canada and the United States of America (CITES 2020). In North America, the furbearing species listed under CITES include the bobcat, Canada lynx, gray wolf, and North American river otter. Although the brown bear (*Ursus arctos*) is also listed as a furbearer by the U.S. Fish and Wildlife Service for CITES purposes (U.S. Fish and Wildlife Service 2020*a*), the American black bear is considered both a big game animal and a furbearer in Canada (Government of Canada 2017), and pelts of the polar bear (*Ursus maritimus*) resulting from regulated sustainable harvest may be traded and legally exported from Canada, none of the bear species will be discussed here. Regardless, all North American furbearing species that are listed under CITES are classified as Appendix II species, which has been described as either (1) a species that is not currently threatened, but may become so if trade controls are not in place, or (2) a species that physically resembles (i.e., look-alike species) one or more listed species and therefore needs to be regulated to effectively control trade of the latter. However, the concept of look-alike species (e.g., the bobcat compared to the rare Iberian lynx [*Lynx pardinus*]) has been challenged based on the opinion that these species do not fall under authority of CITES.

The process for exporting pelts of a CITES-listed species from the United States of America to, for example, an international fur auction first requires an approved program between the relevant state fish and wildlife agency (or the tribal government of a particular jurisdiction) and the U.S. Fish and Wildlife Service. Many state agencies also have state-level tagging programs for furbearing species not listed under CITES. This essentially serves the same purpose: to monitor legally harvested animals, collect additional data on those species, and increase the effectiveness of enforcement of regulations.

In Canada, the export of any wildlife carcass or parts thereof (including meat, untanned pelts, beaver castoreum, or other parts) between provinces or to another country requires a provincial wildlife export permit (Government of Canada 2018). Harvest tags may be part of provincial or territorial harvest reporting regulations (particularly for big game species), but are generally not a requirement for export of furbearer pelts. The provincial export permit system also provides the necessary background paper trail to assure legal harvest origin for issuance of CITES export permits for pelts of furbearer species listed under Appendix II of CITES for pelts exported from Canada. Tagging pelts, collecting data, and the export permit process allow for a paper trail to track the legal harvest of CITES-listed species in North America. Ultimately, these data are used to ensure that trade is not detrimental to the survival of a particular species.

Season Timing and Length

Pelt primeness is a primary consideration of the various pelt characteristics that determine value (e.g., color, size, quality of handling). A fully prime pelt is one where, "...both the guard hairs and the underfur have reached maximum length and density" (Obbard 1987). The annual cycle of pelt primeness varies somewhat by species, latitude, and other factors, but pelts are generally prime between November and March (Stains 1979). Consequently, harvest seasons typically fall within this annual period, although seasons are often similar among many furbearing species to allow fur harvesters flexibility with their activities (Table 23.1). However, when season dates include early

TABLE 23.1
Examples of Harvest Regulations for Selected Furbearing Species in Oregon During 2020–2022[1]

Species Region	Harvest Season	Harvest Limit	Potential Reason for Regulation
Bobcat			
Eastern OR	1 Dec–28 Feb	5	Season length and timing based on pelt primeness; harvest limit based on relatively lower population densities and relatively high harvest effort due to higher pelt value
Western OR	1 Dec–28 Feb	No limit	Season length and timing based on pelt primeness; no harvest limit based on relatively high population densities and relatively low harvest effort due to lower pelt value
American mink, muskrat			
Entire state	15 Nov–31 Mar	No limit	Abundant semi-aquatic species; trappers often target both species simultaneously with same tools and techniques
North American beaver, North American river otter			
Entire state	15 Nov–15 Mar		Often locally abundant; trappers often target both species simultaneously with same tools and techniques, although some species-specific approaches are implemented when required or desired
American badger, coyote, North American porcupine, nutria, striped skunk, Virginia opossum, western spotted skunk, weasel spp.			
Entire state	1 Jan–31 Dec	No limit	Fewer restrictions for species that are invasive, may cause damage (e.g., agricultural damage, livestock depredation, timber production)
Canada lynx, fisher, kit fox, ringtail, sea otter, wolverine			
Entire state	No season	0	Species of conservation concern (e.g., state listed, federally listed) within the state; extirpated or possibly extirpated; very low abundance

Note

[1] Refer to Oregon Department of Fish and Wildlife (2020*a*) for other components of regulations such as area closures, requirements for trapper education and harvest reporting, requirements for licensing requirements, and models of traps and use restrictions. Harvest limit is the maximum number of individuals that may be harvested by an individual fur harvester.

or late periods in pelt primeness, trappers typically avoid capturing those species during the periods when pelt value is expected to be less.

Season timing and length may be refined based on other ecological, enforcement, social, or political factors. For example, some jurisdiction agencies may allow harvest earlier for common furbearing species (e.g., raccoons) that may also be utilized as food. Seasons may also be further refined by activity (e.g., hunting with the aid of dogs, general trapping, use of snares), such that they largely overlap, but certain periods may be activity specific to avoid potential conflicts. In some instances, timing and length of harvest seasons may also be adjusted somewhat to avoid potential conflicts among different consumptive-user groups (e.g., big game hunters, upland gamebird hunters, trappers) as well. For example, the opening of deer season for hunters that use firearms is typically very popular, but is often scheduled prior to when pelts may be prime, so balancing these seasons accordingly often minimizes the potential for any conflicts.

We know of no extensive studies that evaluated the potential effects of relatively minor changes in season length on harvest level. Essentially, season length typically is not used to directly manage annual harvest levels, except in specific situations, e.g., when very brief harvest periods are coupled with harvest limits to ensure sustainable harvest while allowing harvest of lower-density species in a given jurisdiction or management unit.

Hunting Regulations

Legal weapons for hunting furbearers can include a wide variety of handguns, rifles, shotguns (both modern and muzzleloading), and archery equipment (including crossbows). These vary depending on species, safety constraints, and to some extent, popularity among users. Jurisdictions are variable on their allowance for hunting during nighttime. In 2018, 44 or 50 states permitted hunting at night, and 42 of those allowed use of artificial light (Association of Fish and Wildlife Agencies 2019*a*). Hunting raccoons with the aid of dogs and artificial light is often widely accepted in Canada and the United States of America, whereas hunting coyotes with predator calls and artificial light is not always accepted because of safety concerns and concerns that those using lights may be poaching deer (*Odocoileus* spp.). Our experience is that such restrictions may be primarily the result of social influence, as we know of no data that consistently support illegal activity.

In an effort to avoid potential conflicts between hunters using dogs (e.g., upland gamebird hunters) and trappers, some management units may have restrictions on when or where hunters with dogs may hunt, different opening dates for hunting and trapping, and educational efforts for hunters to quickly release dogs in the rare event their dog is captured in a trap. Thus, the relationship between hunters and trappers is an important consideration for managers of furbearer harvest, especially on multiple-use public lands meant for use of both activities. Hunters using dogs are also restricted on when and where they can train dogs and conduct field trials. Training and trialing pose different management challenges because these can occur outside the open hunting seasons for furbearers. Trappers also need to use good judgment when selecting specific trapping locations.

Another regulatory consideration is hunting of animals for damage management. Hunters pursuing animals for damage may be given more flexibility in their activities than those hunting for other reasons. State and provincial jurisdictions do this in an effort to help agricultural producers address livestock and crop depredations during the periods when the damage is occurring. The principles for damage management were articulated by Nagel et al. (1955), especially the concept of removal of a specific offending animal as opposed to attempts at elimination of populations. For example, landowners attempting to mitigate damage may be allowed to hunt at night, or use calls, toxicants, or other methods, whereas those hunting for recreation often may not (Woolsey 1985, Connolly 1988, Connolly 2004, Blom and Connolly 2003, Mitchell et al. 2004).

Harvest Limits

When harvest limits related to furbearers are implemented, they may be categorized as individual limits (i.e., maximum number of a given species harvested by an individual trapper or hunter during a given season) or season limits (i.e., quotas, where a maximum number of individuals of a given species may be harvested within a particular area or jurisdiction by all fur harvesters, and no further harvest is allowed during the remainder of that season if the quota is reached). Harvest regulations for other species (e.g., small game) often use daily and possession limits to manage harvest levels. Many furbearing species are so abundant that harvest limits are not imposed, but annual harvest levels are monitored to help ensure sustainable levels of harvest. When harvest limits for furbearing species are imposed, it is often designed to regulate harvest to better ensure sustainability of a species that is relatively less abundant, has a restricted distribution in a given jurisdiction, or both.

Harvest Data Reporting

Most jurisdictions implement harvest data reporting in the form of either mandatory or voluntary individual reports due soon after the end of most furbearer harvest seasons. For example, jurisdiction agencies acquire data on furbearer harvest levels and harvest effort (e.g., number of days hunted or trapped to estimate catch-per-unit effort [CPUE]) by several means for the purpose of monitoring harvest and potentially implementing regulatory adjustments. The most common type of reporting in the United States of America is by mailed questionnaires (Association of Fish and Wildlife Agencies 2019*a*). Other methods include reports of pelts purchased or sold by licensed fur buyers, total numbers of pelts exported from jurisdiction export permits, numbers of animals tagged with CITES or state-required tags, and individual trapper or hunter reports (Erickson 1982) via either online or mail-in report card (which may be either mandatory or voluntary). In the United States of America, fur-buyer reports from those within the state may not include direct sales of pelts from trappers to out-of-state fur auctions, and animals tagged with CITES tags may not be assignable to a specific year. Fur-dealer reports and CITES-tag reports may at best be considered indices of harvest, whereas mandatory reports from individual fur harvesters and numbers derived from mandatory jurisdictional export permits may provide the most accurate estimates of harvest, depending on compliance rates and other factors. In the United States of America, all available harvest data are acquired from state fish and wildlife agencies, compiled, and organized into an online database, but these should also be considered minimum harvest values (Association of Fish and Wildlife Agencies 2019*c*). Wild furbearer harvest statistics in Canada were formerly collected from provincial and territorial wildlife agencies and compiled by Statistics Canada. This task is now coordinated by Fur Institute of Canada.

Trapping-specific Regulations

Trap and Equipment Restrictions

Some state fish and wildlife agencies have adopted trapping regulations based on results of the Best Management Practices for Trapping program (Association of Fish and Wildlife Agencies 2019*a*), related to the most humane and efficient trap sizes and models. They may place restrictions on certain traps in certain areas because of concerns for potentially capturing pets or domestic animals. For example, foothold and bodygrip traps may not be allowed in areas where people may legally walk dogs, or near other certain areas (e.g., hiking trails, campgrounds, boat-launch ramps) because pets are likely to be present in these areas. Regulations may also be adjusted to help manage damage problems caused by species such as muskrats burrowing into pond dams or river otters depredating fish at aquaculture facilities or hatcheries.

Regulations on Trapping Sets

Agencies may regulate the types of sets that are appropriate for trapping in terrestrial versus aquatic systems. Agencies may also restrict bait usage when there are concerns for potentially capturing non-target species, including pets. However, the use of bait may be warranted, for example, to effectively trap beavers in areas where capture of river otters should be avoided. Set types may be regulated based on recommendations by the Best Management Practices for Trapping program for the target species, or in Canada by recommendations from the guidelines on Best Trapping Practices (Fur Institute of Canada 2014). For example, large-sized bodygrip traps may not be permitted on land unless they are a minimum distance above the ground, or placed inside of a cubby or bucket or within some other form of exclosure. Other restrictions, such as minimum setbacks from the opening of the exclosure, may be required to further reduce accidental captures of non-target species. In some jurisdictions, when using foothold traps for muskrats, only drowning sets may be used. In essence, the reasons for various regulations on trapping sets are just as varied, but often relate to social considerations, concerns about animal welfare, and avoidance of capture of non-target species.

Trap-check Intervals

The maximum amount of time legally allowed between two consecutive events of checking traps for any captures is called the trap-check interval. Regulated intervals currently range from at least once every 24 hours to no required interval, but there are also differences in required intervals for restraining versus lethal traps (or sets) in some jurisdictions (Responsive Management 2016). Some western and northern jurisdictions allow trappers greater intervals for checking traps compared to the rest of Canada and the United States of America, generally to allow for greater distances traveled in these more remote areas, the potential for harsher winter weather, and other factors often associated with access. This situation has routinely resulted in challenges (typically from those opposed to any trapping activities) to decrease trap-check intervals via legislative action, rule-making, and other approaches. Jurisdictions typically set regulations associated with trap-check intervals to balance concerns about animal welfare, logistics associated with travel (e.g., greater distances traveled and effects of inclement weather during winter in northern climates), and other factors.

Registered Traplines

Registered traplines are a means to better regulate harvest levels to ensure sustainability by granting exclusive trapping rights within an area of public land to one trapper and their authorized helpers. Many registered trapline systems originated in the first half of the twentieth century when harvest was unregulated and therefore negatively impacting furbearer populations and in turn, impacting the economies of the trapping communities. In Canada, trapping is more often open to all trappers on any lands which are open to trapping of furbearers in the eastern provinces and in southern and more populated, agricultural-rural landscapes of other provinces. In the western provinces and more remote areas, registered traplines are common.

The holder of a registered trapline is authorized for exclusive right to trap furbearers on a specified area of public land, and manage the furbearer resources of that area accordingly. They are expected to be active on the trapline and manage their harvests sustainably, but they do not otherwise have exclusive right to the land or other natural resources. The trapline holder will have certain mandatory requirements to retain their traplines allocation, which may include minimum harvest effort for certain species (e.g., beaver), mandatory harvest reporting, regulatory compliance (e.g., appropriate use of Agreement on International Humane Trapping Standards—certified traps), and other responsibilities. Failure to fulfill responsibilities could result in license suspension or loss of rights to the trapline and associated cabin, although this would probably occur only under extreme conditions (e.g., abandonment of trapline, conviction of serious non-regulatory compliance). The handling of furbearer trapping by Indigenous trappers varies across the provinces and territories and may be influenced by various treaties and land-claims agreements. Certain trapline lands may be available only to Indigenous communities for trapping by Indigenous trappers in some Canadian jurisdictions, such as the Northwest Territories and Yukon. For more detailed information on registered traplines, see Carmichael (1973), Anderson (1987), Novak (1987*b*), and Berezanski (2004).

Management of Incidental Take

The Endangered Species Act of 1973 (16 U.S.C. §1531 et seq.) defines take as, "*to harass, harm, pursue, hunt, shoot, wound, kill, trap, capture, or collect, or to attempt to engage in any such conduct.*" Incidental take can therefore be described as take while engaged in otherwise lawful activities. Specific to furbearer management, this typically relates to unintentional capture of a wild (e.g., raptors, deer) or domestic (e.g., domestic cat or dog) non-furbearing species or the capture of a protected furbearer species (e.g., Canada lynx or wolverine [*Gulo gulo*] in the coterminous United States of America). Here, we describe regulatory examples and processes designed to minimize incidental captures during otherwise legal harvest activities.

Trappers must minimize incidental captures of species of conservation concern, such as those listed under the U.S. Endangered Species Act. Consider the example of incidental take of Canada lynx during trapping efforts for bobcat and fisher. The Canada lynx is distributed across most of Canada and Alaska, with its southern periphery in very limited areas in the northern coterminous United States of America, including marginal habitat in northern Maine and northeastern Minnesota. Although lynx is currently listed as threatened in the coterminous United States of America, the U.S. Fish and Wildlife Service has proposed delisting due to recovery (U.S. Fish and Wildlife Service 2020*b*). The Maine Department of Inland Fisheries and Wildlife (2020) delineated a lynx protection zone to minimize incidental take, and to increase the probability of releasing captured lynx with minimal or no injuries. Regulations within this zone included several trapping restrictions (e.g., maximum jawspread of foothold traps used on dry land, minimum number of swivels in trap chains for foothold traps, use of lynx exclusion device [see Association of Fish and Wildlife Agencies 2017] on certain bodygrip traps set on dry land), and required any person who captures a lynx to call an established *lynx hotline*. Further, educational brochures have been developed to disseminate information about how to avoid incidental captures of Canada lynx (Golden and Krause 2003) and wolverines (Hiller and White 2013).

Certain species, such as American marten and Canada lynx which are common in more northern areas, are species of conservation concern and may be listed as provincially endangered in some of Canada's eastern provinces. Efforts similar to those used in the United States of America have been implemented to avoid incidental captures, promote release, and encourage sighting reports from trappers and the public. Incidental take of furbearers outside of open seasons or in excess of established harvest limits generally results in the requirement of those animals to be surrendered to the regulating agency if they cannot be released alive.

Numerous other effective regulations are in place to minimize the probability of capturing non-target species, or to increase the probability of being able to release those species with no or minimal injuries. For example, many jurisdictions in Canada and the United States of America require the use of *setbacks*, a regulation that specifies any traps set on public (or private) lands must be a minimum distance from trails, campgrounds, and other areas of high levels of human use, to avoid any potential conflicts (e.g., capturing unleashed dogs). Many Midwestern U.S. states and some Canadian jurisdictions restrict the use of bodygrip traps set in cubbies on dry land to be recessed a certain distance inside the cubby enclosure to avoid capture of dogs (see Association of Fish and Wildlife Agencies 2017; Fig. 23.2). The use of cable restraints or snares on dry land may also be regulated to include minimum sizes for loops, break-away devices, and techniques to minimize capture of deer, moose (*Alces alces*), and other non-target species (e.g., Roy et al. 2005, Association of Fish and Wildlife Agencies 2009, Gardner 2010). A final example is the omnipresent regulation that prohibits either large exposed baits (e.g., carcasses of large animals) or setting traps within a certain distance of such baits to avoid capturing raptors. Several other techniques and approaches are listed in regulations and are used by trappers to minimize incidental captures, or even allow for species-specific sets, such as pan-tension devices on foothold traps, tension devices or trigger modifications on bodygrip traps, use of foot-encapsulating traps, and first and foremost, general avoidance of areas where an incidental capture is likely to occur (e.g., Association of Fish and Wildlife Agencies 2017, White et al. 2021).

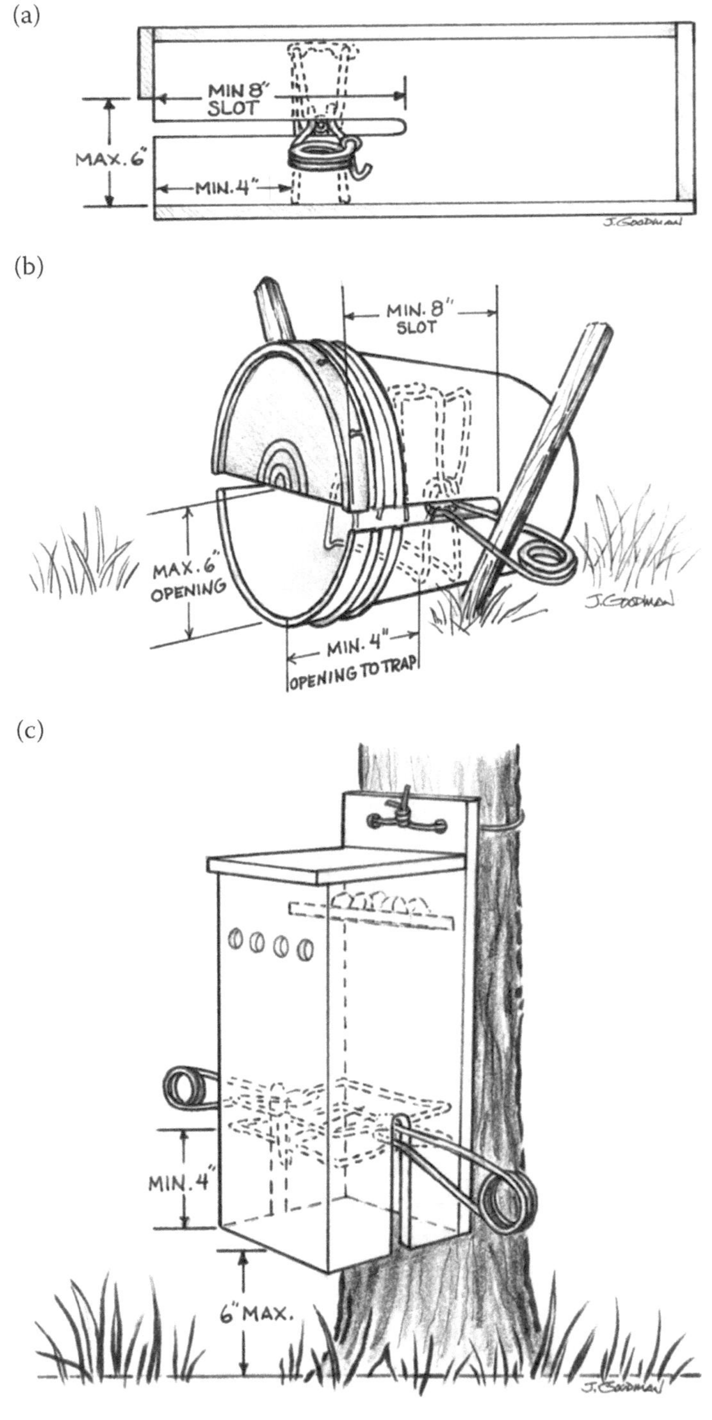

FIGURE 23.2 Several state fish and wildlife agencies require specific dimensions or designs of enclosed cubby sets on dry land that include bodygrip traps, which may include (a) recessed trap, (b) partially restricted entrance, and (c) vertically mounted design. Credit to Joe Goodman, used with permission from Association of Fish and Wildlife Agencies (2017).

There are no accurate and complete data sets associated with number and species of incidental captures of non-furbearing animals. In many instances, trappers simply release non-target captures if the non-target animal appears to have minimal or no injuries, or they report the capture to the jurisdictional fish and wildlife agencies for assistance, especially if legally required. Incidental take of threatened or endangered species can be minimized with proper regulations in place and the use of good judgment by trappers. For example, during trap testing under the Best Management Practices for Trapping program, no individuals of any federally threatened or endangered species (e.g., Canada lynx, San Joaquin kit fox [*Vulpes macrotis mutica*], wolverine) where listed were captured over a 21-year period that included >230,000 trap nights across much of the United States of America (White et al. 2021), despite the potential or actual occupancy of those species in the vicinity.

EVALUATING HARVEST REGULATIONS FOR FURBEARERS

Harvest management of furbearers is typically multi-species management, which increases the complexity of evaluation of regulations. Hunting and trapping seasons are set for species with different biological characteristics, population ecology, and hunter and trapper interests. Economic considerations, such as pelt value and interest in managing damage or nuisance, also play into such decisions. Conversely, big game species, such as bighorn sheep (*Ovis canadensis*), black bear, elk (*Cervus elaphus*), mountain goat (*Oreamnos americanus*), and white-tailed deer, are managed as single species (Diefenbach et al. 2021 [Chapter 22]), which may allow for more straightforward evaluations of harvest regulations.

Harvest management of furbearers has been among the more controversial aspects for jurisdiction agencies. Despite the public attention, data collection and analyses to support informed decision making for furbearer management have often lagged behind that for more popular big game species due to limited resources and the difficulty of monitoring the diverse array of furbearing species, among other factors (Hiller et al. 2018). However, heightened interest in data collection and application of contemporary statistical methods to support defensible decisions (Conroy 2021 [Chapter 1], Cummings and Bernier 2021 [Chapter 10], Runge 2021 [Chapter 7]) should shift this paradigm for furbearer management. Mandatory harvest and activity reporting is an important tool and basic step to better understanding harvest and population trends.

There have been several efforts for developing models to guide furbearer management. Frederick and Cobb (1992) constructed a simulation model of raccoon populations that included hunting and trapping harvest, illegal harvest, disease, and habitat. Most jurisdictions will have harvest data and can obtain habitat data, but will not have disease prevalence or illegal harvest estimates. In spite of this, the model has merit if the comprehensive data needs are obtained. Another model was developed by Thompson et al. (1996) for 23 species of furbearers in New Mexico. This model relied on harvest data and ecological zone to develop a basis for using habitat to estimate harvest of the 23 species. The authors believed that demographic data were needed to properly assess harvest of bobcat and red fox. Neither of these models have seen application in furbearer management programs. In contrast, programs such as Deer Camp (Moen et al. 1986), POP-II (Bartholow 2000), and OnePop (Gross et al. 1973) have been used by agencies for managing big game (e.g., elk, Williams 1991).

THE FUTURE OF FURBEARER HARVEST MANAGEMENT

Social and political challenges to harvest management of furbearers, large carnivores, and other wildlife species will undoubtedly continue to occur through ballot initiatives, litigation, and other means that bypass agency management authority (Hiller et al. 2021*b* [Chapter 2]). Furbearer management is particularly subject to such challenges. We offer some thoughts, in the context of the social and political dynamics described, on what furbearer managers and agency leadership might expect in the future. We also offer thoughts on how to potentially address those issues to ensure the tenets of the North American Model of Wildlife Conservation remain in place, particularly that science is the proper tool to discharge wildlife policy (Organ et al. 2012).

The potential for controversy surrounding harvest management for furbearers creates a current need for increased use of monitoring data collected on furbearers (Cummings and Bernier 2021 [Chapter 10]). Indeed, it becomes increasingly difficult in the current social and political climate to make informed and defensible science-based decisions without collection and utilization of data. Agencies can provide the evidence demanded by stakeholders and the general public when data are available, although we recognize that is only one piece of the puzzle. Financial and logistical constraints will continue to cause disparities in priorities for data collection for furbearers relative to other harvested species. However, conservationists have been working to increase federal funding opportunities for wildlife conservation and management (e.g., Recovering America's Wildlife Act; The Wildlife Society 2020), which may have substantial benefits for monitoring and managing furbearing species.

The decision-making process for harvest management of furbearers is embedded in a social matrix. Thus, wildlife managers need effective communication strategies to demonstrate, understand, and articulate the societal-level benefits and value of sustainable management and use of furbearers (e.g., Kahan et al. 2012), which includes the human societal value that accrues from the management of certain furbearer species that otherwise would create substantial risk, conflict, and damage. Such benefits to local stakeholders have potential to be seen as more relevant to decision makers than the general campaign activities of animal-rights groups that oppose furbearer harvest with substantial resources relative to agencies and consumptive-user groups. Further, although public education efforts are considered beneficial, we should also consider the benefits of targeted communication efforts at smaller scales. For example, a focus on administrators and politicians, those who make decisions and policy, may be the focus for increased efforts. Interestingly, it may be possible that the dissemination of more information about a controversial topic (e.g., climate change) across wide audiences may increase cultural polarization independent of scientific literacy (Kahan et al. 2012). Although the controversy associated specifically with furbearer harvest may not be on the scale of climate change, the humane use of animals could be described as omnipresent.

Global drivers cause concerns about impacts and influences of animal-rights-based interests go beyond local and state-provincial levels to the highest fora of international wildlife conservation, including such venues and critical policy drivers as CITES and the International Union for the Conservation of Nature. For example, the COVID-19 pandemic has caused tremendous global turmoil, concern, and human suffering at the time of this writing, the likes and scale of which the world has not known for many years. As a fuller understanding of the origins of the virus unfolds, it will likely bring quite justifiable and intense scrutiny on the sustainability and safety of use and trade in certain wildlife species and products and the inadequate hygiene and safety protocols and practices of so-called *wet markets*. The outfall of this new scrutiny has already provided considerable opportunity for animal-rights organizations and interests to promote massive generalizations and broad-stroke actions to advance their goal of reducing or banning legitimate, regulated, sustainable wildlife use in the guise of necessary action for public health and safety and biodiversity conservation. Our author team is aware of reports coming out of China as to possible actions to be taken, which may provide an opportunity for overreach of reactionary restrictions on unregulated animal trafficking that could impact regulations for harvest management of furbearers. The concern of furbearer managers in the United States of America is the potential loss of the opportunity to legally harvest furbearers, but also the loss of an effective tool for management of furbearer populations.

Amidst a landslide of controversy generated in Europe during the 1990s, Decker and Batcheller (1993: 153) stated, "It is likely that the import of furs to the [European Union] (75% of the North American fur market) will end in the future." Although that statement was made over a quarter century ago, there are certainly no guarantees that substantial changes to harvest management of furbearers will not occur in the future, even notwithstanding the benefits associated with the Agreement on International Humane Trapping Standards and the Best Management Practices for Trapping programs. These programs have undoubtedly helped guide managers with additional science for informed decision making. However, we believe that we must also increase our monitoring (and other data collection) efforts for furbearers and use those (and existing) data to guide harvest management. Fortunately, the past several years has seemingly resulted in an initiation of a paradigm change toward greater efforts for collection and utilization of data, which is critical to continue.

The future of harvest management for furbearers will mirror efforts for other programs for which the inclusion of social components for regulations, including the formal expression of objectives related to human dimensions, has been critical (Kaemingk et al. 2021 [Chapter 3], Vrtiska 2021 [Chapter 20]). State and provincial agencies will need to consider greater efforts toward furbearer management by increasing budgets and dedicating more personnel to management of these species. Furbearer management programs are often led by agency staff that have multiple primary duties that may too often lead to less support for spending more time on data collection and development of management programs. As a group, furbearers are typically relatively low in priority for funding of research. Also, such efforts must include broader messaging and stakeholder involvement in this dynamic decision-making environment. Perhaps ironically, a program for harvest management that is primarily housed within state and provincial agencies will undoubtedly need to gain increasing coordination at the federal and global levels in light of global dynamics that affect this unique form of harvest management. Regardless of whether fur markets continue to drive management of furbearers, there is an obligation to conduct data-driven management. Evolution of furbearer harvest management must occur at some level to help ensure its persistence and success.

ACKNOWLEDGMENTS

The authors thank D. Berezanski, D. Kay, and A. Walpole for providing input on the registered traplines section, and thoughtful comments from B. White, L. Powell, and one anonymous reviewer. The authors also thank the editors, L. Powell and K. Pope, for meeting the challenges associated with producing a long-overdue tome on such a critical topic as harvest management of fish and wildlife. The authors also thank the various trappers and wildlife professionals whom they have learned from over the years.

24 Harvest Regulations for Inland Recreational Fisheries

R. Scott Gangl

CONTENTS

INTRODUCTION

The history of harvest regulations for inland fisheries is as long and varied as the fisheries profession itself. As the profession advanced, so evolved our understanding of biological processes affecting fish populations, including mortality due to harvest. This evolution of understanding brought periods where regulation philosophies progressed from unrestricted, very restricted, liberalized, to science-based (Redmond 1986, Isermann and Paukert 2010, Rahel 2016).

A fishery is commonly defined as having three components: habitat, humans, and fish. Harvest regulations control the interaction between the human and fish components to ensure long-term sustainability of the fishery. From this interaction, there are two fundamental approaches to regulating harvest of an inland fishery: biological regulations based on the fish component and social regulations for the human component. Fishery managers employ biological regulations when harvest is unsustainable over time, with the intent of influencing biological processes affecting that fish population. For example, biological regulations may be implemented to protect spawning fish

(Quinn 2002, Pierce 2012), increase or maintain the size of the spawning stock (Koupal et al. 2015, Ward 2015, Clower et al. 2018), alter the size structure or density of a fish population (Schneider and Lockwood 2002, Aday and Graeb 2012), or reduce the exploitation rate (Moody-Carpenter et al. 2017, Smith et al. 2018). Regulations serve a social purpose when their intent is to enhance the anglers' experience, such as to create or maintain a trophy or high-quality fishery (Johnson and Bjornn 1975, Simonson and Hewett 1999) or to provide a high-density population to guarantee that anglers, often children, will catch a fish (Eades and Lang 2012).

Whether biological or social, regulations should be science based and objective driven, with properly collected data to suggest that restricting harvest will produce the expected result in the target fishery. Managers need adequate knowledge of their fisheries to know how harvest is affecting population dynamics relative to factors like growth, natural mortality, and recruitment. When data are lacking, outside pressures (e.g., public or political in nature) can persuade managers to impose regulations that are unwarranted (e.g., Thurow and Schill 1994). Such misguided regulations will give anglers the impression that the regulation is needed and makes it difficult to remove in the future. Moreover, regulations based on social pressures without the support of scientific data are simply imposing the beliefs of some anglers on the entire angling public, with neither group seeing any benefit from the regulation. Thus, managers should consider new fishing regulations carefully, by considering the data and potential effects of the regulation and presenting the regulation to the public with transparency on the need and expected result of the regulation. This chapter provides an overview of some of the harvest regulations employed by managers of inland recreational fisheries, briefly reviews the historical application of harvest regulations, and discusses many considerations that managers must make when implementing new regulations.

REGULATION ALTERNATIVES FOR INLAND RECREATIONAL FISHERIES

Creel Limits

Creel limits, also referred to as bag limits, limit the total number of fish an angler can harvest or possess in a given period (Table 24.1). Creel limits are typically set by species, but can also be set by species groups (e.g., sunfish *Lepomis* spp., black bass *Micropterus* spp., crappie *Pomoxis* spp.). Creel limits can be set as a daily limit, a possession or freezer limit, or a season limit.

Daily Creel Limits

Daily creel limits are a principle regulation used by fishery management agencies to regulate harvest. A daily limit is the maximum number of fish an angler can harvest in a 24-hour period. Daily limits serve a dual purpose of regulating harvest and requiring fair allocation of the resource among anglers of varying skill levels. The early setting of daily creel limits was born of necessity, to put limits of harvest on previously unlimited consumption (Redmond 1986). Although early managers factored natural resource production into their regulations, most early daily limits appeared to be arbitrarily set (Fox 1975) to lower harvest from unlimited to a level as low as anglers would accept. As a result, daily limits tended to be lower for fish held in higher esteem by anglers (e.g., walleye *Sander vitreus*, black bass, muskellunge *Esox masquinongy*) than for fish regarded as "rough fish" or undesirable (e.g., common carp *Cyprinus carpio*, buffalo *Ictiobus* spp., suckers *Catostomus* spp.), regardless of their status in the environment. Radomski et al. (2001) examined creel limits across North America, concluding that larger piscivorous fish species had lower creel limits than smaller insectivorous or planktivorous species.

Daily limits do not always regulate total harvest of fish populations, mainly because most anglers rarely catch or harvest a daily limit of fish (e.g., Hartman 1958, Cooper 1970, Snow 1982, Hess 1991, Baccante 1995, Munger and Kraai 1997). This also hinders the distribution of resources among anglers of varying abilities because the most experienced anglers are more likely to catch fish than less avid anglers when catch rates are low. However, creel limits do effectively distribute

TABLE 24.1

Types, Definitions, Rationale, and Some Examples of Creel Limits Commonly Used by Fisheries Managers to Regulate Harvest on Inland Fisheries[1]

Creel Limit	Definition	Rationale
Daily limit	The maximum number of fish an angler can harvest in a day	Promotes equitable harvest among anglers and may limit harvest mortality
Possession limit	The maximum number of fish an angler can have in their possession during trips spanning more than one day	Promotes equitable harvest among anglers
Residence limit	The maximum number of fish an angler can possess at their permanent residence. Often referred to as a *freezer limit*	Rationale not clearly defined, likely promotes equitable harvest among anglers
Season limit	The maximum number of fish an angler can harvest in a season	Limits harvest of fish that cannot sustain high levels of harvest, while providing recreational opportunity

Note

[1] Selected references: bluegill: Jacobson 2005, Rypel 2015, Feiner et al. 2021 [Chapter 16]; crappie: Mosel et al. 2015; yellow perch: Isermann et al. 2007; largemouth bass: Paukert et al. 2007; smallmouth bass: Engel et al. 1999; walleye: Baccante 1995, Munger and Kraai 1997, Beard et al. 2003*a*, Fayram and Schmalz 2006; northern pike: Goeman et al. (1993, Oele et al. 2016; paddlefish: Scarnecchia et al. 2014, 2019; lake sturgeon: Minnesota Department of Natural Resources 2020, Wisconsin Department of Natural Resources 2020; salmon or steelhead: Oregon Department of Fish and Wildlife 2020*b*.

harvest when catch rates are high and individual anglers are able to catch more fish regardless of their ability. Daily creel limits during a "good bite" function to ensure all anglers have an equal opportunity to harvest some fish, and individual anglers are not allowed to harvest more fish than they can reasonably use without wasting the resource. Feiner et al. (2021 [Chapter 16]) reported that reduced daily limits were effective at distributing the harvest of panfish among anglers in Wisconsin.

To effectively reduce harvest, daily limits must be low enough to affect the average angler's harvest. Radomski et al. (2001) suggested that daily limit reductions of 75% would be necessary to reduce harvest by 25% for sunfish and crappie. Substantial reductions in bluegill (*Lepomis macrochirus*) daily limits in Minnesota (Jacobson 2005) and Wisconsin (Rypel 2015) improved size structure in most, but not all, study lakes. Jacobson (2005) suggested a 40% harvest reduction was necessary to see improved bluegill size structure from his study. Isermann et al. (2007) determined that yellow perch (*Perca flavescens*) daily limits in South Dakota would have to be reduced from 25 to 5 fish to reduce harvest by 25% or more on most of the lakes examined, whereas Mosel et al. (2015) determined that limits on black crappie (*Pomoxis nigromaculatus*) and yellow perch would each need to be lowered to ≤10 fish to reduce harvest by 25% in Wisconsin. Beard et al. (2003*b*) reported that a restrictive daily bag limit of two fish did effectively reduce exploitation of walleye in northern Wisconsin lakes. Oele et al. (2016) concluded that restrictive creel limits, combined with high minimum length limits, effectively increased the availability of large northern pike (*Esox lucius*) for Wisconsin anglers.

Some fish management agencies have experimented with liberal bag limits to promote harvest of fish. Redmond (1986) described a period of liberalized regulations in the United States of America during the 1940s–1960s, after studies on some reservoirs suggested that fish populations were largely underfished. In South Dakota, liberal walleye limits have been implemented on Lake

Oahe to reduce predation on the collapsed rainbow smelt (*Osmerus mordax*) prey base (South Dakota Game, Fish and Parks 2019). Though exploitation did increase on Lake Oahe walleye, researchers found that increased natural mortality played a much more prominent role in reducing walleye density (Graeb et al. 2008). Experiments in northern Wisconsin concluded that liberalized limits did not increase harvest of panfish in Spruce Lake (Kempinger 1969) or have a detrimental effect on fish populations in Escanaba Lake (Kempinger et al. 1975). Liberalized limits implemented to reduce population density to increase growth rates (i.e., to reverse effects of "stunting" [Chizinski et al. 2010*b*]) are typically unsuccessful because most anglers do not experience high catch rates when they go fishing, or are uninterested in harvesting small, slow-growing fish (Goeman et al. 1993). If sport fish are subject to liberal daily limits, conflicts can arise when angler groups have preferences that differ from the objectives of population reduction or removal (Paukert et al. 2021 [Chapter 18]).

Although daily limits commonly do not limit harvest for a biological effect on the fishery, they can have a profound effect on the individual angler's perspective. For many anglers, the daily limit for a species is the benchmark they use to define their angling success. Anglers will often strive to catch their limit, and when they do not they may perceive their outing as unsuccessful. As a result, limits that are set close to the catch that the fishery is capable of producing will tend to produce higher satisfaction among anglers than less restrictive limits (Fig. 24.1), even if the actual catch per angler does not change (Fox 1975, Cook et al. 2001).

Enterprising fisheries managers may consider limit reductions as an easy way to increase angler satisfaction, but doing so may lead to unplanned results. Beard et al. (2003*a*) reported that angler effort was less on lakes with a lower daily bag limit for walleye than on lakes with higher limits. They surmised that lakes with higher limits were more attractive to anglers, despite having lower

FIGURE 24.1 Daily and possession limits are set to reduce harvest mortality and require anglers to harvest fish in a fair and equitable manner. Anglers tend to have a higher satisfaction level when they achieve their limit. Photo courtesy of North Dakota Game and Fish Department.

walleye catch rates than lakes with lower bag limits. Fayram and Schmalz (2006) attempted to encourage anglers to harvest small walleyes by restricting harvest of larger fish to one fish over 356 mm daily, but they concluded that the restriction resulted in decreased angler effort, which in turn failed to increase harvest of smaller fish. Most walleye anglers surveyed by Quinn (1992) considered creel limits as an important factor contributing to quality walleye fishing. Conversely, respondents in that survey also rated *inappropriate size or bag limits* as one of the most important factors contributing to low-quality walleye fishing.

Possession and Residence Limits

Possession limits are similar to daily limits in that they limit the total number of fish an angler can have in their possession, typically over the course of a fishing trip of more than one day. Possession limits do not serve a biological purpose, but they do prevent anglers from harvesting more than their fair share of fish. Thus, possession limits serve a social purpose of promoting equitable harvest among anglers. Possession limits are typically set as a multiple (e.g., 1, 2, or 3 times) of the daily limit.

Residence limits, sometimes referred to as *freezer limits*, limit the number of fish that an angler can possess at their permanent domicile. Residence limits vary among states, ranging from one daily limit to unlimited.

Season Limits

The most restrictive type of bag limit is a season limit, which limits the total number of fish that can be harvested in an entire season. Managers generally reserve season limits to control harvest of long-lived species that exhibit late maturation or episodic recruitment, such as paddlefish (*Polyodon spathula,* Scarnecchia et al. 2014, 2019).

Season limits require a more complicated system of harvest monitoring and permitting to implement and enforce. Similar to big-game hunting, tags may be issued to anglers who wish to target the limited species, such as paddlefish in North Dakota or Montana (Scarnecchia et al. 2008) or lake sturgeon (*Acipenser fulvescens*) in Minnesota or Wisconsin (Minnesota Department of Natural Resources 2020, Wisconsin Department of Natural Resources 2020). Fish-harvest tags may be one-time use, and successful anglers may be required to mark the tag and attach it to the fish (Fig. 24.2). Another alternative is a paper tag that is not attached to the fish, but is filled out with harvest information and carried by the angler when they are in possession of the fish. Once the tag is filled, it cannot be used to harvest another fish. In Oregon, for example, anglers may purchase a combination angling tag authorizing them to harvest wild salmon or steelhead (*Oncorhynchus* spp.) annually. They must record each wild salmon harvested, not to exceed their annual limit (Oregon Department of Fish and Wildlife 2020*b*).

Depending on the unique management objectives for the fishery, managers may choose to issue an unlimited number of tags or may limit the number of tags that can be issued each season. Issuing unlimited tags ensures all anglers an opportunity to participate and potentially harvest a fish, while restricting each angler to one tagged fish per year. A system with unlimited tags provides the managing agency with contact information for the angler who purchased the tag, which can be used for harvest monitoring through follow-up surveys. With unlimited participation, other restrictions such as length restrictions or in-season harvest closures may be required to keep harvest below the season total. Limiting the number of tags issued reduces congestion at popular fishing sites and ensures the fishery will remain open through the entire season. Limiting tags requires a special system to allocate those tags (e.g., a lottery), and limits fishing opportunity to just those fortunate to obtain a tag.

Some states require mandatory registration of fish harvested during the season. In Wisconsin and Minnesota for example, anglers are required to purchase a tag to harvest lake sturgeon, and any tagged sturgeon must be reported within 24–48 hours (Minnesota Department of Natural Resources 2020, Wisconsin Department of Natural Resources 2020), whereas in Michigan (Michigan

FIGURE 24.2 Tagging may be required for fish managed with a low season limit, like paddlefish. Photo courtesy of North Dakota Game and Fish Department.

Department of Natural Resources 2020) anglers are not required to tag their fish, but must register any lake sturgeon or muskellunge within 24 hours of harvesting them. Anglers who harvest a fish are required to register the fish with the managing agency within an allotted period. This replaces the act of physically having to tag the fish and serves as a means for the agency to collect important harvest information. The tradeoff of a registration system is the additional staff time and resources required to administer the program.

Length Limits

Length limits, or size limits, provide managers with a mechanism to restrict harvest on specific components of the fish population (Table 24.2). Agencies have used length limits to combat stunting, improve size structure, increase spawning stock, and provide trophy opportunities. Arguably, a length limit is one of the more popular types of harvest regulations because it can limit harvest without inhibiting anglers' opportunity to go fishing; the fisheries literature is replete with studies predicting or evaluating the success, or failure, of various types of length limits on various species of fish (Wilde 1997, Paukert et al. 2001, Paukert et al. 2007).

Length limits rely on biological processes to manipulate size structure, and managers must have appropriate information on fishery population dynamics to implement the correct regulation to meet their objective. Basic processes such as growth, mortality and recruitment factor into designing a length limit. Additional information on angler exploitation and forage abundance can also help predict whether a length limit will produce the desired effect.

Length limits can be popular with some anglers. Often anglers will learn of a length limit successfully applied to a fishery elsewhere, and develop the opinion that the same limit should be applied to their favorite waterbody. This popular opinion can be helpful when implementing a

TABLE 24.2

Types, Definitions, Rationale, and Some Examples of Length Limits Commonly Used by Fisheries Managers to Regulate Harvest on Inland Fisheries[1]

Length Limits	Definition	Rationale
Minimum length limit	Requires release of all fish smaller than the specified length	Protects small fish from harvest until they reach a more desirable size to maximize yield for anglers, or until they reach sexual maturity
Maximum length limit	Requires release of all fish larger than the specified length	Protects large fish from harvest to enhance spawning stock or create trophy fishing opportunities
Slot length limit	Restricts harvest on intermediate segments of the length distribution of a fish population	Protects specific size groups, age groups, or spawning fish
	A *protected slot* limit allows harvest of fish below and above the slot range	A protected slot limit is designed to reduce harvest mortality on specific size or age groups, usually with the intent of enhancing size structure by encouraging harvest fish outside of the slot
	A *harvest slot* limit allows harvest of fish within the slot range and prohibits harvest fish smaller and larger than the slot	Harvest slot limits function as a combination of a minimum and maximum length limit

Note

1 Selected references: largemouth bass: Paragamian 1982, Eder 1984, Gabelhouse 1984*b*, Novinger 1990, Martin 1995, Wilde 1997, Parks and Seidensticker 1998, Garthause and Heidinger 1999, Paukert et al. 2007, Bonds et al. 2008, Dotson et al. 2013, Risley and Johnson 2016, Miranda et al. 2017; smallmouth bass: Paragamian 1984, Hoff 1995, Lyons et al. 1996, Slipke et al. 1998, Engel et al. 1999, Newman and Hoff 2000, Sterling et al. 2019; catfish: Holley et al. 2009, Siddons et al. 2016, Moody-Carpenter et al. 2017; walleye: Serns 1978, Brousseau and Armstrong 1987, Munger and Kraai 1997, Rietveld et al. 1999, Fayram et al. 2001, Stone and Lott 2002, Isermann 2007, Fayram and Treska 2010; yellow perch: Isermann et al. 2007; crappie: Colvin 1991, Webb and Ott 1991, Hale et al. 1999, Bister et al. 2002, Mosel et al. 2015; bluegill: Paukert et al. 2002; muskellunge: Cornelius and Margenau 1999, Simonson and Hewett 1999, Margenau and AveLallemant 2000, Doss et al. 2019; northern pike: Paukert et al. 2001, Pierce 2010, Oele et al. 2016; trout species: Babcock 1971, Caroffino 2013.

regulation that is necessary. However, anglers may look at certain length limits as a panacea for quality fishing and demand limits that are not warranted. Length limits that are improperly applied will not only be ineffective, they may be biologically or sociologically damaging as well.

Minimum Length Limit

A minimum length limit requires that all fish caught smaller than a specified length be returned to the water unharmed (Fig. 24.3), with the objective of increasing abundance of fish larger than the minimum size. The minimum length limit is historically popular with managers who wanted to maximize the yield of fish harvested by anglers (e.g., Ricker 1945), and the regulation is equally popular with anglers who are less concerned with total yield and more of the belief that fish should be "allowed to grow up" because they provide better table fare at sizes above the minimum.

The biological criteria necessary for a minimum length limit to be effective (outlined by Serns [1978] and Brousseau and Armstrong [1987]) include (1) low reproduction; (2) rapid growth, especially of small fish protected by the regulation, (3) low natural mortality, and (4) high angling mortality.

When low reproduction (or poor stocking success) results in a low abundance of young fish, a minimum length limit can protect those fish from harvest, whereas a rapid growth rate ensures the

FIGURE 24.3 When length limits are implemented, anglers are required to accurately measure their catch and release fish that are outside of the allowed harvest range. Photo courtesy of North Dakota Game and Fish Department.

protected fish will quickly grow to a size desirable to anglers. If small fish are very abundant, density-dependent factors can cause growth to slow and create a stockpiling of small fish below the minimum length. When this happens, fish reach harvestable size slowly and are harvested quickly, creating a paucity of fish above the minimum size and not meeting the objective of providing more fish larger than the minimum size for anglers to catch.

The mortality criteria are important from a practical standpoint. Harvest restrictions are only effective if harvest is affecting the population. If angling mortality is not negatively affecting the population, then restricting harvest will have no effect in achieving management objectives (see Sylvia et al. 2021 [Chapter 17]). Likewise, if natural mortality is high, fish are more likely to die of natural causes than being harvested by anglers. Managers should seek to correct the causes of high natural mortality and allow anglers to harvest and utilize fish that may die of natural causes.

Minimum length limits have been applied broadly to fish species across North American, with some states implementing them as a standard statewide regulation for some fish (e.g., Fayram et al. 2001, Pritt et al. 2019, Sterling et al. 2019). Evaluations of minimum length limits generally document mixed success. Wilde (1997) reviewed published length limit evaluations for largemouth bass (*Micropterus salmoides*) across the United States of America, determining that minimum length limits were generally effective at increasing angler catch rates of bass, but did not restructure bass populations to increase the size of bass caught. Lyons et al. (1996) concluded that minimum length limits improved smallmouth bass (*Micropterus dolomieu*) populations in Wisconsin streams with good quality habitat, likely because the quality of habitat helped the population meet the criteria for growth and natural mortality. Isermann (2007) found no direct evidence that minimum length limits improved walleye populations on some Minnesota lakes, and Fayram et al. (2001) reported similar findings in Wisconsin, although the latter study did conclude that the minimum length limit did reduce harvest and exploitation of walleye. Some crappie (*Pomoxis* spp.) fisheries have benefited from minimum length limits, by improved size structure or age structure of the population (Colvin 1991, Webb and Ott 1991) or by increasing the mean size of harvested fish via the regulation directly (Hale et al. 1999).

High minimum length limits have been used to manage trophy fisheries, such as muskellunge or northern pike (e.g., Simonson and Hewett 1999, Margenau and AveLallemant 2000, Oele et al. 2016, Doss et al. 2019), and may provide trophy opportunities for other species such as alligator gar (*Atractosteus spatula*) (Smith et al. 2018), lake trout (*Salvelinus namaycush*) (Lenker et al. 2016), or largemouth bass (Dotson et al. 2013) if those populations meet the criteria to successfully implement a minimum length limit.

Maximum Length Limits

A maximum length limit allows anglers to harvest smaller fish and requires any fish larger than the specified size to be immediately released. Acting as the inverse of the minimum length limit, the maximum length limit is designed to protect larger fish, typically with the intent of either enhancing spawning stock or to create trophy fishing opportunities for anglers.

Like the minimum length limit, a maximum length limit requires certain biological criteria be met to be effective (Brousseau and Armstrong 1987). The criteria include (1) high angling mortality of large fish, and (2) recruitment is limited by the number of spawning fish in the population.

For a maximum length limit to be effective, there needs to be evidence that harvest mortality is having a negative impact on large fish. From a biological standpoint, if the above criteria are met, a maximum length limit may be necessary to ensure a self-sustaining population. However, for fisheries that are maintained through stocking, protecting spawning-sized fish is not necessary for the health of the population, but may be necessary for providing what anglers consider a quality fishery. Pierce (2010) reported that maximum length limits increased the size structure of northern pike in Minnesota. In Manitoba, a 600-mm maximum length limit has maintained a trophy channel catfish (*Ictalurus punctatus*) population in the Red River of the North (Siddons et al. 2016). Where properly applied, maximum length limits may also protect large fish that are sensitive to exploitation, such as alligator gar (Smith et al. 2018), lake trout (Lenker et al. 2016), or trophy largemouth bass (Dotson et al. 2013).

One variation of the maximum size limit is the *one-over* limit, which allows the harvest of one fish over the maximum size (Isermann and Parsons 2011). This regulation is popular with some anglers who believe the maximum size limit is necessary to protect large fish, but also want the opportunity to harvest a trophy *fish of a lifetime* if they are successful. Although it can limit the harvest of larger fish at times when they are vulnerable to anglers (e.g., during spring spawning migrations), this compromise to allow anglers to harvest one large fish can make the regulation much less effective throughout most of the season because most anglers do not catch or harvest multiple large fish during a single trip (Jacobson 1994; see example 1). A modified bag of one walleye greater than 356 mm limit was implemented on select Wisconsin lakes with the expectation that increased harvest of smaller fish would increase growth rates, although this effect was not observed during the three-year evaluation period (Fayram and Schmalz 2006).

DECISION EXAMPLE: USING CREEL SURVEY INFORMATION TO ASSESS THE POTENTIAL EFFECTS OF A *ONE-OVER* LENGTH REGULATION FOR WALLEYE.

Anglers commonly request length limits on fisheries based on their perception that the regulation will improve size structure of their favorite fish population. This perception may be the result of having a positive experience somewhere else where a regulation was applied, or just based on the opinion that too many anglers are keeping too many fish to sustain the fishery. For example, requests for maximum size limits or one-over size limits are often accompanied by complaints of anglers excessively harvesting big fish. In North Dakota, the argument of too many anglers "keeping limits of big walleye" is often accompanied with a request for a one-over size limit for walleye that would allow harvest of only one fish over 508 mm (20 in). However, small sample sizes for larger fish from routine netting surveys sometimes makes it difficult evaluate the biological impacts (e.g., changes in mortality or exploitation rates) a regulation like this may have on a fishery.

Lake Oahe, a Missouri River reservoir, spans the border of North Dakota and South Dakota. In 2009, netting surveys on Upper Lake Oahe in North Dakota revealed a very robust walleye size structure, with a relative stock density for preferred- to memorable-size walleye of 19; that is, 19% of captured walleye longer than 250 mm (10 in) were between 510 and 630 mm (20 and 25 in). This score was the highest value documented on record for the Upper Lake Oahe walleye population. Meanwhile, angling catch rates during the summer of 2009 were very good, providing a great opportunity for anglers to "excessively" harvest fish over 508 mm long if they were so inclined, since there was no length limit on North Dakota's portion of Lake Oahe at the time.

A creel survey conducted that summer by the North Dakota Game and Fish Department (Brooks and Fryda 2011) provided information on catch and harvest by anglers. Individual anglers were interviewed and their harvested fish were measured, providing information on the sizes of fish harvested by individual anglers (or angler parties when all fish were combined in a livewell). Because a one-over regulation allows an angler to harvest one fish over the regulated size, Department biologists used the information to determine the proportion of anglers who actually harvested more than one walleye over that size. They compared the unregulated harvest to what estimated harvest would have been if a regulation had been in effect requiring anglers to release walleye over 508, 559, or 610 mm long.

From April–October 2009, anglers harvested an estimated 360,000 walleye, with 46,000 of those fish greater than 508 mm long, 9000 longer than 559 mm, and 2100 longer than 610 mm. During the survey, creel clerks interviewed nearly 4500 anglers, none of whom harvested more than one walleye over 610 mm long. When single-angler and multiple-angler parties were combined, only 4.4% of anglers harvested more than one walleye over 508 mm long, and 0.4% harvested more than one over 559 mm (Fig. 24.4).

The perception that many anglers were "keeping limits of big fish" was not corroborated by the creel survey. In fact, anglers harvested large walleye in a lower proportion than were present in the population, and a very low proportion of anglers actually harvested more than one large walleye in a fishing trip. From this simple exercise, Department biologists concluded that a *one-over 508 mm* regulation would have had negligible effect on preventing harvest of large walleye.

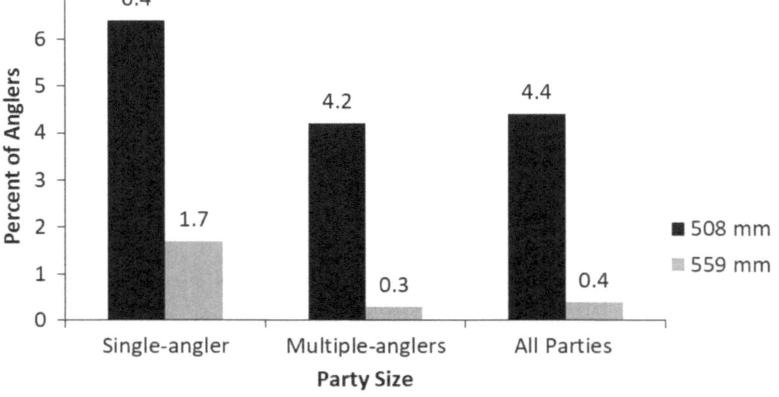

FIGURE 24.4 Percent of anglers interviewed at Lake Oahe, North Dakota, who harvested more than one fish over 508 mm (20 in.) and 559 mm (22 in.).

Slot Length Limits

Slot length limits provide managers with an option to restrict harvest on intermediate segments of the length frequency for a fish population. Martin (1958) was likely the first to suggest a slot limit restricting harvest of 300- to 380-mm (12- to 15-in) largemouth bass from farm ponds in Virginia, although it would be decades later before bass slot limits were applied and evaluated on a more widespread basis (e.g., Eder 1984, Gabelhouse 1985, Novinger 1990). Slot length limits have been applied to restructure the length distribution of a population, encourage harvest of small, slow-growing fish, protect specific size or age groups of fish, and protect spawning fish. The criteria for a successful slot limit (Brousseau and Armstrong 1987) include: (1) consistent natural reproduction; (2) slow growth, especially of small fish; (3) high natural mortality of small fish; and (4) high angling effort.

The first three criteria are interrelated. When natural reproduction is consistent, the density of small fish can increase to a point where growth slows. When fish are growing so slowly that it takes a longer time to reach a size desirable to anglers, natural mortality on small fish can be greater than angling mortality. This can lead to an undesirable situation of stunting, where consistent natural reproduction continually adds new fish to the population, and those fish are dying of natural causes before they can be utilized by anglers. Slot limits are designed to encourage harvest of small fish to decrease density and increase growth rates. When successfully applied, fish will grow through the protected slot range, increasing the number of larger fish available to anglers above the slot. Success of a slot limit strongly depends on whether anglers are willing to harvest small fish. When considering a slot limit with the objective of reducing the density of small fish, managers should consider if anglers are likely to shift their harvest to smaller fish. Regulations designed to reduce the density of small fish have failed where anglers were not willing to harvest small fish (Goeman et al. 1993 , Parks and Seidensticker 1998).

Anglers who hear the term *slot limit* sometimes confuse a *protected slot limit*, a length range protected from harvest, with a *harvest slot limit*, for which only fish within the length range can be harvested. Published evaluations of harvest slot limits are rare (Gwinn et al. 2013). Koupal et al. (2015) reported that a restrictive regulation allowing the harvest of two walleye between 381 and 508 mm (harvest slot) and one walleye greater than 711 mm (one-over) appeared to increase the abundance of female walleye broodstock in Sherman Reservoir, Nebraska. Gwinn et al. (2013) recommended the use of harvest slots to maximize both harvest numbers and trophy catch potential, which is likely to occur if the criteria for a harvest slot limit are met. Different from a protected slot limit, a harvest slot is simply a combined minimum length limit with a maximum length limit, and the combined criteria for both would apply to the successful implementation of a harvest slot length limit. However, harvest slot limits are often requested by anglers who believe that there is an optimal size range for harvesting fish for the table. Absent any biological justification, these requests tend to be social in nature.

Managers should also consider indirect effects that a regulation change may have on other fish populations and trophic interactions (Johnson and Martinez 1995). Changes in target population size structure or abundance can have measurable effects on other fish populations through predation or inter-specific competition. Though harvest regulations are often intended to restore a target population to a healthier balance, some managers have used length limits with the sole purpose of influencing non-target populations (Willis 1989). The use of slot limits on largemouth bass to improve bluegill and crappie, as well as largemouth bass, has been repeated by numerous managers since the summary by Gabelhouse (1984*b*). Other authors have examined the potential impacts that harvest regulations on predators would have on forage fish populations. Johnson et al. (1992) modeled how a minimum size limit on walleye in Lake Mendota (Wisconsin) coupled with stocking would contribute to a reduction in yellow perch recruitment through increased predation. Luecke et al. (1994) examined multiple slot length limits for lake trout in Flaming Gorge Reservoir (Wyoming-Utah) to predict forage demands under each possible regulation.

Seasons

Fishing seasons protect fish at specified times by restricting when anglers are allowed to target or harvest fish (Table 24.3). Historically, fishing seasons were among some of the first regulations imposed by regulating agencies (Redmond 1986, Paukert et al. 2007). Early seasons were typically closed during spawning to allow many species of fish to reproduce undisturbed by anglers. As with many early fishing regulations, over time managers realized that closed seasons did little to limit the overall harvest. No biological purpose is served, for example, to protect spawning fish in populations with little or no natural reproduction. The period of liberalization from 1940–1960 described by Redmond (1986) encompasses a time when many agencies relaxed seasonal restrictions. There are few published studies detailing the effects (positive or negative) of these relaxed seasons, and many regulating agencies currently allow year-round angling for most species of fish.

Fishing seasons do offer valuable protection, however, for certain species of fish during spawning or at times when fish are concentrated and more vulnerable to anglers. The need for a closed season for black bass species has been debated at length (Quinn 2002, Ontario Ministry of Natural Resources 2006). Nest-building and guarding fish, like centrarchids, tend to be more vulnerable to angling than broadcast-spawning fish that do not guard their nests or offspring. Even

TABLE 24.3
Definitions, Rationale, and Some Examples of Regulations for Seasons, Daily Closures, Closed Areas, and Catch-and-Release That May Be Used by Fisheries Managers to Regulate Harvest on Inland Fisheries

Regulation	Definition	Rationale	Selected References
Seasons	Prohibits fishing during specific times of the year	Protect fish from harvest during seasons when they're most vulnerable. Steeped in tradition, seasons have both biological and social implications	Hall 2009, Paukert et al. 2007, Quinn 2002
Daily closures	Prohibits fishing at specific times of the day (e.g., night closures)	Protects fish during times of the day they may be vulnerable to harvest, or may extend the fishing season if a harvest cap is in place. May also be used to promote public safety and reduce poaching activities	Cooke et al. 2017
Closed areas	Prohibits fishing in some areas (aka, sanctuaries), while allowing angling in adjacent areas	May protect vulnerable fish (e.g., spawning or overwinter aggregations) from harvest or disturbance, protect angler safety, or facilitate enforcement activities	Rietveld et al. 1999, Kocovsky and Carline 2001, Saunders et al. 2002, Suski and Cooke 2007, Williams et al. 2011, Funnell 2012, South Dakota Game, Fish and Parks 2019
Catch-and-release regulations	Requires release of all fish covered under regulation	Completely eliminates harvest while providing angling opportunity where recreational value of catching fish ranks higher than food value	Barnhart and Roelofs 1977, Hunt 1981, Anderson and Nehring 1984, Barnhart 1989, Cooke and Schramm 2007

catch-and-release angling for nesting bass can impair offspring survival and recruitment due to the reduced ability of male bass to effectively guard their nest after being played to exhaustion (Kieffer et al. 1995, Philipp et al. 1997, Cooke et al. 2000). Although some anglers consider closed seasons important to allow female fish to spawn without being harvested, it is important to note that protecting nesting fish during spawning may be more important to keep nest-guarding males from being disturbed than to protect or enhance the number of egg-laying females. Closed seasons may do little to reduce harvest of female fish, when you consider that those fish are vulnerable to harvest throughout the open season.

Closed seasons can reduce harvest of fish that are concentrated at particular times, such as seasonal migrations into tributaries. Managers cognizant of these seasonal concentrations may want to limit harvest if they expect it will have a negative impact on the population. Alternately, concentrations of fish at certain times may give some anglers (e.g., shore anglers) the opportunity to catch fish that they do not have otherwise, and this opportunity should be weighed against the need to protect the resource when considering a closed season.

Fishing seasons differ from length or creel limits by limiting fishing effort rather than the size or number of fish that can be harvested. In theory, by limiting the days available to anglers to harvest fish, closed seasons should restrict the total harvest. However, it is not clear whether this occurs, or whether it simply results in more intense harvest over a shorter period when the season is open (Rietveld et al. 1999).

A social reason for closed seasons is steeped in the tradition of the annual *fishing opener*. Anglers eagerly anticipate opening day for their fishing season, and opening day events have become a popular way for local governments to advertise and promote enjoyment of the resource. For some put-and-take fisheries, such as stocked trout ponds, opening day can provide a time of intense harvest to maximize the utilization rate of stocked fish (Hicks et al. 1983).

Daily Closures

Daily closures are not commonly used, although closing fishing during certain hours of the day, or days of the week, can provide a useful tool to regulate harvest, assist enforcement of rules, or extend season length when harvest quotas or caps are used to manage a fishery (Table 24.3).

Night fishing closures have been used to limit harvest for nocturnal fish species more vulnerable to angling at night. In Minnesota, the Mille Lacs walleye fishery is managed with a total allowable catch shared between state licensed anglers and tribal fishers. Managers at Mille Lacs have used nightly closures as one of several regulations to help limit mortality of walleye, thus extending the length of the fishing season before the harvest limit is reached (Cooke et al. 2017).

For some fisheries, night fishing closures are used to curb illegal harvest. White sturgeon (*Acipenser transmontanus*) are highly valued for their flesh and eggs, and in the Fraser River in British Columbia they have been regulated as a catch-and-release only fishery since the early 1990s, but their nocturnal feeding habits make them a prime target for poaching under the cover of darkness. Beginning in 2015, nighttime fishing closures were enacted on the lower Fraser, lower Pitt, and Harrison rivers in British Columbia to help manage the sturgeon fishery, to increase human safety, and to reduce poaching of sturgeon after dark (Cooke et al. 2017).

Closed Areas

Closed areas limit where anglers can fish, rather than when they can fish (Table 24.3). Sometimes referred to as sanctuaries or protected areas, closed areas have become widely used to limit harvest in marine fisheries (e.g., Hall 2009). Freshwater protected areas are generally considered a means to conserve native species and habitats (Saunders et al. 2002, Suski and Cooke 2007, Williams et al. 2011) by limiting disturbances to aquatic ecosystems, but rarely are closed areas used with a sole purpose of regulating harvest in freshwater fisheries. Closed areas have been used to assist with stock recovery and protect spawning areas on the Great Lakes (Sztramko 1985, Madenjian and DeSorcie 1999). On smaller inland fisheries, closed areas are used frequently as a tool for protecting angler safety by limiting their access to unsafe areas, such as lowhead dams, and facilitating enforcement activities by allowing angling to occur where their activities can be easily monitored.

Closed areas may be a useful tool to manage harvest of fish in instances where managers need to protect aggregations of fish (e.g., immediately below dams), but do not want to limit angler opportunity. Managers who want to limit stressors on nesting bass may choose to close spawning areas to fishing rather than close the spring season (Funnell 2012). This would eliminate harvest of bass in those areas while still allowing anglers to fish in adjacent areas. Closed areas are used seasonally in Ontario to protect fish from harvest or disturbance in spawning areas or aggregating areas (Rietveld et al. 1999). Seasonal closed areas have been used to protect fish from barotrauma-related mortality due to catch-and-release fishing when fish concentrate in deep areas, such as on the Missouri River system in South Dakota (South Dakota Game, Fish and Parks 2019). Closed areas may also be used to propagate brood fish for hatchery operations. Pymatuning Sanctuary in Pennsylvania is one such area, managed as a source of walleye eggs for the Pennsylvania Fish and Boat Commission. Kocovsky and Carline (2001) described the unexploited walleye population in Pymatuning Sanctuary as having a high density, large size structure, slow growth, and relatively low mortality.

Catch-and-Release Only

Catch-and-release only regulations requiring release of all fish caught are generally applied to limit harvest in situations where recreational value of catching fish is ranked higher than harvesting fish for food (Table 24.3). Essentially, catch-and-release regulations are a strict form of creel limit that prohibits harvest or possession of any fish, typically by species or by water body. However, catch-and-release fishing encompasses a specific mindset and philosophy of fisheries management, and can be treated as a special regulation all its own. In the mid-twentieth century, during the period when several agencies were moving toward less restrictive harvest limits for warmwater species (Redmond 1986), Hazzard (1952) first suggested the idea of a regulation requiring release of all trout caught by anglers. The National Park Service followed the advice of the *Hazzard Plan* of trout management, and instituted catch-and-release regulations for trout in several National Parks across the United States of America in the 1950s and 1960s (Wallis 1971, Kulp and Moore 2005). Dubbed the *Fish-for-Fun* program, it was well received by trout anglers, and marked the beginning of a paradigm shift as managers began looking at regulations to improve the quality of the fishing experience, rather than to simply maximize yield and harvest (Larkin 1977). Another fisheries management paradigm was changing around this same time, as managers taking a more holistic approach began managing for wild fish rather than relying on stocked fish. Catch-and-release regulations became important tools for wild trout fisheries (Richardson and Hamre 1984), reducing the need for stocking to replace harvested fish.

Popularity of catch-and-release fishing spawned decades of research on the effects of such regulations, as well as side effects, on fish populations and other related topics (e.g., Barnhart and Roelofs 1977, Barnhart 1989, Cooke and Schramm 2007). Population-level responses have been mixed. Though some researchers have documented positive increases in abundance, catch rates, and size structure (e.g., Hunt 1981, Anderson and Nehring 1984), others have documented no population response to catch-and-release regulations (e.g., Detar et al. 2014).

Like most harvest regulations, catch-and-release restrictions are not a panacea, and there are some fundamental points to consider for a catch-and-release regulation to be useful (Barnhart 1989).

1. Habitat should be suitable for the target species with low natural mortality and rapid growth rates. This is important to prevent density-dependent factors from causing growth to slow if abundance increases as a response to elimination of fishing mortality. If natural mortality is

higher than desired, managers should decide whether reducing fishing mortality to zero is important to reduce total mortality of the population, or whether anglers should be allowed to harvest some fish to utilize rather than letting them die of natural causes.

2. Angler desires and behavior should be compatible with the regulation. Catch-and-release regulations may be appropriate when the sport value of catching a fish is greater than the value of harvesting fish for food, and anglers who support the regulation will be more likely to comply. Angler behavior will play an important role in the success of the regulation. If anglers already exhibit a high voluntary-release rate, a catch-and-release regulation may not show any positive effects. Alternately, if anglers exhibit a high rate of illegal harvest, either intentionally or unintentionally, that harvest may mask any benefits from the regulation.
3. Fishing pressure is high, and other regulations are not adequate to protect or enhance the resource (Fig. 24.5).
4. Fish longevity is long enough that any released fish will have a chance of surviving, growing, and being caught again. The basic premise of releasing any fish is that it will survive long enough to carry out a necessary life function, such as reproduction, or will be caught again and enjoyed by another angler. Much of the research on catch-and-release fishing has focused on factors affecting mortality of released fish (Cooke and Suski 2005). But occasionally, fish that survive can be less likely to take an angler's bait and be caught again (Arlinghaus et al. 2017*b*, Fedele 2017). Askey et al. (2006) documented a sharp decline in catch rates of rainbow trout (*Oncorhynchus mykiss*) exposed to heavy fishing pressure in a catch-and-release situation in British Columbia. It is not clear whether this behavior change was a short-term change from stress of being caught, or whether it persisted to a long-term decline in trout catchability.

FIGURE 24.5 Catch-and-release fishing may be required on waters with intense fishing pressure, reducing the need for frequent supplemental stocking. Photo courtesy of North Dakota Game and Fish Department.

FATE OF RELEASED FISH

The principal assumption for any fishing regulation is that released fish have a high survival rate, and that any post-release mortality is low enough to be considered negligible. This is not always true, and fate of released fish is something that managers should consider when evaluating effects of regulations that require anglers to release fish. A large body of research on catch-and-release fishing has been conducted on the fate of fish released by anglers (Muoneke and Childress 1994, Cooke and Schramm 2007). Although mortality rate and causes of mortality vary widely from one fishery to another, Cooke and Wilde (2007) categorized several factors that affect the fate of released fish. These include gear (i.e., hook size and type), bait (live vs. artificial), and angling practice (e.g., fighting time, landing technique, air exposure and handling time, hook removal technique, short-term retention, and fishing location, depth and water temperature). These factors alone, or in combination, can affect the survival of released fish. Additionally, species-specific physiological differences can alter the magnitudes of these effects among species (Cooke and Suski 2005).

Although many factors have been found to contribute to post-release mortality, some generalizations suggest that mortality increases when fish are deeply hooked, and mortality is lowered when fish can be quickly and efficiently released (Schill 1996, Lamansky and Meyer 2016). Management agencies may consider some of the tools or tactics summarized by Brownscombe et al. (2017) as regulation options to reduce mortality of released fish. Fish hooked in the mouth tend to have a much lower mortality rate (e.g., Pauley and Thomas 1993, Persons and Hirsch 1994, Lindsay et al. 2004, Wilde and Pope 2008); thus, regulations can require anglers to use certain types of gear that increase the likelihood of mouth-hooked fish. These can include requiring use of artificial lures rather than live or organic baits (e.g., Payer et al. 1989, Pauley and Thomas 1993, Schisler and Bergersen 1996), restrictions on hook type (Meka 2004), or restrictions on hook size (Cooke et al. 2005). Wilde et al. (2003) suggested that lure-size restrictions may be a valuable tool to reduce catch of nontarget-sized fish, thus reducing mortality of fish that may need to be released by a related length restriction. Requiring the use of barbless hooks can reduce the time it takes to remove the hook, thus shortening the handling time that the fish is out of water (Cooke et al. 2001, Meka 2004).

Environmental conditions, such as water temperature or capture depth, also contribute to catch-and-release fish mortality. However, it is difficult to regulate how deep or the water temperatures at which anglers fish. One alternative to reduce seasonal catch-and-release mortality is to close fishing when water temperatures are particularly warm, or when fish are known to inhabit deep water. Alternately, if additional mortality is acceptable, agencies may choose to relax restrictions during periods when released fish are likely to succumb to mortality, allowing those fish to be harvested and utilized. For example, walleye minimum length limits are relaxed on some Missouri River reservoirs in South Dakota during July and August when water temperatures are high; during the winter months when walleye are caught from deeper areas of the lake length limits are again relaxed and anglers are required to keep the first four walleye they catch (South Dakota Game, Fish and Parks 2019).

Other factors contributing to the fate of released fish may be outside the realm of regulatory control and depend more on the individual angler. Practices such as fighting time, landing procedure, handling, and air exposure time all contribute to fish mortality (Cooke and Wilde 2007), but may not be effectively regulated. Agencies may use their information resources to educate anglers on practices to reduce mortality of fish that must be released due to a harvest regulation.

In addition to mortality, Cooke and Schramm (2007) emphasized the need to evaluate sublethal effects in catch-and-release studies to better understand long-term effects fish experience from being caught and released. Exertion that a fish experiences while being caught typically elicits a physiological response that can lead to changes in behavior, growth, and fitness. When fishing pressure is intense, the cumulative individual fish stressors can lead to population-level responses.

Some authors have reported evolutionary changes to surviving fish populations when anglers selectively harvest fish with certain traits, such as aggressiveness (e.g., Philipp et al. 2015, Leclerc et al. 2017, Festa-Bianchet and Arlinghaus 2021 [Chapter 12]). Harvest regulations that result in selective harvest may exacerbate fishing-induced evolution (Wszola and Fontaine 2021 [Chapter 13]); alternately, harvest regulations may be used to reduce harvest of fish that are vulnerable at certain life stages, such as largemouth bass males guarding nests (Philipp et al. 2015).

IMPLEMENTING REGULATIONS

Social Considerations

To effectively set harvest regulations, fishery managers need an understanding of the interactions between fish populations and anglers. Oftentimes, fishery managers trained as biologists focus on the dynamic functions of a fish population, like growth and mortality, to create fisheries attractive to anglers. Though having an understanding of those functions will give managers a basis to select biologically significant regulations, this *build it and they will come* philosophy neglects to fully consider the motivations that drive anglers to be satisfied with a fishing outing. With growing concerns over declining angler participation (U.S. Fish and Wildlife Service and U.S. Department of Commerce 2016, Burkett and Winkler 2019), more effort has been directed to understanding the motivations driving angler participation (Gruntorad and Chizinski 2021 [Chapter 4]). Angler populations tend to be comprised of diverse of angler types, each with specific preferences, fishing practices, and motivations (Johnston et al. 2015*a*). When it comes to participation, customer satisfaction leads to customer loyalty (Gruntorad and Chizinski 2021 [Chapter 4]), and fishery managers should consider how harvest regulations will affect angler satisfaction (Beardmore et al. 2015), and ultimately, angler participation.

Rather than simply managing for biological yields, such as number or pounds of fish harvested, by combining social considerations with biological rationale into harvest regulations, managers can provide *optimum social yield* for their resource and constituents (Johnston et al. 2015*a*). This play on the traditional optimum sustainable yield (Roedel 1975) requires novel approaches to setting harvest regulations, such as the suggestion by Kaemingk et al. (2021 [Chapter 3]) to view anglers as an apex predator in a complex predator–prey relationship, rather than viewing them as a homogenous group that will react in a predictable manner.

Selecting New Regulations

When a new regulation or a regulation change is considered, managers should approach the process pragmatically. They should have adequate knowledge of their fisheries resource to know if harvest is a driving factor in shaping the fish population in question, and some expectation that regulating harvest will provide a biological or social benefit. Population modeling has become a popular tool for managers to simulate the effects of various regulations (Radomski and Goeman 1996). Simulations of varying complexity have been used to estimate the effects of various harvest regulations on fish populations (e.g., Beamesderfer and North 1995, Beard et al. 1997, Maceina et al. 1998, Allen and Pine 2000, Post et al. 2003, Gwinn and Allen 2010, Allen et al. 2013, Bonvechio et al. 2014,Lenker et al. 2016, Ayllón et al. 2018, Moreau and Matthias 2018, Smith et al. 2018, Wszola and Fontaine 2021 [Chapter 13]). Simulating the effects of any potential regulation can frame expectations of success, and when this information is presented to stakeholders it helps them understand and accept the various options being considered. Computer programs are available that managers can use to automate population modeling, such as Fisheries Analysis and Modeling Simulator (Slipke and Maceina 2014), R (Ogle 2016), or Excel (Haddon 2011).

Once a biological problem is identified, managers should gather input from stakeholders to determine how the problem and potential solutions will affect them. Most management agencies have some internal process for proposing new regulations, soliciting input from stakeholders, and implementing regulations (e.g., Johnson and Martinez 1995; Fig. 24.6). Although soliciting feedback from anglers can be relatively benign, any alternative regulation will affect different stakeholders in different ways, and the process can become very complicated in short order. Managers may want to employ some form of structured decision making (Irwin et al. 2011) to formally incorporated stakeholder input in the decision-making process. Structured decision making provides an inclusive process that increases transparency and promotes consensus building for management (Bain 1987, Irwin et al. 2011). Most importantly, because resources are managed in trust for the benefit of the public, including stakeholders in the decision-making process ultimately leads to greater buy-in from anglers and increases trust (Robinson et al. 2021 [Chapter 9], Runge 2021 [Chapter 7]).

Evaluating Regulations

Often the science needed to manage the resource comes from the experience of management itself (Lester et al. 2003), and any new harvest regulation should be based on objectives with an evaluation plan. Objectives need to be specific and measurable, so that evaluation efforts can reveal whether the regulation was effective or not. Regulations with biological objectives (e.g., to increase catch rates, cause shifts in size structure, improve spawning stock) can be measured through routine sampling. In the absence of sampling data, regulation requests will likely be more of a social nature. For example, managers might receive complaints from anglers who are catching fewer fish, or smaller fish, than they would like. If they suspect harvest may be contributing to these complaints, they may still be able to set some objectives for the fishery, but in this case the

FIGURE 24.6 Public meetings are often used during the regulation setting process to present data and receive feedback from stakeholders on proposed regulations. Photo courtesy of North Dakota Game and Fish Department.

objectives should address the problem directly, such as increasing the catch rate for anglers or improving the size structure in the recreational creel. In any case, by having specific objectives, the evaluation process can be a learning opportunity whereby experience produces new insight on the fishery that managers may not have possessed before the regulation.

In addition to being measurable, the metrics used to evaluate regulation objectives must also be meaningful (Isermann 2007). A meaningful metric is one that can be directly influenced by the regulation change, and not a secondary effect. For example, an increase in the average size of fish harvested by anglers may be due to a shift in angler behavior rather than an actual change in the size structure of a fish population. Anglers complying with a minimum length limit are automatically required to harvest larger fish than they may have otherwise kept before the limit was in place (Isermann 2007). Changes in angler catch rates may be more directly related to forage abundance and habitat conditions than changes of the target fish species due to a regulation.

The time required to effectively evaluate a regulation can vary substantially by fishery. Life-history variations, population dynamics, harvest intensity, and angler effort can all affect how long it takes for a regulation change to manifest a measurable population response. Allen and Pine (2000) modeled recruitment variability effects on minimum length limit evaluations, reporting that evaluation periods of less than five years were insufficient to detect population level changes in white crappie (*Pomoxis annularis*) and largemouth bass managed with minimum length limits, unless the length limit was exceptionally restrictive. They also surmised that the lack of population responses to length limits summarized by Wilde (1997) may have been partially due to environmental variability (including recruitment variability), though most of the studies examined used five or fewer years of data in their evaluations. Similarly, Isermann (2007) suggested that longer evaluation periods of eight to ten years before and after implementing a minimum length limit were necessary to overcome the masking effects of recruitment variability on walleye population changes due to the regulation. Jacobson (2005) discussed the need for a prolonged evaluation period to overcome effects of non-independence in pre- and post-regulation samples, and Pierce (2010) demonstrated the necessity for an extended evaluation period to detect population changes in long-lived species such as northern pike. A good rule of thumb is there should be sufficient time during the evaluation to allow generations of fish that were not exploited before the regulation to cycle through the population post-regulation. Näslund et al. (2005) and Erhardt and Scarnecchia (2014) reported that the full effects of restrictive fishing regulations for European grayling (*Thymallus thymallus*) and bull trout (*Salvelinus confluentus*), respectively, were not apparent until the regulation had been in place approximately as long as the expected lifespan of the regulated species. Thus, short-lived, fast-growing fish would require a shorter evaluation period than long-lived, slow-growing fish.

Angler Behavior

Harvest regulations influence which fish are caught or harvested, and likewise influence angler behavior (Johnston et al. 2015*a*). When anglers change their behavior, it can affect the fishery in unexpected ways. Some anglers may view regulations as attractive and will select a fishing destination where they perceive a regulation will enhance their likelihood of catching fish (Scrogin et al. 2004). Alternately, anglers who are interested in harvesting fish may be deterred from fishing waters with restrictive harvest regulations, opting to select waters with less-restrictive regulations. Dorr et al. (2002) determined that a reduction in daily bag limits for crappies in Sardis Lake, Mississippi would reduce angler trips and related expenditures by roughly 33%. In Minnesota, where anglers place a high value on being able to catch and harvest walleye, Carlin et al. (2012) determined that low bag limits reduced lake preference among anglers. Although difficult to quantify exactly how regulations may alter angler behavior, managers should be cognizant of this and monitor for shifts in angler effort on regulated lakes such as those observed by Beard et al. (2003*a*) or Fayram and Schmalz (2006). If changes in angler effort are substantial enough, they may reduce or eliminate effectiveness of the regulation on the target lake or may require more restrictive regulations on lakes where anglers redirect more effort.

Self-regulating behaviors can also influence the outcome of a harvest regulation. As anglers have become less harvest oriented over time, angler attitudes have shifted more to fishing purely for sport or recreational purposes. For some fish species groups (e.g., bass, muskellunge, or trout), voluntary catch and release has become a common practice among anglers. Simonson and Hewett (1999) related voluntary release of muskellunge in Wisconsin to the reduced harvest of legal-sized muskellunge, despite increased angler effort and consistent catch rates, and Fayram (2003) suggested that voluntary release was as important as regulated release for managing Wisconsin muskellunge fisheries. Most agencies that manage black bass in the United States of America and Canada have noted increases in voluntary catch and release (Noble 2002), with several agencies indicating that voluntary catch and release reduced the need for harvest regulations. High voluntary release rates have been identified as factors contributing to the failure of length limits for largemouth bass (Miranda et al. 2017) and ineffective catch-and-release regulations for brook trout (Detar et al. 2014). Catch and release is so prominent in some fisheries that Sass and Shaw (2020) called it one of the biggest challenges facing inland fisheries management in the twenty-first century. Anglers may also tend to practice self-imposed length limits, preferring to harvest specific sizes of fish deemed to make the best table fare, and releasing smaller or larger fish (e.g., Colvin 2002, Paukert et al. 2002, Isermann et al. 2005, Holley et al. 2009, Chizinski et al. 2014*a*).

Managers need to consider angler behavior and the influence local angler preferences and ethics may have on harvest regulation decisions. Most self-regulating behaviors will negate the need for regulations that serve a biological purpose to lower harvest. However, anglers who practice self-regulating behaviors will often present regulation requests for social reasons, believing their practices should be required for all anglers. Data from routine fish population sampling and angler surveys will be important to engage in discussions with anglers on socially based regulations.

COMPLIANCE AND ENFORCEMENT

Angler compliance with harvest regulations is integral to the success of the regulation (Gigliotti and Taylor 1990, Pierce and Tomcko 1998). There may be several reasons anglers harvest fish illegally, whether intentional or not. According to Gruntorad et al. (2021 [Chapter 4]), three basic explanations why anglers may partake in illegal behavior include:

1. *Willful defiance*. Regulations that are implemented without approval or buy-in from anglers may result in a higher rate of intentional noncompliance by defiant anglers who protest the regulation.
2. *Frustration*. In some cases, restrictive length limits may tempt frustrated anglers to break the rules if harvestable fish are in short supply and protected-sized fish make up the bulk of the catch. Sullivan (2002) documented that illegal harvest of walleye increased sharply when catch rates were low.
3. *Ignorance*. Unintentional noncompliance by anglers who are unaware or confused by a regulation is probably more common than intentional noncompliance. Page and Radomski (2006) compared noncompliance to anglers' awareness of regulations in Minnesota, determining that unaware-angler parties were more likely to be noncompliant than aware-angler parties. Angler awareness was lower for fisheries that had recent changes in regulations and for fisheries with complex regulations. Measurement error by anglers can lead to unintentional noncompliance in fisheries regulated with length limits (Page et al. 2004), as well as post-mortem changes in length (e.g., Blackwell et al. 2003, Page et al. 2004).

Keeping regulations simple and easy to understand can increase angler compliance, mainly by reducing unintentional noncompliance. Rather than enacting lake-specific regulations, managers could consider managing groups of lakes with similar physical and biological characteristics on a watershed or ecoregion level (e.g., Lester et al. 2003). Managers should also resist changing regulations frequently to ensure that anglers can retain regulation knowledge.

Lastly, regulations that are biologically or socially significant will increase angler buy-in and likely lead to increased compliance. When it comes to enforcement, superfluous regulations create unnecessary complexity and increase the likelihood of anglers accidentally violating regulations. Schill and Kline (1995) documented 75% of violations committed by anglers on their study streams were accidental. On further inquiry, many of those anglers indicated that they attempted to comply with all laws, but had made a mistake. If anglers who attempted to comply with all regulations, but make a mistake occasionally, are written citations for their accidental noncompliance of a regulation that is unnecessary, it will create animosity between anglers and regulatory agencies (Schill and Kline 1995).

Preventing or lowering noncompliance through enforcement alone is not practical (Fig. 24.7). Many agencies have staff or budget limitations, and enforcement officers cannot possibly encounter every angler at every water in their jurisdiction. Sullivan (2002) reported that enforcement officers patrolled his study lakes every ten days. Rather than relying on enforcement to increase compliance, managers should work to minimize noncompliance, particularly accidental noncompliance, through education and enacting simple and biologically and socially significant regulations.

FIGURE 24.7 Enforcement activities are important for ensuring that anglers are complying with regulations. Just as important are the social interactions between anglers and enforcement officers who represent the managing agencies. Photo courtesy of North Dakota Game and Fish Department.

SUMMARY

A diverse set of regulation types are available to managers for controlling harvest on inland recreational fisheries. Although the basic utility has not changed over time, experimental application of past regulations has provided information managers need to refine their approach to selecting, implementing, and evaluating the success of new regulations going forward. Harvest regulations are only effective when harvest is affecting the fish population or the angling experience. Thus, managers need to possess the necessary biological and social data on their fishery to identify potential problems and solutions to those problems, which may require modeling various harvest regulations.

Harvest regulations should enhance both the fish population and the fishing experience for anglers when applied correctly. To achieve this, managers are encouraged to keep regulations to a minimum, only implement regulations that are deemed necessary, resist frequent changes to regulations, and strive to devise regulations that are easy for anglers to understand and maximize angling opportunities. Using a combined approach of social input and biological data allows development of regulations that are defensible to most stakeholders and have a high likelihood of successfully accomplishing what the regulation is intended to produce.

ACKNOWLEDGMENTS

The North Dakota Game and Fish Department and Federal Aid in Sportfish Restoration provided resources for this work. Early reviews of this chapter by G. Power, R. Wilson, V. Travnichek, G. Scholten, and K. Pope provided substantial improvements to the final draft.

Literature Cited

Abbott, J. K., and A. C. Haynie. 2012. What are we protecting? Fisher behavior and the unintended consequences of spatial closures as a fishery management tool. Ecological Applications 22:762–777.

Abbott-Jamieson, S., and P. M. Clay. 2010. The long voyage to including sociocultural analysis in NOAA's National Marine Fisheries Service. Marine Fisheries Review 72:14–33.

Abraham, K. F., R. L. Jefferies, and R. T. Alisauskas. 2005. The dynamics of landscape change and snow geese in mid-continent North America. Global Change Biology 11:841–855.

Adams, K. P., and R. J. Hamilton. 2011. Management history. Pages 355–377 *in* D. G. Hewitt, editor. Biology and management of white-tailed deer. CRC Press, Boca Raton, Florida, USA.

Adams, N. S., J. W. Beeman, and J. H. Eiler, editors. 2012. Telemetry techniques: a user guide for fisheries research. American Fisheries Society, Bethesda, Maryland, USA.

Aday, D. D., and B. D. S. Graeb. 2012. Stunted fish in small impoundments: an overview and management perspective. Pages 215–232 *in* J. W. Neal and D. W. Willis, editors. Small impoundment management in North America. American Fisheries Society, Bethesda, Maryland, USA.

Afton, A. D., and M. G. Anderson. 2001. Declining scaup populations: a retrospective analysis of long-term population and harvest survey data. The Journal of Wildlife Management 65:781–796.

Ahrens, R. N. M., M. S. Allen, C. Walters, and R. Arlinghaus. 2020. Saving large fish through harvest slots outperforms the classical minimum-length limit when the aim is to achieve multiple harvest and catch-related fisheries objectives. Fish and Fisheries 21:483–510.

Ahti, P. A., A. Kuparinen, and S. Uusi-Heikkila. 2020. Size does matter–the eco-evolutionary effects of changing body size in fish. Canadian Journal of Fisheries and Aquatic Sciences 28:311–324.

Ajzen, I., and M. Fishbein. 1980. Understanding attitudes and predicting behavior. Prentice Hall, Englewood Cliffs, New Jersey, USA.

Akaike, H. 1973a. Information theory and an extension of the maximum likelihood principle. Pages 267–281 *in* B. N. Petran and F. Csaaki, editors. International Symposium on information theory, 2nd edition. Akadeemiai Kiadi, Budapest.

Akaike, H. 1973b. Maximum likelihood identification of Gaussian autoregressive moving average models. Biometrika 60:255–265.

Albayrak, T., and M. Caber. 2013. Penalty-reward-contrast analysis: a review of its application in customer satisfaction research. Total Quality Management & Business Excellence 24:1288–1300.

Ali, M., A. Nicieza, and R. J. Wooton. 2003. Compensatory growth in fishes: a response to growth depression. Fish and Fisheries 4:147–190.

Alig, R. J., J. D. Kline, and M. Lichtenstein. 2004. Urbanization on the US landscape: looking ahead in the 21st century. Landscape and Urban Planning 69:219–234.

Alisauskas, R. T. 1998. Winter range expansion and relationships between landscape and morphometrics of midcontinent lesser snow geese. The Auk 115:851–862.

Alisauskas, R. T. 2002. Arctic climate, spring nutrition, and recruitment in midcontinent lesser snow geese. The Journal of Wildlife Management 66:181–193.

Alisauskas, R. T., C. D. Ankney, and E. E. Klaas. 1988. Winter diets and nutrition of midcontinental lesser snow geese. The Journal of Wildlife Management 52:403–414.

Alisauskas, R. T., and T. W. Arnold. 1994. American coot. Pages 127–143 *in* T. C. Tacha and C. E. Braun, editors. Migratory shore and upland game bird management in North America. International Association of Fish and Wildlife Agencies, Washington, D.C., USA.

Alisauskas, R. T., T. W. Arnold, J. O. Leafloor, D. L. Otis, and J. S. Sedinger. 2014. Lincoln estimates of mallard (*Anas platyrhynchos*) abundance in North America. Ecology and Evolution 4:132–143.

Alisauskas, R. T., K. L. Drake, and J. D. Nichols. 2009. Filling a void: abundance estimation of North American populations of arctic geese using hunter recoveries. Pages 463–489 *in* D. L. Thomson, E. G. Cooch, and M. J. Conroy, editors. Modeling demographic processes in marked populations. Springer, New York, New York, USA.

Alisauskas, R. T., R. F. Rockwell, K. W. Dufour, E. G. Cooch, G. Zimmerman, K. L. Drake, J. O. Leafloor, T. J. Moser, and E. T. Reed. 2011. Harvest, survival, and abundance of midcontinent lesser snow geese relative to population reduction efforts. Wildlife Monographs 179:1–42.

Allan, J. D., R. Abell, Z. E. B. Hogan, C. Revenga, B. W. Taylor, R. L. Welcomme, and K. Winemiller. 2005. Overfishing of inland waters. BioScience 55:1041–1051.

Allen, C. R., G. S. Cumming, A. S. Garmestani, P. D. Taylor, and B. H. Walker. 2011. Managing for resilience. Wildlife Biology 17:337–349.

Allen, C. R., J. J. Fontaine, K. L. Pope, and A. S. Garmestani. 2011. Adaptive management for a turbulent future. Journal of Environmental Management 92:1339–1345.

Allen, G. T., D. F. Caithamer, and M. Otto. 1999. A review of the status of greater and lesser scaup in North America. U.S. Fish and Wildlife Service, Division of Migratory Bird Management, Laurel, Maryland, USA.

Allen, M. S., R. N. M. Ahrens, M. J. Hansen, and R. Arlinghaus. 2013. Dynamic angling effort influences the value of minimum-length limits to prevent recruitment overfishing. Fisheries Management and Ecology 20:247–257.

Allen, M. S., L. E. Miranda, and R. E. Brock. 1998. Implications of compensatory and additive mortality to the management of selected sportfish populations. Lakes and Reservoirs: Science, Policy and Management for Sustainable Use 3:67–79.

Allen, M. S., and W. E. Pine, III. 2000. Detecting fish population responses to a minimum length limit: effects of variable recruitment and duration of evaluation. North American Journal of Fisheries Management 20:672–682.

Allen, M. S., C. J. Walters, and R. Myers. 2008. Temporal trends in largemouth bass mortality, with fishery implications. North American Journal of Fisheries Management 28:418–427.

Allen, T., K. A. Murray, C. Zambrana-Torrelio, S. S. Morse, C. Rondinini, M. Di Marco, N. Breit, K. J. Olival, and P. Daszak. 2017. Global hotspots and correlates of emerging zoonotic diseases. Nature Communications 8:1124.

Allendorf, F. W., P. R. England, G. Luikart, P. A. Ritchie, and N. Ryman. 2008. Genetic effects of harvest on wild animal populations. Trends in Ecology & Evolution 23:327–337.

Allendorf, F. W., and J. J. Hard. 2009. Human-induced evolution caused by unnatural selection through harvest of wild animals. Proceedings of the National Academy of Sciences 106:9987–9994.

Alós, J., A. Campos-Candela, and R. Arlinghaus. 2019. A modelling approach to evaluate the impact of fish spatial behavioural types on fisheries stock assessment. ICES Journal of Marine Science 76:489–500.

Alós, J., R. Palmer, and R. Arlinghaus. 2012. Consistent selection towards low activity phenotypes when catchability depends on encounters among human predators and fish. PLoS One 7:e48939.

Alós, J., M. Palmer, R. Rossello, and R. Arlinghaus. 2016. Fast and behavior-selective exploitation of a marine fish targeted by anglers. Scientific Reports 6:38093.

Amstrup, S. C., T. L. McDonald, and B. F. J. Manly, editors. 2005. Handbook of capture-recapture analysis. Princeton University Press, Princeton, New Jersey, USA.

Andelt, W. F., R. L. Phillips, R. H. Schmidt, and R. B. Gill. 1999. Trapping furbearers: an overview of the biological and social issues surrounding a public policy controversy. Wildlife Society Bulletin 27:53–64.

Andersen, K. H., L. Marty, and R. Arlinghaus. 2018. Evolution of boldness and life history in response to selective harvesting. Canadian Journal of Fisheries and Aquatic Sciences 75:271–281.

Andersen, K. H., and M. Pedersen. 2010. Damped trophic cascades driven by fishing in model marine ecosystems. Proceedings of the Royal Society B: Biological Sciences 277:795–802.

Andersen, O., H. K. Wam, A. Mysterud, and B. P. Kaltenborn. 2014. Applying typology analyses to management issues: deer harvest and declining hunter numbers. The Journal of Wildlife Management 78:1282–1292.

Anderson, D. R. 1975. Optimal exploitation strategies for an animal population in a Markovian environment: a theory and an example. Ecology 56:1281–1297.

Anderson, D. R. 1985. Constrained optimal exploitation: a quantitative theory. Pages 105–116 *in* S. L. Beasom and S. F. Roberson, editors. Game harvest management. Caesar Kleberg Wildlife Research Institute, Kingsville, Texas, USA.

Anderson, D. R., and K. P. Burnham. 1976. Population ecology of the mallard. VI. The effect of exploitation on survival. U.S. Fish and Wildlife Service, Patuxent Wildlife Research Center, Resource Publication 128, Laurel, Maryland, USA.

Anderson, D. R., K. P. Burnham, J. D. Nichols and M. J. Conroy. 1987. The need for experiments to understand population dynamics of American black ducks. Wildlife Society Bulletin 15:282–284.

Anderson, D. R. and C. J. Henny. 1972. Population ecology of the mallard. I. A review of previous studies and the distribution and migration from breeding areas, volume 105. U.S. Fish and Wildlife Service Resource Publication, Laurel, Maryland, USA.

Anderson, E. M., and M. J. Lovallo. 2003. Bobcat and lynx. Pages 759–786 *in* G. A. Feldhamer, B. C. Thompson, and J. A. Chapman, editors. Wild mammals of North America: biology, management, and conservation. Johns Hopkins University Press, Baltimore, Maryland, USA.

Anderson, E. W., and M. W. Sullivan. 1993. The antecedents and consequences of customer satisfaction for firms. Marketing Science 12:125–143.

Anderson, R. M., and R. B. Nehring. 1984. Effects of a catch-and-release regulation on a wild trout population in Colorado and its acceptance by anglers. North American Journal of Fisheries Management 4:257–265.

Anderson, R. O. 1980. The role of length limits in ecological management. Pages 41–45 *in* S. Gloss and B. Shupp, editors. Practical fisheries management: more with less in the 1980s. New York Chapter of the American Fisheries Society, Workshop proceedings, Ithaca, New York, USA.

Anderson, S. B. 1987. Wild furbearer management in eastern Canada. Pages 1039–1048 *in* M. Novak, J. A. Baker, M. E. Obbard, and B. Malloch, editors. Wild furbearer management and conservation in North America. Ontario Trappers Association, North Bay, Ontario, Canada.

Andreasen, A. M., K. M. Stewart, J. S. Sedinger, C. W. Lackey, and J. P. Beckmann. 2018. Survival of cougars caught in non-target foothold traps and snares. The Journal of Wildlife Management 82:906–917.

Ankney, C. D. 1996. An embarrassment of riches: too many geese. The Journal of Wildlife Management 60:217–223.

Ankney, C. D., D. G. Dennis, and R. C. Bailey. 1987. Increasing mallards, decreasing American black ducks: coincidence or cause and effect? The Journal of Wildlife Management 51:523–529.

Ankney, C. D., D. G. Dennis, and R. C. Bailey. 1989. Increasing mallards, decreasing American black ducks–no evidence for cause and effect: a reply. The Journal of Wildlife Management 53:1072–1075.

Anthony, K. R. N., P. A. Marshall, A. Abdulla, R. Beeden, C. Bergh, R. Black, C. M. Eakin, E. T. Game, M. Gooch, N. A. J. Graham, A. Green, S. F. Heron, R. van Hooidonk, C. Knowland, S. Mangubhai, N. Marshall, J. A. Maynard, P. McGinnity, E. McLeod, Peter. J. Mumby, M. Nyström, D. Obura, J. Oliver, H. P. Possingham, R. L. Pressey, G. P. Rowlands, J. Tamelander, D. Wachenfeld, and S. Wear. 2015. Operationalizing resilience for adaptive coral reef management under global environmental change. Global Change Biology 21:48–61.

Apollionio, M., R. Andersen, and R. Putnam. 2010. European ungulates and their management in the 21st century. Cambridge University Press, Cambridge, England, United Kingdom.

Araya, A. C., and J. A. Dubovsky. 2008. Temporal distribution of harvested mid-continent Sandhill cranes within the Central Flyway states during the 1997-2001 hunting seasons. Pages 50–57 *in* Proceedings of the Tenth North American Crane Workshop. North American Crane Working Group, 7 February-10 February 2006, Zacatecas City, Zacatecas, Mexico.

Arlinghaus, R. 2006a. On the apparently striking disconnect between motivation and satisfaction in recreational fishing: the case of catch orientation of German anglers. North American Journal of Fisheries Management 26:592–605.

Arlinghaus, R. 2006b. Overcoming human obstacles to conservation of recreational fishery resources, with emphasis on central Europe. Environmental Conservation 33:46–59.

Arlinghaus, R. 2007. Voluntary catch-and-release can generate conflict within the recreational angling community: a qualitative case study of specialized carp, *Cyprinus carpio*, angling in Germany. Fisheries Management and Ecology 14:161–171.

Arlinghaus, R., J. Alós, B. Beardmore, K. Daedlow, M. Dorow, M. Fujitani, D. Hühn, W. Haider, L. M. Hunt, B. M. Johnson, F. Johnston, T. Klefoth, S. Matsumura, C. Monk, T. Pagel, J. R. Post, T. Rapp, C. Riepe, H. Ward, and C. Wolter. 2017a. Understanding and managing freshwater recreational fisheries as complex adaptive social-ecological systems. Reviews in Fisheries Science & Aquaculture 25:1–41.

Arlinghaus, R., M. Bork, and E. Fladung. 2008. Understanding the heterogeneity of recreational anglers across an urban-rural gradient in a metropolitan area (Berlin, Germany), with implications for fisheries management. Fisheries Research 92:53–62.

Arlinghaus, R., S. J. Cooke, J. Lyman, D. Policansky, A. Schwab, C. Suski, S. G. Sutton, and E. B. Thorstad. 2007. Understanding the complexity of catch-and-release in recreational fishing: an integrative synthesis of global knowledge from historical, ethical, social, and biological perspectives. Reviews in Fisheries Science 15:75–167.

Arlinghaus, R., S. J. Cooke, S. G. Sutton, A. J. Danylchuk, W. Potts, K. de M. F. Freire, J. Alós, E. T. da Silva, I. G. Cowx, and R. van Anrooy. 2016a. Recommendations for the future of recreational fisheries to prepare the social-ecological system to cope with change. Fisheries Management and Ecology, 23:177–186.

Arlinghaus, R., K. L. Laskowski, J. Alós, T. Klefoth, C. T. Monk, S. Nakayama, and A. Schröder. 2017b. Passive gear-induced timidity syndrome in wild fish populations and its potential ecological and managerial implications. Fish and Fisheries 18:360–373.

Arlinghaus, R., K. Lorenzen, B. M. Johnson, S. J. Cooke, and I. G. Cowx. 2016b. Management of freshwater fisheries: addressing habitat, people and fishes. Freshwater Fisheries Ecology 1:557–579.

Arlinghaus, R., S. Matsumura, and U. Dieckmann. 2009. Quantifying selection differentials caused by recreational fishing: development of modeling framework and application to reproductive investment in pike (*Esox lucius*). Evolutionary Applications 2:335–355.

Arlinghaus, R., A. Schwab, C. Riepe, and T. Teel. 2012. A primer on anti-angling philosophy and its relevance for recreational fisheries in urbanized societies. Fisheries 37:153–164.

Arlinghaus, R., and T. Mehner. 2004. A management-orientated comparative analysis of urban and rural anglers living in a metropolis (Berlin, Germany). Environmental Management 33:331–344.

Árnason, E., U. B. Hernandez, and K. Kristinsson. 2009. Intense habitat-specific fisheries-induced selection at the molecular pan I locus predicts imminent collapse of a major cod fishery. PLoS One 4:e5529.

Arnett, E. B., and R. Southwick. 2015. Economic and social benefits of hunting in North America. International Journal of Environmental Studies 72:734–745.

Arnold, T. W. 2021. Individual heterogeneity in annual survival: quantifying the "doomed surplus." *in* K. L. Pope and L. A. Powell, editors. Harvest of fish and wildlife: new paradigms for sustainable management. CRC Press, Boca Raton, Florida, USA.

Arnold, T. W., A. D. Afton, M. J. Anteau, D. N. Koons, and C. A. Nicolai. 2016a. Temporal variation in survival and recovery rates of lesser scaup. The Journal of Wildlife Management 80:850–861.

Arnold, T. W., C. N. de Sobrino, and H. M. Specht. 2016b. Annual survival rates of migratory shore and upland game birds. Wildlife Society Bulletin 40:470–476.

Arnold, T. W., and D. W. Howerter. 2012. Effects of radiotransmitters and breeding effort on harvest and survival rates of female mallards. Wildlife Society Bulletin 36:286–290.

Artelle, K. A., J. D. Reynolds, A. Treves, J. C. Walsh, P. C. Paquet, and C. T. Darimont. 2018. Hallmarks of science missing from North American wildlife management. Science Advances 4:eaao0167.

Arterburn, J. E., D. J. Kirby, and C. R. Berry. 2002. A survey of angler attitudes and biologist opinions regarding trophy catfish and their management. Fisheries 27:10–21.

Askey, P. J., S. A. Richards, J. R. Post, and E. A. Parkinson. 2006. Linking angling catch rates and fish learning under catch-and-release regulations. North American Journal of Fisheries Management 26:1020–1029.

Asmyhr, L., T. Willebrand, and M. Hörnell-Willebrand. 2012. Successful adult willow grouse are exposed to increased harvest risk. The Journal of Wildlife Management 76:940–943.

Association of Fish and Wildlife Agencies. 2009. Modern snares for capturing mammals: definitions, mechanical attributes, and use considerations. https://www.fishwildlife.org/application/files/5515/2002/6134/Modern_Snares_final.pdf. Accessed 4 April 2020.

Association of Fish and Wildlife Agencies. 2014. Case study: state and federal management of the bobcat in the context of the Convention on International Trade in Endangered Species of Wild Fauna and Flora (CITES). https://www.fishwildlife.org/application/files/6215/1388/9531/bobcatwebsite.pdf. Accessed 21 February 2020.

Association of Fish and Wildlife Agencies. 2015a. Case study: state management of paddlefish in the context of the Convention on International Trade in Endangered Species of Wild Fauna and Flora (CITES). https://www.fishwildlife.org/application/files/2915/1388/9530/paddlefishwebsite.pdf. Accessed 29 March 2020.

Association of Fish and Wildlife Agencies. 2015b. Trap use, furbearers trapped, and trapper characteristics in the United States in 2015. https://www.fishwildlife.org/application/files/3115/2106/4349/FINAL_AFWA_Trap_Use_Report_2015_ed_2016.pdf. Accessed 3 March 2020.

Association of Fish and Wildlife Agencies. 2017. Bodygrip traps on dryland: a guide to responsible use. https://www.fishwildlife.org/application/files/9215/2106/2322/AFWA_Bodygrip_2017_final_compressed.pdf. Accessed 4 April 2020.

Association of Fish and Wildlife Agencies. 2019a. Furbearer management and best management practices for trapping. https://www.fishwildlife.org/afwa-inspires/furbearer-management. Accessed 25 February 2020, 4 March 2020.

Association of Fish and Wildlife Agencies. 2019b. State wildlife actions plans. https://www.fishwildlife.org/afwa-informs/state-wildlife-action-plans. Accessed 20 February 2020.

Association of Fish and Wildlife Agencies. 2019c. U.S. furbearer harvest statistics database 1970–2017. https://www.fishwildlife.org/afwa-inspires/furbearer-management. Accessed 4 April 2020.

Association of Fish and Wildlife Agencies, and Arizona Game and Fish. 2017. The state conservation machine. https://www.fishwildlife.org/application/files/3615/1853/8699/The_State_Conservation_Machine-FINAL.pdf. Accessed 27 June 2020.

Ayllón, D., G. G. Nicola, B. Elvira, and A. Almodóvar. 2019. Optimal harvest regulations under conflicting tradeoffs between conservation and recreational fishery objectives. Fisheries Research 216:47–58.

Ayllón, D., S. F. Railsback, A. Almodóvar, G. G. Nicola, S. Vincenzi, B. Elvira, and V. Grimm. 2018. Eco-evolutionary responses to recreational fishing under different harvest regulations. Ecology and Evolution 8:9600–9613.

Babcock, W. H. 1971. Effect of a size limit regulation on the trout fishery in Trappers Lake, Colorado. Transactions of the American Fisheries Society 100:50–54.

Baccante, D. 1995. Assessing catch inequality in walleye angling fisheries. North American Journal of Fisheries Management 15:661–665.

Bacheler, N. M., J. A. Buckel, J. E. Hightower, L. M. Paramore, and K. H. Pollock. 2009. A combined telemetry-tag return approach to estimate fishing and natural mortality rates of an estuarine fish. Canadian Journal of Fisheries and Aquatic Sciences 66:1230–1244.

Bain, M. B. 1987. Structured decision making in fisheries management: trout fishing regulations on the Au Sable River, Michigan. North American Journal of Fisheries Management 7:475–481.

Baldassarre, G. A., and E. G. Bolen. 2006. Waterfowl ecology and management, 2nd edition. Krieger Publishing Company, Malabar, Florida, USA.

Balkcom, G. D., P. R. Garrettson, and P. I. Padding. 2010. Predicting wood duck harvest rates in eastern North America. The Journal of Wildlife Management 74:1575–1579.

Balmaki, B., and A. Barati. 2006. Harvesting status of migratory waterfowl in northern Iran: a case study from Gilan Province. Pages 868–869 *in* G. Boere, C. Galbraith, and D. Stroud, editors. Waterbirds around the world. The Stationery Office, Edinburgh, Scotland.

Baranov, F. I. 1918. On the question of the biological basis of fisheries. Nauch. Issled. Ikhtiol. Inst. Izv. 1: 81–28 (in Russian).

Barker, R. J., J. E. Hines, and J. D. Nichols. 1991. Effect of hunting on annual survival of grey ducks in New Zealand. The Journal of Wildlife Management 55:260–265.

Barneche, D. R., D. R. Robertson, C. R. White, and D. J. Marshall. 2018. Fish reproductive-energy output increases disproportionately with body size. Science 360:642–645.

Barnhart, R. A. 1989. Symposium review: catch-and-release fishing, a decade of experience. North American Journal of Fisheries Management 9:74–80.

Barnhart, R. A., and T. D. Roelofs, editors. 1977. Catch-and-release fishing as a management tool. California Cooperative Fishery Research Unit, Arcata, California, USA.

Barrett, M. W., G. Proulx, and N. Jotham. 1988. Wild fur industry under challenge: the Canadian response. Transactions of the North American Wildlife and Natural Resources Conference 53:180–190.

Barro, S. C., and M. J. Manfredo. 1996. Constraints, psychological investment, and hunting participation: development and testing of a model. Human Dimensions of Wildlife 1:42–61.

Barrows, C. W., M. B. Swartz, W. L. Hodges, M. F. Allen, J. T. Rotenberry, B. Li, T. A. Scott, and X. Chen. 2005. A framework for monitoring multiple-species conservation plans. The Journal of Wildlife Management 69:1333–1345.

Bartholomew, A., and J. A. Bohnsack. 2005. A review of catch-and-release angling mortality with implications for no-take reserves. Reviews in Fish Biology and Fisheries 15:129–154.

Bartholow, J. 2000. Pop-II v1.2.5 documentation. Fossil Creek Software, Fort Collins, Colorado, USA.

Bartzen, B. A., and K. W. Dufour. 2017. Northern pintail (*Anas acuta*) survival, recovery, and harvest rates derived from 55 years of banding in Prairie Canada, 1960-2014. Avian Conservation and Ecology 12(2):7.

Baskett, T. S., and M. W. Sayre. 1993. Characteristics and importance. Pages 1–6 *in* T. S. Baskett, M. W. Sayre, R. E. Tomlinson, and R. E. Mirarchi, editors. Ecology and management of the mourning dove. Stackpole Books, Harrisburg, Pennsylvania, USA.

Bass Anglers Sportsman Society. 2016. Northwest no limits controversy update. https://www.bassmaster.com/conservation-news/northwest-no-limits-controversy-update. Accessed 1 January 2020.

Bass Anglers Sportsman Society. 2018. Washington updates bass regulations. https://www.bassmaster.com/conservation-news/washington-updates-bass-regulations. Accessed 1 January 2020.

Batavia, C., M. P. Nelson, C. T. Darimont, P. C. Paquet, W. J. Ripple, and A. D. Wallach. 2018. The elephant (head) in the room: a critical look at trophy hunting. Conservation Letters 12(1): e12565.

Batcheller, G. R., M. C. Bambery, L. Bies, T. Decker, S. Dyke, D. Guynn, M. McEnroe, M. O'Brien, J. F. Organ, S. J. Riley, and G. Roehm. 2010. The public trust doctrine: implications for wildlife

management and conservation in the United States and Canada. The Wildlife Society, Technical Review 10-01, Bethesda, Maryland, USA.
Batcheller, G. R., T. A. Decker, D. A. Hamilton, and J. F. Organ. 2000. A vision for the future of furbearer management in the United States. Wildlife Society Bulletin 28:833–840.
Batcheller, G. R., T. A. Decker, and R. P. Lanka. 2018. Public trust doctrine and the legal basis for state wildlife management. Pages 24–38 *in* T. J. Ryder, editor. State wildlife management and conservation. Johns Hopkins University Press, Baltimore, Maryland, USA.
Bates, D., M. Maechler, B. Bolker, and S. Walker. 2015. Fitting linear mixed-effects models using lme4. Journal of Statistical Software 67:1–48.
Batt, B. D. J., editor. 1997. Arctic ecosystems in peril: report of the Arctic Goose Habitat Working Group. Arctic Goose Joint Venture Special Publication, U.S Fish and Wildlife Service, Washington, D.C., USA, and Canadian Wildlife Service, Ottawa, Ontario, Canada.
Beamesderfer, R. C. 2000. Managing fish predators and competitors: deciding when intervention is effective and appropriate. Fisheries 25:18–23.
Beamesderfer, R. C., and J. A. North. 1995. Growth, natural mortality, and predicted response to fishing for largemouth bass and smallmouth bass populations in North America. North American Journal of Fisheries Management 15:688–704.
Beamish, R. J., G. A. McFarlane, and A. Benson. 2006. Longevity overfishing. Progress in Oceanography 68:289–302.
Beard, T. D., Jr., S. P. Cox, and S. R. Carpenter. 2003a. Impacts of daily bag limit reductions on angler effort in Wisconsin walleye lakes. North American Journal of Fisheries Management 23:1283–1293.
Beard, T. D., Jr., M. T. Drake, J. E. Breck, and N. A. Nate. 1997. Effects of simulated angling regulations on stunting in bluegill populations. North American Journal of Fisheries Management 17:525–532.
Beard, T. D., Jr., P. W. Rasmussen, S. Cox, and S. R. Carpenter. 2003b. Evaluation of a management system for a mixed walleye spearing and angling fishery in northern Wisconsin. North American Journal of Fisheries Management 23:481–491.
Beardmore, B., M. Dorow, W. Haider, and R. Arlinghaus. 2011a. The elasticity of fishing effort response and harvest outcomes to altered regulatory policies in eel (*Anguilla anguilla*) recreational angling. Fisheries Research 110:136–148.
Beardmore, B., W. Haider, L. M. Hunt, and R. Arlinghaus. 2011b. The importance of trip context for determining primary angler motivations: are more specialized anglers more catch-oriented than previously believed? North American Journal of Fisheries Management 31:861–879.
Beardmore, B., W. Haider, L. M. Hunt, and R. Arlinghaus. 2013. Evaluating the ability of specialization indicators to explain fishing preferences. Leisure Sciences 35:273–292.
Beardmore, B., L. M. Hunt, W. Haider, M. Dorow, and R. Arlinghaus. 2015. Effectively managing angler satisfaction in recreational fisheries requires understanding the fish species and the anglers. Canadian Journal of Fisheries and Aquatic Sciences 72:500–513.
Beardmore, B. and C. Harris. 2016. Effects of changing license structures and increasing license fees on participation and revenue. A review of the literature. https://www.perc.org/2000/11/09/an-economic-guide-to-state-wildlife-management/ Accessed 12 July 2020.
Beasom, S. L., and S. F. Roberson, editors. Game harvest management. Caesar Kleberg Wildlife Research Institute, Kingsville, Texas, USA.
Beaudreau, A. H., M. N. Chan, and P. A. Loring. 2018. Harvest portfolio diversification and emergent conservation challenges in an Alaskan recreational fishery. Biological Conservation 222:268–277.
Béchet, A., J. F. Giroux, and G. Gauthier. 2004a. The effects of disturbance on behaviour, habitat use and energy of spring staging snow geese. Journal of Applied Ecology 41:689–700.
Béchet, A., J. F. Giroux, G. Gauthier, J. D. Nichols, and J. E. Hines. 2003. Spring hunting changes the regional movements of migrating greater snow geese. Journal of Applied Ecology 40:553–564.
Béchet, A., A. Reed, N. Plante, J.-F. Giroux, and G. Gauthier. 2004b. Estimating the size of the greater snow goose population. The Journal of Wildlife Management 68:639–649.
Bellman, R. 1957. Dynamic programming. Princeton University Press, Princeton, New Jersey, USA.
Bellrose, F. C. 1959. Lead poisoning as a mortality factor in waterfowl populations. Bulletin of the Illinois Natural History Survey 27:235–288.
Belore, M., A. Briggs, A. Cook, M. DuFour, M. Faust, M. Hosack, C. Knight, J. Robinson, B. Schmidt, and S. Thomas. 2020. Report of the Lake Erie Yellow Perch Task Group. Report to Standing Technical Committee, Lake Erie Committee, Great Lakes Fishery Commission, Ann Arbor, U.S.A.
Belsare, A. V., and C. M. Stewart. 2020. OvCWD: an agent-based modeling framework for informing chronic wasting disease management in white-tailed deer populations. Ecological Solutions and Evidence 1:e12017.

Bender, L. C., P. E. Fowler, J. A. Bernatowicz, J. L. Musser, and L. E. Stream. 2002. Effects of open-entry spike-bull, limited-entry branched-bull harvesting on elk composition in Washington. Wildlife Society Bulletin 30:1078–1084.

Bender, L. C., and P. J. Miller. 1999. Effects of elk harvest strategy on bull demographics and herd composition. Wildlife Society Bulletin 27:1032–1037.

Ben-Haim, Y. 2001. Information-gap decision theory: decisions under severe uncertainty. Academic Press, San Diego, California, USA.

Bennear, L. S., R. N. Stavins, and A. F. Wagner. 2005. Using revealed preferences to infer environmental benefits: evidence from recreational fishing licenses. Journal of Regulatory Economics 28:157–179.

Bennett, N. J., R. Roth, S. C. Klain, K. Chan, P. Christie, D. A. Clark, G. Cullman, D. Curran, T. J. Durbin, G. Epstein, A. Greenberg, M. P. Nelson, J. Sandlos, R. Stedman, T. L. Teel, R. Thomas, D. Veríssimo, and C. Wyborn. 2017. Conservation social science: understanding and integrating human dimensions to improve conservation. Biological Conservation 205:93–108.

Benson, D. A. 1968. The Canada migratory game bird hunting and related surveys. Transactions of the Federal-Provincial Wildlife Conference 32:46–50.

Benson, D. A. 1970. Species of water fowl and age and sex ratios of ducks harvested in Canada during the 1969-79 season. Canadian Wildlife Service, Progress Notes Number 19, Ottawa, Canada.

Benson, D. E., and D. J. Decker. 2001. Why people hunt: a theoretical framework. Transactions of the North American Wildlife and Natural Resources Conference 66:140–153.

Berezanski, D. 2004. Trapping comes of age: the registered trapline system of Manitoba. Selected Papers of the Biennial Rupert's Land Colloquium 11:85–97.

Berger, A. M., M. L. Jones, Y. Zhao, and J. R. Bence. 2012. Accounting for spatial population structure at scales relevant to life history improves stock assessment: the case for Lake Erie walleye *Sander vitreus*. Fisheries Research 115–116:44–59.

Berkes, F., C. Folke, and J. Colding. 2000. Linking social and ecological systems: management practices and social mechanisms for building resilience. Cambridge University Press, Cambridge, England, United Kingdom.

Bernier, C., and M. B. Adler. 2012. Vermont furbearer management newsletter. Vermont Fish & Wildlife Department, Springfield, Vermont, USA.

Bernthal, M. J., T. Koesters, K. Ballouli, and M. T. Brown. 2015. Motivations and fan engagement related to professional bass fishing spectatorship. Sport Marketing Quarterly 24: 6.

Bettoli, P. W., G. D. Scholten, and W. C. Reeves. 2007. Protecting paddlefish from overfishing: a case history of the research and regulatory process. Fisheries 32:390–397.

Beverton, R. J. H., and S. J. Holt. 1957. On the dynamics of exploited fish populations. Chapman and Hall, London, England, United Kingdom (facsimile reprint 1993).

Bhandari, P., R. C. Stedman, A. E. Luloff, J. C. Finley, and D. R. Diefenbach. 2006. Effort versus motivation: factors affecting antlered and antlerless deer harvest success in Pennsylvania. Human Dimensions of Wildlife 11:423–436.

Biederbeck, H. H., M. C. Boulay, and D. H. Jackson. 2001. Effects of hunting regulations on bull elk survival and age structure. Wildlife Society Bulletin 29:1271–1277.

Bieg, C., K. S. McCann, and J. M. Fryxell. 2017. The dynamical implications of human behaviour on a social-ecological harvesting model. Theoretical Ecology 10:341–354.

Biro, P. A., and J. R. Post. 2008. Rapid depletion of genotypes with fast growth and bold personality traits from harvested fish populations. Proceedings of the National Academy of Sciences 105:2919–2922.

Biro, P. A., and J. A. Stamps. 2008. Are animal personality traits linked to life-history productivity? Trends in Ecology & Evolution 23:361–368.

Bishop, C. J., G. C. White, D. J. Freddy, and B. E. Watkins. 2005. Effect of limited antlered harvest on mule deer sex and age ratios. Wildlife Society Bulletin 33:662–668.

Bissell, S. J., M. D. Duda, and K. C. Young. 1998. Recent studies on hunting and fishing participation in the United States. Human Dimensions of Wildlife 3:75–80.

Bissonette, J. A., C. A. Kassar, and L. J. Cook. 2008. Assessment of costs associated with deer-vehicle collisions: human death and injury, vehicle damage, and deer loss. Human-Wildlife Conflicts 2:17–27.

Bister, T. J., D. W. Willis, A. D. Knapp, and T. R. St. Sauver. 2002. Evaluation of a 23-cm minimum length limit for black and white crappies in a small South Dakota impoundment. North American Journal of Fisheries Management 22:1364–1368.

Blackwell, B. G., M. J. Hubers, and R. G. Losco. 2003. Postharvest length changes of angler-caught walleyes. North American Journal of Fisheries Management 23:770–778.

Blackwell, B. G., T. M. Kaufman, and T. S. Moos. 2019. Angler exploitation of an unexploited walleye population in the northern Great Plains. Fisheries Research 216:59–64.

Blais, A., and S. Dion. 1990. Are bureaucrats budget maximizers? The Niskanen model & its critics. Polity 22:655–674.

Blanchard, J. L., S. Jennings, R. Law, M. D. Castle, P. McCloghrie, M.-J. Rochet, and E. Benoît. 2009. How does abundance scale with body size in coupled size-structured food webs? Journal of Animal Ecology 78:270–280.

Bleier N., R. Lehoczki, D. Újváry, L. Szemethy, and S. Csányi. 2012. Relationships between wild ungulates density and crop damage in Hungary. Acta Theriologica 57:351–359.

Blohm, R. J. 1989. Introduction to harvest–understanding surveys and season setting. Pages 118–133 *in* Proceedings of the sixth international waterfowl symposium. Ducks Unlimited, Memphis, Tennessee, USA.

Blom, F. S., and G. Connolly. 2003. Inventing and reinventing the sodium cyanide ejectors: a technical history of coyote getters and M-44s in predator damage control. U. S. Department of Agriculture, Animal and Plant Health Inspection Service, National Wildlife Research Center, Research Report 03-02, Fort Collins, Colorado, USA.

Blomberg, E. J. 2015. The influence of harvest timing on greater sage-grouse survival: a cautionary perspective. The Journal of Wildlife Management 79:695–703.

Boggess, E. K., G. R. Batcheller, R. G. Linscombe, J. W. Greer, M. Novak, S. B. Linhart, D. W. Erickson, A. W. Todd, D. C. Juve, and D. A. Wade. 1990. Traps, trapping, and furbearer management: a review. The Wildlife Society, Technical Review 90-1, Bethesda, Maryland, USA.

Bolam, F. C., M. J. Grainger, K. L. Mengersen, G. B. Stewart, W. J. Sutherland, M. C. Runge, and P. J. McGowan. 2019. Using the Value of Information to improve conservation decision making. Biological Reviews 94:629–647.

Bolen, E. G. 2000. Waterfowl management: yesterday and tomorrow. The Journal of Wildlife Management 64:323–335.

Bonds, C. C., J. B. Taylor, and J. Leitz. 2008. Practices and perceptions of Texas anglers regarding voluntary release of largemouth bass and slot length limits. Pages 219–230 *in* M. S. Allen, S. Sammons, and M. J. Maceina, editors. Balancing fisheries management and water uses for impounded river systems. American Fisheries Society, Symposium 62, Bethesda, Maryland, USA.

Bonnot, T. W., J. H. Schulz, and J. J. Millspaugh. 2011. Factors affecting mourning dove harvest in Missouri. Wildlife Society Bulletin 35:76–84.

Bonvechio, T. F., B. R. Bowen, J. M. Wixson, and M. S. Allen. 2014. Exploitation and length limit evaluation of largemouth bass in three Georgia small impoundments. Journal of the Southeastern Association of Fish and Wildlife Agencies 1:33–41.

Bonvechio, T. F., and J. J. Rydell. 2016. Use of a female-only stocking strategy to establish a trophy largemouth bass fishery in a Georgia small impoundment. Journal of the Southeastern Association of Fish and Wildlife Agencies 3:136–143.

Boomer, G. S., F. A. Johnson, M. D. Koneff, T. A. Sanders, and R. E. Trost. 2007. A process to determine scaup regulatory alternatives. https://www.fws.gov/migratorybirds/pdf/management/AHM/ScaupScopingDocument_ProcessToDetermineScaupRegulatoryAlternatives.pdf. Accessed 24 June 2020.

Boomer, G. S., G. S. Zimmerman, N. L. Zimpfer, P. R. Garrettson, M. D. Koneff, T. A. Sanders, K. D. Magruder, and J. A. Royle. 2013. Band reporting probabilities for mallards recovered in the United States and Canada. The Journal of Wildlife Management 77:1059–1066.

Boschung, Jr., H. T., and R. L. Mayden. 2004. Fishes of Alabama. Smithsonian Institution Press, Washington, D.C., USA.

Boulanger, J. R., and P. D. Curtis. 2016. Efficacy of surgical sterilization for managing overabundant suburban white-tailed deer. Wildlife Society Bulletin 40:727–735.

Boulanger, J. R., P. D. Curtis, E. G. Cooch, and A. J. DeNicola.2012a. Sterilization as an alternative deer control technique: a review. Human-Wildlife Interactions 6:273–282.

Boulanger, J. R., G. R. Goff, and P. D. Curtis. 2012b. Use of "earn-a-buck" hunting to manage local deer overabundance. Northeastern Naturalist 19:159–172.

Boulding, W., A. Kalra, R. Staelin, and V. A. Zeithaml. 1993. A dynamic process model of service quality: from expectations to behavioral intentions. Journal of Marketing Research 30:7–27.

Bowles, E., K. Marin, S. Mogensen, P. MacLeod, and D. J. Fraser. 2020. Size reductions and genomic changes within two generations in wild walleye populations: associated with harvest? Evolutionary Applications 13:1128–1144.

Boyce, M. S. 1981. Beaver life-history responses to exploitation. Journal of Applied Ecology 18:749–753.

Boyce, M. S., T. Coulson, J. R. Heffelfinger, and P. R. Krausman. 2019. Mountain sheep management must use representative data: a reply to Festa-Bianchet (2019). The Journal of Wildlife Management 83:9–11.

Boyce, M. S., A. R. E. Sinclair, and G. C. White.1999. Seasonal compensation of predation and harvesting. Oikos 87:419–426.

Boyd, H., G. E. J. Smith, and F. G. Cooch. 1982. The lesser snow geese of the eastern Canadian arctic: their status during 1964-1979 and their management from 1982-1990. Canadian Wildlife Service, Occasional Paper Number 46, Ottawa, Canada.

Boyd, R. J., and J. F. Lipscomb. 1976. An evaluation of yearling bull elk hunting restrictions in Colorado. Wildlife Society Bulletin 4:3–10.

Bozek, M. A., D. A. Baccante, and N. P. Lester. 2011. Walleye and sauger life history. Pages 1—69 *in* B. A. Barton, editor. Biology, management, and culture of walleye and sauger. American Fisheries Society, Bethesda, Maryland, USA.

Bradshaw, L., R. H. Holsman, J. Petchenik, and T. Finger. 2019. Meeting harvest expectations is key for duck hunter satisfaction. Wildlife Society Bulletin 43:102–111.

Braun, C. E., N. J. Silvy, T. S. Baskett, and R. E. Tomlinson. 1993. Research and management needs. Pages 507–513 *in* T. S. Baskett, M. W. Sayre, R. E. Tomlinson, and R. E. Mirarchi, editors. Ecology and management of the mourning dove. Stackpole Books, Harrisburg, Pennsylvania, USA.

Breffle, W. S., E. R. Morey, R. D. Rowe, D. M. Waldman, and S. M. Wytinck.1999. Recreational fishing damages from fish consumption advisories in the waters of Green Bay. Stratus Consulting, Boulder, Colorado, USA.

Breisjøberget, J. I., T. Storaas, and M. Odden. 2017. Ptarmigan hunting restrictions: effects on hunters' opinions and harvest. The Journal of Wildlife Management 81:1179–1186.

Brett, J. R., and N. R. Glass. 1973. Metabolic rates and critical swimming speeds of sockeye salmon (*Oncorhynchus nerka*) in relation to size and temperature. Journal of the Fisheries Board of Canada 30:379–387.

Bromley, R. G. 1996. Characteristics and management implications of the spring waterfowl harvest in the western Canadian Arctic, Northwest Territories. Arctic 49:70–85.

Brook, B. W., N. S. Sodhi, and C. J. A. Bradshaw. 2008. Synergies among extinction drivers under global change. Trends in Ecology & Evolution 23:453–460.

Brooks, L., and D. Fryda. 2011. Angler use and sport fishing catch survey on the Missouri River and Lake Oahe, North Dakota. North Dakota Game and Fish Department, Project F-2-R-55, Study 4, Number 7, Bismarck, North Dakota, USA.

Brøseth, H., and H. C. Pedersen. 2000. Hunting effort and game vulnerability studies on a small scale: a new technique combining radio-telemetry, GPS and GIS. Journal of Applied Ecology 37:182–190.

Brousseau, C. S., and E. R. Armstrong. 1987. The role of size limits in walleye management. Fisheries 12(1):2–5.

Brown, D. T., J. A. Rice, C. D. Suski, and D. D. Aday. 2015. Dispersal patterns of coastal largemouth bass in response to tournament displacement. North American Journal of Fisheries Management 35:431–439.

Brown, G. M., Jr., J. Hammack, and M. F. Tillman. 1976. Mallard population dynamics and management models. The Journal of Wildlife Management 40:542–555.

Brown, P. J. 2008. Perspectives on the past and future of human dimensions of fish and wildlife. Pages 1–14 *in* M. J. Manfredo and B. J. Perry, editors. Wildlife and society: the science of human dimensions. Island Press, Washington, D.C., USA.

Brown, T. L., D. J. Decker, S. J. Riley, J. W. Enck, T. B. Lauber, P. D. Curtis, and G. F. Mattfeld. 2000. The future of hunting as a mechanism to control white-tailed deer populations. Wildlife Society Bulletin 28:797–807.

Brownie, C., D. R. Anderson, K. P. Burnham, and D. R. Robson. 1985. Statistical inference from band recovery data–a handbook, 2nd edition. U.S. Fish and Wildlife Service, Resource Publication 156, Washington, D.C., USA.

Brownscombe, J. W., A. J. Danylchuk, J. M. Chapman, L. F. G. Gutowsky, and S. J. Cooke. 2017. Best practices for catch-and-release recreational fisheries–angling tools and tactics. Fisheries Research 186:693–705.

Brunke, K. D., and K. M. Hunt. 2007. Comparison of two approaches for the measurement of waterfowl hunter satisfaction. Human Dimensions of Wildlife 12:443–457.

Brunke, K. D., and K. M. Hunt. 2008. Mississippi waterfowl hunter expectations, satisfaction, and intentions to hunt in the future. Human Dimensions of Wildlife 13:317–328.

Bruskotter, J. T. 2013. The predator pendulum revisited: social conflict over wolves and their management in the western United States. Wildlife Society Bulletin 37:674–679.

Bryan, H. M., J. E. G. Smits, L. Koren, P. C. Paquet, K. E. Wynne-Edwards, and M. Musiani. 2015. Heavily hunted wolves have higher stress and reproductive steroids than wolves with lower hunting pressure. Functional Ecology 29:347–356.

Bubenik, A. B. 1985. Reproductive strategies in cervids. Pages 367–373 *in* P. F. Fennessy and K. R. Drew, editors. Biology of deer production. Royal Society of New Zealand, Bulletin 22, Wellington, New Zealand.

Buchanan, T. 1983. Toward an understanding of variability in satisfactions within activities. Journal of Leisure Research 15:39–51.

Bull, P., and B. Peyton. 2001. The perceptions of Michigan deer hunters regarding quality deer management (QDM) and related issues. Michigan Department of Natural Resources, East Lansing, Michigan, USA.

Bunch, A. J., R. S. Greenlee, and E. M. Brittle. 2018. Blue catfish density and biomass in a tidal tributary in coastal Virginia. Northeastern Naturalist 25:333–340.

Bunnefeld, N., D. Baines, D. Newborn, and E. J. Milner-Gulland. 2009. Factors affecting unintentional harvesting selectivity in a monomorphic species. Journal of Animal Ecology 78:485–492.

Bunnefeld, N., E. Hoshino, and E. J. Milner-Gulland. 2011a. Management strategy evaluation: a powerful tool for conservation? Trends in Ecology & Evolution 26:441–447.

Bunnefeld, N., D. C. Reuman, D. Baines, and E. J. Milner-Gulland. 2011b. Impact of unintentional selective harvesting on the population dynamics of red grouse. Journal of Animal Ecology 80:1258–1268.

Büntgen, U., J. D. Galvan, A. Mysterud, P. J. Krusic, L. Hülsmann, H. Jenny, J. Senn, and K. Bollmann. 2018. Horn growth variation and hunting selection of the Alpine ibex. Journal of Animal Ecology 87:1069–1079.

Burgner, R. L. 1991. Life history of sockeye salmon (*Oncorhynchus nerka*). Pages 1–118 *in* G. Groot, editor. Pacific salmon life histories. University of British Columbia Press, Vancouver, British Columbia, Canada.

Burkett, E. M., and R. L. Winkler. 2019. Recreational fishing participation trends in Upper Great Lakes states: an age-period-cohort analysis. Human Dimensions of Wildlife 24:95–97.

Burnham, K. P., and D. R. Anderson. 1984. Tests of compensatory vs additive hypotheses of mortality in mallards. Ecology 65:105–112.

Burnham, K. P., and D. R. Anderson. 2002. Model selection and multimodel inference: a practical information-theoretic approach. Springer, New York, New York, USA.

Burnham, K. P., D. R. Anderson, G. C. White, C. Brownie, and K. H. Pollack. 1987. Design and analysis methods for fish survival experiments based on release-recapture data. American Fisheries Society, Monograph 5, Bethesda, Maryland, USA.

Burnham, K. P., G. C. White, and D. R. Anderson. 1984. Estimating the effect of hunting on annual survival rates of adult mallards. The Journal of Wildlife Management 48:350–361.

Burns, R. C., and A. R. Graefe. 2006. Service quality measures: recreationists' perceptions of US Pacific Northwest National Forests. World Leisure Journal 48:40–51.

Buskirk, S. W., A. S. Harestad, M. G. Raphael, and R. A. Powell, editors. 1994. Martens, sables, and fishers: biology and conservation. Comstock, Ithaca, New York, New York, USA.

Bylander, C. B. 2016. R3 Toolkit: a guide to recruitment, retention and reactivation. Minnesota Department of Natural Resources, St. Paul, Minnesota, USA. https://files.dnr.state.mn.us/fish_wildlife/outreach/r3/r3_toolkit.pdf. Accessed 3 August 2020.

Byrne, R., and M. Dunfee. 2018. Evolution and current use of the outdoor recreation adoption model. Council to Advance Hunting and the Shooting Sports, Washington, D.C., USA.

California Fish and Game Commission. 2015. Staff summary for August 4–5, 2015, bobcat trapping. Obtained by request from the California Fish and Game Commission on 24 February 2020.

California Legislative Information. 2019. AB-273 fur-bearing and nongame mammals: recreational and commercial fur trapping: prohibition. http://leginfo.legislature.ca.gov/faces/billTextClient.xhtml?bill_id=201920200AB273. Accessed 21 February 2020.

Callegaro, M., and Y. Yang. 2018. The role of surveys in the era of "big data." Pages 175–192 *in* D. L. Vannette and J. A. Krosnick, editors. The Palgrave Handbook of Survey Research. Palgrave Macmillan, Cham.

Calvert, A. M., R. T. Alisauskas, and G. C. White. 2017. Annual survival and seasonal hunting mortality of midcontinent snow geese. The Journal of Wildlife Management 81:1009–1020.

Calvert, A. M., and G. Gauthier. 2005. Effects of exceptional conservation measures on survival and seasonal hunting mortality in greater snow geese. Journal of Applied Ecology 42:442–452.

Camp, E. V., R. N. M. Ahrens, C. Crandall, and K. Lorenzen. 2018. Angler travel distances: implications for spatial approaches to marine recreational fisheries governance. Marine Policy 87:263–274.

Camp, E. V., M. A. Kaemingk, R. N. M. Ahrens, W. M. Potts, W. E. Pine, O. L. F. Weyl, and K. L. Pope. 2020. Resilience management for conservation of inland recreational fisheries. Frontiers in Ecology and Evolution 7:498.

Campbell, H. 1985. Harvesting wildlife: past, present, and future. Pages 3–14 *in* S. L. Beasom and S. F. Roberson, editors. Game harvest management. Caesar Kleberg Wildlife Research Institute, Kingsville, Texas, USA.

Campbell, H., D. K. Martin, P. E. Ferkovich, and B. K. Harris. 1973. Effects of hunting and some other environmental factors on scaled quail in New Mexico. Wildlife Monographs 34:3–49.

Canadian Wildlife Service Waterfowl Committee. 2018. Migratory birds regulations in Canada, July 2018. 2018/2019 and 2019/2020 hunting seasons. Environment and Climate Change Canada, CWS Migratory Birds Regulatory Report Number 51, Ottawa, Canada.

Canadian Wildlife Service Waterfowl Committee. 2020a. Population status of migratory game birds in Canada. November 2019. Environment and Climate Change Canada, CWS Migratory Birds Regulatory Report Number 52, Ottawa, Canada.

Canadian Wildlife Service Waterfowl Committee. 2020b. Proposals to amend the Canadian migratory birds regulations. December 2019. Consultation document. Hunting seasons 2020- 2021 and 2021-2022. Environment and Climate Change Canada, CWS Migratory Birds Regulatory Report Number 53, Ottawa, Canada.

Carey, M. P., B. L. Sanderson, T. A. Friesen, K. A. Barnas, and J. D. Olden. 2011. Smallmouth bass in the Pacific Northwest: a threat to native species; a benefit for anglers. Reviews in Fisheries Science 19:305–315.

Carlin, C., S. A. Schroeder, and D. C. Fulton. 2012. Site choice among Minnesota walleye anglers: the influence of resource conditions, regulations, and catch orientation on lake preference. North American Journal of Fisheries Management 32:299–312.

Carlson, A. J., and D. A. Isermann. 2010. Mandatory catch and release and maximum length limits for largemouth bass in Minnesota: is exploitation still a relevant concern? North American Journal of Fisheries Management 30:209–220.

Carlson, A. K., W. W. Taylor, M. R. Cronin, M. J. Eaton, L. E. Eckert, M. A. Kaemingk, A. J. Reid, and A. Trudeau. 2020. A social–ecological odyssey in fisheries and wildlife management. Fisheries 45:238–243.

Carlson, D. M., and P. S. Bonislawsky. 1981. The paddlefish (*Polyodon spathula*) fisheries of the Midwestern United States. Fisheries 6:17–22.

Carmichael, R. G. 1973. Innovation and enterprise: a history of fur conservation in northern Manitoba, 1935-1948. University of Manitoba, Unpublished report, Winnipeg, Manitoba, Canada.

Caroffino, D. C. 2013. Angler compliance with lake trout length limit regulations in Great Lakes waters. North American Journal of Fisheries Management 33:1203–1209.

Carothers, A. D. 1973. The effects of unequal catchability on Jolly-Seber estimates. Biometrics 29:79–100.

Carpenter, L. H. 2000. Harvest management goals. Pages 192–213 *in* S. Demarais and P. R. Krausman, editors. Ecology and management of large mammals in North America. Prentice Hall, Upper Saddle River, New Jersey, USA.

Carpenter, L. H., and R. B. Gill. 1987. Antler point regulations: the good, the bad, and the ugly. Proceedings of the Western Association of Fish and Wildlife Agencies 67:94–107.

Carruthers, T. R., K. Dabrowska, W. Haider, E. A. Parkinson, D. A. Varkey, H. Ward, M. K. McAllister, T. Godin, B. Van Poorten, P. J. Askey, K. L. Wilson, L. M. Hunt, A. Clarke, E. Newton, C. Walters, and J. R. Post. 2018. Landscape-scale social and ecological outcomes of dynamic angler and fish behaviours: processes, data, and patterns. Canadian Journal of Fisheries and Aquatic Sciences 76:970–988.

Casazza, M. L., J. L. Yee, M. R. Miller, D. L. Orthmeyer, D. R. Yparraguirre, R. L. Jarvis, and C. T. Overton. 2005. Evaluation of current population indices for band-tailed pigeons. Wildlife Society Bulletin 33:606–615.

Caswell, H. 2001. Matrix population models: construction, analysis, and interpretation, 2nd edition. Sinauer Associates, Inc., Sunderland, Massachusetts, USA.

Caughley, G. 1977. Analysis of vertebrate populations. Wiley and Sons, Ltd. New York, New York, USA.

Caughley, G. 1985. Harvesting of wildlife: past, present, and future. Pages 3–14 *in* S. L. Beasom and S. F. Roberson, editors. Game harvest management. Caesar Kleberg Wildlife Research Institute, Kingsville, Texas, USA.

Central Flyway Webless Migratory Game Bird Technical Committee. 2018. Management guidelines for the mid-continent population of Sandhill cranes. Unpublished report, Denver, Colorado, USA.

Central, Mississippi, and Pacific Flyway Councils. 2015. Management plan for midcontinent greater white-fronted geese. Unpublished report, Denver, Colorado, USA.

Cha, W., and R. T. Melstrom. 2018. Catch-and-release regulations and paddlefish angler preferences. Journal of Environmental Management 214:1–8.

Chadès, I., E. McDonald-Madden, M. A. McCarthy, B. Wintle, M. Linkie, and H. P. Possingham. 2008. When to stop managing or surveying cryptic threatened species. Proceedings of the National Academy of Sciences 105:13936–13940.

Chapman, J. K., L. G. Anderson, C. L. A. Gough, and A. R. Harris. 2016. Working up an appetite for lionfish: a market-based approach to manage the invasion of *Pterois volitans* in Belize. Marine Policy 73:256–262.

Charnov, E. L. 1976. Optimal foraging, the marginal value theorem. Theoretical Population Biology 9:129–136.

Charnov, E. L. 1993. Life history invariants: some explorations of symmetry in evolutionary ecology. Oxford University Press, Oxford, England, United Kingdom.

Chatfield, C., and G. Goodhardt. 1970. The beta-binomial model for consumer purchasing behaviour. The Royal Statistical Society Series C-Applied Statistics 19:240–250.

Child, K. N., D. A. Aitken, R. V. Rea, and R. A. Demarchi. 2010. Potential vulnerability of bull moose in central British Columbia to three antler-based hunting regulations. Alces 46:113–121.

Chiyo, P. I., V. Obanda, and D. K. Korir. 2015. Illegal tusk harvest and the decline of tusk size in the African elephant. Ecology and Evolution 5:5216–5229.

Chizinski, C. J., D. R. Martin, K. L. Hurley, and K. L. Pope. 2014a. Self-imposed length limits in recreational fisheries. Fisheries Research 155:83–89.

Chizinski, C. J., D. R. Martin, K. L. Pope, T. J. Barada, and J. J. Schuckman. 2014b. Angler effort and catch within a spatially complex system of small lakes. Fisheries Research 154:172–178.

Chizinski, C. J., K. L. Pope, and G. R. Wilde. 2010a. A modelling approach to evaluate potential management actions designed to increase growth of white perch in a high-density population. Fisheries Management and Ecology 17:262–271.

Chizinski, C. J., K. L. Pope, G. R. Wilde, and R. E. Strauss. 2010b. Implications of stunting on morphology of freshwater fishes. Journal of Fish Biology 76:564–579.

Ciuti, S., T. B. Murphy, D. G. Paton, A. D. McDevitt, M. Musiani, and M. S. Boyce. 2012. Human selection of elk behavioural traits in a landscape of fear. Proceedings of the Royal Society B 279:4407–4416.

Clark, C. W. 1990. Mathematical bioeconomics: the optimal management of renewable resources, 2nd edition. John Wiley and Sons, Inc., New York, New York, USA.

Clark, C. W. 2006. The worldwide crisis in fisheries: economic models and human behavior. Cambridge University Press, Cambridge, England, United Kingdom.

Clark, W. R. 1987. Effects of harvest on annual survival of muskrats. The Journal of Wildlife Management 51:265–272.

Clausen, K. K., T. K. Christensen, O. M. Gundersen, and J. Madsen. 2017. Impact of hunting along the migration corridor of pink-footed geese *Anser brachyrhynchus*–implications for sustainable harvest management. Journal of Applied Ecology 54:1563–1570.

Clawson, M., and J. L. Knetsch. 2013. Economics of outdoor recreation. John Hopkins Press for Resources for the Future, Baltimore, Maryland, USA.

Clemen, R. T. 1996. Making hard decisions: an introduction to decision analysis, 2nd edition. Duxbury Press, Pacific Grove, California, USA.

Clower, K., E. Evarts, M. Hennen, and C. Kavanaugh. 2018. Lake Vermilion Fisheries Management Plan 2017-2020. Minnesota Department of Natural Resources Special Publication 183. St Paul, Minnesota, USA.

Cohen, B. S., E. H. Leone, E. P. Garrison, C. R. Morea, S. M. Shea, K. V. Miller, M. J. Cherry, J. H. Stickles, and J. D. Kelly. 2016. Evaluation of antler-based selective harvest criteria on harvest and antler size of male white-tailed deer in Florida. Journal of the Southeastern Association of Fish and Wildlife Agencies 3:203–209.

Colby, S., and J. M. Ortman. 2015. Projections of the size and composition of the U.S. population: 2014 to 2060. U.S. Census Bureau, Washington, D.C., USA.

Cole, R. A., and F. A. Ward. 1994. Optimum fisheries management policy: angler opportunity versus angler benefit. North American Journal of Fisheries Management 14:22–33.

Coleman, T. S. 2019. Evaluation of a shoreline rotenone application to control largemouth bass *Micropterus salmoides* recruitment in small impoundments. Thesis, Auburn University, Alabama, USA.

Collier, B. A., and D. Krementz. 2007. Uncertainty in age-specific harvest estimates and consequences for white-tailed deer management. Ecological Modelling 201:194–204.

Colvin, M. A. 1991. Evaluation of minimum-size limits and reduced daily limits on the crappie populations and fisheries in five large Missouri reservoirs. North American Journal of Fisheries Management 11:585–597.

Colvin, M. A. 2002. Population and fishery characteristics of white bass in four large Missouri reservoirs. North American Journal of Fisheries Management 22:677–689.

Committee on the Status of Endangered Wildlife in Canada. 2019. Canadian wildlife species at risk, October 2019. https://wildlife-species.canada.ca/species-risk-registry/virtual_sara/files/species/CanadianWildlifeSpeciesAtRisk-2019.pdf. Accessed 25 February 2020.

Confer, J. L., and M. V. Moore. 1987. Interpreting selectivity indices calculated from field data or conditions of prey replacement. Canadian Journal of Fisheries and Aquatic Sciences 44:1529–1533.

Conn, P. B., and W. L. Kendall. 2004. Evaluating mallard adaptive management models with time series. The Journal of Wildlife Management 68:1065–1081.

Connelly, J. W., J. H. Gammonley, and J. M. Peek. 2005. Harvest management. Pages 658–690 *in* C. E. Braun, editor. Techniques for wildlife investigations and management. The Wildlife Society, Bethesda, Maryland, USA.

Connelly, J. W., K. P. Reese, E. O. Garton, and M. L. Commons-Kemner. 2003. Response of greater sage-grouse *Centrocercus urophasianus* populations to different levels of exploitation in Idaho, USA. Wildlife Biology 9:335–340.

Connolly, G. 1988. M-44 cyanide ejectors in the animal damage control program, 1976-1986. Proceedings of the Vertebrate Pest Conference 13:220–225.

Connolly, G. 2004. Development and use of compound 1080 in coyote control 1944-1972. Proceedings of the Vertebrate Pest Conference 21:221–239.

Conover, D. O., and S. B. Munch. 2002. Sustaining fisheries yields over evolutionary time scales. Science 297:94–96.

Conover, D. O., S. B. Munch, and S. A. Arnott. 2009. Reversal of evolutionary downsizing caused by selective harvest of large fish. Proceedings of the Royal Society B: Biological Sciences 276:2015–2020.

Conover, M. R., J. B. Dinkins, and R. E. Ruzicka. 2015. Consequences of hunter harvest, winter weather, and increasing population size on survival of non-migratory Canada geese in Connecticut. The Journal of Wildlife Management 79:1239–1245.

Conroy, M. J. 2021. Some perspectives on the development of a paradigm for modern harvest management. *in* K. L. Pope and L. A. Powell, editors. Harvest of fish and wildlife: new paradigms for sustainable management. CRC Press, Boca Raton, Florida, USA.

Conroy, M. J., G. G. Barnes, R. W. Bethke, and T. D. Nudds. 1989. Increasing mallards, decreasing American black ducks-no evidence for cause and effect: a comment. The Journal of Wildlife Management 53:1065–1071.

Conroy, M. J., and W. W. Blandin. 1984. Geographic and temporal differences in band reporting rates for American black ducks. The Journal of Wildlife Management 48:23–36.

Conroy, M. J., and R. T. Eberhardt. 1983. Variation in survival and recovery rates of ring-necked ducks. The Journal of Wildlife Management 47:127–137.

Conroy, M. J., C. J. Fonnesbeck, and N. L. Zimpfer. 2005. Modeling regional waterfowl harvest rates using Markov chain Monte Carlo. The Journal of Wildlife Management 69:77–90.

Conroy, M. J., M. W. Miller, and J. E. Hines. 2002. Identification and synthetic modeling of factors affecting American black duck populations. Wildlife Monographs 150:1–67.

Conroy, M. J., and J. T. Peterson. 2013. Decision making in natural research management: a structured, adaptive approach. John Wiley & Sons, Chichester, West Sussex, United Kingdom.

Conroy, M. J., M. C. Runge, J. D. Nichols, K. W. Stodola, and R. J. Cooper. 2011. Conservation in the face of climate change: the roles of alternative models, monitoring, and adaptation in confronting and reducing uncertainty. Biological Conservation 144:1204–1213.

Convention on International Trade in Endangered Species of Wild Fauna and Flora. 2020. What is CITES? https://www.cites.org/eng/disc/what.php. Accessed 27 January 2020.

Converse, S. J., C. T. Moore, M. J. Folk, and M. C. Runge. 2013. A matter of tradeoffs: reintroduction as a multiple objective decision. The Journal of Wildlife Management 77:1145–1156.

Converse, S. J., K. J. Shelley, S. Morey, J. Chan, A. LaTier, C. Scafidi, D. T. Crouse, and M. C. Runge. 2011. A decision-analytic approach to the optimal allocation of resources for endangered species consultation. Biological Conservation 144:319–329.

Cooch, E., E. Cam, and W. Link. 2002. Occam's shadow: levels of analysis in evolutionary ecology–where to next? Journal of Applied Statistics 29:19–48.

Cooch, E. G., M. Guillemain, G. S. Boomer, J. D. Lebreton, and J. D. Nichols. 2014. The effect of harvest on waterfowl populations. Wildfowl 4:220–276.

Cook, J. G. 2002. Nutrition and food. Pages 259–349 *in* D. E. Toweill and J. W. Thomas, editors. North American elk: ecology and management, 2nd edition. Smithsonian Institute, Washington, D.C., USA.

Cook, M. A., M. N. Peterson, M. Colter Chitwood, D. Palmer, C. S. DePerno, and K. Gross. 2015. Evaluating deer hunters' support for hunting deer with dogs. Human Dimensions of Wildlife 20:174–181.

Cook, M. F., T. J. Goeman, P. J. Radomski, J. A. Younk, and P. C. Jacobson. 2001. Creel limits in Minnesota: a proposal for change. Fisheries 26(5):19–26.

Cooke, S. J., B. L. Barthel, C. D. Suski, M. J. Siepker, and D. P. Philipp. 2005. Influence of circle hook size on hooking efficiency, injury, and size selectivity of bluegill with comments on circle hook conservation benefits in recreational fisheries. North American Journal of Fisheries Management 25:211–219.

Cooke, S. J., and I. G. Cowx. 2004. The role of recreational fishing in global fish crises. BioScience 54:857–859.

Cooke, S. J., R. J. Lennox, S. D. Bower, A. Z. Horodysky, M. K. Treml, E. Stoddard, L. A. Donaldson, and A. J. Danylchuk. 2017. Fishing in the dark: the science and management of recreational fisheries at night. Bulletin of Marine Science 93:519–538.

Cooke, S. J., R. J. Lennox, B. Cantrell, and A. J. Danylchuk. 2020. Micro-fishing as an emerging form of recreational angling: research gaps and policy considerations. Fisheries 45:517–521.

Cooke, S. J., D. P. Philipp, K. M. Dunmall, and J. F. Schreer. 2001. The influence of terminal tackle on injury, handling time, and cardiac disturbance of rock bass. North American Journal of Fisheries Management 21:333–342.

Cooke, S. J., D. P. Philipp, J. F. Schreer, and R. S. McKinley. 2000. Locomotory impairment of nesting male largemouth bass following catch-and-release angling. North American Journal of Fisheries Management 20:968–977.

Cooke, S. J., and H. L. Schramm. 2007. Catch-and-release science and its application to conservation and management of recreational fisheries. Fisheries Management and Ecology 14:73–79.

Cooke S. J., J. F. Schreer, D. H. Wahl and D. P. Philipp. 2002. Physiological impacts of catch-and-release angling practices on largemouth and smallmouth bass. Pages 489–512 *in* D. P. Philipp and M. S. Ridgeway, editors. Black bass: ecology, conservation, and management. American Fisheries Society, Symposium 31, Bethesda, Maryland, USA.

Cooke, S. J., and C. D. Suski. 2005. Do we need species-specific guidelines for catch-and-release recreational angling to effectively conserve diverse fishery resources? Biodiversity and Conservation 14:1195–1209.

Cooley, H. S., R. B. Wielgus, G. M. Koehler, H. S. Robinson, and B. T. Maletzke. 2009. Does hunting regulate cougar populations? A test of the compensatory mortality hypothesis. Ecology 90:2913–2921.

Cooke, S. J., and G. R. Wilde. 2007. The fate of fish released by recreational anglers. Pages 181–234 *in* S. J. Kennelly, editor. By-catch reduction in the world's fisheries. Springer publishing, Dordrecht, The Netherlands.

Cooper, E. L. 1970. Management of trout streams. Pages 153–162 *in* N. G. Benson, editor. A century of fisheries in North America. American Fisheries Society, Special Publication Number 7. Washington, D.C., USA.

Cooper, J. F. 1823. The pioneers, volume 2 .Wiley, New York, USA.

Cooper, J. F. 1827. The prairie. Carey, Lee, and Carey.

Cormack, R. M. 1964. Estimates of survival from the sighting of marked animals. Biometrika 51:429–438.

Cornelius, R. R., and T. L. Margenau. 1999. Effects of length limits on muskellunge in Bone Lake, Wisconsin. North American Journal of Fisheries Management 19:300–308.

Cornicelli, L., D. C. Fulton, M. D. Grund, and J. Fieberg. 2011. Hunter perceptions and acceptance of alternative deer management regulations. Wildlife Society Bulletin 35:323–329.

Council to Advance Hunting and the Shooting Sports. 2016. National hunting and shooting sports action plan: strategies for recruiting, retaining, and reactivating hunting and shooting sports participants. Association of Fish and Wildlife Agencies, and U.S. Fish and Wildlife Service, Washington, D.C., USA.

Cox, R. R., Jr., and A. D. Afton. 1997. Use of habitats by female northern pintails wintering in southwestern Louisiana. The Journal of Wildlife Management 61:435–443.

Cox, R. R., Jr., and A. D. Afton. 2000. Predictable interregional movements by female northern pintails during winter. Waterbirds 23:258–269.

Craig, G. R., G. C. White, and J. H. Enderson. 2004. Survival, recruitment, and rate of population change of the peregrine falcon population in Colorado. The Journal of Wildlife Management 68:1032–1038.

Crawford, D. W., E. L. Jackson, and G. Godbey. 1991. A hierarchical model of leisure constraints. Leisure Sciences 13:309–320.

Crissey, W. F. 1971. Some thoughts on wildlife research and management objectives. Wildlife Society News 134:27–28.

Crosby, A. D., R. D. Elmore, and D. M. Leslie. 2013. Northern bobwhite response to habitat restoration in eastern Oklahoma. Wildlife Society Bulletin 37:733–740.

Crosmary, W.-G., A. J. Loveridge, H. Ndaimani, S. Lebel, V. Booth, S. D. Côté, and H. Fritz. 2013. Trophy hunting in Africa: long-term trends in antelope horn size. Animal Conservation 16:648–660.

Crow, M. S., T. O. Shelley, and P. B. Stretesky. 2013. Camouflage-collar crime: an examination of wildlife crime and characteristics of offenders in Florida. Deviant Behavior 34:635–652.

Cruz, J., C. Thomson, J. P. Parkes, I. Gruner, D. M. Forsyth. 2017. Long-term impacts of an introduced ungulate in native grasslands: Himalayan tahr (*Hemitragus jemlahicus*) in New Zealand's Southern Alps. Biological Invasions 19:339–349.

Cummings, J. W., and C. Bernier. 2021. Structured decision making provides insight when selecting population-monitoring programs. *in* K. L. Pope and L. A. Powell, editors. Harvest of fish and wildlife: new paradigms for sustainable management. CRC Press, Boca Raton, Florida, USA.

Cummings, J. W., M. J. Hague, D. A. Patterson, and R. M. Peterman. 2011. The impact of different performance measures on model selection for Fraser River sockeye salmon. North American Journal of Fisheries Management 31:323–334.

Curtis, P. D., and J. R. Hauber. 1997. Public involvement in deer management decisions: consensus versus consent. Wildlife Society Bulletin 25:399–403.

Dabrowska, K., W. Haider, and L. Hunt. 2014. Examining the impact of fisheries resources and quality on licence sales. Journal of Outdoor Recreation and Tourism 5:58–67.

Dahlgren, D. K., E. J. Blomberg, C. A. Hagen, and R. D. Elmore. 2021. Upland game bird harvest management. *in* K. L. Pope and L. A. Powell, editors. Harvest of fish and wildlife: new paradigms for sustainable management. CRC Press, Boca Raton, Florida, USA.

Daigle, J. J., R. M. Muth, R. R. Zwick, and R. J. Glass. 1998. Sociocultural dimensions of trapping: a factor analytic study of trappers in six northeastern states. Wildlife Society Bulletin 26:624–625.

Dale, F. H. 1951. The refuge in pheasant management. The Journal of Wildlife Management 15:337–346.

Danylchuk, S. E., A. J. Danylchuk, S. J. Cooke, T. L. Goldberg, J. Koppelman, and D. P. Philipp. 2007. Effects of recreational angling on the post-release behavior and predation of bonefish (*Albula vulpes*): the role of equilibrium status at the time of release. Journal of Experimental Marine Biology and Ecology 346:127–133.

Dappen, P., J. Merchant, I. Ratcliffe, and C. Robbins. 2007. Delineation of 2005 land use patterns for the State of Nebraska Department of Natural Resources. https://calmit.unl.edu/pdf/2005_Landuse_FinalReport.pdf Accessed 1 September 2020.

Darimont, C. T., S. M. Carlson, M. T. Kinnison, P. C. Paquet, T. E. Reimchen, and C. C. Wilmers. 2009. Human predators outpace other agents of trait change in the wild. Proceedings of the National Academy of Sciences 106:952–954.

Darimont, C. T., and K. R. Child. 2014. What enables size-selective trophy hunting of wildlife? PLoS One 9:e103487.

Darwin, C. 1859. On the Origin of Species. John Murray, London, England, United Kingdom.

Dasmann, R. F. 1964. Wildlife biology. John Wiley and Sons, New York, New York, USA.

Davies, A. L., R. Bryce, and S. M. Redpath. 2013. Use of multicriteria decision analysis to address conservation conflicts: multicriteria decision analysis. Conservation Biology 27:936–944.

Davis, A. J., B. Leland, M. Bodenchuk, K. C. VerCauteren, and K. M. Pepin. 2018. Costs and effectiveness of damage management of an overabundant species (*Sus scrofa*) using aerial gunning. Wildlife Research 45:696–705.

Davis, M. L., J. Berkson, D. Steffen, and M. K. Tilton. 2007. Evaluation of accuracy and precision of downing population reconstruction. The Journal of Wildlife Management 71:2297.

Dawe, K. L., E. M. Bayne, and S. Boutin. 2014. Influence of climate and human land use on the distribution of white-tailed deer (*Odocoileus virginianus*) in the western boreal forest. Canadian Journal of Zoology 92:353–363.

De Marchi, B. 2003. Public participation and risk governance. Science and Public Policy 30:171–176.

De Roos, A. M., D. S. Boukal, and L. Persson. 2006. Evolutionary regime shifts in age and size at maturation of exploited fish stocks. Proceedings of the Royal Society B: Biological Sciences 273:1873–1880.

De Roos, A. M., and L. Persson. 2002. Size-dependent life-history traits promote catastrophic collapses of top predators. Proceedings of the National Academy of Sciences 99:12907–12912.

De Roos, A. M., T. Schellekens, T. van Kooten, K. van de WolfShaar, D. Claessen, and L. Persson. 2007. Food-dependent growth leads to overcompensation in stage-specific biomass when mortality increases: the influence of maturation versus reproduction regulation. The American Naturalist 170:E59—E76.

Deblinger, R. D., W. A. Woytek, and R. R. Zwick. 1998. Demographics of voting on the 1996 Massachusetts ballot referendum. Human Dimensions of Wildlife 4:40–55.

deCalesta, D. S. 2017. Achieving and maintaining sustainable white-tailed deer density with adaptive management. Human-Wildlife Interactions 11:99–111.

Decker, D., C. Smith, A. Forstchen, D. Hare, E. Pomeranz, C. Doyle-Capitman, K. Schuler, and J. Organ. 2016. Governance principles for wildlife conservation in the 21st century. Conservation Letters 9:290–295.

Decker, D. J., T. L. Brown, and R. Gutierrez. 1980. Further insights into the multiple-satisfactions approach for hunter management. Wildlife Society Bulletin 8:323–331.

Decker, D. J., T. L. Brown, and W. F. Siemer. 2001. Human dimensions of wildlife management in North America. The Wildlife Society, Bethesda, Maryland, USA.

Decker, D. J., and N. A. Connelly. 1989. Motivations for deer hunting: implications for antlerless deer harvest as a management tool. Wildlife Society Bulletin 17:455–463.

Decker, D. J., and N. A. Connelly. 1990. The need for hunter education in deer management: insights from New York. Wildlife Society Bulletin 18:447–452.

Decker, D. J., C. C. Krueger, R. A. Baer, Jr., B. A. Knuth, and M. E. Richmond. 1996. From clients to stakeholders: a philosophical shift for fish and wildlife management. Human Dimensions of Wildlife 1:70–82.

Decker, D. J., S. J. Riley, and W. F. Siemer. 2012. Human dimensions of wildlife management. Pages 1–14 *in* D. J. Decker, S. J. Riley, and W. F. Siemer, editors. Human dimensions of wildlife management, 2nd edition. Johns Hopkins University Press, Baltimore, Maryland, USA.

Decker, D. J., R. A. Smolka, Jr., J. O'Pezio, and T. L. Brown. 1985. Social determinants of black bear management for the Northern Catskill Mountains. Pages 239–247 *in* S. L. Beasom and S. F. Roberson, editors. Game harvest management. Caesar Kleberg Wildlife Research Institute, Kingsville, Texas, USA.

Decker, T. A., and G. R. Batcheller. 1993. Furbearer management in transition: challenges for the future. Northeast Wildlife 50:153–157.

Degnbol, P., H. Gislason, S. Hanna, S. Jentoft, J. Raakjær Nielsen, S. Sverdrup-Jensen, and D. Clyde Wilson. 2006. Painting the floor with a hammer: technical fixes in fisheries management. Marine Policy 30:534–543.

Deines, A., E. Peterson, D. Boeckner, J. Boyle, A. Keighley, J. Kogut, J. Lubben, R. Rebarber, R. Ryan, B. Tenhumberg, S. Townley, and A. J. Tyre. 2007. Robust population management under uncertainty for structured population models. Ecological Applications 17:2175–2183.

DelGiudice, G. D., M. R. Riggs, P. Joly, and W. Pan. 2002. Winter severity, survival, and cause-specific mortality of female white-tailed deer in north-central Minnesota. The Journal of Wildlife Management 66:698–717.

DeLong, J. P., and T. M. Luhring. 2018. Size-dependent predation and correlated life history traits alter eco-evolutionary dynamics and selection for faster individual growth. Population Ecology 60:9–20.

Demarais, S., K. V. Miller, and H. A. Jacobson. 2000. White-tailed deer. Pages 601–628 *in* S. Demarais and P. R. Krausman, editors. Ecology and management of large mammals in North America. Prentice Hall, Upper Saddle River, New Jersey, USA.

Demarais, S., and B. K. Strickland. 2011. Antlers. Pages 107–145 *in* D. G. Hewitt, editor. Biology and management of white-tailed deer. CRC Press, Boca Raton, Florida, USA.

Demarais, S. and B. K. Strickland. 2017. Strategic Harvest System: How to Break Through the Buck Management Glass Ceiling. Independently Published.

Demarais, S., B. K. Strickland, and L. E. Castle. 2005. Antler regulation effects on white-tailed deer on Mississippi public hunting areas. Proceedings of the Southeastern Association of Fish and Wildlife Agencies 59:1–9.

Deng, W.-J., Y.-F. Kuo, and W.-C. Chen. 2008. Revised importance-performance analysis: three-factor theory and benchmarking. The Service Industries Journal 28:37–51.

DeNicola, A. J., D. R. Etter, and T. Almendinger. 2008. Demographics of non-hunted white-tailed deer populations in suburban areas. Human-Wildlife Conflicts 2:102–109.

Department of Interior, U.S. Fish and Wildlife Service. 2019. Migratory bird hunting; proposed 2020-21 migratory game bird hunting regulations (preliminary) with requests for Indian tribal proposals. Government Publishing Office, Code of Federal Regulations 84(199):55120–55129.

Deroba, J. J., and J. R. Bence. 2008. A review of harvest policies: understanding relative performance of control rules. Fisheries Research 94:210–223.

Detar, J., D. Kristine, T. Wagner, and T. Greene. 2014. Evaluation of catch-and-release regulations on brook trout in Pennsylvania streams. North American Journal of Fisheries Management 34:49–56.

Devers, P. K., A. J. Roberts, S. Knoche, P. I. Padding, and R. Raftovich. 2017. Incorporating human dimensions objectives into waterfowl habitat planning and delivery. Wildlife Society Bulletin 41:405–415.

Devers, P. K., D. F. Stauffer, G. W. Norman, D. E. Steffen, D. M. Whitaker, J. D. Sole, T. J. Allen, S. L. Bittner, D. A. Buehler, J. W. Edwards, D. E. Figert S. T. Friedhoff, W. W. Guiliano, C. A. Harper, W. K. Ido, R. L. Kirkpatrick, M. H. Seamster, H. A. Spiker, Jr., D. A. Swanson, and B. C. Tefft. 2007. Ruffed grouse population ecology in the Appalachian Region. Wildlife Monographs 168:1–36.

DeVos, J. C., Jr., D. L. Shroufe, and V. C. Supplee. 1998. Managing wildlife by ballot initiative: the Arizona experience. Human Dimensions of Wildlife 3:60–66.

Di Minin, E., N. Leader-Williams, and C. J. A. Bradshaw. 2016. Banning trophy hunting will exacerbate biodiversity loss. Trends in Ecology & Evolution 39:99–102.

Dickman, A., R. Cooney, P. J. Johnson, M. P. Louis, and D. Roe. 2019. Trophy hunting bans imperil biodiversity. Science 365:874.

Diefenbach, D. R., G. L. Alt, B. D. Wallingford, C. S. Rosenberry, and E. S. Long. 2019. Effect of male age structure on reproduction in white-tailed deer. The Journal of Wildlife Management 83:1368–1376.

Diefenbach, D. R., J. C. Finley, A. Luloff, R. C. Stedman, C. B. Swope, H. C. Zinn, and G. J. San Julian. 2005. Bear and deer hunter density and distribution on public land in Pennsylvania. Human Dimensions of Wildlife 10:201–212.

Diefenbach, D. R., W. M. Knox, and C. S. Rosenberry. 2021. The future of managing ungulate species: white-tailed deer as a case study. *in* K. L. Pope and L. A. Powell, editors. Harvest of fish and wildlife: new paradigms for sustainable management. CRC Press, Boca Raton, Florida, USA.

Diefenbach, D. R., and W. L. Palmer. 1997. Deer management: marketing the science. Wildlife Society Bulletin 25:378–381.

Diefenbach, D. R., W. L. Palmer, and W. K. Shope. 1997. Attitudes of Pennsylvania sportsmen towards managing white-tailed deer to protect the ecological integrity of Pennsylvania forests. Wildlife Society Bulletin 25:244–251.

Diefenbach, D. R., and S. M. Shea. 2011. Managing white-tailed deer: eastern North America. Pages 481–500 *in* D. G. Hewitt, editor. Biology and management of white-tailed deer. CRC Press, Boca Raton, Florida, USA.

Diekert, F. K. 2012. Growth overfishing: the race to fish extends to the dimension of size. Environmental and Resource Economics 52:549–572.

Diekert, F. K., A. Richter, I. M. Rivrud, and A. Mysterud. 2016. How constraints affect the hunter's decision to shoot a deer. Proceedings of the National Academy of Sciences 113:14450–14455.

Dietz, T., and P. C. Stern, editors. 2008. Public participation in environmental assessment and decision making. National Academic Press, Washington, D.C., USA.

Dillon, J. C., D. J. Schill, and D. M. Teuscher. 2000. Relative return to creel of triploid and diploid rainbow trout stocked in eighteen Idaho streams. North American Journal of Fisheries Management 20:1–9.

Dinges, A. J. 2013. Participation in the light goose conservation order and effects on behavior and distribution of waterfowl in the Rainwater Basin of Nebraska. Thesis, University of Missouri, Columbia, Missouri, USA.

Dinges, A. J., E. B. Webb, and M. P. Vrtiska. 2015. Effects of the Light Goose Conservation Order on non-target waterfowl distribution during spring migration. Wildlife Biology 21:88–97.

Dinges, A. J., E. B. Webb, M. P. Vrtiska, C. H. Nilon, and S. A. Wilhelm-Stanis. 2014. Migratory bird hunter opinions regarding potential management strategies for controlling light goose populations. Wildlife Society Bulletin 38:728–733.

Dizard, J. 1999. Going wild: hunting, animal rights, and the contested meaning of nature. University of Massachusetts Press, Amherst, Massachusetts, USA.

Doerr, M. L., J. B. McAninch, and E. P. Wiggers. 2001. Comparison of 4 methods to reduce white-tailed deer abundance in an urban community. Wildlife Society Bulletin 29:1105–1113.

Dolin, E. J. 2008. Leviathan: the history of whaling in America. W.W. Norton & Company, New York, New York, USA.

Dolton, D. D. 1993. The call-count survey: historic development and current procedures. Pages 233–252 *in* T. S. Baskett, M. W. Sayre, R. E. Tomlinson, and R. E. Mirarchi, editors. Ecology and management of the mourning dove. Stackpole Books, Harrisburg, Pennsylvania, USA.

Dorendorf, R. R., P. J. Fix, and L. R. Prugh. 2016. Motivations of fur trappers in interior Alaska. Human Dimensions of Wildlife 21:522–537.

Dorr, B., I. A. Munn, and K. O. Meals. 2002. A socioeconomic and biological evaluation of current and hypothetical crappie regulations in Sardis Lake, Mississippi: an integrated approach. North American Journal of Fisheries Management 22:1376–1384.

Doss, S. S., B. R. Murphy, L. Castello, J. A. Williams, J. Copeland, and Vi. J. DiCenzo. 2019. Field evaluation and simulation modeling of length limits and their effects on fishery quality for muskellunge in the New River, Virginia. North American Journal of Fisheries Management 39:3–16.

Dotson, J. R., M. S. Allen, J. A. Kerns, and W. F. Pouder. 2013. Utility of restrictive harvest regulations for trophy largemouth bass management. North American Journal of Fisheries Management 33:499–507.

Douhard, M., M. Festa-Bianchet, F. Pelletier, J. M. Gaillard, and C. Bonenfant. 2016. Changes in horn size of Stone's sheep over four decades correlate with trophy hunting pressure. Ecological Applications 26:309–321.

Downing, R. 1980. Vital statics of animal populations. Wildlife management technique manual. The Wildlife Society, Bethesda, Maryland, USA.

Drewien, R. C., J. T. Herbert, T. W. Aldrich, and S. H. Bouffard. 1999. Detecting trumpeter swans harvested in tundra swan hunts. Wildlife Society Bulletin 27:95–102.

Driscoll, M. T., J. L. Smith, and R. A. Myers. 2007. Impact of tournaments on the largemouth bass population at Sam Rayburn Reservoir, Texas. North American Journal of Fisheries Management 27:425–433.

Driver, B. L. 1985. Specifying what is produced by management of wildlife by public agencies. Leisure Sciences 7:281–295.

Driver, B. L., and R. C. Knopf. 1976. Temporary escape: one product of sport fisheries management. Fisheries 1(2):21–29.

Driver, B. L., and R. C. Knopf. 1977. Personality, outdoor recreation, and expected consequences. Environment and Behavior 9:169–193.

Driver, B. L., C. Phillips, E. P. Bergersen, and C. C. Harris. 1984. Using angler preference data in defining types of sport fisheries to manage. Transactions of the North American Wildlife and Natural Resources Conference 49:82–90.

Dubovsky, J. A. 2019a. Status and harvests of Sandhill cranes: mid-Continent, Rocky Mountain, Lower Colorado River Valley and eastern populations. U.S. Fish and Wildlife Service, Administrative Report, Lakewood, Colorado, USA.

Dubovsky, J. A., compiler. 2019b. Central Flyway harvest and population survey data book. U.S. Fish and Wildlife Service, Denver, Colorado, USA.

Duda, M. D., S. J. Bissel, and K. C. Young. 1996. Factors related to hunting and fishing participation in the United States. Transactions of North American Wildlife and Natural Resources Conference 61:324–337.

Dufour, K. W., C. D. Ankney, and P. J. Weatherhead. 1993. Condition and vulnerability to hunting among mallards tagging at Lake St. Clair, Ontario. The Journal of Wildlife Management 57:209–215.

Dugger, K. M., B. D. Dugger, and L. H. Fredrickson. 1999. Annual survival rates of female hooded mergansers and wood ducks in southeastern Missouri. Wilson Bulletin 111:1–6.

Duncan, M. I., A. E. Bates, N. C. James, and W. M. Potts. 2019. Exploitation may influence the climate resilience of fish populations through removing high performance metabolic phenotypes. Scientific Reports 9:11437.

Dunfee, M., A. Forstchen, E. Haubold, M. Humpert, J. Newmark, J. Sumners, and C. Smith., editors. 2019. Fish and wildlife relevancy roadmap (v1.0): enhanced conservation through broader engagement. Association of Fish and Wildlife Agencies, and The Wildlife Management Institute, Washington, D.C., USA.

Dunlap, T. R. 1988. Sport hunting and conservation, 1880-1920. Environmental Review 12:51–60.

Dunlop, E. S., A. M. Eikeset, and N. C. Stenseth. 2015. From genes to populations: how fisheries-induced evolution alters stock productivity. Ecological Applications 25:1860–1868.

Dunlop, E. S., Z. S. Feiner, and T. O. Höök. 2018. Potential for fisheries-induced evolution in the Laurentian Great Lakes. Journal of Great Lakes Research 44:735—747.

Dziurzynski, L., E. Wadsworth, and D. McCarthy. 2014. BTYD: implementing buy 'til you die models. https://CRAN.R-project.org/package=BTYD. Accessed 6 November 2020.

Eades, R. T., and T. J. Lang. 2012. Community fishing ponds. Pages 351–372 *in* J. W. Neal and D. W. Willis, editors. Small impoundment management in North America. American Fisheries Society, Bethesda, Maryland, USA.

Ebeling-Schuld, A. M., and C. T. Darimont. 2017. Online hunting forums identify achievement as prominent among multiple satisfactions. Wildlife Society Bulletin 41:523–529.

Eberhardt, L. L. 1982. Calibrating an index by using removal data. The Journal of Wildlife Management 46:734.

Edeline, E., S. M. Carlson, L. C. Stige, I. J. Winfield, J. M. Fletcher, J. B. James, T. O. Haugen, L. A. Vøllestad, and N. C. Stenseth. 2007. Trait changes in a harvested population are driven by a dynamic tug-of-war between natural and harvest selection. Proceedings of the National Academy of Sciences 104:15799–15804.

Edeline, E., A. Le Rouzic, I. J. Winfield, J. M. Fletcher, J. B. James, N. Chr. Stenseth, and L. A. Vøllestad. 2009. Harvest-induced disruptive selection increases variance in fitness-related traits. Proceedings of the Royal Society B: Biological Sciences 276:4163–4171.

Eder, S. 1984. Effectiveness of an imposed slot length limit of 12.0-14.9 inches on largemouth bass. North American Journal of Fisheries Management 4:469–478.

Edwards, G. P., Jr., R. M. Neumann, R. P. Jacobs, and E. B. O'Donnell. 2004. Factors related to mortality of black bass caught during small club tournaments in Connecticut. North American Journal of Fisheries Management 24:801–810.

Edwards, W. 1971. Social utilities. Pages 119–129 *in* Decision and risk analysis: powerful new tools for management. Proceedings of the sixth triennial symposium, June 19–20, Engineering Economy

Division, America Society for Engineering Education, American Institute of Industrial Engineers, Annapolis, Maryland, USA.

Edwards, W., and F. Barron. 1994. SMARTS and SMARTER: improved simple methods for multiattribute utility measurement. Organizational Behavior and Human Decision Processes 60:306–325.

Ehrenberg, A. 1959. The pattern of consumer purchases. The Royal Statistical Society Series C-Applied Statistics 8:26–41.

Eikeset, A. M., E. S. Dunlop, M. Heino, G. Storvik, N. C. Stenseth, and U. Dieckmann. 2016. Roles of density-dependent growth and life history evolution in accounting for fisheries-induced trait changes. Proceedings of the National Academy of Sciences 113:15030–15035.

Eikeset, A. M., A. Richter, E. S. Dunlop, U. Dieckmann, and N. C. Stenseth. 2013. Economic repercussions of fisheries-induced evolution. Proceedings of the National Academy of Sciences 110:12259–12264.

Elden, R. C., W. V. Bevill, P. I. Padding, J. E. Frampton, and D. L. Shroufe. 2002. A history of the development of the harvest information program. Pages 7–16 *in* J. M. Ver Steeg and R. C. Elden, compilers. Harvest information program: evaluation and recommendations. International Association of Fish and Wildlife Agencies, Migratory Shore and Upland Game Bird Working Group, Ad Hoc Committee on HIP, Washington, D.C., USA.

Eliason, S. 2012. From the King's deer to a capitalist commodity: a social historical analysis of the poaching law. International Journal of Comparative and Applied Criminal Justice 36:133–148.

Eliason, S. L. 2020. Poaching, social conflict, and the public trust: some critical observations on wildlife crime. Capitalism Nature Socialism 31:110–126.

Eliason, S. L. 2021. A place to hunt: some observations on access to wildlife resources in the western United States. Human Dimensions of Wildlife 26: in press.

Elliott, D. S. 1991. Hunter ethics: A look at hunter behavior and hunter education in the US and Canada. Izaak Walton League of America and the U.S. Fish and Wildlife Service, Arlington, VA.

Ellison, L. N. 1991. Shooting and compensatory mortality in tetraonids. Ornis Scandinavica 22:229–240.

Embke, H. S., A. L. Rypel, S. R. Carpenter, G. G. Sass, D. Ogle, T. Cichosz, J. Hennessy, T. E. Essington, and M. J. Vander Zanden. 2019. Production dynamics reveal hidden overharvest of inland recreational fisheries. Proceedings of the National Academy of Sciences 116:24676–24681.

Enberg, K., C. Jorgensen, E. S. Dunlop, M. Heino, and U. Dieckmann. 2009. Implications of fisheries-induced evolution for stock rebuilding and recovery. Evolutionary Applications 2:394–414.

Enberg, K., C. Jørgensen, E. S. Dunlop, O. Varpe, D. S. Boukal, L. Baulier, S. Eliassen, and M. Heino. 2012. Fishing-induced evolution of growth: concepts, mechanisms and the empirical evidence. Marine Ecology 33:1–25.

Enck, J. W., and T. L. Brown. 2008a. 2007 statewide deer hunter survey: participation during the '06 seasons, opinions about hot-button issues, and trends in characteristics of hunters. Cornell University, Human Dimensions Research Unit Series Number 08-5, Ithaca, New York, USA.

Enck, J. W., and T. L. Brown. 2008b. Deer hunters' assessment of antler restrictions in Wildlife Management Units (WMUs) 3C, 3H, and 3K during the 2007 hunting season. Cornell University, Human Dimensions Research Unit Series Number 08-7, Ithaca, New York, USA.

Enck, J. W., and D. J. Decker. 2011. Hunters' experiences with and attitudes about antler restrictions in Wildlife Management Units 3C, 3J, 3H, and 3K. Cornell University, Human Dimensions Research Unit Series Number 11-3, Ithaca, New York, USA.

Enck, J. W., D. J. Decker, and T. L. Brown. 2000. Status of hunter recruitment and retention. Wildlife Society Bulletin 28:817–824.

Enck, J. W., D. J. Decker, S. J. Riley, J. F. Organ, L. H. Carpenter, and W. F. Siemer. 2006. Integrating ecological and human dimensions in adaptive management of wildlife-related impacts. Wildlife Society Bulletin 34:698–705.

Enck, J. W., R. C. Stedman, and D. J. Decker. 2011. Final report: statewide deer hunter survey-2010. Cornell University, Human Dimensions Research Unit Series Number 11–1, Ithaca, New York, USA.

Enck, J. W., B. L. Swift, and D. J. Decker. 1993. Reasons for decline in duck hunting: insights from New York. Wildlife Society Bulletin 21:10–21.

Enck, J. W., and H. A. Van Den Berg. 2007. Developing a conceptual model of duck hunters' experiences to help task forces understand reasons underlying preferences for season dates. Cornell University, Human Dimensions Research Unit Series Number 07-13, Ithaca, New York, USA.

Engel, S., M. H. Hoff, and S. P. Newman. 1999. Evaluating a smallmouth bass slot length and daily bag limit on Nebish Lake, Wisconsin. Wisconsin Department of Natural Resources Research Report 181, Madison, Wisconsin, USA.

Engeman, R. M., B. U. Constantin, S. A. Shwiff, H. T. Smith, J. Woolard, J. Allen, and J. Dunlap. 2007. Adaptive and economic management methods for feral hog control in Florida. Human-Wildlife Conflicts 1:178–185.

Engeman, R. M., A. B. Shiels, and C. S. Clark. 2018. Objectives and integrated approaches for the control of brown tree snakes: an updated overview. Journal of Environmental Management 219:115–124.

Enticott, G. 2011. Techniques of neutralising wildlife crime in rural England and Wales. Journal of Rural Studies 27:200–208.

Environment and Climate Change Canada. 2018. National harvest survey. https://www.canada.ca/en/environment-climate-change/services/bird-surveys/waterfowl/national-harvest.html. Accessed 8 July 2020.

Environmental Reporting BC. 2012. Grizzly bear population status in B.C. State of Environment Reporting, Ministry of Environment, British Columbia, Canada. http://www.env.gov.bc.ca/soe/indicators/plants-and-animals/grizzly-bears.html. Accessed 25 February 2020.

Erhardt, J. M., and D. L. Scarnecchia. 2014. Population changes after 14 years of harvest closure on a migratory population of bull trout in Idaho. North American Journal of Fisheries Management 34:482–492.

Erickson, D. W. 1982. Estimating and using furbearer harvest information. Pages 53–65 *in* G. C. Sanderson, editor. Midwest furbearer management. Kansas Chapter of The Wildlife Society, Wichita, Kansas, USA.

Erickson, G. L., J. R. Heffelfinger, and J. H. Ellenberger. 2003. Potential effects of hunting and hunt structure on mule deer abundance and demographics. Pages 119–138 *in* J. C. deVos, Jr., M. R. Conover, and N. E. Headrick, editors. Mule deer conservation: issues and management strategies. Berryman Institute Press, Utah State University, Logan, Utah, USA.

Erisman, B. E., L. G. Allen, J. T. Claisse, D. J. Pondella, E. F. Miller, and J. H. Murray. 2011. The illusion of plenty: hyperstability masks collapses in two recreational fisheries that target fish spawning aggregations. Canadian Journal of Fisheries and Aquatic Sciences 68:1705–1716.

Errington, P. L. 1945. Some contributions of a fifteen-year local study of the northern bobwhite to a knowledge of population phenomena. Ecological Monographs 15:1–34.

Errington, P. L. 1946. Predation and vertebrate populations. Quarterly Review of Biology 21:144–177.

Errington, P. L. 1954. The special responsiveness of minks to epizootics in muskrat populations. Ecological Monographs 24:377–393.

Errington, P. L., and F. N. Hammerstrom, Jr. 1936. Bob-white winter survival on experimentally shot and unshot areas. Iowa State College Journal of Science 9:625–639.

Escobar, O. 2017. Pluralism and democratic participation: what kind of citizen are citizens invited to be? Contemporary Pragmatism 14:416–438.

Etter, D. R., T. R. Van Deelen, D. R. Ludwig, S. N. Kobal, and R. E. Warner. 2000. Management of white-tailed deer in Chicago, Illinois forest preserves. Proceedings of the Vertebrate Pest Conference, 19. http://dx.doi.org/10.5070/V419110290. Accessed 27 June 2020.

European Commission. 1991. Council regulation (EEC) no. 3254/91. http://eur-lex.europa.eu/legal-content/EN/TXT/PDF/?uri=CELEX:31991R3254&from=EN. Accessed 3 March 2020.

European Commission. 1998a. Agreement on international humane trapping standards between the European Community, Canada and the Russian Federation. http://fur.ca/wp-content/uploads/2015/09/AIHTS-Copy-of-Agreement.pdf. Accessed 4 March 2020.

European Commission. 1998b. Agreed minute and annex: standards for the humane trapping of specified terrestrial and semiaquatic mammals. http://ec.europa.eu/environment/biodiversity/animal_welfare/hts/pdf/l_21919980807en00260037.pdf. Accessed 4 March 2020.

Evans, T. 2019. The amazing rise—and shocking fall—of Indiana's caviar king. Indy Star, Indianapolis, Indiana, USA. March 21. https://www.indystar.com/story/news/investigations/2019/03/21/amazing-rise-and-shocking-fall-indianas-caviar-king/2859237002/. Accessed 29 March 2020.

Everett, M. W. 2014. Measuring flow in Michigan youth firearm deer hunters and implications for hunter recruitment. Human Dimensions of Wildlife 19:215–217.

Fader, P., B. Hardie, and K. Lee. 2005. "Counting your customers" the easy way: an alternative to the Pareto/NBD model. Marketing Science 24:275–284.

Fader, P. S., and B. G. S. Hardie. 2009. Probability models for customer-base analysis. Journal of Interactive Marketing 23:61–69.

Fader, P. S., B. G. S. Hardie, and J. Shang. 2010. Customer-base analysis in a discrete-time noncontractual setting. Marketing Science 29:1086–1108.

Failing, L., R. Gregory, and P. Higgins. 2013. Science, uncertainty, and values in ecological restoration: a case study in structured decision-making and adaptive management. Restoration Ecology 21:422–430.

Failing, L., G. Horn, and P. Higgins. 2004. Using expert judgment and stakeholder values to evaluate adaptive management options. Ecology and Society 9:13.

Fascione, N., A. Delach, and M. E. Smith, editors. 2004. People and predators: from conflict to coexistence. Island Press, Washington, D.C., USA.

Fauchald, P. 2010. Predator-prey reversal: a possible mechanism for ecosystem hysteresis in the North Sea? Ecology 91:2191–2197.

Fausch, K. D. 2008. A paradox of trout invasions in North America. Biological Invasion 10:685–701.

Fayram, A. H. 2003. A comparison of regulatory and voluntary release of muskellunge and walleyes in Northern Wisconsin. North American Journal of Fisheries Management 23:619–624.

Fayram, A. H., S. W. Hewett, S. J. Gilbert, S. D. Plaster, and T. D. Beard. 2001. Evaluation of a 15-inch minimum length limit for walleye angling in Northern Wisconsin. North American Journal of Fisheries Management 21:816–824.

Fayram, A. H., and P. J. Schmalz. 2006. Evaluation of a modified bag limit for walleyes in Wisconsin: effects of decreased angler effort and lake selection. North American Journal of Fisheries Management 26:606–611.

Fayram, A. H., and T. Treska. 2010. An evaluation of the efficacy of a 14-18 inch slot limit for walleyes in Northern Wisconsin. Wisconsin Department of Natural Resource, Research Report 190, Madison, Wisconsin, USA.

Fedele, A. D. 2017. Influences of catch-and-release angling on fish avoidance behavior. Thesis, University of Nebraska, Lincoln, Nebraska, USA.

Fedler, A. J., and R. B. Ditton. 1994. Understanding angler motivations in fisheries management. Fisheries 19:6–13.

Feiner, Z. F., S. C. Chong, C. T. Knight, T. E. Lauer, M. V. Thomas, J. T. Tyson, and T. O. Höök. 2015. Rapidly shifting maturation schedules following reduced commercial harvest in a freshwater fish. Evolutionary Applications 8:724–737.

Feiner, Z. S., A. W. Latzka, and M. H. Wolter. 2021. What to exploit when you're exploiting: angling rates and size selection responses to changing bag limits. *in* K. L. Pope and L. A. Powell, editors. Harvest of fish and wildlife: new paradigms for sustainable management. CRC Press, Boca Raton, Florida, USA.

Feiner, Z. S., A. W. Latzka, M. H. Wolter, L. D. Eslinger, and G. R. Hatzenbeler. 2020a. Assessing the rage against the machines: do ice anglers' electronics improve catch and harvest rates? Fisheries 45:327–333.

Feiner, Z. S., M. H. Wolter, and A. W. Latzka. 2020b. "I will look for you, I will find you, and I will [harvest] you": persistent hyperstability in Wisconsin's recreational fishery. Fisheries Research 230:105679.

Festa-Bianchet, M. 2003. Exploitative wildlife management as a selective pressure for the life-history evolution of large mammals. Pages 191–207 *in* M. Festa-Bianchet and M. Apollonio, editors. Animal behavior and wildlife conservation. Island Press, Washington, D.C., USA.

Festa-Bianchet, M. 2017. When does selective hunting select, how can we tell, and what should we do about it? Mammal Review 47:76–81.

Festa-Bianchet, M., and R. Arlinghaus. 2021. How harvest regulations can affect the evolutionary impacts of recreational harvests on fish and mammals. *in* K. L. Pope and L. A. Powell, editors. Harvest of fish and wildlife: new paradigms for sustainable management. CRC Press, Boca Raton, Florida, USA.

Festa-Bianchet, M., and S. D. Côté. 2008. Mountain goats: ecology, behavior and conservation of a mountain ungulate. Island Press, Washington, D.C., USA.

Festa-Bianchet, M., and A. Mysterud. 2018. Hunting and evolution: theory, evidence, and unknowns. Journal of Mammalogy 99:1281–1292.

Festa-Bianchet, M., F. Pelletier, J. T. Jorgenson, C. Feder, and A. Hubbs. 2014. Decrease in horn size and increase in age of trophy sheep in Alberta over 37 years. The Journal of Wildlife Management 78:133–141.

Fincel, M. J., C. M. Longhenry, and D. A. James. 2015. Effects of a protected slot limit on smallmouth bass size structure and angler harvest. Lake and Reservoir Management 31:180–189.

Fink, D., T. Auer, A. Johnston, V. Ruiz-Gutierrez, W. M. Hochachka, and S. Kelling. 2019. Modeling avian full annual cycle distribution and population trends with citizen science data. Ecological Applications 30(3):e02056.

Fisher, M. R. 1997. Segmentation of the angler population by catch preference, participation, and experience: a management-oriented application of recreation specialization. North American Journal of Fisheries Management 17:1–10.

Fleegle, T. A., C. S. Rosenberry, and B. D. Wallingford. 2013. Use of citizen advisory committees to direct deer management in Pennsylvania. Wildlife Society Bulletin 37:129–136.

Fleming, K. K., M. G. Brasher, D. D. Humburg, M. J. Petrie, and G. J. Soulliere. 2016. Derivation of non-breeding duck population abundance objectives to inform regional conservation planning. North American Waterfowl Management Plan Science Support Team Technical Report 2016-01, Washington, D.C., USA.

Floyd, M. F., and J. H. Gramann. 1997. Experience-based setting management: implications for market segmentation of hunters. Leisure Sciences 19:113–128.

Floyd, M. F., K. J. Shinew, F. A. McGuire, and F. P. Noe. 1994. Race, class, and leisure activity preferences: marginality and ethnicity revisited. Journal of Leisure Research 26:158–173.

Folke, C., S. Carpenter, B. Walker, M. Scheffer, T. Elmqvist, L. Gunderson, and C. S. Holling. 2004. Regime shifts, resilience, and biodiversity in ecosystem management. Annual Review of Ecology, Evolution, and Systematics 35:557–581.

Folke, C., T. Hahn, P. Olsson, and J. Norberg. 2005. Adaptive governance of social-ecological systems. Annual Review of Environment and Resources 30:441–473.

Follis, T. B. 1972. Reproduction and hematology of the Cache elk herd. Dissertation, Utah State University, Logan, Utah, USA.

Foote, C. J., G. S. Brown, and C. C. Wood. 1997. Spawning success of males using alternative mating tactics in sockeye salmon. Canadian Journal of Fisheries and Aquatic Sciences 54: 1785–1795.

Fournier, D., and C. P. Archibald. 1982. A general theory for analyzing catch at age data. Canadian Journal of Fisheries and Aquatic Sciences 39:1195–1207.

Fowler, D. N., E. B. Webb, and M. P. Vrtiska. 2020. Condition bias of decoy-harvested light geese during the conservation order. The Journal of Wildlife Management 84:33–44.

Fox, A. C. 1975. Effects of traditional harvest regulations on bass populations and fishing. Pages 392–407 *in* R. H. Stroud and H. Clepper, editors. Black bass biology and management. Sport Fishing Institute, Washington, D.C., USA.

Francis, C. M., J. R. Sauer, and J. R. Serie. 1998. Effect of restrictive harvest regulations on survival and recovery rates of American black ducks. The Journal of Wildlife Management 62:1544–1557.

Fraser, N. H., N. B. Metcalfe, and J. E. Thorpe. 1993. Temperature-dependent switch between diurnal and nocturnal foraging in salmon. Proceedings of the Royal Society of London. Series B: Biological Sciences 252:135–139.

Frederick, R. B., and D. T. Cobb. 1992. Computer simulation of furbearer population dynamics. Pages 911–921 *in* D. R. McCullough and R. H. Barrett, editors. Wildlife 2001: populations. Elsevier Science, Essex, United Kingdom.

Frederick, S. W., and R. M. Peterman. 1995. Choosing fisheries harvest policies: when does uncertainty matter? Canadian Journal of Fisheries and Aquatic Sciences 52:291–306.

Freyfogle, E. T. , D. D. Goble, and T. Wildermuth. 2019. Wildlife Law, 2nd edition. Island Press, Washington, DC, USA.

Friedlander, A., K. Poepoe, K., Poepoe, K. Helm, P. Bartram, J. Maragos, and I. Abbott. 2002. Application of Hawaiian traditions to community-based fishery management. Proceedings of the International Coral Reef Symposium 9:813–815.

Friedlander, A., K. Poepoe, K. Helm, P. Bartram, J. Maragos, and I. Abbott. 2002. Application of Hawaiian traditions to community-based fishery management. Proceedings of the Ninth International Coral Reef Symposium 2:813–815.

Fry, F. E. J. 1949. Statistics of a lake trout fishery. Biometrics 5:17–67.

Frye, B. 2006. Deer wars: science, tradition, and the battle over managing whitetails in Pennsylvania. The Pennsylvania State University Press, University Park, Pennsylvania, USA.

Fugère, V., and A. P. Hendry. 2018. Human influences on the strength of phenotypic selection. Proceedings of the National Academy of Sciences 115:10070–10075.

Fuller, A. K., D. J. Decker, M. V. Schiavone, and A. B. Forstchen. 2020. Ratcheting up rigor in wildlife management decision making. Wildlife Society Bulletin 44:29–41.

Fuller, A. K., J. C. Stiller, W. F. Siemer, and K. A. Perkins. 2021. Engaging hunters in selecting duck season dates using decision science: problem framing, objective setting, devising management alternatives. in K. L. Pope and L. A. Powell, editors. Harvest of fish and wildlife: new paradigms for sustainable management. CRC Press, Boca Raton, Florida, USA.

Fuller, M. 2016. Deer hunting in the United States: demographics and trends. Addendum to the 2011 national survey of fishing, hunting, and wildlife-associated recreation. Report 2011–10, U.S. Fish and Wildlife Service, Falls Church, Virginia, USA.

Fuller, R. J., and R. M. A. Gill. 2001. Ecological impacts of increasing numbers of deer in British woodland. Forestry 74:193–199.

Fulton, D. C., and K. Hundertmark. 2004. Peer-reviewed articles assessing the effects of a selective harvest system on moose hunters' behaviors, beliefs, and satisfaction. Human Dimensions of Wildlife 9:1–16.
Fulton, E. A., A. D. M. Smith, D. C. Smith, and I. E. van Putten. 2011. Human behaviour: the key source of uncertainty in fisheries management. Fish and Fisheries 12:2–17.
Funk, J. L. 1974. Symposium on overharvest and management of largemouth bass in small impoundments. American Fisheries Society, North Central Division, Special Publication Number 3, Bethesda, Maryland, USA.
Funnell, E. 2012. The smallmouth bass in Ontario. Biodiversity Branch, Ontario Ministry of Natural Resources. Peterborough, Ontario, Canada.
Fur Commission USA. 2020. New research looks beyond retail sales. https://furcommission.com/new-research-reveals-the-global-fur-trade-is-worth-as-much-as-wi-fi/. Accessed 3 March 2020.
Fur Information Council of America. 2020. FICA facts. https://www.fur.org/fica-facts/. Accessed 3 March 2020.
Fur Institute of Canada. 2014. Best trapping practices. https://fur.ca/wp-content/uploads/2015/10/Best-Trapping-Practices.pdf. Accessed 14 April 2020.
Fur Institute of Canada. 2019. AIHTS: humane trapping standards and animal welfare. https://fur.ca/fur-trapping/humane-trapping-standards-and-animal-welfare/. Accessed 25 February 2020, 4 March 2020.
Gabelhouse, D. 1985. Slot length limits: an overview. Kansas Wildlife 42(5):22–25.
Gabelhouse, D. W., Jr. 1984a. A length-categorization system to assess fish stocks. North American Journal of Fisheries Management 4:273–285.
Gabelhouse, D. W., Jr. 1984b. An assessment of largemouth bass slot length limits in five Kansas lakes. Kansas Fish and Game Commission, Federal Aid in Sportfish Restoration project FW-9-P-3, Emporia, Kansas, USA.
Gabelhouse, D. W., Jr., and D. W. Willis. 1986. Biases and utility of angler catch data for assessing size structure and density of largemouth bass. North American Journal of Fisheries Management 6:481–489.
Gaillard, J. M., M. Festa-Bianchet, D. Delorme, and J. Jorgenson. 2000a. Body mass and individual fitness in female ungulates: bigger is not always better. Proceedings of the Royal Society of London B 267:471–477.
Gaillard, J.-M., M. Festa-Bianchet, and N. G. Yoccoz. 1998. Population dynamics of large herbivores: variable recruitment with constant adult survival. Trends in Ecology & Evolution 13:58–63.
Gaillard, J.-M., M. Festa-Bianchet, N. G. Yoccoz, A. Loison, and C. Toïgo. 2000b. Temporal variation in fitness components and population dynamics of large herbivores. Annual Review of Ecology and Systematics 31:367–393.
Galea, S., and M. Tracy. 2007. Participation rates in epidemiologic studies. Annals of Epidemiology 17:643–653.
Gamborg, C., and F. S. Jensen. 2016. Wildlife value orientations among hunters, landowners, and the general public: a Danish comparative quantitative study. Human Dimensions of Wildlife 21:328–344.
Gamelon, M., A. Besnard, J. M. Gaillard, S. Servanty, E. Baubet, S. Brandt, and O. Gimenez. 2011. High hunting pressure selects for earlier birth date: wild boar as a case study. Evolution 65:3100–3112.
Gammonley, J. H., M. P. Vrtiska, T. L. Shaffer, D. J. Witter, J. L. Hansen, M. A. Johnson, K. Kraai, F. NcNew, T. Mitchusson, D. Morrison, J. L. Richardson, L. Roberts, M. Szymanski, and S. Vaa. 2010. Evaluation of species-specific duck hunting regulations in the Central Flyway. Central Flyway Council report. Denver, Colorado, USA.
Gangl, R. S. 2021. Harvest regulations for inland recreational fisheries. *in* K. L. Pope and L. A. Powell, editors. Harvest of fish and wildlife: new paradigms for sustainable management. CRC Press, Boca Raton, Florida, USA.
Gardner, C. L. 2010. Reducing non-target moose capture in wolf snares. Alces 46:167–182.
Garthause, M. W., and R. C. Heidinger. 1999. Response of largemouth bass and bluegill to length limits in Lake of Egypt. Transactions of the Illinois State Academy of Science 92:79–87.
Gates, C., and K. Aune. 2008. *Bison bison*. IUCN Red List of Threatened Species. International Union for Conservation of Nature, Gland, Switzerland.
Gauthier, G., J.-F. Giroux, A. Reed, A. Be´chet, and L. Be´langer. 2005. Interactions between land use, habitat use, and population increase in greater snow geese: what are the consequences for natural wetlands? Global Change Biology 11:856–868.
Gaynor, A., J. Frawley, and K. S. Máñez. 2016. 'Slim female records the same old story': Newspapers, gender, and recreational fishing in Australia, 1957–2000. Geoforum 77:114–123.
Geis, A. D. 1972. Role of banding data in migratory bird population studies. Pages 213–228 *in* Population ecology of migratory birds. Department of Interior, Bureau of Sport Fisheries and Wildlife, Wildlife Research Report 2, Washington, D.C., USA.

Geis, A. D., and W. F. Crissey. 1969. Effect of restrictive hunting regulations on canvasback and redhead harvest rates and survival. The Journal of Wildlife Management 33:860–866.

Geis, A. D., and R. D. Taber. 1963. Measuring hunting and other mortality. Pages 284–298 *in* H. S. Mosby, editor. Wildlife investigational techniques. The Wildlife Society, Washington, D.C., USA.

Geist, V. 1991. The big bull in the breeding biology of elk. Pages 6–9 *in* A. G. Christensen, L. J. Lyon, and T. N. Lonner, compilers. Proceedings of a symposium on elk vulnerability. Montana State University, Bozeman, Montana, USA.

Geist, V., S. P. Mahoney, and J. F. Organ. 2001. Why hunting has defined the North American model of wildlife conservation. Transactions of the North American Wildlife and Natural Resources Conference 66:175–185.

Geist, V., and J. F. Organ. 2004. The public trust foundation of the North American model of wildlife conservation. Northeast Wildlife 58:49–56.

Gelman, A., and D. B. 1996. Markov chain Monte Carlo methods in biostatistics. Statistical Methods in Medical Research 5:339–355.

Gendron, M. H., and A. C. Smith. 2019. National harvest survey web site. Canadian Wildlife Service, National Wildlife Research Centre, Bird Populations Monitoring, Ottawa, Ontario, Canada. http://www.ec.gc.ca/reom-mbs/enp-nhs/index.cfm?do=def&lang=e. Accessed 25 June 2020.

George, R. R., R. E. Tomlinson, R. W. Engel-Wilson, G. L., Waggerman, and A. G. Spratt. 1994. White-winged dove. Pages 29–58 *in* T. C. Tacha and C. E. Braun, editors. Migratory shore and upland game bird management in North America. International Association of Fish and Wildlife Agencies, Washington, D.C., USA.

Gigliotti, L. M. 2000. A classification scheme to better understand satisfaction of Black Hills deer hunters: the role of harvest success. Human Dimensions of Wildlife 5:32–51.

Gigliotti, L. M., and D. J. Decker. 1992. Human dimensions in wildlife management education: pre-service opportunities and in-service needs. Wildlife Society Bulletin 20:8–14.

Gigliotti, L. M., and E. C. Metcalf. 2016. Motivations of female black hills deer hunters. Human Dimensions of Wildlife 21:371–378.

Gigliotti, L. M., and W. W. Taylor. 1990. The effect of illegal harvest on recreational fisheries. North American Journal of Fisheries Management 10:106–110.

Gilbert, F. F. 2000. Considerations in managing wildlife populations for sport. Wildlife Society Bulletin 28:459–463.

Giles, B. G., and C. S. Findlay. 2004. Effectiveness of a selective harvest system in regulating deer populations in Ontario. The Journal of Wildlife Management 68:266–277.

Gill, R. 1990. Monitoring the status of European and North American cervids. The Global Environment Monitoring System Information Series 8, United Nations Environment Programme, Nairobi, Kenya.

Gimenez, O., E. Cam, and J.-M. Gaillard. 2018. Individual heterogeneity and capture-recapture models: what, why and how? Oikos 127:664–686.

Gimenez, O., and R. Choquet. 2010. Individual heterogeneity in studies on marked animals using numerical integration: capture-recapture mixed models. Ecology 91:951–957.

Glass, R. J., and T. A. More. 1992. Satisfaction, valuation, and views toward allocation of Vermont goose hunting opportunities. U.S. Department of Agriculture, Forest Service, Northeastern Forest Experiment Station, Research Paper NE-668, Radnor, Pennsylvania, USA.

Glover, R., and T. Baskett. 1984. Socioeconomic profiles of Missouri deer poachers: management applications. Transactions of the North American Wildlife & Natural Resources Conference 49:104–111.

Goeman, T. J., P. D. Spencer, and R. B. Pierce. 1993. Effectiveness of liberalized bag limits as management tools for altering northern pike population size structure. North American Journal of Fisheries Management 13:621–624.

Golden, H., and T. Krause. 2003. How to avoid incidental take of lynx while trapping or hunting bobcats and other furbearers. https://files.dnr.state.mn.us/recreation/hunting/trapping/avoidlynx.pdf. Accessed 4 April 2020.

Goodwin, P., and G. Wright. 2004. Decision analysis for management judgment, 3rd edition. John Wiley & Sons, Chichester, West Sussex, United Kingdom.

Goodwin, P., and G. Wright. 2009. Decision analysis for management judgment, 4th edition. John Wiley & Sons, Chichester, West Sussex, United Kingdom.

Gosselin, J., A. Zedrosser, J. E. Swenson, and F. Pelletier. 2015. The relative importance of direct and indirect effects of hunting mortality on the population dynamics of brown bears. Proceedings of the Royal Society B 282:20141840.

Government of Canada. 2017. American black bear: non-detriment finding. https://www.canada.ca/en/environment-climate-change/services/convention-international-trade-endangered-species/non-detriment-findings/american-black-bear.html. Accessed 14 April 2020.

Government of Canada. 2018. About WAPPRIITA: wild species protection and trade. https://www.canada.ca/en/environment-climate-change/services/environmental-enforcement/acts-regulations/wild-species-protection.html. Accessed 14 April 2020.

Graeb, B. D. S., S. R. Chipps, D. W. Willis, J. P. Lott, R. P. Hanten, W. Nelson-Stastny, and J. W. Erickson. 2008. Walleye response to rainbow smelt population decline and liberalized angling regulations in a Missouri River reservoir. Pages 275–292 *in* M. S. Allen, S. Sammons, and M. J. Maceina, editors. Balancing fisheries management and water uses for impounded river systems. American Fisheries Society, Symposium 62, Bethesda, Maryland, USA.

Graham, K. A. 2019. Modeling waterfowl hunter and Ducks Unlimited member behavior using mark-recapture methods. Thesis, University of Nebraska, Lincoln, Nebraska, USA.

Graham, K. A., N. B. Price, V. K. Jones, J. J. Fontaine, and C. J. Chizinski. 2021. Marketing and ecological models to predict permit-purchasing behavior of sportspersons. *in* K. L. Pope and L. A. Powell, editors. Harvest of fish and wildlife: new paradigms for sustainable management. CRC Press, Boca Raton, Florida, USA.

Grahn, G. L. 1969. NBD model of repeat-purchase loyalty: an empirical investigation. Journal of Marketing Research 6:72–78.

Grams, A. S. 2018. Motivations, specializations, identities, and constraints associated with hunting in Nebraska. University of Nebraska-Lincoln, Lincoln, Nebraska, USA. https://digitalcommons.unl.edu/natresdiss/269/.

Gray, B. T. and R. M. Kaminski. 1994. Illegal waterfowl hunting in the Mississippi Flyway and recommendations for alleviation. Wildlife Monographs 127:1–60.

Great Lakes Fishery Commission. 2007. A joint strategic plan for management of the Great Lakes fisheries (adopted in 1997 and supersedes 1981 original). Great Lakes Fishery Commission, Miscellaneous Publications 2007-01, Ann Arbor, Michigan, USA.

Great Lakes Fishery Commission. 2019. Great lakes databases. http://www.glfc.org/great-lakes-databases.php. Accessed 10 July 2020.

Gregory, R., T. McDaniels, and D. Fields. 2001. Decision aiding, not dispute resolution: creating insights through structured environmental decisions. Journal of Policy Analysis and Management 20:415–432.

Gregory, R. S. 2017. The troubling logic of inclusivity in environmental consultations. Science, Technology & Human Values 42:144–165.

Gregory, R. S., L. Failing, M. Harstone, G. Long, T. L. McDaniels, and D. Ohlson. 2012. Structured decision making: a practical guide to environmental management choices. Wiley Blackwell, Chichester, West Sussex, United Kingdom.

Gregory, R. S., and R. L. Keeney. 2002. Making smarter environmental management decisions. Journal of the American Water Resources Association 38:1601–1612.

Grimmelikhuijsen, S. 2010. Transparency of public decision-making: towards trust in local government? Policy and Internet 2:5–35.

Groepper, S. R., M. P. Vrtiska, L. A. Powell, and S. E. Hygnstrom. 2012. Evaluation of September hunting seasons on resident Canada geese in Nebraska. Wildlife Society Bulletin 36:524–530.

Gross, J. E., J. E. Roelle, and G. L. Williams. 1973. Program OnePop and information processor: a systems modeling and communication project. Colorado State University, Colorado Cooperative Wildlife Research Unit, Fort Collins, Colorado, USA.

Grosslein, R. 1980. Ideas at work: the Minnesota voluntary adult hunter education program. Pages 56–59 *in* J. Lorenz, editor. Proceedings of the international conference on outdoor ethics. Izaak Walton League of America Gaithersburg, Maryland, USA.

Gruntorad, M. P., and C. J. Chizinski. 2020. Constraints to hunting and harvesting elk in a landscape dominated by private land. Wildlife Biology 2020.

Gruntorad, M. P., and C. J. Chizinski. 2021. Hunter and angler behavior in harvest management. in K. L. Pope and L. A. Powell, editors. Harvest of fish and wildlife: new paradigms for sustainable management. CRC Press, Boca Raton, Florida, USA.

Gruntorad, M. P., J. J. Lusk, M. P. Vrtiska, and C. J. Chizinski. 2020. Identifying factors influencing hunter satisfaction across hunting activities in Nebraska. Human Dimensions of Wildlife 25:215–231.

Gruver, B. J., D. C. Guynn, Jr., and H. A. Jacobson. 1984. Simulated effects of harvest strategy on reproduction in white-tailed deer. The Journal of Wildlife Management 48:535–541.

Guckian, M. L., A. J. Danylchuk, S. J. Cooke, and E. M. Markowitz. 2018. Peer pressure on the riverbank: assessing catch-and-release anglers' willingness to sanction others' (bad) behavior. Journal of Environmental Management 219:252–259

Gude, J. A., J. A. Cunningham, J. T. Herbert, and T. Baumeister. 2012. Deer and elk hunter recruitment, retention, and participation trends in Montana. The Journal of Wildlife Management 76:471–479.

Gulland, J. A. 1983. Fish stock assessment: a manual of basic methods. Wiley, New York, New York, USA.

Gulsby, W. D., C. H. Killmaster, J. W. Bowers, and K. V. Miller. 2019. Effects of restrictive harvest criteria on antler size of hunter-harvested male white-tailed deer and hunting opportunity. Wildlife Society Bulletin 43:213–221.

Guthery, F. S. 2012. On some founding ideas of quailology and their propounders. Proceedings of the National Quail Symposium 7:50.

Guthery, F. S., A. K. Crews, J. J. Lusk, R. N. Chapman, and M. Sams. 2004. Effects of bag limits on bobwhite hunters and harvest. The Journal of Wildlife Management 68:1095–1103.

Guthery, F. S., M. J. Peterson, and R. R. George. 2000. Viability of northern bobwhite populations. The Journal of Wildlife Management 64:646–662.

Guynn, D. E., and J. L. Schmidt. 1984. Managing deer hunters on private lands in Colorado. Wildlife Society Bulletin 12:12–19.

Gwinn, D. C., and M. S. Allen. 2010. Exploring population-level effects of fishery closures during spawning: an example using largemouth bass. Transactions of the American Fisheries Society 139:626–634.

Gwinn, D. C., M. S. Allen, F. D. Johnston, P. Brown, C. R. Todd, and R. Arlinghaus. 2013. Rethinking length-based fisheries regulations: the value of protecting old and large fish with harvest slots. Fish and Fisheries 16:259–281.

Haag, J. 2020. New mountain lion data for the Pine Ridge. http://magazine.outdoornebraska.gov/2020/05/new-mountain-lion-data-for-the-pine-ridge/. Accessed 15 June 2020.

Haddon, M. 2011. Modelling and quantitative methods in fisheries, 2nd edition. CRC Press, Boca Raton, Florida, USA.

Hale, R. S., M. E. Lundquist, R. L. Miller, and R. W. Petering. 1999. Evaluation of a 254-mm minimum length limit on crappies in Delaware Reservoir, Ohio. North American Journal of Fisheries Management 19:804–814.

Hall, D. L., G. J. Bonnaffons, and R. M. Jackson. 1989. The relationship of enforcement, courts and sentencing to compliance with waterfowl hunting regulations. Transactions of the North American Wildlife and Natural Resources Conference 54:341–354.

Hall, M. 1991. Citizen task force on deer management. Page 195 *in* P.D. Curtis, M.J. Fargione, J.E. Caslick, editors (with editorial assistance of C. Rundle). Proceedings of the fifth eastern wildlife damage control conference. New York State Department of Environmental Conservation, Albany, New York, USA.

Hall, S. 2009. Area and time restrictions. Pages 196–219 *in* K. L. Cochrane and S. M. Garcia, editors. A fishery managers handbook, 2nd edition. Wiley-Blackwell, United Kingdom.

Halliday, T. 1980. The extinction of the passenger pigeon *Ectopistes migratorius* and its relevance to contemporary conservation. Biological Conservation 17:157–162.

Hamilton, D. A., B. Roberts, G. Linscombe, N. R. Jotham, H. Noseworthy, and J. L. Stone. 1998. The European Union's wild fur regulation: a battle of politics, cultures, animal rights, international trade and North America's wildlife policy. Transactions of the North American Wildlife and Natural Resources Conference 63:572–588.

Hammitt, W. E., C. D. McDonald, and M. E. Patterson. 1990. Determinants of multiple satisfaction for deer hunting. Wildlife Society Bulletin 18:331–337.

Hammond, J. S., R. L. Keeney, and H. Raiffa. 1999. Smart choices: a practical guide to making better life decisions. Harvard Business School Press, Boston, Massachusetts, USA.

Hampton, J. O., and K. Teh-White. 2019. Animal welfare, social license, and wildlife use industries. The Journal of Wildlife Management 83:12–21.

Hansen, G. J. A., S. R. Carpenter, J. W. Gaeta, J. M. Hennessy, and M. J. Vander Zanden. 2015a. Predicting walleye recruitment as a tool for prioritizing management actions. Canadian Journal of Fisheries and Aquatic Sciences 72:661–672.

Hansen, G. J. A., J. S. Read, J. F. Hansen, and L. A. Winslow. 2017a. Projected shifts in fish species dominance in Wisconsin lakes under climate change. Global Change Biology 23:1463–1476.

Hansen, H. O. 2017. Global fur retail value. https://fur.ca/wp-content/uploads/2017/09/Global_Fur_Retail_July_20172.pdf. Accessed 3 March 2020.

Hansen, J., and M. H. Wolter. 2017. A 10-year strategic plan for managing Wisconsin's panfish. Wisconsin Department of Natural Resources, Administrative Report, Madison, Wisconsin, USA.

Hansen, J. F., G. G. Sass, J. W. Gaeta, G. A. Hansen, D. A. Isermann, J. Lyons, and M. J. Vander Zanden. 2015b. Largemouth bass management in Wisconsin: intraspecific and interspecific implications of abundance increases. Pages 193–206 *in* M. D. Tringali, J. M. Long, T. W. Birdsong, and M. S. Allen, editors. Black bass diversity: multidisciplinary science for conservation. America Fisheries Society, Symposium 82, Bethesda, Maryland, USA.

Hansen, K. 2007. Bobcat: master of survival. Oxford University Press, New York, New York, USA.

Hansen, L. 2011. Extensive management. Pages 409–452 *in* D. G. Hewitt, editor. Biology and management of white-tailed deer. CRC Press, Boca Raton, Florida, USA.

Hansen, L. P., and J. Beringer. 1997. Managed controlled hunts to control white-tailed deer populations on urban public areas in Missouri. Wildlife Society Bulletin 25:484–487.

Hansen, L. P., E. B. Flinn, J. A. Sumners, X. Gao, and J. J. Millspaugh. 2017b. Effects of an antler point restriction on harvest of white-tailed deer in Missouri. Wildlife Society Bulletin 41:516–522.

Hansen, L. P., J. A. Sumners, R. Reitz, Y. Bian, X. Gao, and J. J. Millspaugh. 2018. Effects of an antler point restriction on deer hunter perceptions and satisfaction in Missouri. Wildlife Society Bulletin 42:607–615.

Hansen, M. J., M. A. Bozek, J. R. Newby, S. P. Newman, and M. D. Staggs. 1998. Factors affecting recruitment of walleyes in Escanaba Lake, Wisconsin, 1958–1996. North American Journal of Fisheries Management 18:764–774.

Hanson, K. C., C. T. Hasler, S. J. Cooke, C. D. Suski, and D. P. Philipp. 2008. Intersexual variation in the seasonal behaviour and depth distribution of a freshwater temperate fish, the largemouth bass. Canadian Journal of Zoology 86:801–811.

Hanson, L. B., M. S. Mitchell, J. B. Grand, D. B. Jolley, B. D. Sparklin, and S. S. Ditchkoff. 2009. Effect of experimental manipulation on survival and recruitment of feral pigs. Wildlife Research 36:185–191.

Harmon, B. S., D. R. Martin, C. J. Chizinski, and K. L. Pope. 2018. Variation in angler distribution and catch rates of stocked rainbow trout in a small reservoir. PLoS One 13:e0190745.

Harms, T. M., and S. J. Dinsmore. 2021. Direct and indirect effects of harvest regulations on wildlife populations. *in* K. L. Pope and L. A. Powell, editors. Harvest of fish and wildlife: new paradigms for sustainable management. CRC Press, Boca Raton, Florida, USA.

Harper, C. A., C. E. Shaw, J. M. Fly, and J. T. Beaver. 2012. Attitudes and motivations of Tennessee deer hunters toward quality deer management. Wildlife Society Bulletin 36:277–285.

Harrop, S. R. 1998. The agreements on international humane trapping standards-background, critique and the texts. Journal of International Wildlife Law and Policy 1:387–394.

Hartman, W. L. 1958. Estimation of catch and related statistics of the stream rainbow trout fishery of the Finger Lakes region. New York Fish and Game Journal 5:205–212.

Hartman, W. L. 1972. Lake Erie: effects of exploitation, environmental changes and new species on the fishery resource. Journal Fisheries Research Board of Canada 29:899–912.

Haugen, M. T., L. A. Powell, and M. P. Vrtiska. 2014. Assessment of spatial changes of duck harvest in the Central Flyway, 1997-2011. Wildfowl 64:75–90.

Haugen, M. T., L. A. Powell, M. P. Vrtiska, and K. L. Pope. 2015. The effects of harvest regulations on behaviors of duck hunters. Human Dimensions of Wildlife 20:15–29.

Haugen, M. T., M. P. Vrtiska, and L. A. Powell. 2017. Variation in number of ducks harvested among hunters in the Central Flyway. Journal of the Southeastern Association of Fish and Wildlife Agencies 4:80–84.

Hauser, C. E., and H. P. Possingham. 2008. Experimental or precautionary? Adaptive management over a range of time horizons. Journal of Applied Ecology 45:72–81.

Hauser, C. E., D. Southwell, J. J. Lahoz-Monfort, L. Rumpff, J. Benshemesh, T. Burnard, R. van Hespen, J. Wright, B. Wintle, and M. Bode. 2019. Adaptive management informs conservation and monitoring of Australia's threatened malleefowl. Biological Conservation 233:31–40.

Hautaluoma, J., and P. J. Brown. 1978. Attributes of the deer hunting experience: a cluster-analytic study. Journal of Leisure Research 10:271–287.

Havera, S. P., R. M. Whitton, and R. T. Shealy. 1992. Blood lead and ingested and embedded shot in diving ducks during spring. The Journal of Wildlife Management 56:539–545.

Hawkins, A. S., R. C. Hanson, H. K. Nelson, and H. M. Reeves, editors. 1984. Flyways: pioneering waterfowl management in North America. U.S. Fish and Wildlife Service, Washington, D.C., USA.

Hawkins, T. O. 2017. A case study analysis of collaborative conservation: restoring bighorn sheep to the Santa Catalina Mountains. Dissertation, Prescott College, Prescott, Arizona, USA.

Hayes, D. B., W. W. Taylor, and P. A. Soranno. 1999. Natural lakes and large impoundments. Pages 589–621 *in* C. C. Kohler and W. A. Hubert, editors. Inland fisheries management in North America, 2nd edition. American Fisheries Society, Bethesda, Maryland, USA.

Hayes, J. 2018. Hunters and anglers support license fee increases in poll. Pittsburgh Post-Gazette. https://www.post-gazette.com/life/outdoors/2018/11/01/Pennsylvania-Fish-Boat-Commission-hunters-anglers-license-fee-John-Arway/stories/201811010033. Accessed 9 June 2020.

Hayslette, S. E., J. B. Armstrong, and R. E. Mirarchi. 2001. Mourning dove hunting in Alabama: motivations, satisfactions, and sociocultural influences. Human Dimensions of Wildlife 6:81–95.

Hazel, K. L., E. E. Langenau, Jr., and R. L. Levine. 1990. Dimensions of hunting satisfaction: multiple-satisfactions of wild turkey hunting. Leisure Sciences 12:383–393.

Hazzard, A. S. 1952. Better trout fishing–and how. Sports Afield 128(2):17–19, 95–96.

Healy, W. M., and S. M. Powell. 1999. Wild turkey harvest management: biology, strategies, and techniques. U.S. Fish and Wildlife Service Biological Technical Publication, BTP-R5001-1999, Washington, D.C., USA

Heberlein, T. A. 2004. Fire in the Sistine Chapel: how Wisconsin responded to chronic wasting disease. Human Dimensions of Wildlife 9:165–179.

Heberlein, T. A., and W. F. Kuentzel. 2002. Too many hunters or not enough deer? Human and biological determinants of hunter satisfaction and quality. Human Dimensions of Wildlife 7:229–250.

Heffelfinger, J. R., V. Geist, and W. Wishart. 2013. The role of hunting in North American wildlife conservation. International Journal of Environmental Studies 70:399–413.

Hefley, T. J., A. J. Tyre, and E. E. Blankenship. 2013. Statistical indicators and state–space population models predict extinction in a population of bobwhite quail. Theoretical Ecology 6:319–331.

Heino, J. 2011. A macroecological perspective of diversity patterns in the freshwater realm. Freshwater Biology 56:1703–1722.

Heino, M., B. Díaz Pauli, and U. Dieckmann. 2015. Fisheries-induced evolution. Annual Review of Ecology, Evolution, and Systematics 46:461–480.

Heino, M., and U. Dieckmann. 2008. Detecting fisheries-induced life-history evolution: an overview of the reaction-norm approach. Bulletin of Marine Science 83:69–93.

Hendee, J. C. 1969. Rural-urban differences reflected in outdoor recreation participation. Journal of Leisure Research 1:333–341.

Hendee, J. C. 1974. A multiple-satisfaction approach to game management. Wildlife Society Bulletin 2:104–113.

Hendee, J. C., and R. J. Burdge. 1974. The substitutability concept: implications for recreation research and management. Journal of Leisure Research 6:157–162.

Hendry, A. P., T. J. Farrugia, and M. T. Kinnison. 2008. Human influences on rates of phenotypic change in wild animal populations. Molecular Ecology 17:20–29.

Hengeveld, P. E., and M. Festa-Bianchet. 2011. Harvest regulations and artificial selection on horn size in male bighorn sheep. The Journal of Wildlife Management 75:189–197.

Henny, C. J., and K. P. Burnham. 1976. A reward band study of mallards to estimate reporting rates. The Journal of Wildlife Management 40:1–14.

Hepp, G. R., R. J. Blohm, R. E. Reynolds, J. E. Hines, and J. D. Nichols. 1986. Physiological condition of autumn-banded mallards and its relationship to hunting vulnerability. The Journal of Wildlife Management 50:177–183.

Hess, T. B. 1991. The impact of daily creel limits on sport fish harvest in Georgia. Proceedings of the Annual Conference of the Southeastern Association of Fish and Wildlife Agencies, White Sulphur Springs, West Virginia 45:282–287.

Hessenauer, J.-M., J. Vokoun, J. Davis, R. Jacobs, and E. O'Donnell. 2018. Size structure suppression and obsolete length regulations in recreational fisheries dominated by catch-and-release. Fisheries Research 200:33–42.

Hessenauer, J.-M., J. C. Vokoun, C. D. Suski, J. Davis, R. Jacobs, and E. O'Donnell. 2015. Differences in the metabolic rates of exploited and unexploited fish populations: a signature of recreational fisheries induced evolution? PLoS One 10:e0128336.

Heusmann, H. W. 1999a. Special hunting seasons and resident Canada goose populations. Wildlife Society Bulletin 27:456–464.

Heusmann, H. W. 1999b. Let's get rid of the midwinter waterfowl inventory in the Atlantic Flyway. Wildlife Society Bulletin 27:559–565.

Heusmann, H. W., and J. E. McDonald. 2002. Distribution of wood duck harvest in the Atlantic and Mississippi Flyways in relation to hunting season length. Wildlife Society Bulletin 30:666–674.

Hewitt, J. A., and D. K. Brown. 2000. Agency costs in environmental not-for-profits. Public Choice 103:163–183.

Hicks, C. E., L. C. Belusz, D. J. Witter, and P. S. Haverland. 1983. Application of angler attitudes and motives to management strategies at Missouri's trout parks. Fisheries 8(5):2–7.

Hightower, J. E., and J. E. Harris. 2017. Estimating fish mortality rates using telemetry and multistate models. Fisheries 42:210–219.

Hilaire, L. J., D. E. Benson, and K. P. Burnham. 1998. Evaluation of home-study versus teacher-taught hunter education: expanding learning opportunities. Wildlife Society Bulletin 26:56–67.

Hilborn, R., and C. J. Walters. 1992. Quantitative fisheries stock assessment: choice, dynamics, and uncertainty. Routledge, Chapman and Hall, New York, New York, USA.

Hiller, T. L., R. D. Applegate, and M. S. O'Brien. 2021a. Harvest management of furbearers. *in* K. L. Pope and L. A. Powell, editors. Harvest of fish and wildlife: new paradigms for sustainable management. CRC Press, Boca Raton, Florida, USA.

Hiller, T. L., R. D. Applegate, and L. A. Powell. 2021b. The social and political context of harvest management. *in* K. L. Pope and L. A. Powell, editors. Harvest of fish and wildlife: new paradigms for sustainable management. CRC Press, Boca Raton, Florida, USA.

Hiller, T. L., L. A. Powell, T. D. McCoy, and J. J. Lusk. 2009. Long-term agricultural land-use trends in Nebraska, 1866–2007. Great Plains Research 19:225–237.

Hiller, T. L., and H. B. White. 2013. How to avoid incidental take of wolverine during regulated trapping activities. Association of Fish and Wildlife Agencies, Washington, D.C., USA.

Hiller, T. L., B. White, and J. Erb. 2018. State management of furbearing animals. Pages 116–134 *in* T. J. Ryder, editor. State wildlife conservation and management. Johns Hopkins University Press, Baltimore, Maryland, USA.

Hilling, C. D., A. J. Bunch, R. S. Greenlee, D. J. Orth, and Y. Jiao. 2018. Natural mortality and size structure of introduced blue catfish in Virginia tidal rivers. Journal of the Southeastern Association of Fish and Wildlife Agencies 5:30–38.

Hilling, C. D., Y. Jiao, A. J. Bunch, R. S. Greenlee, J. D. Schmitt, and D. J. Orth. In press. Growth dynamics of invasive blue catfish in four subestuaries of the Chesapeake Bay, USA. North American Journal of Fisheries Management.

Hindman, L. J., K. M. Dickson, W. F. Harvey, IV, and J. R. Serie. 2004. Atlantic Flyway Canada geese: new perspectives in goose management. Pages 12–21 *in* T. J. Moser, K. C. VerCauteren, R. D. Lien, K. F. Abraham, D. E. Andersen, J. G. Bruggink, J. M. Coluccy, D. A. Graber, J. O. Leafloor, D. R. Luukkonen, and R. E. Trost, editors. Proceedings of the 2003 International Canada Goose Symposium, Milwaukee, Wisconsin, USA.

Hindman, L. J., W. F. Harvey, IV, J. B. Hestbeck, and W. E. Rhodes. 1998. Effect of restrictive hunting regulations on harvest and survival of Canada goose in Maryland. Pages 151–158 *in* D. H. Rusch, M. D. Samuel, D. D. Humburg, and B. D. Sullivan, editors. Biology and management of Canada geese. Proceedings of the International Canada Goose Symposium, Milwaukee, Wisconsin, USA.

Hinrichs, M. P. 2019. Hunter and angler motivations, preferences, and barriers to waterfowl hunting in the central United States. Thesis, University of Nebraska, Lincoln, Nebraska, USA.

Hinrichs, M. P., N. B. Price, M. P. Gruntorad, K. L. Pope, J. J. Fontaine, and C. J. Chizinski. 2020. Understanding sportsperson retention and reactivation through license purchasing behavior. Wildlife Society Bulletin 44:383–390.

Hinrichs, M. P., M. P. Vrtiska, M. A. Pegg, and C. J. Chizinski. 2021. A comparison of motivations of hunters and anglers in the central United States. Human Dimensions of Wildlife DOI:10.1080/10871209.2020.1858208

Hixon, M. A., D. W. Johnson, and S. M. Sogard. 2014. BOFFFFs: on the importance of conserving old-growth age structure in fishery populations. ICES Journal of Marine Science 71:2171–2185.

Hobaugh, W. C. 1984. Habitat use by snow geese wintering in southeast Texas. The Journal of Wildlife Management 48:1085–1096.

Hoenig, J. M., N. J. Barrowman, and K. H. Pollock. 1998. Multiyear tagging studies incorporating fishing effort. Canadian Journal of Fisheries and Aquatic Sciences 55:1466–1467.

Hoff, M. H. 1995. Comparisons of the effects of 10-inch, 8-inch, and no minimum length limits on the smallmouth bass population and fishery in Nebish Lake, Wisconsin. North American Journal of Fisheries Management 15:95–102.

Hogg, J. T., and S. H. Forbes. 1997. Mating in bighorn sheep: frequent male reproduction via a high-risk "unconventional" tactic. Behavioral Ecology and Sociobiology 41:33–48.

Hokanson, K. E. F. 1977. Temperature requirements of some percids and adaptations to the seasonal temperature cycle. Journal of the Fisheries Research Board of Canada 34:1524–1550.

Holbrook, II, J. A. 1975. Bass fishing tournaments. Pages 404–414 *in* R. H. Stroud and H. Clepper, editors. Black bass biology and management. Sport Fishing Institute, Washington, D.C., USA.

Holland, S. M., and R. B. Ditton. 1992. Fishing trip satisfaction: a typology of anglers. North American Journal of Fisheries Management 12:28–33.

Holley, M. P., M. D. Marshall, and M. J. Maceina. 2009. Fishery and population characteristics of blue catfish and channel catfish and potential impacts of minimum length limits on the fishery in Lake Wilson, Alabama. North American Journal of Fisheries Management 29:1183–1194.

Hollins, J., D. Thambithurai, B. Koeck, A. Crespel, D. M. Bailey, S. J. Cooke, J. Lindström, K. J. Parsons, and S. S. Killen. 2018. A physiological perspective on fisheries-induced evolution. Evolutionary Applications 11:561–576.

Holsman, R. H. 2000. Goodwill hunting? Exploring the role of hunters as ecosystem stewards. Wildlife Society Bulletin 28:808–816.

Holsman, R. H., J. Petchenik, and E. E. Cooney. 2010. CWD after "the Fire": six reasons why hunters resisted Wisconsin's eradication effort. Human Dimensions of Wildlife 15:180–193.

Holt, R. E., and C. Jørgensen. 2015. Climate change in fish: effects of respiratory constraints on optimal life history and behaviour. Biology Letters 11:20141032.

Hörnberg, S. 2001. Changes in population density of moose (*Alces alces*) and damage to forests in Sweden. Forest Ecology and Management 149:141–151. DOI: 10.1016/S0378-1127(00)00551-X

Houde, E. 2006. A fisheries ecosystem plan for the Chesapeake Bay. Pages 1-12 *in* National Oceanic and Atmospheric Administration Chesapeake Bay Fisheries Ecosystem Advisory Panel. Fisheries ecosystem planning for Chesapeake Bay. American Fisheries Society, Trends in Fisheries Management 3, Bethesda, Maryland.

Hoyle, J. A., and A. Keast. 1987. The effect of prey morphology and size on handling time in a piscivore, the largemouth bass (*Micropterus salmoides*). Canadian Journal of Zoology 65:1972–1977.

Hrubes, D., I. Ajzen, and J. Daigle. 2001. Predicting hunting intentions and behavior: an application of the theory of planned behavior. Leisure Sciences 23:165–178.

Huang, I. B., J. Keisler, and I. Linkov. 2011. Multi-criteria decision analysis in environmental sciences: ten years of applications and trends. Science of the Total Environment 409:3578–3594.

Huang, J. L., M. Liu, and N. A. Bowling. 2015. Insufficient effort responding: examining an insidious confound in survey data. Journal of Applied Psychology 100:828.

Hubbard, J., and R. C. Mannell. 2001. Testing competing models of the leisure constraint negotiation process in a corporate employee recreation setting. Leisure Sciences 23:145–163.

Hudson, P. J., and A. P. Dobson. 2001. Harvesting unstable populations: red grouse *Lagopus lagopus scoticus* (Lath.) in the United Kingdom. Wildlife Biology 7:189–195.

Hudson, P. J., D. Newborn, and P. A. Robertson. 1997. Geographical and seasonal patterns of mortality in red grouse *Lagopus lagopus scoticus* populations. Wildlife Biology 3:79–88.

Humburg, D. D., T. W. Aldrich, S. Baker, G. Costanzo, J. H. Gammonley, M. A. Johnson, B. Swift, and D. Yparraguirre. 2000. Adaptive harvest management: has anything really changed? Transactions of the North American Wildlife and Natural Resources Conference 65:78–93.

Hunt, L., W. Haider, and K. Armstrong. 2002. Understanding the fish harvesting decisions by anglers. Human Dimensions of Wildlife 7:75–89.

Hunt, L. M., R. Arlinghaus, N. Lester, and R. Kushneriuk. 2011. The effects of regional angling effort, angler behavior, and harvesting efficiency on landscape patterns of overfishing. Ecological Applications 21:2555–2575.

Hunt, L. M., E. Camp, B. van Poorten, and R. Arlinghaus. 2019. Catch and non-catch-related determinants of where anglers fish: a review of three decades of site choice research in recreational fisheries. Reviews in Fisheries Science & Aquaculture 27:261–286.

Hunt, L. M., W. Haider, and B. Bottan. 2005. Accounting for varying setting preferences among moose hunters. Leisure Sciences 27:297–314.

Hunt, L. M., S. G. Sutton, and R. Arlinghaus. 2013. Illustrating the critical role of human dimensions research for understanding and managing recreational fisheries within a social-ecological system framework. Fisheries Management and Ecology 20:111–124.

Hunt, R. L. 1981. A successful application of catch and release regulations on a Wisconsin trout stream. Wisconsin Department of Natural Resources Technical Bulletin Number 119, Madison, Wisconsin, USA.

Hutchings, J. A. 1996. Spatial and temporal variation in the density of northern cod and a review of hypotheses for the stock's collapse. Canadian Journal of Fisheries and Aquatic Sciences 53:943–962.

Hutchings, J. A. 2000. Collapse and recovery of marine fishes. Nature 406:882–885.

Hutchings, J. A. 2005. Life history consequences of overexploitation to population recovery in Northwest Atlantic cod (*Gadus morhua*). Canadian Journal of Fisheries and Aquatic Sciences 62:824–832.

Hutchings, J. A. 2009. Avoidance of fisheries-induced evolution: management implications for catch selectivity and limit reference points. Evolutionary Applications 2:324–334.

Hutchings, J. A., W. Ardren, B. T. Barlaup, E. J. Bergman, K. D. Clarke, L. A. Greenberg, C. Lake, J. Piironen, P. Sirois, L. E. Sundt-Hansen, and D. J. Fraser. 2019. Life-history variability and conservation status of landlocked Atlantic salmon: an overview. Canadian Journal of Fisheries and Aquatic Sciences 76:1697–1708.

Hutchings, J. A., and A. Kuparinen. 2020. Implications of fisheries-induced evolution for population recovery: refocusing the science and refining its communication. Fish and Fisheries 21:453–464.

Hutchings, J. A., and J. D. Reynolds. 2004. Marine fish population collapses: consequences for recovery and extinction risk. BioScience 54:297.

Hygnstrom, S., D. Drake, T. R. Van Deelen, and S. Vantassel. 2014. Managing overabundant white-tailed deer: is it time to consider regulated commercial harvest? Outlooks on Pest Management 25:11–16.

Idaho Department of Fish and Game. 2003. Nez Perce license valid only on reservation. https://idfg.idaho.gov/press/nez-perce-license-valid-only-reservation. Accessed 26 October 2020.

International Association of Fish and Wildlife Agencies. 2005. Potential costs of losing hunting and trapping as wildlife management methods. https://www.fishwildlife.org/application/files/1115/2002/5933/PCLHuntingAndTrapping.pdf. Accessed 14 April 2020.

Invasive Catfish Task Force. 2014. Final report of the sustainable fisheries goal implementation team. https://www.chesapeakebay.net/documents/Updated_2014_ICTF_Report.pdf. Accessed 14 October 2020.

Irwin, B. J., M. J. Wilberg, M. L. Jones, and J. R. Bence. 2011. Applying structured decision making to recreational fisheries management. Fisheries 36:113–122.

Isermann, D. A. 2007. Evaluating walleye length limits in the face of population variability: case histories from western Minnesota. North American Journal of Fisheries Management 27:551–568.

Isermann, D. A., and J. sssB. Maxwell. 2013. Temporal and regional trends in black bass release rates in Minnesota. North American Journal of Fisheries Management 33:344–350.

Isermann, D. A., and B. G. Parsons. 2011. Harvest regulations and sampling. Pages 403–422 *in* B. A. Barton, editor. Biology, management and culture of walleye and sauger. American Fisheries Society, Bethesda, Maryland, USA.

Isermann, D. A., and C. P. Paukert. 2010. Regulating harvest. Pages 185–212 *in* W. A. Huber and M. C. Quist, editors. Inland fisheries management in North America, 3rd edition. American Fisheries Society, Bethesda, Maryland, USA.

Isermann, D. A., D. W. Willis, B. G. Blackwell, and D. O. Lucchesi. 2007. Yellow perch in South Dakota: population variability and predicted effects of creel limit reductions and minimum length limits. North American Journal of Fisheries Management 27:918–931.

Isermann, D. A., D. W. Willis, D. O. Lucchesi, and B. G. Blackwell. 2005. Seasonal harvest, exploitation, size selectivity, and catch preferences associated with winter yellow perch anglers on South Dakota lakes. North American Journal of Fisheries Management 25:827–840.

Iso-Ahola, S. E. 1980. The social psychology of leisure and recreation. W. C. Brown Company Publishers, Dubuque, Iowa, USA.

Iverson, S. A., E. T. Reed, R. J. Hughes, and M. R. Forbes. 2014. Age and breeding stage-related variation in the survival and harvest of temperate-breeding Canada geese in Ontario. The Journal of Wildlife Management 78:24–34.

Jackson, E. L. 1991. Leisure constraints/constrained leisure: special issue introduction. Journal of Leisure Research 23:279–285.

Jackson, E. L. 2000. Will research on leisure constraints still be relevant in the twenty-first century? Journal of Leisure Research 32:62–68.

Jackson, E. L., D. W. Crawford, and G. Godbey. 1993. Negotiation of leisure constraints. Leisure Sciences 15:1–11.

Jacobson, P. C. 1994. Population dynamics of large walleye in Big Sand Lake. Minnesota Department of Natural Resources, Investigational Report 436, St Paul, Minnesota, USA.

Jacobson, P. C. 2005. Experimental analysis of a reduced daily bluegill limit in Minnesota. North American Journal of Fisheries Management 25:203–210.

Jacobson, S. K., and M. D. McDuff. 1998. Training idiot savants: the lack of human dimensions in conservation biology. Conservation Biology 12:263–267.

Jagnow, C. P., R. C. Stedman, A. Luloff, G. J. San Julian, J. C. Finley, and J. Steele. 2006. Why landowners in Pennsylvania post their property against hunting. Human Dimensions of Wildlife 11:15–26.

Jakus, P. M., M. Downing, M. S. Bevelhimer, and J. M. Fly. 1997. Do sportfish consumption advisories affect reservoir anglers' site choice? Agricultural and Resource Economics Review 26:196–204.

Jenkins, R. E., and N. M. Burkhead. 1994. Freshwater fishes of Virginia. American Fisheries Society, Bethesda, Maryland, USA.

Jennings, C. A., and S. J. Zigler. 2009. Biology and life history of paddlefish in North America: an update. Pages 1–22 *in* C. P. Paukert and G. D. Scholten, editors. Paddlefish management, propagation, and conservation in the 21st Century: building from 20 years of research and management. American Fisheries Society, Symposium 66, Bethesda, Maryland, USA.

Johnsgard, P. A. 1973. Grouse and quails of North America. University of Nebraska Press, Lincoln, Nebraska, USA.

Johnsgard, P. A. 1983. Grouse of the world. University of Nebraska Press, Lincoln, Nebraska, USA.

Johnson, B. M., R. Arlinghaus, and P. J. Martinez. 2009. Introduced species—are we doing all we can to stem the tide of illegal fish stocking? Fisheries 34:389–394.

Johnson, B. M., C. Luecke, R. S. Stewart, M. D. Staggs, S. J. Gilbert, and J. F. Kitchell. 1992. Forecasting effects of harvest regulations and stocking of walleyes on prey fish communities in Lake Mendota, Wisconsin. North American Journal of Fisheries Management 12:797–807.

Johnson, B. M., and P. J. Martinez. 1995. Selecting harvest regulations for recreational fisheries: opportunities for research/management cooperation. Fisheries 20(10):22–29.

Johnson, D. H., J. D. Nichols, M. J. Conroy, and L. M. Cowardin. 1988. Some considerations in modeling the mallard life cycle. Pages 9–20 *in* M. W. Weller, editor. Waterfowl in winter. University of Minnesota Press, Minneapolis, Minnesota, USA.

Johnson, F. A. 2006. Adaptive harvest management and double-loop learning. Transactions of the North American Wildlife and Natural Resources Conference 71:23–39.

Johnson, F. A., G. S. Boomer, B. K. Williams, J. D. Nichols, and D. J. Case. 2015. Multilevel learning in the adaptive management of waterfowl harvests: 20 years and counting. Wildlife Society Bulletin 39:9–19.

Johnson, F. A., and D. J. Case. 2000. Adaptive regulation of waterfowl harvest: lessons learned and prospects for the future. Transactions of the North American Wildlife and Natural Resources Conference 65:94–108.

Johnson, F. A., and J. A. Dubvosky. 2001. Framework-date extensions for duck hunting in the United States: projected impacts and coping with uncertainty. U.S. Fish and Wildlife Service, Administrative Report, Washington, D.C., USA.

Johnson, F. A., J. E. Hines, F. Montalbano, III, and J. D. Nichols. 1973. Effects of liberalized harvest regulations on wood ducks in the Atlantic Flyway. Wildlife Society Bulletin 14:383–388.

Johnson, F. A., G. H. Jensen, J. Madsen, and B. K. Williams. 2014. Uncertainty, robustness, and the value of information in managing an expanding Arctic goose population. Ecological Modelling 273:186–199.

Johnson, F. A., W. L. Kendall, and J. A. Dubovsky. 2002. Conditions and limitations on learning in the adaptive management of mallard harvests. Wildlife Society Bulletin 30:176–185.

Johnson, F. A., and C. T. Moore. 1996. Harvesting multiple stocks of ducks. The Journal of Wildlife Management 60:551–559.

Johnson, F. A., C. T. Moore, W. L. Kendall, J. A. Dubovsky, D. F. Caithamer, J. R. Kelley, Jr., and B. K. Williams. 1997. Uncertainty and the management of mallard harvests. The Journal of Wildlife Management 61:202–216.

Johnson, F. A., B. K. Williams, J. D. Nichols, J. E. Hines, W. L. Kendall, G. W. Smith, and D. F. Caithamer. 1993. Developing an adaptive management strategy for harvesting waterfowl in North America. Transactions of the North American Wildlife and Natural Resources Conference 58:565–583.

Johnson, F. A., G. S. Zimmerman, M. T. Huang, P. I. Padding, G. D. Balkcom, M. C. Runge, and P. K. Devers. 2019. Multi-species duck harvesting using dynamic programming and multi-criteria decision analysis. Journal of Applied Ecology 56:1447–1459.

Johnson, K. N., R. L. Johnson, D. Edwards, and C. A. Wheaton. 1993. Public participation in wildlife management: opinions from public meetings and random surveys. Wildlife Society Bulletin 21:218–225.

Johnson, T. H., and T. C. Bjornn. 1975. Evaluation of angling regulations in management of cutthroat trout. Idaho Fish and Game Department, Job Performance Report, Project F-59-R-6, Boise, Idaho, USA.

Johnston, C. A. 2014. Agricultural expansion: land use shell game in the U.S. northern plains. Landscape Ecology 29:81–95.

Johnston, F. D., M. S. Allen, B. Beardmore, C. Riepe, T. Pagel, D. Hühn, and R. Arlinghaus. 2018. How ecological processes shape the outcomes of stock enhancement and harvest regulations in recreational fisheries. Ecological Applications 28:2033–2054.

Johnston, F. D., R. Arlinghaus, and U. Diekmann. 2012. Fish life history, angler behaviour and optimal management of recreational fisheries. Fish and Fisheries 14:554–579.

Johnston, F. D., R. Arlinghaus, and U. Dieckmann. 2015a. Fish life history, angler behavior, and optimal management of recreational fisheries. Interim Report IR-12-044, International Institute for Applied Systems Analysis, Laxenburg, Austria.

Johnston, F. D., B. Beardmore, and R. Arlinghaus. 2015b. Optimal management of recreational fisheries in the presence of hooking mortality and noncompliance—predictions from a bioeconomic model incorporating a mechanistic model of angler behavior. Canadian Journal of Fisheries and Aquatic Sciences 72:37–53.

Jolly, G. M. 1965. Explicit estimates from capture-recapture data with both death and immigration-stochastic model. Biometrika 52:225–247.

Jones, M. L., M. J. Catalano, L. K. Peterson, and A. M. Berger. 2016. Stakeholder-centered development of a harvest control rule for Lake Erie walleye *Sander vitreus*. Pages 163–183 *in* C. Edwards and D. Dankel, editors. Management science in fisheries: an introduction to simulation-based methods. Routledge, United Kingdom.

Jørgensen, C., and R. E. Holt. 2013. Natural mortality: its ecology, how it shapes fish life histories, and why it may be increased by fishing. Journal of Sea Research 75:8–18.

Jorgensen, C. F., L. A. Powell, J. J. Lusk, A. A. Bishop, and J. J. Fontaine. 2014. Assessing landscape constraints on species abundance: does the neighborhood limit species response to local habitat conservation programs? PLoS One 9:e99339.

Jotham, N., and R. L. Phillips. 1994. Developing international trap standards, a progress report. Proceedings of the Vertebrate Pest Conference 16:308–310.

Jun, J., and G. T. Kyle. 2012. Gender identity, leisure identity, and leisure participation. Journal of Leisure Research 44:353–378.

Kaemingk, M. A., C. N. Bender, C. J. Chizinski, A. J. Bunch, and K. L. Pope. 2021. Temporal invariance of social-ecological catchments. Ecological Applications 31(2):e02272.

Kaemingk, M. A., C. J. Chizinski, K. L. Hurley, and K. L. Pope. 2018. Synchrony—an emergent property of recreational fisheries. Journal of Applied Ecology 55:2986–2996.

Kaemingk, M. A., K. L. Hurley, C. J. Chizinski, and K. L. Pope. 2020. Harvest-release decisions in recreational fisheries. Canadian Journal of Fisheries and Aquatic Sciences 77:194–201.

Kaemingk, M. A., M. P. Vrtiska, K. L. Hurley, and L. A. Powell. 2021. Linking social and ecological components to spatial and temporal components of harvest. *in* K. L. Pope and L. A. Powell, editors. Harvest of fish and wildlife: new paradigms for sustainable management. CRC Press, Boca Raton, Florida, USA.

Kahan, D. M., E. Peters, M. Wittlin, P. Slovic, L. L. Ouellette, D. Braman, and G. Mandel. 2012. The polarizing impact of science literacy and numeracy on perceived climate change risks. Nature Climate Change 2:732–735.

Kahler, J. S., G. J. Roloff, and M. L. Gore. 2013. Poaching risks in community-based natural resource management. Conservation Biology 27:177–186.

Kahneman, D., and A. Tversky. 2000. Choices, values, and frames. Cambridge University Press, Cambridge, Massachusetts, USA.

Kaltenborn, B. P., and O. Andersen. 2009. Habitat preferences of ptarmigan hunters in Norway. European Journal of Wildlife Research 55:407–413.

Kamakura, W. A., M. Wedel, F. De Rosa, and J. A. Mazzon. 2003. Cross-selling through database marketing: a mixed data factor analyzer for data augmentation and prediction. International Journal of Research in Marketing 20:45–65.

Kaminski, D. J., T. M. Harms, and J. M. Coffey. 2019. Using spotlight observations to predict resource selection and abundance for white-tailed deer. The Journal of Wildlife Management 83:1565–1580.

Kandoth, C., J. Leskie, D. Roberts, J. Powers, and A. Riviello. 2010. An assessment of New Jersey hunter opinion on expanding antler point restriction (APR) regulations in deer management zones 28, 30, 31, 34, and 47. New Jersey Division of Wildlife, Hampton, New Jersey, USA.

Kanstrup, N. 2006. Sustainable harvest of waterbirds: a global review. Pages 98–106 *in* G. Boere, C. Galbraith, and D. Stroud, editors. Waterbirds around the world. The Stationery Office, Edinburgh, Scotland.

Karns, G. R., J. T. Bruskotter, and R. J. Gates. 2015. Explaining hunting participation in Ohio: a story of changing land use and new technology. Human Dimensions of Wildlife 20:484–500.

Karns, G. R., R. J. Gates, S. N. Matthews, J. T. Bruskotter, J. C. McCoy, and M. J. Tonkovich. 2016. Factors influencing spatial heterogeneity of female white-tailed deer harvest dynamics. Wildlife Society Bulletin 40:758–763.

Katz, R., and S. Sterrett. 2019. How a decision-analysis tool helped one scientist couple make some tough career choices. Nature 572:275–276.

Kayle, K., K. Oldenburg, C. Murray, J. Francis, and J. Markham. 2015. Lake Erie walleye management plan 2015-2019. Great Lakes Fishery Commission, Ann Arbor, Michigan, USA.

Keeney, R. L. 1992. Value-focused thinking: a path to creative decision making. Harvard University Press, Cambridge, Massachusetts, USA.

Keeney, R. L. 2004. Framing public policy decisions. International Journal of Technology, Policy, and Management 4:95–115.

Keeney, R. L., and R. S. Gregory. 2005. Selecting attributes to measure the achievement of objectives. Operations Research 53:1–11.

Keeney, R. L., and H. Raiffa. 1976. Decisions with multiple objectives: preference and value tradeoffs. John Wiley & Sons, New York, New York, USA.

Keiling, T. D., M. J. Louison, and C. D. Suski. 2020. Behavioral phenotype does not predict habitat occupancy or angling capture of largemouth bass (*Micropterus salmoides*). Canadian Journal of Zoology 98:399–409.

Kellert, S. R. 1980. American attitudes toward and knowledge of animals: an update. International Journal for the Study of Animal Problems 1:87–119.

Kelly, J. D., and J. E. Hurst. 2016. Delineation of management zones for buck harvest decision making. New York State Department of Environmental Conservation, Albany, New York, USA.

Kempinger, J. J. 1969. Experimental management of Spruce Lake, a small bog lake in northeastern Wisconsin. Wisconsin Department of Natural Resources, Research Report Number 40, Madison, Wisconsin, USA.

Kempinger, J. J., W. S. Churchill, G. R. Priegel, and L. M. Christenson. 1975. Estimate of abundance, harvest, and exploitation of the fish populations of Escanaba Lake, Wisconsin, 1946-69. Wisconsin Department of Natural Resources, Technical Bulletin Number 84, Madison, Wisconsin, USA.

Keogh George, S. 2016. Pink camouflage: reshaping the gendered nature of hunting in the twenty-first century. Loisir et Société/Society and Leisure 39:481–499.

Kerns, J. A., M. S. Allen, J. R. Dotson, and J. E. Hightower. 2015. Estimating regional fishing mortality for freshwater systems: a Florida largemouth bass example. North American Journal of Fisheries Management 35:681–689.

Kerns, J. A., M. S. Allen, and J. E. Hightower. 2016. Components of mortality within a black bass high-release recreational fishery. Transactions of the American Fisheries Society 145:578–588.

Kerr, G. N., and W. Abell. 2016. What are they hunting for? Investigating heterogeneity among sika deer (*Cervus nippon*) hunters. Wildlife Research 43:69–79.

Kéry, M., and M. Schaub. 2012. Bayesian population analysis using WinBUGS. Academic Press, Boston, Massachusetts, USA.

Kieffer, J. D., M. R. Kubacki, F. J. S. Phelan, D. P. Philipp, and B. L. Tufts. 1995. Effects of catch-and-release angling on nesting male smallmouth bass. Transactions of the American Fisheries Society 124:70–76.

Kilpatrick, H. J., and A. M. LaBonte. 2003. Deer hunting in a residential community: the community's perspective. Wildlife Society Bulletin 31:340–348.

Kilpatrick, H. J., A. M. Labonte, and J. S. Barclay. 2007. Acceptance of deer management strategies by suburban homeowners and bowhunters. The Journal of Wildlife Management 71:2095–2101.

Kilpatrick, H. J., A. M. LaBonte, and J. A. Seymore. 2002. A shotgun-archery deer hunt in a residential community: evaluation of hunt strategies and effectiveness. Wildlife Society Bulletin 30:478–486.

Kilpatrick, H. J., and W. D. Walter. 1999. A controlled archery deer hunt in a residential community: cost, effectiveness, and recovery rates. Wildlife Society Bulletin 27:115–123.

Kindsvater, H. K., and E. P. Palkovacs. 2017. Predicting eco-evolutionary impacts of fishing on body size and trophic role of Atlantic cod. Copeia 105:475–482.

Kirby, R. E., J. E. Hines, and J. D. Nichols. 1983. Afternoon closure of hunting and recovery rates of mallards banded in Minnesota. The Journal of Wildlife Management 47:209–213.

Klefoth, T., C. Skov, A. Kuparinen, and R. Arlinghaus. 2017. Toward a mechanistic understanding of vulnerability to hook-and-line fishing: boldness as the basic target of angling-induced selection. Evolutionary Applications 10:994–1006.

Knoche, S., and F. Lupi. 2012. The economic value of publicly accessible deer hunting land. The Journal of Wildlife Management 76:462–470.

Knoche, S., and F. Lupi. 2016. Demand for fishery regulations: effects of angler heterogeneity and catch improvements on preferences for gear and harvest restrictions. Fisheries Research 181:163–171.

Knox, W. M. 2011. The antler religion. Wildlife Society Bulletin 35:45–48.

Kocovsky, P. M., and R. F. Carline. 2001. Dynamics of the unexploited walleye population of Pymatuning Sanctuary, Pennsylvania, 1997-1998. North American Journal of Fisheries Management 21:178–187.

Koeck, B., L. Zavorka, D. Aldven, J. Näslund, R. Arlinghaus, P.-O. Thörnqvist, S. Winberg, B. T. Björnsson, and J. I. Johnsson. 2019. Angling selects against active and stress-resilient phenotypes in rainbow trout. Canadian Journal of Fisheries and Aquatic Sciences 76:320–333.

Koen, E. L., J. Bowman, and C. S. Findlay. 2007. Fisher survival in Eastern Ontario. The Journal of Wildlife Management 71:1214–1219.

Kokko, H. 2001. Optimal and suboptimal use of compensatory responses to harvesting: timing of hunting as an example. Wildlife Biology 7:141–150.

Konisky, D. M., and T. C. Beierle. 2001. Innovations in public participation and environmental decision making: examples from the Great Lakes region. Society & Natural Resources 14:815–826.

Koons, D. N., T. W. Arnold, and M. Schaub. 2017. Understanding the demographic drivers of realized population growth rates. Ecological Applications 27:2102–2115.

Koons, D. N., L. M. Aubry, and R. F. Rockwell. 2019. Liberalized harvest regulations have not affected overabundant snow geese in northern Manitoba. The Condor 2:1–13.

Kosack, J. 1995. The Pennsylvania Game Commission 1895–1995. Pennsylvania Game Commission, Harrisburg, Pennsylvania, USA.

Koupal, K. D., J. D. Katt, C. W. Schoenebeck, and B. E. Eifert. 2015. Sex-specific changes in walleye abundance, size structure and harvest following implementation of regulation to protect broodstock. Journal of Fish and Wildlife Management 6:448–455.

Krause, T. 2007. NTA trapping handbook: a guide for better trapping, 2nd edition. National Trapper's Association, Bedford, Indiana, USA.

Krementz, D. G., M. J. Conroy, J. E. Hines, and H. F. Percival. 1987. Sources of variation in survival and recovery rates of American black ducks. The Journal of Wildlife Management 51:689–700.

Krementz, D. G., M. J. Conroy, J. E. Hines, and H. F. Percival. 1988. The effects of hunting on survival rates of American black ducks. The Journal of Wildlife Management 52:214–226.

Krementz, D. G., J. E. Hines, and D. R. Luukkonen. 2003. Survival and recovery rates of American woodcock in Michigan. The Journal of Wildlife Management 67:398–407.

Kron, J. 2012. Red state, blue city: how the urban-rural divide is splitting America. November 30, 2012, The Atlantic.

Kruuk, L. E. B. 2017. A new explanation for unexpected evolution in body size. PLoS Biology 15:e2001832.

Krymkowski, D. H., R. E. Manning, and W. A. Valliere. 2014. Race, ethnicity, and visitation to national parks in the United States: tests of the marginality, discrimination, and subculture hypotheses with national-level survey data. Journal of Outdoor Recreation and Tourism 7:35–43.

Kucharski, A. J., T. W. Russell, C. Diamond, Y. Liu, J. Edmunds, S. Funk, and R. M. Eggo. 2020. Early dynamics of transmission and control of covid-19: a mathematical modelling study. The Lancet Infectious Diseases 5:553–558.

Kuehn, D., V. Luzadis, and M. Brincka. 2013. An analysis of the factors influencing fishing participation by resident anglers. Human Dimensions of Wildlife 18:322–339.

Kulp, M. A., and S. E. Moore. 2005. A case history in fishing regulations in Great Smoky Mountains National Park: 1934-2004. North American Journal of Fisheries Management 25:510–524.

Kumar, V., and J. A. Petersen. 2012. Statistical methods in customer relationship management. John Wiley & Sons, Chichester, West Sussex, United Kingdom.

Kumar, V., B. Rajan, R. Venkatesan, and J. Lecinski. 2019. Understanding the role of artificial intelligence in personalized engagement marketing. California Management Review 61:135–155.

Kuparinen, A., A. Boit, F. S. Valdovinos, H. Lassaux, and N. D. Martinez. 2016. Fishing-induced life-history changes degrade and destabilize harvested ecosystems. Scientific Reports 6:22245.

Kuparinen, A., and J. A. Hutchings. 2012. Consequences of fisheries-induced evolution for population productivity and recovery potential. Proceedings of the Royal Society B: Biological Sciences 279:2571–2579.

Kuparinen, A., and J. A. Hutchings. 2017. Genetic architecture of age at maturity can generate divergent and disruptive harvest-induced evolution. Philosophical Transactions of the Royal Society B: Biological Sciences 372:20160035.

Kuparinen, A., and M. Festa-Bianchet. 2017. Harvest-induced evolution: insights from aquatic and terrestrial systems. Philosophical Transactions of the Royal Society B 372:20160036.

Kuparinen, A., and J. Merilä. 2007. Detecting and managing fisheries-induced evolution. Trends in Ecology & Evolution 22:652–659.

Kurt, F., G. B. Hartl, and R. Tiedemann. 1995. Tuskless bulls in Asian elephant Elephas maximus. History and population genetics of a man-made phenomenon. Acta Theriologica Supplement 3:125–143.

Kuznetsova, A., P. B. Brockhoff, and R. H. B. Christensen. 2016. lmerTest: tests in linear mixed effects models. https://CRAN.R-project.org/package=lmerTest. Accessed 24 September 2020.

Kuzyk, G., C. Procter, I. Hatter, and D. Jury. 2011. Utilizing antler point restrictions for mule deer to maximize hunter opportunity in southern British Columbia. Pages 37–45 *in* Proceedings of the 9th western states and provinces deer and elk workshop. New Mexico Department of Game and Fish, Santa Fe, New Mexico, USA.

Kvalnes, T., B.-E. Saether, H. Haanes, K. H. Røed, S. Engen, and E. J. Solberg. 2016. Harvest-induced phenotypic selection in an island population of moose, *Alces alces*. Evolution 70:1486–1500.

Kwak, T. J., and M. G. Henry. 1995. Largemouth bass mortality and related causal factors during live-release fishing tournaments on a large Minnesota lake. North American Journal of Fisheries Management 15:621–630.

Laborde, L. P., F. C. Rohwer, M. D. Kaller, and L. A. Reynolds. 2014. Surveying Louisiana waterfowl hunters: open web and random mail surveys produce similar responses to attitudinal questions. Wildlife Society Bulletin 38:821–826.

Ladd, W. N., Jr., J. C. Bartonek, K. E. Gamble, and J. R. Serie. 1989. Experiences with special harvest management strategies for ducks. Transactions of the North American Wildlife and Natural Resources Conference 54:552–565.

Ladhari, R. 2009. Service quality, emotional satisfaction, and behavioural intentions. Managing Service Quality: An International Journal 19:308–331.

Lamansky, J. A., Jr., and K. A. Meyer. 2016. Air exposure time of trout released by anglers during catch and release. North American Journal of Fisheries Management 36:1018–1023.

Lande, U. S., I. Herfindal, M. H. Finne, and L. Kastdalen. 2010. Use of hunters in wildlife surveys: does hunter and forest grouse habitat selection coincide? European Journal of Wildlife Research 56:107–115.

Landsman, S. J., H. J. Wachelka, C. D. Suski, and S. J. Cooke. 2011. Evaluation of the physiology, behaviour, and survival of adult muskellunge (*Esox masquinongy*) captured and released by specialized anglers. Fisheries Research 110:377–386.

Langeland, A., and B. Jonsson. 1990. Management of stunted populations of Arctic char (*Salvelinus alpinus*) and brown trout (*Salmo trutta*) in Norway. Pages 396–405 *in* W. L. T. van Densen, B. Steinmetz, and R. H. Hughes, editors. Management of freshwater fisheries. European Inland Fisheries Advisory Commission, Norway.

Larkin, P. A. 1977. An epitaph for the concept of maximum sustained yield. Transactions of the American Fisheries Society 106:1–11.

Larson, G. L., S. E. Moore, and D. C. Lee. 1986. Angling and electrofishing for removing nonnative rainbow trout from a stream in a national park. North American Journal of Fisheries Management 6:580–585.

Larson, L. R., D. J. Decker, R. C. Stedman, W. F. Siemer, M. S. Baumer, and J. W. Enck. 2013. Hunter recruitment and retention: a framework for research and action. Cornell University, Human Dimensions Research Unit Series Number 13-04, Ithaca, New York, USA.

Larson, L. R., R. C. Stedman, D. J. Decker, W. F. Siemer, and M. S. Baumer. 2014. Exploring the social habitat for hunting: toward a comprehensive framework for understanding hunter recruitment and retention. Human Dimensions of Wildlife 19:105–122.

Lasharr, T. N., R. A. Long, J. R. Heffelfinger, V. C. Bleich, P. R. Krausman, R. T. Bowyer, J. M. Shannon, R. W. Klaver, C. Brewer, M. Cox, A. A. Holland, A. Hubbs, C. P. Lehman, J. D. Muir, B. Sterling, and K. B. Monteith. 2019. Hunting and mountain sheep: do current harvest practices affect horn growth? Evolutionary Applications 12:1823–1836.

Lauber, T. B., and T. L. Brown. 2000. Hunting access on private lands in Dutchess County. Cornell University, Human Dimensions Research Unit Series Number 00-12, Ithaca, New York, USA.

Lauber, T. B., D. J. Decker, K. M. Leong, L. C. Chase, and T. M. Schusler. 2012. Stakeholders engagement in wildlife management. Pages 139–156 *in* D. J. Decker, S. J. Riley, and W. F. Siemer, editors. Human dimensions of wildlife management, 2nd edition. Johns Hopkins University Press, Baltimore, Maryland, USA.

Lauber, T. B., and B. A. Knuth. 1997. Fairness in moose management decision-making: the citizens' perspective. Wildlife Society Bulletin 25:776–787.

Lauber, T. B., R. C. Stedman, D. J. Decker, and B. A. Knuth. 2009. Using State Wildlife Action Plans to achieve your conservation goals through collaboration. Cornell University, Human Dimension Research Unit Series Number 09-5, Ithaca, New York, USA.

Laurian, C., J. P. Ouellet, R. Courtois, L. Breton, and S. St-Onge. 2000. Effects of intensive harvesting on moose reproduction. Journal of Applied Ecology 37:515–531.

Law, R. 2000. Fishing, selection, and phenotypic evolution. ICES Journal of Marine Science 57:659–668.

Law, R., and M. J. Plank. 2018. Balanced harvesting could reduce fisheries-induced evolution. Fish and Fisheries 19:1078–1091.

Lawler, E. E. 1973. Motivation in work organizations. Brooks/Cole, Monterey, California, USA.

Le Luyer, J., M. Laporte, T. D. Beacham, K. H. Kaukinen, R. E. Withler, J. S. Leong, E. B. Rondeau, B. F. Koop, and L. Bernatchez. 2018. Parallel epigenetic modifications induced by hatchery rearing in a Pacific salmon. Proceedings of the National Academy of Sciences 114:12964–12969.

Leafloor, J. O., T. J. Moser, and B. Batt. 2012. Evaluation of special management measures for midcontinent lesser snow geese and Ross's geese. Arctic Goose Joint Venture Special Publication, Washington, D.C., USA.

Lebreton, J.-D. 2005. Dynamical and statistical models for exploited populations. Australian & New Zealand Journal of Statistics 47:49–63.

Lebreton J.-D., K. P. Burnham, J. Clobert, and D. R. Anderson. 1992. Modeling survival and testing biological hypotheses using marked animals: a unified approach with case studies. Ecological Monographs 62:67–118.

Leclerc, M., A. Zedrosser, and F. Pelletier. 2017. Harvesting as a potential selective pressure on behavioural traits. Journal of Applied Ecology 54:1941–1945.

Lee, J., A. R. Graefe, and R. C. Burns. 2004. Service quality, satisfaction, and behavioral intention among forest visitors. Journal of Travel & Tourism Marketing 17:73–82.

Lee, S.-H., A. R. Graefe, and C.-L. Li. 2007. The effects of specialization and gender on motivations and preferences for site attributes in paddling. Leisure Sciences 29:355–373.

Lefebvre, J., G. Gauthier, J. F. Giroux, A. Reed, E. T. Reed and L. Bélanger. 2017. The greater snow goose (*Anser caerulescens atlanticus*): managing an overabundant population. Ambio 46:262–274.

Legagneux, P., P. Inchausti, F. Bourguemestre, F. Latraube, and V. Bretagnolle. 2009. Effect of predation risk, body size and habitat characteristics on emigration decisions in mallards. Behavioral Ecology 20:186–194.

Lenker, M. A., B. C. Weidel, O. P. Jensen, and C. T. Solomon. 2016. Developing recreational harvest regulations for an unexploited lake trout population. North American Journal of Fisheries Management 36:385–397.

Lennox, R. J., M. Falkegård, L. A. Vøllestad, S. J. Cooke, and E. B. Thorstad. 2016. Influence of harvest restrictions on angler release behaviour and size selection in a recreational fishery. Journal of Environmental Management 176:139–148.

Lenth, R. V. 2019. emmeans: estimated marginal means, aka least-squares means. R. https://CRAN.R-project.org/package=emmeans. Accessed 24 September 2020.

Leong, K. M., D. J. Decker, and T. B. Lauber. 2012. Stakeholders as beneficiaries of wildlife management. Pages 26–40 *in* D. J. Decker, S. J. Riley, and W. F. Siemer, editors. Human dimensions of wildlife management, 2nd edition. Johns Hopkins University Press, Baltimore, Maryland, USA.

Leopold, A. 1939. Game management. Charles Scribner's Sons, New York, New York, USA.

Leopold, A. 1968. A Sand County almanac, and sketches here and there. Oxford University Press, New York, New York, USA.

Leopold, A., L. K. Sowls and D. L. Spencer. 1947. A survey of over-populated deer ranges in the United States. The Journal of Wildlife Management 11:162–177.

Lercel, B. A., R. M. Kaminski, and R. R. Cox, Jr. 1999. Mate loss in winter affects reproduction of mallards. The Journal of Wildlife Management 63:621–629.

Leslie, P. H. 1945. On the use of matrices in certain population mathematics. Biometrika 33:183–212.

Lester, N. P., T. R. Marshall, K. Armstrong, W. I. Dunlop, and B. Ritchie. 2003. A broad-scale approach to management of Ontario's recreational fisheries. North American Journal of Fisheries Management 23:1312–1328.

Lester, N. P., B. J. Shuter, P. Venturelli, and D. Nadeau. 2014. Life-history plasticity and sustainable exploitation: a theory of growth compensation applied to walleye management. Ecological Applications 24:38—54.

Levin, S., T. Xepapadeas, A.-S. Crépin, J. Norberg, A. de Zeeuw, C. Folke, T. Hughes, K. Arrow, S. Barrett, G. Daily, P. Ehrlich, N. Kautsky, K.-G. Mäler, S. Polasky, M. Troell, J. R. Vincent, and B. Walker. 2013. Social-ecological systems as complex adaptive systems: modeling and policy implications. Environment and Development Economics 18:111–132.

Lilleeng, M. S., S. J. Hegland, K. Rydgren, and S. R. Moe. 2016. Red deer mediate spatial and temporal plant heterogeneity in boreal forests. Ecological Research 31:777–784.

Lim, J., K. Lim, J. Heinrichs, K. Al-Aali, A. Aamir, and M. Qureshi. 2018. The role of hospital service quality in developing the satisfaction of the patients and hospital performance. Management Science Letters 8:1353–1362.

Lima, S. L. 2002. Putting predators back into behavioral predator-prey interactions. Trends in Ecology & Evolution 17:70–75.

Lima, S. L., and P. A. Bednekoff. 1999. Temporal variation in danger drives antipredator behavior: the predation risk allocation hypothesis. The American Naturalist 153:649–659.

Lin, H., K. F. Robinson, M. L. Jones, and L. Walter. 2019. Using structured decision making to overcome scale mismatch challenges in barrier removal for watershed restoration. Fisheries 44:545–550.

Lincoln, F. C. 1935. The waterfowl flyways of North America. U.S. Department of Agriculture, Circular 342, Washington, D.C., USA.

Lindberg, M. S., J. S. Sedinger, and J.-D. Lebreton. 2013. Individual heterogeneity in black brant survival and recruitment with implications for harvest dynamics. Ecology and Evolution 3:4045–4056.

Lindmark, M., J. Ohlberger, M. Huss, and A. Gårdmark. 2019. Size-based ecological interactions drive food web responses to climate warming. Ecology Letters 22:778–786.

Lindqvist, S., S. Camilla, T. Bjärstig, and E. Kvastegård. 2014. The changing role of hunting in Sweden : from subsistence to ecosystem stewardship? Alces 50:35–51.

Lindsay, R. B., R. K. Schroeder, K. R. Kenaston, R. N. Toman, and M. A. Buckman. 2004. Hooking mortality by anatomical location and its use in estimating mortality of spring chinook salmon caught and released in a river sport fishery. North American Journal of Fisheries Management 24:367–378.

Ling, S., and E. J. Milner-Gulland. 2008. When does spatial structure matter in models of wildlife harvesting? Journal of Applied Ecology 45:63–71.

Link, J. S. 2002. Ecological considerations in fisheries management: when does it matter? Fisheries 27:10–17.

Logan, K. A., L. L. Sweanor, and M. G. Hornocker. 2001. Desert puma: evolutionary ecology and conservation of an enduring carnivore. Island Press, Washington, D.C., USA.

Loison, A., M. Festa-Bianchet, J.-M. Gaillard, J. T. Jorgenson, and J.-M. Jullien. 1999. Age-specific survival in five populations of ungulates: evidence of senescence. Ecology 80:2539–2554.

Lone, K., L. E. Loe, T. Gobakken, J. D. Linnell, J. Odden, J. Remmen, and A. Mysterud. 2014. Living and dying in a multi-predator landscape of fear: roe deer are squeezed by contrasting pattern of predation risk imposed by lynx and humans. Oikos 123:641–651.

Lone, K., L. E. Loe, E. L. Meisingset, I. Stamnes, and A. Mysterud. 2015. An adaptive behavioural response to hunting: surviving male red deer shift habitat at the onset of the hunting season. Animal Behaviour 102:127–138.

Long, J. M., M. S. Allen, W. F. Porak, and C. D. Suski. 2015. A historical perspective of black bass management in the United States. Pages 99–122 *in* M. D. Tringali, J. M. Long, T. W. Birdsong, and M. S. Allen, editors. Black bass diversity: multidisciplinary science for conservation. America Fisheries Society, Symposium 82, Bethesda, Maryland, USA.

Losey, R. J., T. Nomokonova, and O. I. Goriunova. 2008. Fishing ancient Lake Baikal, Siberia: inferences from the reconstruction of harvested perch (*Perca fluviatilis*) size. Journal of Archaeological Science 35:577–590.

Louison, M. J., J. A. Stein, and C. D. Suski. 2019. The role of social network behavior, swimming performance, and fish size in the determination of angling vulnerability in bluegill. Behavioral Ecology and Sociobiology 73:139.

Lubben, J., D. Boeckner, R. Rebarber, S. Townley, and B. Tenhumberg. 2009. Parameterizing the growth-decline boundary for uncertain population projection models. Theoretical Population Biology 75:85–97.

Lubben, J., B. Tenhumberg, A. Tyre, and R. Rebarber. 2008. Management recommendations based on matrix projection models: the importance of considering biological limits. Biological Conservation 141:517–523.

Lubow, B. 1995. SDP: generalized software for solving stochastic dynamic optimization problems. Wildlife Society Bulletin 23:738–742.

Lueck, M. 2000. An Economic Guide to State Wildlife Management. https://www.perc.org/2000/11/09/an-economic-guide-to-state-wildlife-management/ Accessed 13 August 2020.

Luecke, C., T. C. Edwards, Jr., M. W. Wengert, Jr., S. Brayton, and R. Schneidervin. 1994. Simulated changes in lake trout yield, trophies, and forage consumption under various slot limits. North American Journal of Fisheries Management 14:14–21.

Luellen, D. R., M. J. LaGuardia, T. D. Tuckey, M. C. Fabrizio, G. W. Rice, and R. C. Hale. 2018. Assessment of legacy and emerging contaminants in an introduced catfish and implications for the fishery. Environmental Science and Pollution Research 25:28355–28366.

Lyach, R., and M. Čech. 2019. Differences in fish harvest, fishing effort, and angling guard activities between urban and natural fishing grounds. Urban Ecosystems 22:975–987.

Lyman, J. 2002. Cultural values and change: catch and release in Alaska's sport fisheries. Page 29–36 *in* Jon A. Lucy and Anne L. Studholme, editors. Catch and release in marine recreational fisheries. American Fisheries Society, Symposium 30, Bethesda, Maryland, USA.

Lyons, J., P. D. Kanehl, and D. M. Day. 1996. Evaluation of a 356-mm minimum-length limit for smallmouth bass in Wisconsin streams. North American Journal of Fisheries Management 16:952–957.

Lyons, J., A. L. Rypel, J. F. Hansen, and D. C. Rowe. 2017. Fillet weight and fillet yield: new metrics for the management of panfish and other consumption-oriented recreational fisheries. North American Journal of Fisheries Management 37:550–557.

Lyons, J. E., M. C. Runge, H. P. Laskowski, and W. L. Kendall. 2008. Monitoring in the context of structured decision-making and adaptive management. The Journal of Wildlife Management 72:1683–1692.

Lyons, T., L. A. Powell, and M. P. Vrtiska. 2020. Effects of liberalizing harvest regulations on Canada goose (*Branta canadensis*) demography in Nebraska. Journal of Fish and Wildlife Management 11:in press.

MacArthur, R. H., and E. O. Wilson. 1967. The theory of island biogeography. Princeton University Press, Princeton, New Jersey, USA.

MacCrimmon, H. R., and W. H. Robbins. 1979. Suitability of jaw and Atkins streamer tags for marking smallmouth bass, *Micropterus dolomieu*. Transactions of the American Fisheries Society 108:499–501.

Macdonald, H. F., and K. J. Boyle. 1997. Effect of a statewide sport fish consumption advisory on open-water fishing in Maine. North American Journal of Fisheries Management 17:687–695.

Maceina, M. J., P. W. Bettoli, S. D. Finely, and V. J. DiCenzo. 1998. Analysis of the sauger fishery with simulated effects of a minimum size limit in the Tennessee River of Alabama. North American Journal of Fisheries Management 18:66–75.

Maceina, M. J., J. Boxrucker, D. L. Buckmeier, R. S. Gangl, D. O. Lucchesi, D. A. Isermann, J. R. Jackson, and P. J. Martinez. 2007. Current status and review of freshwater fish aging procedures used by state and provincial fisheries agencies with recommendations for future directions. Fisheries 32:329–340.

Maceina, M. J., and S. M. Sammons. 2015. Stocking threadfin shad to enhance largemouth bass populations in two Alabama ponds. Journal of the Southeastern Associated Fish and Wildlife Agencies 2:28–34.

Maceina, M. J., S. M. Sammons, and R. P. Phelps. 2016. Evaluation of stocking all female largemouth bass (*Micropterus salmoides*) in Alabama (USA) ponds. Natural Resources 7:315–325.

Madden, J. R., E. J. Langley, M. A. Whiteside, C. E. Beardsworth, and J. O. van Horik. 2018. The quick are the dead: pheasants that are slow to reverse a learned association survive for longer in the wild. Philosophical Transactions of the Royal Society B: Biological Sciences 373:20170297.

Madden, J. R., and M. A. Whiteside. 2014. Selection on behavioural traits during 'unselective' harvesting means that shy pheasants better survive a hunting season. Animal Behaviour 87:129–135.

Madenjian, C. P., and T. J. DeSorcie. 1999. Status of lake trout rehabilitation in the Northern Refuge of Lake Michigan. North American Journal of Fisheries Management 19:658–669.

Madenjian, C. P., J. T. Tyson, R. L. Knight, M. W. Kershner, and M. J. Hansen. 1996. First-year growth, recruitment, and maturity of walleyes in western Lake Erie. Transactions of the American Fisheries Society 125:821–830.

Madin, E. M., J. S. Madin, and D. J. Booth. 2011. Landscape of fear visible from space. Scientific Reports 1:14.

Madsen, J., K. K. Clausen, T. K. Christensen, and F. A. Johnson. 2016. Regulation of the hunting season as a tool for adaptive harvest management—first results for pink-footed geese *Anser brachyrhynchus*. Wildlife Biology 22:204–208.

Maine Department of Inland Fisheries and Wildlife. 2020. Lynx protection zone and trap restrictions. https://www.maine.gov/ifw/hunting-trapping/trapping-laws/lynx-protection.html Accessed 4 April 2020.

Mainguy, J., S. D. Côté, M. Festa-Bianchet, and D. W. Coltman. 2009. Father-offspring phenotypic correlations suggest intralocus sexual conflict for a fitness-linked trait in a wild sexually dimorphic mammal. Proceedings of the Royal Society of London B 276:4067–4075.

Malison, J. A., L. S. Procarione, T. P. Barry, A. R. Kapuscinski, and T. B. Kayes. 1994. Endocrine and gonadal changes during the annual reproductive cycle of the freshwater teleost, *Stizostedion vitreum*. Fish Physiology and Biochemistry 13:473–484.

Manfredo, M. J. 1984. The comparability of onsite and offsite measures of recreation needs. Journal of Leisure Research 16:245–249.

Manfredo, M. J., B. L. Driver, and P. J. Brown. 1983. A test of concepts inherent in experience based setting management for outdoor recreation areas. Journal of Leisure Research 15:263–283.

Manfredo, M. J., B. L. Driver, and M. A. Tarrant. 1996a. Measuring leisure motivation: a meta-analysis of the recreation experience preference scales. Journal of Leisure Research 28:188–213.

Manfredo, M. J., P. J. Fix, T. L. Teel, J. Smeltzer, and R. Kahn. 2004. Assessing demand for big-game hunting opportunities: applying the multiple-satisfaction concept. Wildlife Society Bulletin 32:1147–1155.

Manfredo, M. J., D. C. Fulton, and C. L. Pierce. 1997. Understanding voter behavior on wildlife ballot initiatives: Colorado's trapping amendment. Human Dimensions of Wildlife 2:22–39.

Manfredo, M. J., and R. A. Larson. 1993. Managing for wildlife viewing recreation experiences: an application in Colorado. Wildlife Society Bulletin 21:226–236.

Manfredo, M. J., L. Sullivan, A. W. Don Carlos, A. M. Dietsch, T. L. Teel, A. D. Bright, and J. Bruskotter. 2018. America's wildlife values: the social context of wildlife management in the U.S. Colorado State University, Department of Human Dimensions of Natural Resources, Fort Collins, Colorado, USA. https://www.fishwildlife.org/application/files/9915/4049/1625/AWV_-_National_Final_Report.pdf. Accessed 20 August 2020.

Manfredo, M. J., T. L. Teel, and A. D. Bright. 2003. Why are public values toward wildlife changing? Human Dimensions of Wildlife 8:287–306.

Manfredo, M. J., J. J. Vaske, and L. Sikorowski. 1996b. Human dimensions of wildlife management. Pages 53–72 *in* A. Ewert, editor. Natural resource management: the human dimension. Westview Press, Boulder, Colorado, USA.

Manfredo, M. J., and H. C. Zinn. 2008. Population change and its implications for wildlife management in the New West: a case study of Colorado. Human Dimensions of Wildlife 1:62–74.

Manjerovic, M. B., M. L. Green, N. Mateus-Pinilla, and J. Novakofski. 2014. The importance of localized culling in stabilizing chronic wasting disease prevalence in white-tailed deer populations. Preventive Veterinary Medicine 113:139–145.

Manville, R. H. 1966. The extinct sea mink, with taxonomic notes. Proceedings of the United States National Museum 122(3584):1–12.

Margenau, T. L., and S. P. AveLallemant. 2000. Effects of a 40-inch minimum length limit on muskellunge in Wisconsin. North American Journal of Fisheries Management 20:986–993.

Marks, S. A. 1991. Southern hunting in black and white: nature, history, and ritual in a Carolina community. Princeton University Press, Princeton, New Jersey, USA.

Martin, A. M., M. Festa-Bianchet, D. W. Coltman, and F. Pelletier. 2016. Demographic drivers of age-dependent sexual selection. Journal of Evolutionary Biology 29:1437–1446.

Martin, C., P.-P. Pastoret, B. Brochier. M.-F. Humblet, and C. Saegerman. 2011. A survey of the transmission of infectious diseases/infections between wild and domestic ungulates in Europe. Veterinary Research 42:70.

Martin, C. C. 1995. Evaluation of slot length limits for largemouth bass in two Delaware ponds. North American Journal of Fisheries Management 15:713–7719.

Martin, C. L. 2016. How nature, culture and legal calendars influence the calendrical timing of consumer behaviour. Journal of Customer Behaviour 15:337–368.

Martin, D. R., C. J. Chizinski, and K. L. Pope. 2015. Reservoir area of influence and implications for fisheries management. North American Journal of Fisheries Management 35:185–190.

Martin, D. R., and K. L. Pope. 2011. Luring anglers to enhance fisheries. Journal of Environmental Management 92:1409–1413.

Martin, E. M., and S. M. Carney. 1977. Population ecology of the mallard: IV. A review of duck hunting regulations, activity, and success, with special reference to the mallard. U.S. Fish and Wildlife Service, Resource Publication 130, Washington, D.C., USA.

Martin, F. W., R. S. Pospahala, and J. D. Nichols. 1979. Assessment and population management of North American migratory birds. Pages 187–239 *in* J. Cairns, Jr., G. P. Patil, and W. E. Walters, editors. Environmental biomonitoring, assessment, prediction, and management-certain case studies and related quantitative issues. International Cooperative Publishing House, Fairland, Maryland, USA.

Martin, R. G. 1958. More fish from farm ponds. Virginia Wildlife 19(11):10–12.

Martinez-Padilla, J., S. M. Redpath, M. Zeineddine, and F. Mougeot. 2014. Insights into population ecology from long-term studies of red grouse *Lagopus lagopus scoticus*. Journal of Animal Ecology 83:85–98.

Martorelli-Barcelo, M., A. Campos-Candela, and J. Alós. 2018. Fitness consequences of fish circadian behavioural variation in exploited marine environments. PeerJ 6:e4814.

Maryland Department of Natural Resources. 2020. Junior hunt days. https://dnr.maryland.gov/huntersguide/Pages/JrHunters.aspx. Accessed 27 October 2020.

Matsumura, S., R. Arlinghaus, and U. Dieckmann. 2011. Assessing evolutionary consequences of size-selective recreational fishing on multiple life-history traits, with an application to northern pike (*Esox lucius*). Evolutionary Ecology 25:711–735.

Matsumura, S., B. Beardmore, W. Haider, U. Dieckmann, and R. Arlinghaus. 2019. Ecological, angler, and spatial heterogeneity drive social and ecological outcomes in an integrated landscape model of freshwater recreational fisheries. Reviews in Fisheries Science & Aquaculture 27:170–197.

Mattsson, B. J., M. C. Runge, J. H. Devries, G. S. Boomer, J. M. Eadie, D. A. Haukos, J. P. Fleskes, D. N. Koons, W. E. Thogmartin, and R. G. Clark. 2012. A modeling framework for integrated harvest and habitat management of North American waterfowl: case-study of northern pintail metapopulation dynamics. Ecological Modelling 225:146–158.

Matzler, K., and E. Sauerwein. 2002. The factor structure of customer satisfaction: an empirical test of the importance grid and the penalty-reward-contrast analysis. International Journal of Service Industry Management 13:314–332.

Maunder, M. N. 2012. Evaluating the stock-recruitment relationship and management reference points: application to summer flounder (*Paralichthys dentatus*) in the U.S. mid-Atlantic. Fisheries Research 125–126:20–26.

Mawdsley, J. R., J. F. Organ, D. J. Decker, A. B. Forstchen, R. J. Regan, S. J. Riley, M. S. Boyce, J. E. McDonald, C. Dwyer, and S. P. Mahoney. 2018. Artelle et al. (2018) miss the science underlying North American wildlife management. Science Advances 4:eaat8281.

Mawson, P. R., J. O. Hampton, and B. Dooley. 2016. Subsidized commercial harvesting for cost-effective wildlife management in urban areas: a case study with kangaroo sharpshooting. Wildlife Society Bulletin 40:251–260.

McCauley, R. W., and D. M. Kilgour. 1990. Effect of air temperature on growth of largemouth bass in North America. Transactions of the American Fisheries Society 119:276–281.

McCrea, R. S., B. J. T. Morgan, and D. J. Cole. 2013. Age-dependent mixture models for recovery data on animals marked at unknown age. Journal of the Royal Statistical Society: Series C (Applied Statistics) 62:101–113.

McCullough, D. R. 1980. The George Reserve deer herd: population ecology of a K-selected species. University of Michigan Press, Ann Arbor, Michigan, USA.

McCullough, D. R. 1996. Spatially structured populations and harvest theory. The Journal of Wildlife Management 60:1–9.

McDaniels, T., H. Longstaff, and H. Dowlatabadi. 2006. A value-based framework for risk management decisions involving multiple scales: a salmon aquaculture example. Environmental Science & Policy 9:423–438.

McDougall, M. B., and C. L. Amundson. 2017. Harvest dynamics and annual survival of mallards and grey ducks. The Journal of Wildlife Management 81:449–460.

McGinnis, M. D., and E. Ostrom. 2014. Social-ecological system framework: initial changes and continuing challenges. Ecology and Society 19:30.

McGowan, C. P. 2013. A structured approach to incidental take decision making. Environmental Management 41:241–250.

McGowan, C. P., D. R. Smith, J. D. Nichols, J. E. Lyons, J. Sweka, K. Kalasz, L. J. Niles, R. Wong, J. Brust, M. Davis, and B. Spear. 2015. Implementation of a framework for multi-species, multi-objective adaptive management in Delaware Bay. Biological Conservation 191:759–769.

McGrath, D. J., T. M. Terhune, and J. A. Martin. 2018. Northern bobwhite foraging response to hunting. The Journal of Wildlife Management 82:966–976.

McGuigan, L. 2017. The hunting industry: exploring the marriage of consumerism, sport hunting, and commercial entertainment. Journal of Consumer Culture 17:910–930.

McIlhenny, E. A. 1932. The blue goose in its winter home. Auk 49:279–306.

McLellan, B., and V. Banci. 1999. Status and management of the brown bear in Canada. Pages 46–50 *in* C. Servheen, S. Herrero, and B. Peyton, editors. Bears: status survey and conservation action plan. International Union for Conservation of Nature and Natural Resources/Species Survival Commission Bear and Polar Bear Specialists Groups, International Union for Conservation of Nature, Gland, Switzerland.

McMullin, S. L., V. DiCenzo, R. Essig, C. Bonds, R. L. DeBruyne, M. A. Kaemingk, M. E. Mather, C. Myrick, Q. E. Phelps, T. M. Sutton, and J. R. Triplett. 2016. Are we preparing the next generation of fisheries professionals to succeed in their careers? A survey of AFS members. Fisheries 41:436–449.

Meade, A. W., and S. B. Craig. 2012. Identifying careless responses in survey data. Psychological Methods 17:437–455.

Mech, L. D. 1978. Report of the working group on bobcat, lynx, and river otter, to the Endangered Species Scientific Authority, 28 March 1978. U.S. Endangered Species Scientific Authority, Falls Church, Virginia, USA.

Mehmood, S., D. Zhang, and J. Armstrong. 2003. Factors associated with declining hunting license sales in Alabama. Human Dimensions of Wildlife 8:243–262.

Meka, J. M. 2004. The influence of hook type, angler experience, and fish size on injury rates and the duration of capture in an Alaskan catch-and-release rainbow trout fishery. North American Journal of Fisheries Management 24:1309–1321.

Melstrom, R. T. 2018. How do license prices affect participation in recreational fishing? Evidence from Pennsylvania. Human Dimensions of Wildlife 23:273–283.

Melstrom, R. T. 2021. Expanding participation in recreational fishing through multiple license markets. *in* K. L. Pope and L. A. Powell, editors. Harvest of fish and wildlife: new paradigms for sustainable management. CRC Press, Boca Raton, Florida, USA.

Mendoza, G. A., and H. Martins. 2006. Multi-criteria decision analysis in natural resource management: a critical review of methods and new modelling paradigms. Forest Ecology and Management 230:1–22.

Merrill, K., and A. Graefe. 1998. The relationship between activity specialization and preferences for setting and route attributes of selected rock climbers. Pages 40–43 in H. G. Vogelsong, editor. Proceedings of the 1997 northeastern recreation research symposium. U.S. Forest Service, Northeastern Forest Experiment Station, General Technical Report NE-421, Newtown Square, Pennsylvania, USA.

Messinger, L. N., E. F. Stuber, C. J. Chizinski, and J. J. Fontaine. 2019. Mortality, perception, and scale: understanding how predation shapes space use in a wild prey population. PLoS One 14:e0222272.

Messmer, T. A. 2000. The emergence of human-wildlife conflict management: turning challenges into opportunities. International Biodeterioration and Biodegradation 45:97–102.

Metcalf, A. L., E. C. Metcalf, K. Khumalo, J. Gude, Q. Kujala, and M. S. Lewis. 2017. Public wildlife management on private lands: reciprocity, population status, and stakeholders' normative beliefs. Human Dimensions of Wildlife 22:564–582.

Metcalf, E. C., A. R. Graefe, N. E. Trauntvein, and R. C. Burns. 2015. Understanding hunting constraints and negotiation strategies: a typology of female hunters. Human Dimensions of Wildlife 20:30–46.

Michaels, S., and A. J. Tyre. 2012. How indeterminism shapes ecologists' contributions to managing socio-ecological systems. Conservation Letters 5:289–295.

Michigan Department of Natural Resources. 2020. 2020 Michigan fishing guide. Online fishing regulations guide. https://www.michigan.gov/documents/dnr/FishingGuide2020_684742_7.pdf. Accessed 4 August 2020.

Miller, C. A., L. K. Campbell, J. A. Yeagle, and R. J. Williams. 2002. Results of studies of Illinois hunters, landowners, and participants in Access Illinois. Illinois Natural History Survey, Human Dimensions Program Report SR-02-01, Champaign, Illinois, USA.

Miller, C. A., and A. R. Graefe. 2000. Degree and range of specialization across related hunting activities. Leisure Sciences 22:195–204.

Miller, C. A., and A. R. Graefe. 2001. Effect of harvest success on hunter attitudes toward white-tailed deer management in Pennsylvania. Human Dimensions of Wildlife 6:189–203.

Miller, C. A., and J. Vaske. 2003. Individual and situational influences on declining hunter effort in Illinois. Human Dimensions of Wildlife 8:263–276.

Miller, K. V. and R. L. Marchington. 2007. Quality whitetails, the why and how of quality deer management. Stackpole Books, Mechanicsburg, Pennsylvania, USA.

Miller, K. V., L. Muller, and S. Demarais. 2003. White-tailed deer. Pages 906–930 *in* G. A. Feldhamer, B. C. Thompson, and J. A. Chapman, editors. Wild mammals of North America, 2nd edition. John Hopkins University Press, Baltimore, Maryland, USA.

Miller, L. R., and D. J. Herrig. 2001. Creel survey of Brushy Creek, 2001. Iowa Department of Natural Resources, Des Moines, Iowa, USA.

Miller, S. D. 2007. Distinct population segments and grizzly bear delisting in Yellowstone: a response to Rosen. Ursus 18:117–123.

Miller, S. W. 1998. The biopolitics of Mississippi Valley population management: the Wisconsin perspective. Pages 467–474 *in* D. H. Rusch, M. D. Samuel, D. D. Humburg, and B. D. Sullivan, editors. Biology and management of Canada geese. Proceedings International Canada Goose Symposium, Milwaukee, Wisconsin, USA.

Miller, T. W. 2015. Marketing data science: modeling techniques in predictive analytics with R and Python. Pearson, Old Tappan, New Jersey, USA.

Millspaugh, J. J., M. S. Boyce, D. R. Diefenbach, L. P. Hansen, K. Kammermeyer, and J. R. Skalski. 2007. An evaluation of the SAK model as applied in Wisconsin. https://dnr.wi.gov/topic/wildlifehabitat/documents/SAKAuditFinalReport.pdf. Accessed 27 June 2020.

Millspaugh, J. J., J. R. Skalski, R. L. Townsend, D. R. Diefenbach, M. S. Boyce, L. P. Hansen, and K. Kammermeyer. 2009. An evaluation of sex-age-kill (SAK) model performance. The Journal of Wildlife Management 73:442–451.

Milner, J. M., C. Bonenfant, A. Mysterud, J.-M. Gaillard, S. Csányi, N. C. Stenseth. 2006. Temporal and spatial development of red deer harvesting in Europe: biological and cultural factors. Journal of Applied Ecology 43:721–734.

Milner, J. M., E. B. Nilsen, and H. P. Andreassen. 2007. Demographic side effects of selective hunting in ungulates and carnivores. Conservation Biology 21:36–47.

Mincher, B. J. 2002. Harvest as a component of Greater Yellowstone Ecosystem grizzly bear management. Wildlife Society Bulletin 30:1287–1292.

Miner, J. G. and J. Bart. 1989. Woodcock abundance and hunter success: regional trends from hunter surveys. Wildlife Society Bulletin 17:258–263.

Ministry of Water, Land, and Air Protection. 2002. Grizzly bears in British Columbia: ecology, conservation and management. https://www2.gov.bc.ca/gov/content/environment/plants-animals-ecosystems/wildlife/wildlife-conservation/grizzly-bear. Accessed 25 February 2020.

Minnesota Department of Natural Resources. 2018. Lake Vermilion fisheries management plan 2017-2022. Minnesota Department of Natural Resources, Special Publication 183, St. Paul, Minnesota, USA.

Minnesota Department of Natural Resources. 2020. Minnesota fishing regulations: effective March 2020 through February 28, 2021. Minnesota Department of Natural Resources, St. Paul, Minnesota, USA.

Minnis, D. L. 1998. Wildlife policy-making by the electorate: an overview of citizen-sponsored ballot measures on hunting and trapping. Wildlife Society Bulletin 26:75–83.

Miranda, L. E., M. E. Colvin, A. C. Shamaskin, L. A. Bull, T. Holman, and R. Jones. 2017. Length limits fail to restructure a largemouth bass population: a 28-year case history. North American Journal of Fisheries Management 37:624–632.

Mississippi Flyway Council. 1996. Mississippi Flyway giant Canada goose management plan. U.S. Fish and Wildlife Service, Mississippi Flyway Council Technical Section, Fort Snelling, Minnesota, USA.

Mitchell, B. R., M. M. Jaeger, and R. H. Barrett. 2004. Coyote depredation management: current methods and research needs. Wildlife Society Bulletin 32:1209–1218.

Mitchell, M. S., H. Cooley, J. A. Gude, J. Kolbe, J. J. Nowak, K. M. Proffitt, S. N. Sells, and M. Thompson. 2018. Distinguishing values from science in decision making: setting harvest quotas for mountain lions in Montana. Wildlife Society Bulletin 42:13–21.

Mitchell, M. S., J. A. Gude, N. J. Anderson, J. M. Ramsey, M. J. Thompson, M. G. Sullivan, V. L. Edwards, C. N. Gower, J. F. Cochrane, E. R. Irwin, and T. Walshe. 2013. Using structured decision making to manage disease risk for Montana wildlife. Wildlife Society Bulletin 37:107–114.

Moa, P. F., L. F. Eriksen, and E. B. Nilsen. 2017. Harvest regulations and implementation uncertainty in small game harvest management. Frontiers in Ecology and Evolution 5:118.

Moen, A. N., C. W. Severinghaus, and R. A. Moen. 1986. Deer CAMP computer-assisted management program operating manual and tutorial. CornerBrook Press, Lansing, New York, USA.

Monk, C. T. 2019. Mining the behavioural reality of fish-fisher interactions to understand vulnerability to hook-and-line fishing. Dissertation, Humboldt University, Berlin, Germany.

Monk, C. T., and R. Arlinghaus. 2018. Eurasian perch, *Perca fluviatilis*, spatial behaviour determines vulnerability independent of angler skill in a whole-lake reality mining experiment. Canadian Journal of Fisheries and Aquatic Sciences 75:417–428.

Monteith, K. L., R. A. Long, V. C. Bleich, J. R. Heffelfinger, P. R. Krausman, and R. T. Bowyer. 2013. Effects of harvest, culture, and climate on trends in size of horn-like structures in trophy ungulates. Wildlife Monographs 183:1–26.

Moody-Carpenter, C. J., G. G. Sass, L. D. Frankland, E. K. Bollinger, and R. E. Colombo. 2017. Demographics and interstate management implications for flathead catfish in the lower Wabash River, Illinois and Indiana. North American Journal of Fisheries Management 37:1265–1273.

Moon, J. A., D. A. Haukos, and W. C. Conway. 2017. Seasonal survival of adult female mottled ducks. The Journal of Wildlife Management 81:461–469.

Moore, J. L., and M. C. Runge. 2012. Combining structured decision making and value-of-information analyses to identify robust management strategies. Conservation Biology 26:810–820.

Moreau, C. M., and B. G. Matthias. 2018. Using limited data to identify optimal bag and size limits to prevent overfishing. North American Journal of Fisheries Management 38:747–758.

Morgan, M., and H. Yun. 2018. Perception of Asian carp as a possible food source among Missouri anglers. Human Dimensions of Wildlife 23:491–498.

Morina, D. L., C. P. Hansen, L. P. Hansen, and J. J. Millspaugh. 2021. The efficacy of antler harvest regulations in meeting management objectives. *in* K. L. Pope and L. A. Powell, editors. Harvest of fish and wildlife: new paradigms for sustainable management. CRC Press, Boca Raton, Florida, USA.

Morita, K. 2019. Earlier migration timing of salmonids: an adaptation to climate change or maladaptation to the fishery? Canadian Journal of Fisheries and Aquatic Sciences 76:475–479.

Mosel, K. J., D. A. Isermann, and J. F. Hansen. 2015. Evaluation of daily creel and minimum length limits for black crappie and yellow perch in Wisconsin. North American Journal of Fisheries Management 35:1–13.

Munger, C. R., and J. E. Kraai. 1997. Evaluation of length and bag limits for walleyes in Meredith Reservoir, Texas. North American Journal of Fisheries Management 17:438–445.

Munn, I. A., A. Hussain, S. Spurlock, and J. E. Henderson. 2010. Economic impact of fishing, hunting, and wildlife-associated recreation expenditures on the southeast U.S. regional economy: an input-output analysis. Human Dimensions of Wildlife 15:433–449.

Munro, R. E., and C. F. Kimball. 1982. Population ecology of the mallard. VII. Distribution and derivation of the harvest. U.S. Fish and Wildlife Service Resource Publication 147.

Muoneke, M. I., and W. M. Childress. 1994. Hooking mortality: a review for recreational fisheries. Reviews in Fisheries Science 2:123–156.

Musiani, M., T. Myhly, C. Callaghan, C. C. Gates, M. E. Smith, S. Stone, and E. Tosoni. 2004. Wolves in rural agricultural areas of western North America: conflict and conservation. Pages 51–80 *in* N. Fascione, A. Delach, and M. E. Smith, editors. People and predators: from conflict to coexistence. Island Press, Washington, D.C., USA.

Muth, R. M. 2000. On the destiny of deer camps and duck blinds: the rise of the animal rights movement and the future of wildlife conservation. Wildlife Society Bulletin 28:841–851.

Muth, R. M., and J. F. Bowe. 1998. Illegal harvest of renewable natural resources in North America: toward a typology of the motivations for poaching. Society & Natural Resources 11:9–24.

Myers, R. A., J. A. Hutchings, and N. J. Barrowman. 1997. Why do fish stocks collapse? The example of cod in Atlantic Canada. Ecological Applications 7:91–106.

Myers, R. A., J. Taylor, M. S. Allen, and T. F. Bonvechio. 2008. Temporal trends in voluntary release of largemouth bass. North American Journal of Fisheries Management 28:428–433.

Mysterud, A., and R. Bischof. 2010. Can compensatory culling offset undesirable evolutionary consequences of trophy hunting? Journal of Animal Ecology 79:148–160.

Mysterud, A., T. Coulson, and N. C. Stenseth. 2002. The role of males in the dynamics of ungulate populations. Journal of Animal Ecology 71:907–915.

Mysterud, A., O. Strand, and C. M. Rolandsen. 2019. Efficacy of recreational hunters and marksmen for host culling to combat chronic wasting disease in reindeer. Wildlife Society Bulletin 43:683–692.

Nagel, W. O., F. W. Sampson, and A. Brohn. 1955. Predator control: why and how. Missouri Department of Conservation, Jefferson City, Missouri, USA.

Naidoo, R., L. C. Weaver, R. W. Diggle, G. Matongo, G. Stuart-Hill, and C. Thouless. 2016. Complementary benefits of tourism and hunting to communal conservancies in Namibia. Conservation Biology 30:628–638.

Nakashima, D., L. Prott, and P. Bridgewater. 2000. Tapping into the world's wisdom. *UNESCO Sources* 125(5):12.

Näslund, I., F. Nordwall, T. Eriksson, D. Hannersjö, and L.-O. Eriksson. 2005. Long-term responses of a stream-dwelling grayling population to restrictive fishing regulations. Fisheries Research 72:323–332.

Natcher, D. C. 2009. Subsistence and the social economy of Canada's Aboriginal North. Northern Review 30:83–98.

Natcher, D. C., L. Felt, K. Chaulk, A. Procter, and the Nunatsiavut Governments. 2011. Monitoring the domestic harvest of migratory birds in Nunatsiavut, Labrador. Arctic 64:362–366.

National Flyway Council and Wildlife Management Institute. 2006. National duck hunter survey, 2005 national report. D. J. Case and Associates, Minnewaska, Indiana, USA.

National Oceanic and Atmospheric Administration. 2019. Population assessments. https://www.fisheries.noaa.gov/topic/population-assessments. Accessed 11 September 2020.

National Oceanic and Atmospheric Administration. 2020a. Fisheries Office of Science and Technology, commercial landings query. https://foss.nmfs.noaa.gov/apexfoss/f?p=215:200. Accessed 14 October 2020.

National Oceanic and Atmospheric Administration. 2020b. Stock assessment model descriptions. National Oceanic and Atmospheric Administration, Fisheries. https://www.fisheries.noaa.gov/insight/stock-assessment-model-descriptions. Accessed 26 June 2020.

National Oceanic and Atmospheric Administration. 2020c. Tropical cyclone climatology. https://www.nhc.noaa.gov/climo/. Accessed 15 June 2020.

Nebraska Game and Parks Commission. 2020. Mountain lions in Nebraska. http://outdoornebraska.gov/mountainlions/. Accessed 15 June 2020.

Neckles, H. A., J. E. Lyons, G. R. Guntenspergen, W. G. Shriver, and S. C. Adamowicz. 2015. Use of structured decision making to identify monitoring variables and management priorities for salt marsh ecosystems. Estuaries and Coasts 38:1215–1232.

Needham, M. D., J. J. Vaske, M. P. Donnelly, and M. J. Manfredo. 2007. Hunting specialization and its relationship to participation in response to chronic wasting disease. Journal of Leisure Research 39:413–437.

Needle, D. B., V. C. Burnell, M. J. Forzán, E. J. Dubovi, K. L. Schuler, C. Bernier, N. A. Hollingshead, J. C. Ellis, B. A. Stevens, P. Tate, E. Anis, and R. P. Wilkes. 2019. Infection of eight mesocarnivores in New Hampshire and Vermont with a distinct clade of canine distemper virus in 2016–2017. Journal of Veterinary Diagnostic Investigation 31:562–567.

Neely, B. C., S. F. Steffen, S. T. Lynott, and J. D. Koch. 2015. Review of paddlefish management in Kansas from 1972 to 2013 and implications for future conservation. Journal of the Southeastern Association of Fish and Wildlife Agencies 2:20–27.

Nepal, V., and M. C. Fabrizio. 2019. High salinity tolerance of invasive blue catfish suggests potential for further range expansion in the Chesapeake Bay region. PLoS One 14(11):e0224770.

Nepal, V., and M. C. Fabrizio. 2020. Density-dependence mediates the effects of temperature on growth of juvenile blue catfish in nonnative habitats. Transactions of the American Fisheries Society 149:108–120.

New York State Department of Environmental Conservation. 2011. Management plan for white-tailed deer in New York State 2012-2016. Bureau of Wildlife, Division of Fish, Wildlife & Marine Resources, New York State Department of Environmental Conservation, Albany, New York, USA.

New York State Department of Environmental Conservation. 2019. Waterfowl season setting process. https://www.dec.ny.gov/outdoor/40737.html#:~:text=2019%2D2024%20Duck%20Season%20Structures&text=Western%20Zone%20%2D%20Beginning%20the%203rd,%2D%20Jan%205%2C%202020). Accessed 9 October 2020.

New York State Department of Environmental Conservation. 2020. White-tailed deer. https://www.dec.ny.gov/animals/6965.html. Accessed 4 February 2020.

Newig, J. 2007. Does public participation in environmental decisions lead to improved environmental quality? Towards an analytical framework. Communication, Cooperation, Participation (International Journal of Sustainability Communication) 1:51–71.

Newman, S. P., and M. H. Hoff. 2000. Evaluation of a 16-inch minimum length limit for smallmouth bass in Pallette Lake, Wisconsin. North American Journal of Fisheries Management 20:90–99.

Ng, E. L., J. P. Fredericks, and M. C. Quist. 2016. Population dynamics and evaluation of alternative management strategies for nonnative lake trout in Priest Lake, Idaho. North American Journal of Fisheries Management 36:40–54.

Nichols, J. D. 1991. Responses of North American duck populations to exploitation. Pages 498–525 *in* C. M. Perrins, J.-D. Lebreton, G. Hirons, editors. Bird population studies: their relevance to conservation and management. Oxford University Press, Oxford, England, United Kingdom.

Nichols, J. D. 2000. Evolution of harvest management for North American waterfowl: selective pressures and preadaptations for adaptive harvest management. Transactions of the North American Wildlife and Natural Resources Conference 65:65–77.

Nichols, J. D., M. J. Conroy, D. R. Anderson, and K. P. Burnham. 1984a. Compensatory mortality in waterfowl populations: a review of the evidence and implications for research and management. Transactions of the North American Wildlife and Natural Resources Conference 49:535–554.

Nichols, J. D., and J. E. Hines. 1983. The relationship between harvest and survival rates of mallards. A straightforward approach with partitioned data sets. The Journal of Wildlife Management 43:334–348.

Nichols, J. D., J. E. Hines, and K. H. Pollock. 1984b. Effects of permanent trap response in capture probability on Jolly-Seber capture-recapture model estimates. The Journal of Wildlife Management 48:289–294.

Nichols, J. D., and F. A. Johnson. 1989. Evaluation and experimentation with duck management strategies. Transactions of the North American Wildlife and Natural Resources Conference 54:566–593.

Nichols, J. D., F. A. Johnson, and B. K. Williams. 1995. Managing North American waterfowl in the face of uncertainty. Annual Review of Ecology and Systematics 26:177–199.

Nichols, J. D., and W. L. Kendall. 1995. The use of multi-state capture-recapture models to address questions in evolutionary ecology. Journal of Applied Statistics 22:835–846.

Nichols, J. D., M. D. Koneff, P. J. Heglund, M. G. Knutson, M. E. Seamans, J. E. Lyons, J. M. Morton, M. T. Jones, G. S. Boomer, and B. K. Williams. 2011. Climate change, uncertainty, and natural resource management. The Journal of Wildlife Management 75:6–18.

Nichols, J. D., M. C. Runge, F. A. Johnson, and B. K. Williams. 2007. Adaptive harvest management of North American waterfowl populations: a brief history and future prospects. Journal of Ornithology 148:S343–S349.

Nichols, J. D., S. L. Stokes, J. E. Hines, and M. J. Conroy. 1982. Additional comments on the assumption of homogeneous survival rates in modern bird banding estimation models. The Journal of Wildlife Management 46:953–962.

Nicolaisen, M., K. Thorsen, and S. H. Eriksen. 2012. Jump into the void? Factors related to a preferred retirement age: gender, social interests, and leisure activities. The International Journal of Aging and Human Development 75:239–271.

Nielsen, C. K., W. F. Porter, and H. B. Underwood. 1997. An adaptive management approach to controlling suburban deer. Wildlife Society Bulletin 25:470–477.

Nilsson, J., H. Flink, and P. Tibblin. 2019. Predator-prey role reversal may impair the recovery of declining pike populations. Journal of Animal Ecology 88:927–939.

Nilsson, P. A., and C. Bronmark. 2000. Prey vulnerability to a gape-size limited predator: behavioural and morphological impacts on northern pike piscivory. Oikos 88:539–546.

Noble, R. L. 2002. Reflections on 25 years of progress in black bass management. Pages 419–431 *in* D. P. Philipp and M. S. Ridgeway, editors. Black bass: ecology, conservation, and management. American Fisheries Society, Symposium 31, Bethesda, Maryland, USA.

Nobel, R. L., and W. T. Jones. 1999. Managing fisheries with regulations. Pages 455–477 *in* C. Kohler and W. A. Hubert, editors. Inland fisheries management in North America, 2nd edition. American Fisheries Society, Bethesda, Maryland, USA.

North American Waterfowl Management Plan. 2018. North American Waterfowl Management Plan (NAWMP) update: connecting people, waterfowl, and wetlands. U.S. Department of Interior,

Washington, D.C., USA, Environment and Climate Change, Gatineau, Quebec, Canada, and Secretariat of Environment and Natural Resources, Mexico City, Mexico.

Norton, A. S., D. R. Diefenbach, C. S. Rosenberry, and B. D. Wallingford. 2013. Incorporating harvest rates into the sex-age-kill model for white-tailed deer: PASAK model evaluation. The Journal of Wildlife Management 77:606–615.

Novak, M. 1987a. Traps and trap research. Pages 941–969 *in* M. Novak, J. A. Baker, M. E. Obbard, and B. Malloch, editors. Wild furbearer management and conservation in North America. Ontario Trappers Association, North Bay, Ontario, Canada.

Novak, M. 1987b. Wild furbearer management in Ontario. Pages 1049–1061 *in* M. Novak, J. A. Baker, M. E. Obbard, and B. Malloch, editors. Wild furbearer management and conservation in North America. Ontario Trappers Association, North Bay, Ontario, Canada.

Novinger, G. D. 1984. Observations on the use of size limits for black basses in large impoundments. Fisheries 9(4):2–6.

Novinger, G. D. 1987. Evaluation of a 15.0-inch minimum length limit on largemouth bass and spotted bass catches at Table Rock Lake, Missouri. North American Journal of Fisheries Management 7:260–272.

Novinger, G. D. 1990. Slot length limits for largemouth bass in small private impoundments. North American Journal of Fisheries Management 10:330–337.

Nowell, K., and P. Jackson. 1996. Wild cats: status survey and conservation action plan. IUCN/SSC Cat Specialist Group. Burlington Press, Cambridge, England, United Kingdom.

Noyes, J. H., B. K. Johnson, L. D. Bryant, S. L. Findholt, and J. W. Thomas. 1996. Effects of bull age on conception dates and pregnancy rates of cow elk. The Journal of Wildlife Management 60:508–517.

Nugent, G., and K. Mawhinney. 1987. Recreational hunters' views on fallow deer management in the Blue Mountains, Otago. New Zealand Journal of Forestry 32:32–35.

O'Dea, A., M. L. Shaffer, D. R. Doughty, T. A. Wake, and F. A. Rodriguez. 2014. Evidence of size-selective evolution in the fighting conch from prehistoric subsistence harvesting. Proceedings of the Royal Society B: Biological Sciences 281:20140159.

Obbard, M. E. 1987. Fur grading and pelt identification. Pages 717–826 *in* M. Novak, J. A. Baker, M. E. Obbard, and B. Malloch, editors. Wild furbearer management and conservation in North America. Ontario Trappers Association, North Bay, Ontario, Canada.

Odell, D. 2000. A sportsman's task force for establishing waterfowl seasons. Transactions of the North American Wildlife and Natural Resources Conference 65:310–314.

Oele, D. L., A. L. Rypel, J. Lyons, P. Cunningham, and T. Simonson. 2016. Do higher size and reduced bag limits improve northern pike size structure in Wisconsin lakes? North American Journal of Fisheries Management 36:982–994.

Office of Legislative Auditor. 2016. Evaluation report: Department of Natural Resources: deer population management. Minnesota Office of Legislative Auditor, St. Paul, Minnesota, USA. https://www.auditor.leg.state.mn.us/ped/pedrep/deermanagement.pdf. Accessed 27 June 2020.

Office of the Auditor General of British Columbia. 2017. An independent audit of grizzly bear management, October 2017, revised 10/31/2017. https://www.bcauditor.com/sites/default/files/publications/reports/FINAL_Grizzly_Bear_Management.pdf. Accessed 25 February 2020.

Ogle, D. H. 2016. Introductory fisheries analysis with R. CRC Press, Boca Raton, Florida, USA.

Oh, C.-O., R. B. Ditton, B. Gentner, and R. Riechers. 2005. A stated preference choice approach to understanding angler preferences for management options. Human Dimensions of Wildlife 10:173–186.

Oh, H., and S. C. Parks. 1996. Customer satisfaction and service quality: a critical review of the literature and research implications for the hospitality industry. Hospitality Research Journal 20:35–64.

Ólafsdóttir, R., and M. C. Runnström. 2013. Assessing hiking trails condition in two popular tourist destinations in the Icelandic highlands. Journal of Outdoor Recreation and Tourism 3:57–67.

Olin, M., J. Tiainen, M. Rask, M. Vinni, K. Nyberg, and H. Lehtonen. 2017. Effects of non-selective and size-selective fishing on perch populations in a small lake. Boreal Environment Research 22:137–155.

Olsen, E. M., M. Heino, G. R. Lilly, M. J. Morgan, J. Brattey, B. Ernande, and U. Dieckmann. 2004. Maturation trends indicative of rapid evolution preceded the collapse of northern cod. Nature 428:932–935.

Ontario Ministry of Natural Resources. 2006. Regulatory guidelines for managing the recreational fishery for largemouth and smallmouth bass in Ontario. Fisheries Section, Fish and Wildlife Branch, Toronto, Ontario, Canada.

Oregon Department of Fish and Wildlife. 2017. 2017 Oregon cougar management plan. https://www.dfw.state.or.us/wildlife/cougar/docs/2017_Oregon_Cougar_Management_Plan.pdf. Accessed 3 April 2020.

Oregon Department of Fish and Wildlife. 2020a. Oregon furbearer trapping and hunting regulations: July 1, 2020 through June 30, 2022. Oregon Department of Fish and Wildlife, Salem, Oregon, USA.

Oregon Department of Fish and Wildlife. 2020b. Oregon sport fishing regulations: 2020. Oregon Department of Fish and Wildlife, Salem, Oregon, USA.

Organ, J. F., D. J. Decker, S. S. Stevens, T. M. Lama and C. Doyle-Capitman. 2014. Public trust principles and trust administration functions in the North American model of wildlife conservation: contributions of human dimensions research. Human Dimensions of Wildlife 19:407–416.

Organ, J. F., V. Geist, S. P. Mahoney, S. Williams, P. R. Krausman, G. R. Batcheller, T. A. Decker, R. Carmichael, P. Nanjappa, R. Regan, R. A. Medellin, R. Cantu, R. E. McCabe, S. Craven, G. M. Vecellio, and D. J. Decker. 2012. The North American model of wildlife conservation. The Wildlife Society, Technical Review 12-04, Bethesda, Maryland, USA.

Organ, J. F., R. F. Gotie, T. A. Decker, and G. R. Batcheller. 1998. A case study in the sustained use of wildlife: the management of beaver in the northeastern United States. Pages 125–139 *in* H. A. van der Linke and M. H. Danskin, editors. Enhancing sustainability: resources for our future. International Union for Conservation of Nature and Natural Resources, Gland, Switzerland.

Organ, J. F., S. P. Mahoney, and V. Geist. 2013. Born in the hands of hunters: the North American model of wildlife conservation. Pages 69–75 *in* P. R. Krausman and B. D. Leopold, editors. Essential readings in wildlife management and conservation. Johns Hopkins University Press, Baltimore, Maryland, USA.

Ortman, J. M., and C. E. Guarneri. 2009. United States population projections: 2000 to 2050. U.S. Census Bureau, Washington, D.C., USA. https://www.census.gov/content/dam/Census/library/working-papers/2009/demo/us-pop-proj-2000-2050/analytical-document09.pdf. Accessed 9 October 2020.

Osnas, E. E., M. C. Runge, B. J. Mattson, J. Austin, G. S. Boomer, R. G. Clark, P. Devers, J. M. Eadie, E. V. Lonsdorf, and B. G. Tavernia. 2014. Managing harvest and habitat as integrated components. Wildfowl (Special Issue) 4:305–328.

Ostrom, E. 2009. A general framework for analyzing sustainability of social-ecological systems. Science 325:419.

Otis, D. L. 2002. Survival models for harvest management of mourning dove populations. The Journal of Wildlife Management 66:1052–1063.

Otis, D. L. 2006. A mourning dove hunting regulation strategy based on annual harvest statistics and banding data. The Journal of Wildlife Management 70:1302–1307.

Otis, D. L., and G. C. White. 2002. Re-analysis of a banding study to test the effects of an experimental increase in bag limits of mourning doves. Journal of Applied Statistics 29:479–495.

Otis, D. L., and G. C. White. 2004. Evaluation of ultrastructural and random effects band recovery models for estimating relationships between survival and harvest rates in exploited populations. Animal Biodiversity and Conservation 27:157–173.

Otis, K. J., R. R. Piette, J. K. Keppler, and P. W. Rasmussen. 1998. A largemouth bass closed fishery to control an overabundant bluegill population in a Wisconsin lake. Journal of Freshwater Ecology 13:391–403.

Otterstrom, S. M. 2001. Trends in national and regional population concentration in the United States from 1790 to 1990: from the frontier to the urban transformation. Social Science Journal 38:393–407.

Otterstrom, S. M. 2003. Population concentration in United States city-systems from 1790 to 2000: historical trends and current phases. Tijdschrift voor Economische en Sociale Geografie 94:477–495.

Padding, P. I., J.-F. Gobeil, and C. Wentworth. 2006. Estimating waterfowl harvest in North America. Pages 849–852 *in* G. C. Boere, C. A. Galbraith, and D. A. Stroud, editors. Waterbirds around the world. The Stationery Office, Edinburgh, Scotland.

Padding, P. I., G. S. Zimmerman, M. T. Huang, G. D. Balkcom, and P. K. Devers. 2018. A multi-stock adaptive harvest management protocol for informing annual harvest regulations in the Atlantic flyway. 9 March 2018. Unpublished report.

Page, K. S., G. C. Grant, P. Radomski, T. S. Jones, and R. E. Bruesewitz. 2004. Fish total length measurement error from recreational anglers: causes and contribution to noncompliance for the Mille Lacs walleye fishery. North American Journal of Fisheries Management 24:939–951.

Page, K. S., and P. Radomski. 2006. Compliance with sport fishery regulations in Minnesota as related to regulation awareness. Fisheries 31:166–178.

Pahl-Wostl, C. 2009. A conceptual framework for analysing adaptive capacity and multi-level learning processes in resource governance regimes. Global Environmental Change 19:354–365.

Pain, D. J., I. Dickie, R. Green, N. Kanstrup, and R. Cromie. 2019. Wildlife, human and environmental costs of using lead ammunition: an economic review and analysis. Ambio 48:969–988.

Palkovacs, E. P. 2011. The overfishing debate: an eco-evolutionary perspective. Trends in Ecology & Evolution 26:616–617.

Palkovacs, E. P., B. A. Wasserman, and M. T. Kinnison. 2011. Eco-evolutionary trophic dynamics: loss of top predators drives trophic evolution and ecology of prey. PLoS One0 6:e18879.

Pang, A. 2017. Incorporating the effect of successfully bagging big game into recreational hunting: an examination of deer, moose and elk hunting. Journal of Forest Economics 28:12–17.

Papenfuss, J. T., N. Phelps, D. Fulton, and P. A. Venturelli. 2015. Smartphones reveal angler behavior: a case study of a popular mobile fishing application in Alberta, Canada. Fisheries 40:318–327.

Papworth, S. K., J. Rist, L. Coad, and E. J. Milner-Gulland. 2009. Evidence for shifting baseline syndrome in conservation. Conservation Letters 2:93–100.

Paragamian, V. L. 1982. Catch rates and harvest results under a 14.0-inch minimum length limit for largemouth bass in a new Iowa impoundment. North American Journal of Fisheries Management 2:224–231.

Paragamian, V. L. 1984. Angler compliance with a 12.0-inch minimum length limit for smallmouth bass in Iowa streams. North American Journal of Fisheries Management 4:228–229.

Paragi, T. F., S. M. Arthur, and W. B. Krohn. 1994. Seasonal and circadian activity patterns of female fishers, *Martes pennanti*, with kits. The Canadian Field-Naturalist 108:52–57.

Parks, J. O., and E. P. Seidensticker. 1998. Evaluation of a 356- to 457-mm slot length limit for largemouth bass in 5 Texas reservoirs. Proceedings of the Annual Conference of Southeastern Association of Fish and Wildlife Agencies 52:76–85.

Pascoe, S., K. Brooks, T. Cannard, C. M. Dichmont, E. Jebreen, J. Schirmer, and L. Triantafillos. 2014. Social objectives of fisheries management: what are managers' priorities? Ocean and Coastal Management 98:1–10.

Pasko, S., and J. Goldberg. 2014. Review of harvest incentives to control invasive species. Management of Biological Invasions 5:263–277.

Patterson, J. H. 1979. Can ducks be managed by regulations? Experience in Canada. Transactions of the North American Wildlife and Natural Resources Conference 44:130–139.

Patterson, J. H., and R. D. Sparrowe. 1987. Background and objectives on stabilized duck hunting regulations: Canadian and U.S. perspectives. Transactions of the North American Wildlife and Natural Resources Conference 52:177–185.

Paukert, C., M. McInerny, and R. Schultz. 2007. Historical trends in creel limits, length-based limits, and season restrictions for black basses in the United States and Canada. Fisheries 32(2):62–72.

Paukert, C. P., J. A. Klammer, R. B. Pierce, and T. D. Simonson. 2001. An overview of northern pike regulations in North America. Fisheries 26(6):6–13.

Paukert, C. P., E. B. Webb, D. N. Fowler, and C. D. Hilling. 2021. Harvest as a tool to manage populations of undesirable or overabundant fish and wildlife. in K. L. Pope and L. A. Powell, editors. Harvest of fish and wildlife: new paradigms for sustainable management. CRC Press, Boca Raton, Florida, USA.

Paukert, C. P., D. W. Willis, and D. W. Gabelhouse, Jr. 2002. Effect and acceptance of bluegill length limits in Nebraska natural lakes. North American Journal of Fisheries Management 22:1306–1313.

Paul, A. J. 2000. Recruitment dynamics in bull trout (*Salvelinus confluentus*): linking theory and data to species management. Dissertation, University of Calgary, Alberta, Canada.

Paul, A. J., J. R. Post, and J. D. Stelfox. 2003. Can anglers influence the abundance of native and nonnative salmonids in a stream from the Canadian Rocky Mountains? North American Journal of Fisheries Management 23:109–119.

Pauley, G. B., and G. L. Thomas. 1993. Mortality of anadromous coastal cutthroat trout caught with artificial lures and natural bait. North American Journal of Fisheries Management 13:337–345.

Payer, R. D., R. B. Pierce, and D. L. Pereira. 1989. Hooking mortality of walleyes caught on live and artificial baits. North American Journal of Fisheries Management 9:188–192.

Pearse, A. T., G. L. Krapu, and R. R. Cox, Jr. 2012. Spring snow goose hunting influences body composition of waterfowl staging in Nebraska. The Journal of Wildlife Management 76:1393–1400.

Peek, J., B. Dale, H. Hristienko, L. Kantar, K. A. Loyd, S. Mahoney, C. Miller, D. Murray, L. Olver, and C. Soulliere. 2012. Management of large mammalian carnivores in North America. The Wildlife Society, Technical Review 12-1, Bethesda, Maryland, USA.

Pennington, D. D., G. L. Simpson, M. S. McConnell, J. M. Fair, and R. J. Baker. 2013. Transdisciplinary research, transformative learning, and transformative science. BioScience 63:564–573.

Penobscot Nation. 2020. Chapter VII: Inland Fish and Game Regulations. https://www.penobscotnation.org/images/natural-resources/2020DNR/Chapter_.pdf>. Accessed 15 September 2020.

People for the Ethical Treatment of Animals. 2020. What is the distinction between "animal rights" and "animal welfare"? [sic] https://www.peta.org/about-peta/faq/what-is-the-difference-between-animal-rights-and-animal-welfare/. Accessed 27 October 2020.

Péron, G. 2013. Compensation and additivity of anthropogenic mortality: life-history effects and review of methods. Journal of Animal Ecology 82:408–417.

Péron, G., C. A. Nicolai, and D. N. Koons. 2012. Demographic response to perturbations: the role of compensatory density dependence in a North American duck under variable harvest regulations and changing habitat. Journal of Animal Ecology 81:960–969.

Persons, S. E., and S. A. Hirsch. 1994. Hooking mortality of lake trout angled through ice by jigging and setlining. North American Journal of Fisheries Management 14:664–668.

Peterman R., and C. N. Peters. 1998. Decision analysis: taking uncertainties into account in forest resource management. Pages 105–128 *in* V. Sit and B. Taylor, editors. Statistical methods for adaptive management studies. Land management handbook. BC Ministry of Forests, Victoria, British Columbia.

Peterson, J. T., and J. W. Evans. 2003. Quantitative decision analysis for sport fisheries management. Fisheries 28:10–21.

Peterson, M. N., and M. P. Nelson. 2017. Why the North American model of wildlife conservation is problematic for modern wildlife management. Human Dimensions of Wildlife 22:43–54.

Peterson, M. N., H. P. Hansen, M. J. Peterson, and T. R. Peterson. 2011. How hunting strengthens social awareness of coupled human-natural systems. Wildlife Biology in Practice 6:127–143.

Petrides, G. A. 1949. Viewpoints on the analysis of open season sex and age ratios. Transactions of the North American Wildlife Conference 14:391–410.

Philipp, D. P., J. E. Claussen, J. B. Koppelman, J. A. Stein, S. J. Cooke, C. D. Suski, D. H. Wahl, D. A. H. Sutter, and R. Arlinghuas. 2015. Fisheries-induced evolution in largemouth bass: linking vulnerability to angling, parental care, and fitness. Pages 223–234 *in* M. D. Tringali, J. M. Long, T. W. Birdsong, and M. S. Allen, editors. Black bass diversity: multidisciplinary science for conservation. America Fisheries Society, Symposium 82, Bethesda, Maryland, USA.

Philipp, D. P., C. A. Toline, M. F. Kubacki, D. B. F. Philipp, and F. J. S. Phelan. 1997. The impact of catch-and-release angling on the reproductive success of smallmouth bass and largemouth bass. North American Journal of Fisheries Management 17:557–567.

Pierce, C. L., J. K. Ringelman, and M. R. Szymczak. 1996. An investigation of factors affecting waterfowl hunting participation in Colorado. Human Dimensions of Wildlife 1:80–81.

Pierce, R. B. 2010. Long-term evaluation of length limit regulations for northern pike in Minnesota. North American Journal of Fisheries Management 30:412–432.

Pierce, R. B. 2012. Northern pike ecology, conservation, and management history. University of Minnesota Press, Minneapolis, Minnesota, USA.

Pierce, R. B., and C. M. Tomcko. 1998. Angler noncompliance with slot length limits for northern pike in five small Minnesota lakes. North American Journal of Fisheries Management 18:720–724.

Pigeon, G., M. Festa-Bianchet, D. W. Coltman, and F. Pelletier. 2016. Intense selective hunting leads to artificial evolution in horn size. Evolutionary Applications 9:521–530.

Pine, W. E., K. H. Pollock, J. E. Hightower, T. J. Kwak, and J. A. Rice. 2003. A review of tagging methods for estimating fish population size and components of mortality. Fisheries 28:10–23.

Platzer, M. 2016. BTYDplus: probabilistic models for assessing and predicting your customer base. https://CRAN.R-project.org/package=BTYDplus. Accessed 6 November 2020.

Pledger, S., and P. Phillpot. 2008. Using mixtures to model heterogeneity in ecological capture-recapture studies. Biometrical Journal 50:1022–1034.

Pledger, S., and C. J. Schwarz. 2002. Modelling heterogeneity of survival in band-recovery data using mixtures. Journal of Applied Statistics 29:315–327.

Poisson, Y., M. Festa-Bianchet, and F. Pelletier. 2020. Testing the importance of harvest refuges for phenotypic rescue of trophy-hunted populations. Journal of Applied Ecology 57:526–535.

Polacheck, T., N. L. Klaer, C. Millar, and A. L. Preece. 1999. An initial evaluation of management strategies for the southern bluefin tuna fishery. ICES Journal of Marine Science 56:811–826.

Policansky, D. 2002. Catch-and-release recreational fishing: a historical perspective. Pages 74–94 *in* T. J. Pitcher and C. Hollingworth, editors. Recreational fisheries: ecological, economic and social evaluation. Blackwell Science, Oxford, England, United Kingdom.

Pollan, M. 2006. An omnivore's dilemma: a natural history of four meals. Penguin Press, New York, New York, USA.

Pollock, K. H., J. M. Hoenig, and C. M. Jones. 1991. Estimation of fishing and natural mortality when a tagging study is combined with a creel survey or port sampling. Pages 423–434 *in* D. Guthrie, J. M. Hoenig, M. Holliday, C. M. Jones, M. J. Mills, S. A. Moberly, K. H. Pollock, and D. R. Talhelm editors. Creel and angler surveys in fisheries management. American Fisheries Society, Symposium 12, Bethesda, Maryland, USA.

Pollock, K. H., H. Jiang, and J. E. Hightower. 2004. Combining telemetry and fisheries tagging models to estimate fishing and natural mortality rates. Transactions of the American Fisheries Society 133:639–648.

Pollock, K. H., and D. G. Raveling. 1982. Assumptions of modern band-recovery models, with emphasis on heterogeneous survival rates. The Journal of Wildlife Management 46:88–98.

Pope, K. L., C. J. Chizinski, C. L. Wiley, and D. R. Martin. 2016. Influence of anglers' specializations on catch, harvest, and bycatch of targeted taxa. Fisheries Research 183:128–137.

Pope, K. L., M. A. Pegg, N. W. Cole, S. F. Siddons, A. D. Fedele, B. S. Harmon, R. L. Ruskamp, D. R. Turner, and C. C. Uerling. 2016. Fishing for ecosystem services. Journal of Environmental Management 183:408–417.

Pope, K. L., L. A. Powell, B. S. Harmon, M. A. Pegg, and C. J. Chizinski. 2017. Estimating the number of recreational anglers for a given waterbody. Fisheries Research 191:69–75.

Portatz, E. J., J. S. Brown, T. Gallo, C. Anchor, and R. M. Santymire. 2019. Effects of demography and urbanization on stress and body condition of urban white-tailed deer. Urban Ecosystems 22:807–816.

Posewitz, J. 1994. Beyond fair chase. Falcon Press Publishing, Helena, Montana, USA.

Pospahala, R. S., D. R. Anderson, and C. J. Henny. 1974. Population ecology of the mallard. II. Breeding habitat conditions, size of the breeding populations, and production indices. U.S. Fish and Wildlife Service Resource Publication 115.

Post, J. R., C. Mushens, A. Paul, and M. Sullivan. 2003. Assessment of alternative harvest regulations for sustaining recreational fisheries: model development and application to bull trout. North American Journal of Fisheries Management 23:22–34.

Post, J. R., and E. A. Parkinson. 2012. Temporal and spatial patterns of angler effort across lake districts and policy options to sustain recreational fisheries. Canadian Journal of Fisheries and Aquatic Sciences 69:321–329.

Post, J. R., M. Sullivan, S. Cox, N. P. Lester, C. J. Walters, E. A. Parkinson, A. J. Paul, L. Jackson, and B. J. Shuter. 2002. Canada's recreational fisheries: the invisible collapse? Fisheries 27(1):6–17.

Post van der Burg, M., and A. J. Tyre. 2011. Integrating info-gap decision theory with robust population management: a case study using the mountain plover. Ecological Applications 21:303–312.

Poudyal, N. C., S. H. Cho, and D. G. Hodges. 2008. Effects of urban sprawl on hunting participation in the southeastern United States. Southern Journal of Applied Forestry 32:134–138.

Powell, L. 2020. Principles for management of fisheries and wildlife: the manager as decision-maker. Cognella, Inc., San Diego, California, USA.

Powell, L. A., W. R. Clark, and E. E. Klaas. 1998. Distribution and derivation of mallard band recoveries on the Upper Mississippi River, 1961-89. North American Bird Bander 23:1–12.

Powell, L. A., and G. A. Gale. 2015. Estimation of parameters for animal populations: a primer for the rest of us. Caught Napping Publications, Lincoln, Nebraska, USA.

Powell, L. A., J. S. Taylor, J. J. Lusk, and T. W. Matthews. 2011. Adaptive harvest management and harvest mortality of greater prairie-chickens. Studies in Avian Biology 39:329–339.

Powell, L. A., M. P. Vrtiska, and N. Lyman. 2004. Survival rates and recovery distributions of Canada geese banded in Nebraska. Pages 60–65 *in* T. J. Moser, K. C. VerCauteren, R. D. Lien, K. F. Abraham, D. E. Andersen, J. G. Bruggink, J. M. Coluccy, D. A. Graber, J. O. Leafloor, D. R. Luukkonen, and R. E. Trost, editors. Proceedings 2003 International Canada Goose Symposium, Madison, Wisconsin, USA.

Powell, R. A., S. W. Buskirk, and W. J. Zielinski. 2003. Fisher and marten (*Martes pennanti* and *Martes americana*). Pages 635–649 *in* G. A. Feldhamer, B. C. Thompson, and J. A. Chapman, editors. Wild mammals of North America: biology, management, and conservation, 2nd edition. John Hopkins University Press, Baltimore, Maryland, USA.

Powell, R. A., and W. J. Zielinski. 1994. Fisher. Pages 38–73 *in* L. F. Ruggiero, K. B. Aubry, S. W. Buskirk, L. J. Lyon, and W. J. Zielinski, technical editors. The scientific basis for conserving forest carnivores: American marten, fisher, lynx, and wolverine in the western United States. U.S. Forest Service, Rocky Mountain Forest and Range Experiment Station, General Technical Report RM-254, Fort Collins, Colorado, USA.

Pöysä, H. 2004. Ecological basis of sustainable harvesting: is the prevailing paradigm of compensatory mortality still valid? Oikos 104:612–615.

Pracheil, B. M., M. A. Pegg, L. A. Powell, and G. E. Mestl. 2012. Swimways: protecting paddlefish through movement-centered management. Fisheries 37:449–457.

Pradel, R. 1996. Utilization of capture-mark-recapture for the study of recruitment and population growth rate. Biometrics 52:703–709.

Pradel, R. 2005. Multievent: an extension of multistate capture–recapture models to uncertain states. Biometrics 61:442–447.

Presser, L., and W. V. Taylor. 2011. An autoethnography of hunting. Crime, Law and Social Change 55:483–494.

Price, N. B., C. J. Chizinski, J. J. Fontaine, K. L. Pope, M. Rahe, and J. Rawlinson. 2020. An open-sourced, web-based application to improve our ability to understand hunter and angler purchasing behavior from license data. PLoS One 15(10):e0226397.

Price Tack, J. L., C. P. McGowan, S. S. Ditchkoff, W. C. Morse, and O. J. Robinson. 2018. Managing the vanishing North American hunter: a novel framework to address declines in hunters and hunter-generated conservation funds. Human Dimensions of Wildlife 23:515–532.

Pritt, J. J., J. D. Conroy, K. S. Page, S. M. Tyszko, and R. D. Zweifel. 2019. Standard assessments reveal context-dependent responses of crappie populations to a length-based regulation in Ohio reservoirs. North American Journal of Fisheries Management 39:999–1014.

Proffitt, K. M., J. L. Grigg, K. L. Hamlin, and R. A. Garrott. 2009. Contrasting effects of wolves and human hunters on elk behavioral responses to predation risk. The Journal of Wildlife Management 73:345–356.

Pukk, L., A. Kuparinen, L. Järv, R. Gross, and A. Vasemägi. 2013. Genetic and life-history changes associated with fisheries-induced population collapse. Evolutionary Applications 6:749–760.

Punt, A. E., D. S. Butterworth, C. L. de Moor, J. A. De Oliveira, and M. Haddon. 2016. Management strategy evaluation: best practices. Fish and Fisheries 17:303–334.

Quartuch, M. R., R. C. Stedman, D. J. Decker, L. R. Larson, W. F. Siemer, and M. S. Baumer. 2017. Exploring nontraditional pathways into hunting in New York State: implications for recruitment and retention. Human Dimensions of Wildlife 22:391–405.

Quinn, S. 1996. Trends in regulatory and voluntary catch-and-release fishing. Pages 152–162 *in* L. E. Miranda and D. R. DeVries, editors. Multidimensional approaches to reservoir fisheries management. American Fisheries Society, Symposium 16, Bethesda, Maryland, USA.

Quinn, S. 2001. Catch-and-release fishing: a new challenge of managers. Lakeline 3:29–32.

Quinn, S. P. 1992. Angler perspectives on walleye management. North American Journal of Fisheries Management 12:367–378.

Quinn, S. P. 2002. Status of seasonal restrictions on black bass fisheries in Canada and the United States. Pages 455–465 *in* D. P. Philipp and M. S. Ridgway, editors. Black bass: ecology, conservation, and management. American Fisheries Society, Symposium 31, Bethesda, Maryland, USA.

Quinn, T. J., II, and R. B. Deriso. 1999. Quantitative fish dynamics. Oxford University Press, New York, New York, USA.

R Core Team. 2018. R: a language and environment for statistical computing. R Foundation for Statistical Computing, Vienna, Austria. https://www.R-project.org/.

R Core Team. 2017. R: a language and environment for statistical computing. R Foundation for Statistical Computing, Vienna, Austria. http://www.R-project.org/. Accessed 24 September 2020.

R Core Team. 2019. R: a language and environment for statistical computing. R Foundation for Statistical Computing, Vienna, Austria. https://www.R-project.org/.

R Core Team. 2020. R: A language and environment for statistical computing. R Foundation for Statistical Computing, Vienna, Austria. https://www.R-project.org/.

Radomski, P. 2003. Initial attempts to actively manage recreational fishery harvest in Minnesota. North American Journal of Fisheries Management 23:1329–1342.

Radomski, P. J., and T. J. Goeman. 1996. Decision making and modeling in freshwater sport-fisheries management. Fisheries 21(12):14–21.

Radomski, P. J., G. C. Grant, P. C. Jacobson, and M. F. Cook. 2001. Visions for recreational fishing regulations. Fisheries 26(5):7–18.

Raedeke, A. H., J. S. Rikoon, and K. A. Bradley. 1996. Landowner resistance to lease hunting: a sociocultural perspective. Human Dimensions of Wildlife 1:48–59.

Raedeke, K. J., J. J. Millspaugh, and P. E. Clark. 2002. Population characteristics. Pages 449–491 *in* D. E. Toweill and J. W. Thomas, editors. North American elk: ecology and management. Wildlife Management Institute, and Smithsonian Institution Press, Washington, D.C., USA.

Raftovich, R. V., K. K. Fleming, S. C. Chandler, and C. M. Cain. 2019. Migratory bird hunting activity and harvest during the 2017-18 and 2018-19 hunting seasons. U.S. Fish and Wildlife Service, Laurel, Maryland, USA.

Raftovich, R. V., and K. A. Wilkins. 2013. Migratory bird hunting activity and harvest during the 2011-12 and 2012-13 hunting seasons. U.S. Fish and Wildlife Service, Laurel, Maryland, USA.

Rahel, F. J. 2016. Changing philosophies of fisheries management as illustrated by the history of fishing regulations in Wyoming. Fisheries 41:38–48.

Rasmussen, P. W., M. D. Staggs, T. D. Beard, and S. P. Newman. 1998. Bias and confidence interval coverage of creel survey estimators evaluated by simulation. Transactions of the American Fisheries Society 127:469–480.

Ratikainen, I. I., J. A. Gill, T. G. Gunnarsson, W. J. Sutherland, and H. Kokko. 2008. When density dependence is not instantaneous: theoretical developments and management implications. Ecology Letters 11:184–198.

Ray, A. J. 1987. The fur trade in North America: an overview from a historical geographical perspective. Pages 21–30 *in* M. Novak, J. A. Baker, M. E. Obbard, and B. Malloch, editors. Wild furbearer management and conservation in North America. Ontario Trappers Association, North Bay, Ontario, Canada.

Recce, S. 2008. Perpetuating hunter traditions: access to public and private lands. Pages 73–78 *in* J. Nobile and M. D. Duda, editors. Strengthening America's hunting heritage and wildlife conservation in the 21st century: challenges and opportunities. Sporting Conservation Council, White House Conference on North American Wildlife Policy, Washington, D.C., USA.

Redmond, L. C. 1986. The history and development of warmwater fish harvest regulations. Pages 186–195 *in* G. E. Hall and M. J. Van Den Avyle, editors. Reservoir fisheries management: strategies for the 80's. American Fisheries Society, Southern Division, Reservoir Committee, Bethesda, Maryland, USA.

Redpath, T. D., S. J. Cooke, C. D. Suski, R. Arlinghaus, P. Couture, D. H. Wahl, and D. P. Philipp. 2010. The metabolic and biochemical basis of vulnerability to recreational angling after three generations of angling-induced selection in a teleost fish. Canadian Journal of Fisheries and Aquatic Sciences 67:1983–1992.

Reed, E. T., and A. M. Calvert, editors. 2007. Evaluation of the special conservation measures for greater snow geese: report of the Greater Snow Goose Working Group. Arctic Goose Joint Venture Special Publication, Canadian Wildlife Service, Sainte-Foy, Quebec, Canada.

Reed, J. R., and B. G. Parsons. 1999. Angler opinions of bluegill management and related hypothetical effects on bluegill fisheries in four Minnesota lakes. North American Journal of Fisheries Management 19:515–519.

Reed, M. S. 2008. Stakeholder participation for environmental management: a literature review. Biological Conservation 141:2417–2431.

Reed, M. S., S. Vella, E. Challies, J. de Vente, L. Frewer, D. Hohenwallner-Ries, T. Huber, R. K. Neumann, E. A. Oughton, J. Sidoli del Ceno, and H. van Delden. 2018. A theory of participation: what makes stakeholder and public engagement in environmental management work? Restoration Ecology 26:S7–S17.

Reeves, H. M. 1993. Mourning dove hunting regulations. Pages 429–448 *in* T. S. Baskett, M. W. Sayre, R. E. Tomlinson, and R. E. Mirarchi, editors. Ecology and management of the mourning dove. Stackpole Books, Harrisburg, Pennsylvania, USA.

Regehr, E. V., R. R. Wilson, K. D. Rode, M. C. Runge, and H. L. Stern. 2017. Harvesting wildlife affected by climate change: a modelling and management approach for polar bears. Journal of Applied Ecology 54:1534–1543.

Reinartz, W., J. S. Thomas, and G. Bascoul. 2008. Investigating cross-buying and customer loyalty. Journal of Interactive Marketing 22:5–20.

Reis, A. C. 2009. More than the kill: hunters' relationships with landscape and prey. Current Issues in Tourism 12:573–587.

Reisner, M. 1992. Game wars: the undercover pursuit of wildlife poachers. Penguin Books, New York, New York, USA.

Renecker, L. A., and C. C. Schwartz. 1998. Nutrition and energetics. Pages 403–439 *in* W. Franzmann and C. C. Schwartz, editors. Ecology and management of the North American moose. Smithsonian Institution, Washington, D.C., USA.

Responsive Management. 2015. Trap use, furbearers trapped, and trapper characteristics in the United States in 2015. Association of Fish and Wildlife Agencies, Washington, D.C., USA.

Responsive Management. 2016. 2016 summary of furbearer trapping regulation in the United States. Furbearer Conservation Technical Work Group, Association of Fish and Wildlife Agencies, Washington, D.C., USA.

Responsive Management, and National Shooting Sports Foundation. 2017. Hunting, fishing, sport shooting, and archery recruitment, retention, and reactivation: a practitioner's guide. Responsive Management, Harrisonburg, Virginia, USA.

Rexstad, E. A. 1992. Effect of hunting on annual survival of Canada geese in Utah. The Journal of Wildlife Management 56:297–305.

Reynolds, J. D., G. M. Mace, K. H. Redford, and J. G. Robinson, editors. 2001. Conservation of exploited species. Cambridge University Press, Cambridge, England, United Kingdom.

Reynolds, J. H., M. G. Knutson, K. B. Newman, E. D. Silverman, and W. L. Thompson. 2016. A road map for designing and implementing a biological monitoring program. Environmental Monitoring and Assessment 188:399.

Reynolds, R. E. 1987. Breeding duck population, production and habitat surveys, 1979-85. Transactions of the North American Wildlife and Natural Resources Conference 52:186–205.

Rezaei, J., O. Kothadiya, L. Tavasszy, and M. Kroesen. 2018. Quality assessment of airline baggage handling systems using SERVQUAL and BWM. Tourism Management 66:85–93.

Reznick, D. N., J. B. Losos, and J. Travis. 2019. From low to high gear: there has been a paradigm shift in our understanding of evolution. Ecology Letters 22:233–244.

Rhoades, O. K., S. I. Lonhart, and J. J. Stachowicz. 2019. Human-induced reductions in fish predator boldness decrease their predation rates in kelp forests. Proceedings of the Royal Society B 286:20182745.

Rice, M. B., D. A. Haukos, J. A. Dubovsky, and M. C. Runge. 2010. Continental survival and recovery rates of northern pintails using band-recovery data. The Journal of Wildlife Management 74:778–787.

Richardson, F., and R. H. Hamre, editors. 1984. Wild Trout III, proceedings of the symposium. Trout Unlimited, Vienna, Virginia, USA.

Ricker, W. E. 1945. A method of estimating minimum size limits for obtaining maximum yield. Copeia 1945(2):84–94.

Ricker, W. E. 1954. Stock and recruitment. Journal of the Fisheries Research Board of Canada 11:559–623.

Ricker, W. E. 1975. Computation and interpretation of biological statistics of fish populations. Fisheries Research Board of Canada Bulletin 191.

Rider, S. J., D. K. Riecke, and D. L. Scarnecchia. 2019. Proactive harvest management of commercial paddlefish fisheries. Pages 267–297 *in* J. D. Schooley and L. Scarnecchia, editors. Paddlefish: ecological, aquacultural, and regulatory challenges of managing a global resource. American Fisheries Society, Symposium 88. Bethesda, Maryland, USA.

Rieman, B. E., J. T. Peterson, and D. L. Meyers. 2006. Have brook trout (*Salvelinus fontinalis*) displaced bull trout (*Salvelinus confluentus*) along longitudinal gradients in central Idaho streams? Canadian Journal of Fisheries and Aquatic Sciences 62:2431–2442.

Rietveld, H. J., E. W. Armstrong, D. A. Baccante, R. M. Korver, R. Leith, R. A. S. Mathers, T. E. Mosindy, D. M. Reid, and J. Seyler. 1999. Regulatory control of walleye (*Stizostedion vitreum* [Mitchill, 1818]) sport fisheries in Ontario. Report of the Harvest Control Working Group of the Percid Community Synthesis, Ontario Ministry of Natural Resources, Toronto, Ontario, Canada.

Riley, S. J., D. J. Decker, J. W. Enck, P. D. Curtis, T. B. Lauber, and T. L. Brown. 2003. Deer populations up, hunter populations down: implications of interdependence of deer and hunter population dynamics on management. Ecoscience 10:455–461.

Riley S. J., and R. S. Gregory. 2012. Decision making in wildlife management. Pages 101–112 *in* D. J. Decker, S. J. Riley, and W. F. Siemer, editors. Human dimensions of wildlife management, 2nd edition. Johns Hopkins University Press, Baltimore, Maryland, USA.

Ringelman, J. K. 1997. Effects of regulations and duck abundance on duck hunter participation and satisfaction. Transactions of the North American Wildlife and Natural Resources Conference 62:361–376.

Risley, J. T., and R. L. Johnson. 2016. Effectiveness of slot-length limits to maintain an Arkansas trophy largemouth bass fishery characterized by high voluntary release rates by anglers. Journal of the Southeastern Association of Fish and Wildlife Agencies 3:112–120.

Ritchie, H., and M. Roser. 2020. Urbanization. https://ourworldindata.org/urbanization. Accessed 17 April 2020.

Robbins, C. S., D. Bystrak, and P. H. Geissler. 1986. The breeding bird survey: its first fifteen years, 1965-1979. U.S. Fish and Wildlife Service, Resource Publication 157, Washington, D.C., USA.

Robbins, W. H., and H. R. MacCrimmon. 1974. The black bass in America and overseas. Biomanagement and Research Enterprises, Sault Ste. Marie, Canada.

Roberts, A., J. M. Eadie, D. W. Howerter, F. A. Johnson, J. D. Nichols, M. C. Runge, M. P. Vrtiska, and B. K. Williams. 2018. Strengthening links between waterfowl research and management: linking waterfowl research and management. The Journal of Wildlife Management 82:260–265.

Roberts, C. 2007. Mixing modes of data collection in surveys: a methodological review. Discussion Paper (unpublished). http://eprints.ncrm.ac.uk/418/.

Roberts, N. M., and S. M. Crimmins. 2010. Bobcat population status and management in North America: evidence of large-scale population increase. Journal of Fish and Wildlife Management 1:169–174.

Robinson, K. F., D. R. Diefenbach, A. K. Fuller, J. E. Hurst, and C. S. Rosenberry. 2014. Can managers compensate for coyote predation of white-tailed deer? The Journal of Wildlife Management 78:571–579.

Robinson, K. F., M. DuFour, M. Jones, S. Herbst, T. Newcomb, J. Boase, T. Brenden, D. Chapman, J. Dettmers, J. Francis, T. Hartman, P. Kočovský, B. Locke, C. Mayer, and J. Tyson. 2021. Using decision analysis to collaboratively respond to invasive species threats: a case study of Lake Erie grass carp (*Ctenopharyngodon idella*). Journal of Great Lakes Research 47:108–119.

Robinson, K. F., and A. K. Fuller. 2017. Participatory modeling and structured decision making. Pages

83–101 *in* S. Gray, M. Paolisso, R. Jordan, and S. Gray, editors. Environmental modeling with stakeholders: theory, methods, and applications. Springer International Publishing, Cham, Switzerland.

Robinson, K. F., A. K. Fuller, J. E. Hurst, B. L. Swift, A. Kirsch, J. Farquhar, D. J. Decker, and W. F. Siemer. 2016a. Structured decision making as a framework for large-scale wildlife harvest management decisions. Ecosphere 7:e01613.

Robinson, K. F., A. K. Fuller, J. E. Hurst, B. Swift, A. Kirsch, and J. Farquhar. 2016b. A structured decision making approach to white-tailed deer buck harvest management in New York State. New York State Department of Environmental Conservation, Albany, New York, USA.

Robinson, K. F., A. K. Fuller, and M. L. Jones. 2021. Using structured decision making to incorporate ecological and social values into harvest decisions: case studies of white-tailed deer and walleye. *in* K. L. Pope and L. A. Powell, editors. Harvest of fish and wildlife: new paradigms for sustainable management. CRC Press, Boca Raton, Florida, USA.

Robinson, K. F., A. K. Fuller, M. V. Schiavone, B. L. Swift, D. R. Diefenbach, W. F. Siemer, and D. J. Decker. 2017. Addressing wild turkey population declines using structured decision making. The Journal of Wildlife Management 81:393–405.

Robinson, K. F., A. K. Fuller, R. C. Stedman, W. F. Siemer, and D. J. Decker. 2019. Integration of social and ecological sciences for natural resource decision making: challenges and opportunities. Environmental Management 63:565–573.

Robinson, W. B. 1953. Population trends of predators and fur animals in 1080 station areas. Journal of Mammalogy 34:220–227.

Robison, K. K., and D. Ridenour. 2012. Whither the love of hunting? Explaining the decline of a major form of rural recreation as a consequence of the rise of virtual entertainment and urbanism. Human Dimensions of Wildlife 17:418–436.

Röckmann, C., C. Ulrich, M. Dreyer, E. Bell, E. Borodzicz, P. Haapasaari, K. H. Hauge, D. Howell, S. Mäntyniemi, D. Miller, G. Tserpes, and M. Pastoors. 2012. The added value of participatory modelling in fisheries management—what has been learnt? Marine Policy 36:1072–1085.

Rodriguez, S., M. Chitwood, M. Peterson, C. Jensen, B. Gardner, and H. Hansen. 2016. Gender differences in hunter recruitment and dedication in Denmark. Wildlife Biology in Practice 2:1–10.

Roedel, P. M., editor. 1975. Optimum sustainable yield as a concept in fisheries management. American Fisheries Society, Special Publication 9, Bethesda, Maryland, USA.

Rolland, V., J. A. Hostetler, T. C. Hines, H. F. Percival, and M. K. Oli. 2010. Impact of harvest on survival of a heavily hunted game bird population. Wildlife Research 37:392–400.

Roseberry, J. L. 1979. Bobwhite population responses to exploitation: real and simulated. The Journal of Wildlife Management 43:285–305.

Roseberry, J. L., D. C. Autry, W. D. Klimstra, and L. A. Merhoff, Jr. 1969. A controlled deer hunt on Crab Orchard National Wildlife Refuge. The Journal of Wildlife Management 34:23–28.

Roseberry, J. L., and A. Woolf. 1991. A comparative evaluation of techniques for analyzing white-tailed deer harvest data. Wildlife Monographs 117:1–59.

Roseberry, J. L. and W. D. Klimstra. 1974. Differential vulnerability during a controlled deer harvest. Journal of Wildlife Management 38:499–507.

Rosenberry, C. S., and D. R. Diefenbach. 2019. A comparison of spatial variation of deer harvests between small and large management units in Pennsylvania. Wildlife Society Bulletin 43:71–76.

Rosenberry, C. S., J. T. Fleegle, and B. D. Wallingford. 2009. Management and biology of white-tailed deer in Pennsylvania 2009–2018. Pennsylvania Game Commission, Harrisburg, Pennsylvania, USA.

Ross, M. V., R. T. Alisauskas, D. C. Douglas, and D. K. Kellett. 2017. Decadal declines in avian herbivore reproduction: density-dependent nutrition and phenological mismatch in the arctic. Ecology 98:1869–1883.

Ross, M. V., R. T. Alisauskas, D. C. Douglas, D. K. Kellett, and K. L. Drake. 2018. Density-dependent and phenological mismatch effects on growth and survival in lesser snow and Ross's goslings. Journal of Avian Biology 49:1–12.

Rowe, G., T. Horlick-Jones, J. Walls, and N. Pidgeon. 2005. Difficulties in evaluating public engagement initiatives: reflections on an evaluation of the U.K. *GM Nation?* public debate about transgenic crops. Public Understanding of Science 14:331–352.

Roy, L. D., C. Twitchell, and M. Hiltz. 2005. Factors influencing the effectiveness of breakaway snares to capture coyotes and release deer in Alberta. Alberta Research Council, Vegreville, Alberta, Canada.

Royle, J. A., and P. R. Garrettson. 2005. The effect of reward band value on mid-continent mallard band reporting rates. The Journal of Wildlife Management 69:800–804.

Rughetti, M., and M. Festa-Bianchet. 2010. Compensatory growth limits opportunities for artificial selection in Alpine chamois. The Journal of Wildlife Management 74:1024–1029.

Rughetti, M., and M. Festa-Bianchet. 2014. Effects of selective harvest of non-lactating females on chamois population dynamics. Journal of Applied Ecology 51:1075–1084.

Rughetti, M., M. Festa-Bianchet, S. D. Côté, and S. Hamel. 2017. Ecological and evolutionary effects of selective harvest of non-lactating female ungulates. Journal of Applied Ecology 54:1571–1580.

Rummer, J. L. 2007. Factors affecting catch and release (CAR) mortality in fish: insight into CAR mortality in red snapper and the influence of catastrophic decompression. Pages 123–144 *in* W. F. Patterson, III, J. H. Cowan, Jr., G. R. Fitzhugh, and D. L. Nieland, editors. Red snapper: ecology and fisheries in the U.S. Gulf of Mexico. American Fisheries Society, Symposium 60, Bethesda, Maryland, USA.

Runge, M. C. 2021. A decision-analytical framework for developing harvest regulations. *in* K. L. Pope and L. A. Powell, editors. Harvest of fish and wildlife: new paradigms for sustainable management. CRC Press, Boca Raton, Florida, USA.

Runge, M. C., E. Bean, D. R. Smith, and S. Kokos. 2011. Non-native fish control below Glen Canyon Dam—report from a structured decision-making project. U.S. Geological Survey, Open File Report 2011-1012, Reston, Virginia, USA.

Runge M. C., G. S. Boomer, T. A. Sanders, and M. D. Koneff. 2009. Proposal for a derived and adaptive harvest strategy for northern pintails. U.S. Department of the Interior, Fish and Wildlife Service, Division of Migratory Bird Management, Washington, D.C., USA.

Runge, M. C., and S. J. Converse. 2020. Introduction to risk analysis. Pages 149–155 *in* M. C. Runge, S. J. Converse, J. E. Lyons, and D. R. Smith, editors. Structured decision making: case studies in natural resource management. Johns Hopkins University Press, Baltimore, Maryland, USA.

Runge, M. C., S. J. Converse, and J. E. Lyons. 2011. Which uncertainty? Using expert elicitation and expected value of information to design an adaptive program. Biological Conservation 144:1214–1223.

Runge, M. C., S. J. Converse, J. E. Lyons, and D. R. Smith. 2020. Structured decision making: case studies in natural resource management. John Hopkins University Press, Baltimore, Maryland, USA.

Runge, M. C., and F. A. Johnson. 2002. The importance of functional form in optimal control solutions of problems in population dynamics. Ecology 83:1357–1371.

Runge, M. C., F. A. Johnson, M. G. Anderson, M. D. Koneff, E. T. Reed, and S. E. Mott. 2006. The need for coherence between waterfowl harvest and habitat management. Wildlife Society Bulletin 34:1231–1237.

Runge, M. C., W. L. Kendall, and J. D. Nichols. 2004. Exploitation. Pages 303–328 *in* W. J. Sutherland, I. Newton, and R. E. Green, editors. Bird ecology and conservation: a handbook of techniques. Oxford University Press, Oxford, England, United Kingdom.

Runge, M. C., J. R. Sauer, M. L. Avery, B. F. Blackwell, and M. D. Koneff. 2009. Assessing allowable take of migratory birds. The Journal of Wildlife Management 73:556–565.

Rutberg, A. T., and R. E. Naugle. 2008. Population-level effects of immunocontraception in white-tailed deer (*Odocoileus virginianus*). Wildlife Research 35:494–501.

Rypel, A. L. 2015. Effects of a reduced daily bag limit on bluegill size structure in Wisconsin lakes. North American Journal of Fisheries Management 35:388–397.

Rypel, A. L., J. Lyons, J. D. Tober Griffin, and T. D. Simonson. 2016. Seventy-year retrospective on size-structure changes in the recreational fisheries of Wisconsin. Fisheries 41:230–243.

Salinas, R. A., W. H. Stiver, J. L. Corn, S. Lenhart, C. Collins, M. Madden, K. C. Vercauteren, B. Schmit, E. Karsi, A. Odoi, G. Hickling, and H. MaCallum. 2015. An individual-based model for feral hogs in Great Smoky Mountains National Park. Natural Resource Modeling 28:18–36.

Sandercock, B. K., E. B. Nilsen, H. Brøseth, and H. C. Pedersen. 2011. Is hunting mortality additive or compensatory to natural mortality? Effects of experimental harvest on the survival and cause-specific mortality of willow ptarmigan. Journal of Animal Ecology 80:244–258.

Sanderson, G. C., and F. C. Bellrose. 1986. A review of the problem of lead poisoning in waterfowl. Illinois Natural History Survey. Special Publication 4, Champaign, Illinois, USA.

Sands, J. P., and M. D. Pope. 2010. A survey of galliform monitoring programs and methods in the United States and Canada. Wildlife Biology 16:342–356.

Sass, G. G., J. W. Gaeta, M. S. Allen, C. D. Suski, and S. L. Shaw. 2018. Effects of catch-and-release angling on a largemouth bass (Micropterus salmoides) population in a north temperate lake, 2001–2005. Fisheries Research 204:95–x102.

Sass, G. G., and S. L. Shaw. 2020. Catch-and-release influences on inland recreational fisheries. Reviews in Fisheries Science & Aquaculture 28:211–227.

Sauer, J. R., D. K. Niven, J. E. Hines, D. J. Ziolkowski, Jr., K. L. Pardieck, J. E. Fallon, and W. A. Link. 2017a. The North American breeding bird survey, results and analysis 1966-2015, Version 2.07.2017. U.S. Geolgoical Survey, Patuxent Wildlife Research Center, Laurel, Maryland, USA.

Sauer, J. R., K. L. Pardieck, D. J. Ziolkowski, Jr., A. C. Smith, A.-M. R. Hudson, V. Rodriguez, H. Berlanga,

D. K. Niven, and W. A. Link. 2017b. The first 50 years of the North American breeding bird survey. Condor 119:576–593.

Sauer, J. R., W. A. Link, J. E. Fallon, K. L.Pardieck, & D. J. Ziolkowski, Jr. 2013. The North American Breeding Bird Survey 1966-2011: summary analysis and species accounts. North American Fauna 79:1–32.

Saunders, D. L., J. J. Meeuwig, and A. C. J. Vincent. 2002. Freshwater protected areas: strategies for conservation. Conservation Biology 16:30–41.

Sawyer, R. K. 2012. A hundred years of Texas waterfowl hunting. Texas A&M University Press, College Station, Texas, USA.

Sbragaglia, V., C. Gliese, D. Bierbach, A. E. Honsey, S. Uusi-Heikkila, and R. Arlinghaus. 2019. Size-selective harvesting fosters adaptations in mating behaviour and reproductive allocation, affecting sexual selection in fish. Journal of Animal Ecology 88:1343–1354.

Scarnecchia, D. L., Y. Lim, L. F. Ryckman, K. M. Backes, S. E. Miller, R. S. Gangl, B. J. Schmitz. 2014. Virtual population analysis, episodic recruitment, and harvest management of paddlefish with applications to other acipenseriform fishes. Reviews in Fisheries Science & Aquaculture 22:16–35.

Scarnecchia, D. L., L. F. Ryckman, B. J. Schmitz, S. Gangl, W. Wiedenheft, L. L. Leslie, and Y. Lim. 2008. Management plan for North Dakota and Montana paddlefish stocks and fisheries. North Dakota Game and Fish Department and Montana Department of Fish, Wildlife and Parks. Bismarck and Helena, Montana, USA.

Scarnecchia, D. L., J. D. Schooley, K. M. Backes, A. Slominski, S. Dalbey, and Y. Lim. 2019. Paddlefish life history: advances and applications in design of harvest management regulations. Pages 1–28 *in* J. D. Schooley and D. L. Scarnecchia, editors. Paddlefish: ecological, aquacultural, and regulatory challenges of managing a global resource. American Fisheries Society, Symposium 88, Bethesda, Maryland, USA.

Schill, D. J. 1996. Hooking mortality of bait-caught rainbow trout in an Idaho trout stream and a hatchery: implications for special-regulation management. North American Journal of Fisheries Management 16:348–356.

Schill, D. J., and P. A. Kline. 1995. Use of random response to estimate angler noncompliance with fishing regulations. North American Journal of Fisheries Management 15:721–731.

Schisler, G. J., and E. P. Bergersen. 1996. Postrelease hooking mortality of rainbow trout caught on scented artificial baits. North American Journal of Fisheries Management 16:570–578.

Schloesser, R. W., M. C. Fabrizio, R. J. Latour, G. C. Garman, R. S. Greenlee, M. Groves, and J. Gartland. 2011. Ecological role of blue catfish in Chesapeake Bay communities and implications for management. Pages 369–382 *in* P. H. Michaletz and V. H. Travnichek, editors. Conservation, ecology and management of catfish: the second international symposium. American Fisheries Society, Symposium 77, Bethesda, Maryland, USA.

Schmitt, J. D., B. K. Peoples, L. Castello, and D. J. Orth. 2019. Feeding ecology of a generalist consumer: a case study of invasive blue catfish *Ictalurus furcatus* in the Chesapeake Bay. Environmental Biology of Fishes 102:443–465.

Schmittlein, D. C., D. G. Morrison, and R. Colombo. 1987. Counting your customers: who are they and what will they do next? Management Science 33:1–24.

Schneider, J. C., and R. N. Lockwood. 2002. Use of walleye stocking, antimycin treatments, and catch-and-release angling regulations to increase growth and length of stunted bluegill populations in Michigan Lakes. North American Journal of Fisheries Management 22:1041–1052.

Schofield, P., and P. Reeves. 2015. Does the factor theory of satisfaction explain political voting behaviour? European Journal of Marketing 49:968–992.

Schorr, R. A., P. M. Lukacs, and J. A. Gude. 2014. The Montana deer and elk hunting population: the importance of cohort group, license price, and population demographics on hunter retention, recruitment, and population change. The Journal of Wildlife Management 78:944–952.

Schramm, H. L., and G. E. Gilliland. 2015. Achieving high survival of tournament-caught black bass: past efforts and future needs and opportunities. Journal of the Southeastern Association of Fish and Wildlife Agencies 2:50–56.

Schroeder, M. A., and L. A. Robb. 2005. Criteria for gender and age. Pages 303–338 *in* C. E. Braun, editor. Techniques for wildlife investigation and management. 6th edition. The Wildlife Society, Bethesda, Maryland, USA.

Schroeder, S. A., L. Cornicelli, D. C. Fulton, and M. D. Grund. 2014. What predicts support for antler point restrictions? Human Dimensions of Wildlife 19:301–318.

Schroeder, S. A., L. Cornicelli, D. C. Fulton, and S. S. Merchant. 2018. Explicit versus implicit motivations: clarifying how experiences affect turkey hunter satisfaction using revised importance-performance, importance grid, and penalty-reward-contrast analyses. Human Dimensions of Wildlife 23:1–20.

Schroeder, S. A., L. Cornicelli, D. C. Fulton, and S. S. Merchant. 2019a. The influence of motivation versus experience on recreation satisfaction: how appreciative-versus achievement-oriented recreation experience preferences relate to hunter satisfaction. Journal of Leisure Research 50:107–131.

Schroeder, S. A., D. C. Fulton, L. Cornicelli, S. D. Cordts, and J. S. Lawrence. 2019b. Clarifying how hunt-specific experiences affect satisfaction among more avid and less avid waterfowl hunters. Wildlife Society Bulletin 43:455–467.

Schroeder, S. A., D. C. Fulton, and J. S. Lawrence. 2006. Managing for preferred hunting experiences: a typology of Minnesota waterfowl hunters. Wildlife Society Bulletin 34:380–387.

Schroeder, S. A., D. C. Fulton, J. S. Lawrence, and S. D. Cordts. 2013. Identity and specialization as a waterfowl hunter. Leisure Sciences 35:218–234.

Schroeder, S. A., D. C. Fulton, J. S. Lawrence, and S. D. Cordts. 2017. How hunter perceptions of wildlife regulations, agency trust, and satisfaction affect attitudes about duck bag limits. Human Dimensions of Wildlife 22:454–475.

Schulz, J. H., J. J. Millspaugh, A. J. Bermudez, X. Gao, T. W. Bonnot, L. G. Britt, and M. Paine. 2006. Acute lead toxicosis in mourning doves. The Journal of Wildlife Management 70:413–421.

Schulz, J. H., J. J. Millspaugh, B. E. Washburn, G. R. Wester, J. T. Lanigan, III, and J. C. Franson. 2002. Spent-shot availability and ingestion on areas managed for mourning doves. Wildlife Society Bulletin 30:112–120.

Schulz, J. H., J. J. Millspaugh, D. T. Zekor, and B. E. Washburn. 2003. Enhancing sport-hunting opportunities for urbanites. Wildlife Society Bulletin 31:565–573.

Schummer, M. L., J. Simpson, J. B. Davis, B. Shirkey B. and K. E. Wallen. 2020. Balancing waterfowl hunting opportunity and quality to recruit, retain, and reactivate. Wildlife Society Bulletin 44:391–395.

Schwartz, C. C., K. J. Hundertmark, and T. H. Spraker. 1992. An evaluation of selective bull moose harvest on the Kenai Peninsula, Alaska. Alces 28:1–13.

Scrogin, D., K. Boyle, G. Parsons, and A. J. Plantinga. 2004. Effects of regulations on expected catch, expected harvest, and site choice of recreational anglers. American Journal of Agricultural Economics 86:963–974.

Seamans, M. E. 2019a. Band-tailed pigeon population status, 2019. U.S. Department of the Interior, Fish and Wildlife Service, Division of Migratory Bird Management, Laurel, Maryland, USA.

Seamans, M. E. 2019b. Mourning dove population status, 2019. U.S. Department of the Interior, Fish and Wildlife Service, Division of Migratory Bird Management, Laurel, Maryland, USA.

Seamans, M. E., and C. E. Braun. 2016. Estimation of band-tailed pigeon band recovery and population vital rates in Colorado, 1969-1981. Journal of Fish and Wildlife Management 7:369–376.

Seamans, M. E., and R. D. Rau. 2019. American woodcock population status, 2019. U.S. Fish and Wildlife Service, Laurel, Maryland, USA.

Seber, G. A. F. 1982. Estimation of animal abundance and related parameters, 2nd edition. Griffin, London, England, United Kingdom.

Sedinger, J. S., and M. P. Herzog. 2012. Harvest and dynamics of duck populations. The Journal of Wildlife Management 76:1108–1116.

Sedinger, J. S., and E. A. Rexstad. 1994. Do restrictive harvest regulations result in higher survival rates in mallards? A comment. The Journal of Wildlife Management 58:571–577.

Sedinger, J. S., and J. J. Rotella. 2005. Effect of harvest on sage-grouse *Centrocercus urophasianus* populations: what can we learn from the current data? Wildlife Biology 11:371–375.

Sedinger, J. S., G. C. White, S. Espinosa, E. T. Partee, and C. E. Braun. 2010. Assessing compensatory versus additive harvest mortality: an example using greater sage-grouse. The Journal of Wildlife Management 74:326–332.

Seiler, A. 2004. Trends and spatial patterns in ungulate-vehicle collisions in Sweden. Wildlife Biology 10:301–313.

Seng, P. T., D. J. Case, M. Conover, D. J. Decker, J. W. Enck, S. N. Frey, M. Z. Stange, B. Stanton, R. Stedman, C. Thomas, and D. Thorne. 2001. Contributions of hunting to North American culture and society. Transactions of the North American Wildlife and Natural Resources Conference 66:202–231.

Serenari, C., and M. N. Peterson. 2018. Evaluating the cultural fit of hunting and angling among minority sportspersons in North Carolina. Leisure Sciences DOI:10.1080/01490400.2018.1458262.

Serns, S. L. 1978. Effects of a minimum size limit on the walleye population of a northern Wisconsin lake. Pages 390-–397 *in* R. L. Kendall, editor. Selected coolwater fishes of North America. American Fisheries Society, Special Publication 11, Bethesda, Maryland, USA.

Servanty, S., R. Choquet, É. Baubet, S. Brandt, J.-M. Gaillard, M. Schaub, C. Toïgo, J.-D. Lebreton, M. Buoro, and O. Gimenez. 2010. Assessing whether mortality is additive using marked animals: a Bayesian state-space modeling approach. Ecology 91:1916–1923.

Servheen, C. 1999. Status and management of the grizzly bear in the lower 48 United States. Pages 50–54 *in* C. Servheen, S. Herrero, and B. Peyton, editors. Bears: status survey and conservation action plan. International Union for Conservation of Nature and Natural Resources/Species Survival Commission Bear and Polar Bear Specialists Groups, International Union for Conservation of Nature, Gland, Switzerland.

Sexson, M. L., and J. R. Choate. 1981. Historical biogeography of the pronghorn in Kansas. Transactions of the Kansas Academy of Sciences 84:128–133.

Shannon, C. E. 1948. A mathematical theory of communication. Bell System Technical Journal 27:379–423.

Shea, S. M., and R. E. Vanderhoof. 1999. Evaluation of a five-inch regulation for increasing antler size of harvested deer in northwest Florida. Proceedings of the Annual Southeastern Deer Study Group 22:18–19.

Sheaffer, S. E., W. L. Kendall, and E. F. Bowers. 2005. Impact of special early harvest seasons on subarctic-nesting and temperate-nesting Canada geese. The Journal of Wildlife Management 69:1494–1507.

Shelby, B., and J. J. Vaske. 1991. Resource and activity substitutes for recreational salmon fishing in New Zealand. Leisure Sciences 13:21–32.

Shelley, T. O., and M. S. Crow. 2009. The nature and extent of conservation policing: law enforcement generalists or conservation specialists? American Journal of Criminal Justice 34:9–27.

Shirkey, B. T., and R. J. Gates. 2020. Survival, harvest, and Lincoln estimates of wood ducks banded in Ohio. Journal of Fish and Wildlife Management 11:185–195.

Shrestha, S. K., and R. C. Burns. 2016. Integrating constraints to the theory of planned behavior in predicting deer hunting participation. Human Dimensions of Wildlife 21:445–459.

Shuter, B. J., and J. F. Koonce. 1977. A dynamic model of the western Lake Erie walleye (*Stizostedion vitreum vitreum*) population. Journal of the Fisheries Research Board of Canada 34:1972–1982.

Siddons, S. F., M. A. Pegg, N. P. Hogberg, and G. M. Klein. 2016. Age, growth, and mortality of a trophy channel catfish population in Manitoba, Canada. North American Journal of Fisheries Management 36:1368–1374.

Siemer, W. F., J. R. Boulanger, D. J. Decker, and M. S. Baumer. 2014. Activities and satisfactions of fall turkey hunters in New York State. Cornell University, Human Dimensions Research Unit Series Number 14–1, Ithaca, New York, USA.

Siemer, W. F., and T. L. Brown. 1993. Hunter demand for access to public and private land in New York. Cornell University, Human Dimensions Research Unit Series Number 93-5, Ithaca, New York, USA.

Siemer, W. F., and D. J. Decker. 2015. Hunter satisfactions with deer harvest opportunities in New York State. Cornell University, Human Dimensions Research Unit Series Number 15-05, Ithaca, New York, USA.

Siemer, W. F., D. J. Decker, and R. C. Stedman. 2012. Cornell University lands deer hunting program: profile of active participants in 2011. Cornell University, Human Dimensions Research Unit Series Number 12-04, Ithaca, New York, USA.

Siemer, W. F., D. J. Decker, and R. C. Stedman. 2015. Hunter satisfactions with deer harvest opportunities in New York State. Cornell University, Human Dimensions Research Unit Series Number 15-05, Ithaca, New York, USA.

Siemer, W. F., and R. C. Stedman. 2018. Understanding factors that influence hunter preferences for timing of waterfowl hunting seasons. Cornell University, Social Sciences Publication Series 18-4, Ithaca, New York, USA.

Siepker, M. J., K. G. Ostrand, S. J. Cooke, D. Philipp P., and D. H. Wahl. 2007. A review of the effects of catch-and-release angling on black bass, *Micropterus* spp.: implications for conservation and management of populations. Fisheries Management and Ecology 14:91–101.

Sigmon, M. R. 2004. Hunting and posting on private land in America. Duke Law Journal 54:549–585.

Simard, M. A., C. Dussault, J. Huot, and S. D. Cote. 2013. Is hunting an effective tool to control overabundant deer? A test using an experimental approach. The Journal of Wildlife Management 77:254–269.

Simmons, R. H., and E. P. Dvorin. 1977. Public administration: values, policy, and change. Alfred Publishing, Port Washington, New York, USA.

Simon, A. 2009. The use of historic and contemporary justifications for killing wolves by the Palin administration. Capitalism Nature Socialism 20:6–30.

Simonson, T. D., and S. W. Hewett. 1999. Trends in Wisconsin's muskellunge fishery. North American Journal of Fisheries Management 19:291–299.

Skalski, J. R., M. V. Clawson, and J. J. Millspaugh. 2012. Model evaluation in statistical population reconstruction. Wildlife Biology 18:225–234.

Slagle, K., and A. Dietsch. 2018a. National survey of waterfowl hunters: summary report Atlantic Flyway.

Report to the National Flyway Council from the Minnesota Cooperative Fish and Wildlife Research Unit, University of Minnesota and The Ohio State University, St. Paul, Minnesota, USA.

Slagle, K., and A. Dietsch. 2018b. National survey of waterfowl hunters: summary report Central Flyway. Report to the National Flyway Council from the Minnesota Cooperative Fish and Wildlife Research Unit, University of Minnesota and The Ohio State University, St. Paul, Minnesota, USA.

Slagle, K., and A. Dietsch. 2018c. National survey of waterfowl hunters: summary report Mississippi Flyway. Report to the National Flyway Council from the Minnesota Cooperative Fish and Wildlife Research Unit, University of Minnesota and The Ohio State University, St. Paul, Minnesota, USA.

Slagle, K., and A. Dietsch. 2018d. National survey of waterfowl hunters: summary report Pacific Flyway. Report to the National Flyway Council from the Minnesota Cooperative Fish and Wildlife Research Unit, University of Minnesota and The Ohio State University, St. Paul, Minnesota, USA.

Slipke, J. W., and M. J. Maceina. 2014. Fisheries analysis and modeling simulator (FAMS), version 1.64. American Fisheries Society, Bethesda, Maryland, USA.

Slipke, J. W., M. J. Maceina, V. H. Travnichek, and K. C. Weathers. 1998. Effects of a 356-mm minimum length limit on the population characteristics and sport fishery of smallmouth bass in the Shoals Reach of the Tennessee River, Alabama. North American Journal of Fisheries Management 18:76–84.

Small, R. J., J. C. Holzwart, and D. H. Rusch 1991. Predation and hunting mortality of ruffed grouse in central Wisconsin. The Journal of Wildlife Management 55:512–520.

Smith, D. R. 2020. Structured decision making: case studies in natural resource management. Johns Hopkins University Press, Baltimore, Maryland, USA.

Smith, G. 2009. Democratic innovations: designing institutions for citizen participation. Cambridge University Press, Cambridge, England, United Kingdom.

Smith, G. W. 1995. A critical review of the aerial and ground surveys of breeding waterfowl in North America. U. S. Department of the Interior, National Biological Service, Biological Science Report 5, Washington, D.C., USA.

Smith, G. W., and J. A. Dubovsky. 1998. The point system and duck harvest management. Wildlife Society Bulletin 26:333–341.

Smith, G. W., and R. E. Reynolds. 1992. Hunting and mallard survival. The Journal of Wildlife Management 56:306–316.

Smith, J. A., M. B. Brown, L. A. Powell. 2019. Training wildlife biologists for work in the Anthroscape. Pages 447–452 *in* M. I. Goldstein and D. A. DellaSala, editors. Encyclopedia of Worlds Biomes. Elsevier, New York, USA.

Smith, N. G., D. J. Daugherty, J. W. Schlechte, and D. L. Buckmeier. 2018. Modeling the responses of alligator gar populations to harvest under various length-based regulations: implications for conservation and management. Transactions of the American Fisheries Society 147:665–673.

Smith, R. I., R. J. Blohm, S. T. Kelly, R. E. Reynolds, and F. D. Caswell. 1989. Review of databases for managing duck harvests. Transactions of the North American Wildlife and Natural Resources Conference 54:537–544.

Smith, R., and B. Deppa. 2009. Two dimensions of attribute importance. Journal of Consumer Marketing 26:28–38.

Snepenger, D. J., and R. B. Ditton. 1985. A longitudinal analysis of nationwide hunting and fishing indicators: 1955–1980. Leisure Sciences 7:297–319.

Snow, H. E. 1982. Hypothetical effects of fishing regulations in Murphy Flowage, Wisconsin. Wisconsin Department of Natural Resources, Technical Bulletin Number 131, Madison, Wisconsin, USA.

Soper, J. D. 1941. History, range, and home life of the northern bison. Ecological Monographs 11:347–412.

South Dakota Game, Fish and Parks. 2019. Fisheries and aquatic resources adaptive management system, Missouri River fisheries management area, 2019-2023. South Dakota Game, Fish and Parks, Wildlife Division. Pierre, South Dakota, USA.

Springer, P. F., and R. W. Lowe. 1998. Population, distribution and ecology of migrating and wintering Aleutian Canada geese. Pages 425–434 *in* D. H. Rusch, M. D. Samuel, D. D. Humburg, and B. D. Sullivan, editors. Biology and management of Canada geese. Proceedings of the International Canada Goose Symposium, Milwaukee, Wisconsin, USA.

Squibb, R. C., R. E. Denver, J. Kimball J. F., S. T. Davis, and T. D. Bunch. 1991. Ecology of conception in a northern Utah elk herd. Pages 10–12 *in* A. G. Christensen, L. J. Lyon, and T. N. Lonner, compilers. Proceedings of a symposium on elk vulnerability. Montana State University, Bozeman, Montana, USA

Stains, H. J. 1979. Primeness in North American furbearers. Wildlife Society Bulletin 7:120–124.

Stearns, S. C. 1989. Trade-offs in life-history evolution. Functional Ecology 3:259–268.

Stedman, R. C. 2003. Sense of place and forest science: toward a program of quantitative research. Forest Science 49:822–829.

Stedman, R. C., P. Bhandari, A. E. Luloff, D. R. Diefenbach, and J. C. Finley. 2008. Deer hunting on Pennsylvania's public and private lands: a two-tiered system of hunters? Human Dimensions of Wildlife 13:222–233.

Stedman, R. C., and D. J. Decker. 1996. Illuminating an overlooked hunting stakeholder group: nonhunters and their interest in hunting. Human Dimensions of Wildlife 1:29–41.

Stedman, R. C., D. R. Diefenbach, C. B. Swope, J. C. Finley, A. Luloff, H. C. Zinn, G. J. San Julian, and G. A. Wang. 2004. Integrating wildlife and human-dimensions research methods to study hunters. The Journal of Wildlife Management 68:762–773.

Stedman, R. C., and T. A. Heberlein. 2001. Hunting and rural socialization: contingent effects of the rural setting on hunting participation. Rural Sociology 66:599–617.

Stenseth, N. Chr, W. Falck, O. N. Bjørnstad, and C. J. Krebs. 1997. Population regulation in snowshoe hare and Canadian lynx: asymmetric food web configurations between hare and lynx. Proceedings of the National Academy of Sciences 94:5147.

Sterk, M., I. A. van de Leemput, and E. T. Peeters. 2017. How to conceptualize and operationalize resilience in socio-ecological systems? Sustainability Governance 28:108–113.

Sterling E. J., E. Betley, A. Sigouin, A. Gomez, A. Toomey, G. Cullman, C. Malone, A. Pekor, F. Arengo, M. Blair, C. Filardi, K. Landrigan, and A. L. Porzecanski. 2017. Assessing the evidence for stakeholder engagement in biodiversity conservation. Biological Conservation 209:159–171.

Sterling, E. M., K. B. Sterling, Q. E. Phelps, T. D. Ham, and S. J. Tripp. 2019. Smallmouth bass population demographics in Missouri Ozark streams and an evaluation of a statewide 305-mm minimum length limit. North American Journal of Fisheries Management 39:1175–1191.

Stone, C., and J. Lott. 2002. Use of a minimum length limit to manage walleyes in Lake Francis Case, South Dakota. North American Journal of Fisheries Management 22:975–984.

Storm, D. J., C. K. Neilsen, E. M. Schauber, and A. Woolf. 2007. Deer-human conflict and hunter access in an exurban landscape. Human-Wildlife Conflicts 1:53–59.

Stout, R. J., D. J. Decker, B. A. Knuth, J. C. Proud, and D. H. Nelson. 1996. Comparison of three public-involvement approaches for stakeholder input into deer management decisions: a case study. Wildlife Society Bulletin 24:312–317.

Stradtmann, M. L., and J. B. Mcaninch. 1996. Development and implementation of a metro bowhunter resource base. Minnesota Department of Natural Resources, Farmland Wildlife Population and Research Group, Madelia, Minnesota, USA.

Strickland, B. K., and S. Demarais. 2000. Age and regional differences in antlers and mass of white-tailed deer. The Journal of Wildlife Management 64:903–911.

Strickland, B. K., S. Demarais, L. E. Castle, J. W. Lipe, W. H. Lunceford, H. A. Jacobson, D. Frels, and K. V. Miller. 2001. Effects of selective-harvest strategies on white-tailed deer antler size. Wildlife Society Bulletin 29:509–520.

Strickland, M. D., H. J. Harju, K. R. McCaffery, H. W. Miller, L. M. Smith, and R. J. Stoll. 1994. Harvest management. Pages 445–473 *in* T. A. Bookhout, editor. Research and management techniques for wildlife and habitats, The Wildlife Society, Bethesda, Maryland, USA.

Sullivan, C. J., D. A. Isermann, K. E. Whitlock, and J. F. Hansen. 2020. Assessing the potential to mitigate climate-related expansion of largemouth bass populations using angler harvest. Canadian Journal of Fisheries and Aquatic Sciences 77:520–533.

Sullivan, M. G. 2002. Illegal angling harvest of walleye protected by length limits in Alberta. North American Journal of Fisheries Management 22:1053–1063.

Sunde, P., and T. Asferg. 2014. How does harvest size vary with hunting season length? Wildlife Biology 20:176–184.

Suski, C. D., and S. J. Cooke. 2007. Conservation of aquatic resources through the use of freshwater protected areas: opportunities and challenges. Biodiversity and Conservation 16:2015–2029.

Suski, C. D., and D. P. Philipp. 2004. Factors affecting the vulnerability to angling of nesting male largemouth and smallmouth bass. Transactions of the American Fisheries Society 133:1100–1106.

Susskind, L., and J. L. Cruikshank. 1987. Breaking the impasse: consensual approaches to resolving public disputes. Basic Books, New York, New York, USA.

Sutinen, J. G., and R. J. Johnston. 2003. Angling management organizations: integrating the recreational sector into fishery management. Marine Policy 27:471–487.

Sutter, D. A. H., C. D. Suski, D. P. Philipp, T. Klefoth, D. H. Wahl, P. Kersten, S. J. Cooke, and R. Arlinghaus. 2012. Recreational fishing selectively captures individuals with the highest fitness potential. Proceedings of the National Academy of Sciences 109:20960–20965.

Sutton, S. G., J. R. Stoll, and R. B. Ditton. 2001. Understanding anglers' willingness to pay increased fishing license fees. Human Dimensions of Wildlife 6:115–130.

Svedäng, H., and S. Hornborg. 2014. Selective fishing induces density-dependent growth. Nature Communications 5:4152.

Swain, D. P., A. F. Sinclair, and J. M. Hanson. 2007. Evolutionary response to size-selective mortality in an exploited fish population. Proceedings of the Royal Society B: Biological Sciences 274:1015–1022.

Swihart, R. K., J. N. Caudell, J. M. Brooke, and Z. Ma. 2020. A flexible model-based approach to delineate wildlife management units. Wildlife Society Bulletin 44:77–85.

Sylvia, A. 2019. An evaluation of tournament angling impacts on a largemouth bass population using mark-recapture data. Dissertation, Iowa State University, Ames, Iowa, USA.

Sylvia, A., B. Maahs, and M. J. Weber. 2021. Shifting angler harvest behaviors: a case study using large-mouth bass. *in* K. L. Pope and L. A. Powell, editors. Harvest of fish and wildlife: new paradigms for sustainable management. CRC Press, Boca Raton, Florida, USA.

Symes, D., and J. Phillipson. 2009. Whatever became of social objectives in fisheries policy? Fisheries Research 95:1–5.

Syslo, J. M., C. S. Guy, P. E. Bigelow, P. D. Doepke, B. D. Ertel, and T. M. Koel. 2011. Response of non-native lake trout (*Salvelinus namaycush*) to 15 years of harvest in Yellowstone Lake, Yellowstone National Park. Canadian Journal of Fisheries and Aquatic Sciences 68:2132–2145.

Sztramko, L. K. 1985. Effects of a sanctuary on the smallmouth bass fishery of Long Point Bay, Lake Erie. North American Journal of Fisheries Management 5:233–241.

Tamura, A., and K. Nakajima. 2017. Effects of 10 years of fencing under a gap and closed canopy on the regeneration of tree seedlings in an old-growth Japanese fir (*Abies firma*) forest overbrowsed by sika deer. Journal of Forest Research 22:224–232.

Taylor, R. G., J. A. Whittington, and D. E. Haymans. 2001. Catch-and-release mortality rates of common snook in Florida. North American Journal of Fisheries Management 21:70–75.

Taylor, W. W., and C. P. Ferreri, editors. 1999. Great Lakes fisheries policy and management: a binational perspective. Michigan State University Press, East Lansing, Michigan, USA.

Teal Harvest Potential Working Group. 2013. An assessment of the harvest potential of North American teal. https://www.fws.gov/migratorybirds/pdf/surveys-and-data/FinalTealAssessmentReport.pdf. Accessed October 8, 2020.

Teel, T. L., A. A. Dayer, M. J. Manfredo, and A. D. Bright. 2005. Regional results from the research project entitled "Wildlife Values in the West." (Project Rep. No. 58). Project Report for the Western Association of Fish and Wildlife Agencies. Colorado State University, Human Dimensions in Natural Resources Unit, Fort Collins, Colorado, USA. https://www.researchgate.net/publication/287997441_Regional_Results_from_the_Research_Project_Entitled_Wildlife_Values_in_the_West. Accessed 16 February 2020.

Teel, T. L., and M. J. Manfredo. 2010. Understanding the diversity of public interests in wildlife conservation. Conservation Biology 37: 153–164.

Teisl, M. F., K. J. Boyle, and R. E. Record Jr. 1999. License-sales revenues: Understanding angler and hunter reactions to changes in license prices. Human Dimensions of Wildlife 4:1–17.

Tenhumberg, B., A. J. Tyre, A. R. People, and H. P. Possingham. 2004. Do harvest refuges buffer kangaroos against evolutionary responses to selective harvesting? Ecology 85:2003–2017.

Tennessee State Wildlife Action Plan Team. 2015. Tennessee State Wildlife Action Plan 2015. Tennessee Wildlife Resources Agency, Nashville, Tennessee, USA.

Theobald, D. M., J. R. Miller, and N. T. Hobbs. 1997. Estimating the cumulative effects of development on wildlife habitat. Landscape and Urban Planning 39:25–36.

Therkildsen, N. O., A. P. Wilder, D. O. Conover, S. B. Munch, H. Baumann, and S. R. Palumbi. 2019. Contrasting genomic shifts underlie parallel phenotypic evolution in response to fishing. Science 365:487–490.

Thompson, B. D., D. F. Miller, T. A. Doumitt, T. R. Jacobson, and M. L. Munson-McGee. 1996. An ecological framework for monitoring sustainable management of wildlife: a New Mexico furbearer example. U. S. Department of the Interior, National Biological Service, Information and Technology Report 5, Washington, D.C., USA.

Thorbjørnsen, S. H., E. Moland, C. Simpfendorfer, M. Heupel, H. Knutsen, and E. M. Olsen. 2018. Potential of a no-take marine reserve to protect home ranges of anadromous brown trout (*Salmo trutta*). Ecology and Evolution 9:417–426.

Thorne, D. H. 2016. Missouri conservation sales tax. https://olis.leg.state.or.us/liz/2015I1/Downloads/CommitteeMeetingDocument/89604. Accessed 17 April 2020.

Thorson, J. T., S. B. Munch, J. M. Cope, and J. Gao. 2017. Predicting life history parameters for all fishes worldwide. Ecological Applications 27:2262–2276.

Thurow, R. F., and D. J. Schill. 1994. Conflicts in allocation of wild trout resources: an Idaho case history.

Pages 132–140 *in* R. Barnhart, B. Shake, and R. H. Hamre, editors. Wild trout V: wild trout in the 21st century. Trout Unlimited, Vienna, Virginia, USA.

Tilghman, N. G. 1989. Impacts of white-tailed deer on forest regeneration in northwestern Pennsylvania. The Journal of Wildlife Management 53:524–532.

Tillotson, M. D., and T. P. Quinn. 2017. Selection on the timing of migration and breeding: a neglected aspect of fishing-induced evolution and trait change. Fish and Fisheries 19:170–181.

Tilman, D., M. Clark, D. R. Williams, K. Kimmel, S. Polasky, and C. Packer. 2017. Future threats to biodiversity and pathways to their prevention. Nature 546:73–81.

Toïgo, C., J.-M. Gaillard, M. Festa-Bianchet, É. Largo, J. Michallet, and D. Maillard. 2007. Sex- and age-specific survival of the highly dimorphic Alpine ibex: evidence for a conservative life-history tactic. Journal of Animal Ecology 76:679–686.

Towell, W. E. 1979. The role of policy making boards and commissions. Pages 49–54 *in* R. D. Teague and E. Decker, editors. Wildlife conservation principles and practices. The Wildlife Society, Washington, D.C., USA.

Trebitz, A., S. Carpenter, P. Cunningham, B. Johnson, R. Lillie, D. Marshall, T. Martin, R. Narf, T. Pellett, S. Stewart, C. Storlie, and J. Unmuth. 1997. A model of bluegill-largemouth bass interactions in relation to aquatic vegetation and its management. Ecological Modelling 94:139–156.

Trefethen, J. B. 1964. Wildlife management and conservation. D. C. Heath, Boston, Massachusetts, USA.

Trefethen, J. B. 1975. An American crusade for wildlife. Winchester, New York, New York, USA.

Trost, R. E., K. E. Gamble, and D. J. Nieman. 1990. Goose surveys in North America: current procedures and suggested improvements. Transactions of the North American Wildlife and Natural Resources Conference 55:338–349.

Truth About Fur. 2017. Animal welfare vs. animal rights: an important distinction. https://www.truthaboutfur.com/blog/animal-welfare-animal-rights-distinction/#. Accessed 27 October 2020.

Tsehaye, I., M. Catalanoc G. Sass, D. Glover, and B. Roth. 2013. Prospects for fishery-induced collapse of invasive Asian Carp in the Illinois River. Fisheries 38:445–454.

Tsehaye, I., Roth, B. M., and G. G. Sass. 2016. Exploring optimal walleye exploitation rates for northern Wisconsin Ceded Territory lakes using a hierarchical Bayesian age-structured model. Canadian Journal of Fisheries and Aquatic Sciences 73:1413–1433.

Tuckey, T. D., and M. C. Fabrizio. 2019. 2019 Annual report estimating relative juvenile abundance of ecologically important finfish in the Virginia portion of Chesapeake Bay (1 June 2018–30 June 2019). Submitted to Virginia Marine Resources Commission, Newport News, Virginia. Project Number F-104-R-23. https://doi.org/10.25773/8znm-db81. Accessed 14 October 2020.

Turner, M. M., C. S. Deperno, W. Booth, E. L. Vargo, M. C. Conner, and R. A. Lancia. 2016. The mating system of white-tailed deer under Quality Deer Management. The Journal of Wildlife Management 80:935–940.

Tyre, A. J., and S. Michaels. 2011. Confronting socially generated uncertainty in adaptive management. Journal of Environmental Management 92:1365–1370.

Tyre, A. J., and B. Tenhumberg. 2021. Making harvest management decisions robust to uncertainty. *in* K. L. Pope and L. A. Powell, editors. Harvest of fish and wildlife: new paradigms for sustainable management. CRC Press, Boca Raton, Florida, USA.

U.S. Fish and Wildlife Service. 1975. Final environmental statement: issuance of annual regulations permitting the sport hunting of migratory birds. U.S. Fish and Wildlife Service, Washington, D.C., USA.

U.S. Fish and Wildlife Service. 1988. Final supplemental environmental impact statement: issuance of annual regulations permitting the sport hunting of migratory birds. U.S. Fish and Wildlife Service, Washington, D.C., USA.

U.S. Fish and Wildlife Service. 2004. Policies and responsibilities of the migratory bird Program. https://www.fws.gov/policy/720fw1.html. Accessed 31 March 2020.

U.S. Fish and Wildlife Service. 2005. Final environmental impact statement: resident Canada goose management. U.S. Fish and Wildlife Service, Washington, D.C., USA.

U.S. Fish and Wildlife Service. 2007. Final environmental impact statement: light goose management. U.S. Fish and Wildlife Service, Washington, D.C., USA.

U.S. Fish and Wildlife Service. 2013. Final supplemental environmental impact statement: issuance of annual regulations permitting the sport hunting of migratory birds. U.S. Fish and Wildlife Service, Washington, D.C., USA.

U.S. Fish and Wildlife Service. 2014. Mourning dove harvest strategy. U.S. Department of the Interior, Fish and Wildlife Service, Division of Migratory Bird Management, Washington, D.C., USA.

U.S. Fish and Wildlife Service. 2015a. Economic impact of waterfowl hunting in the United States: addendum to the 2011 national survey of fishing, hunting, and wildlife-associated recreation. U.S. Fish and Wildlife Service, Report 2011-6, Washington, D.C., USA.

U.S. Fish and Wildlife Service. 2015b. Grants: funding for migratory bird conservation partnerships. https://www.fws.gov/birds/grants.php. Accessed 20 February 2020.

U.S. Fish and Wildlife Service. 2015c. Wildlife & sport fish restoration program. https://wsfrprograms.fws.gov/Subpages/GrantPrograms/WR/WR_Act.htm. Accessed 10 October 2019.

U.S. Fish and Wildlife Service. 2017a. Historical fishing license data. U.S. Fish and Wildlife Service, Washington, D.C., USA. https://www.fws.gov/wsfrprograms/subpages/licenseinfo/Fishing.htm. Accessed 3 March 2019.

U.S. Fish and Wildlife Service. 2017b. Historical hunting license data. U.S. Fish and Wildlife Service, Washington, D.C., USA. https://www.fws.gov/wsfrprograms/subpages/licenseinfo/hunting.htm. Accessed 3 March 2019.

U.S. Fish and Wildlife Service. 2018a. Adaptive harvest management: 2019 hunting season. U.S. Department of the Interior, Washington, D.C., USA.

U.S. Fish and Wildlife Service. 2018b. Grizzly Bear Recovery Program, 2018 annual report. https://fws.gov/mountain-prairie/es/species/mammals/grizzly/2018GB_AnnualReport_FINAL.pdf. Accessed 25 February 2020.

U.S. Fish and Wildlife Service. 2018c. Migratory bird conservation commission report–fiscal year 2017. U.S. Department of the Interior, Washington, D.C., USA. https://www.fws.gov/refuges/realty/pdf/MBCC_2017.pdf. Accessed 8 June 2020.

U.S. Fish and Wildlife Service. 2018d. Migratory bird joint ventures. https://www.fws.gov/birds/management/bird-conservation-partnership-and-initiatives/migratory-bird-joint-ventures.php. Accessed 20 February 2020.

U.S. Fish and Wildlife Service. 2019a. Adaptive harvest management: 2020 hunting season. U.S. Department of the Interior, Washington, D.C., USA.

U.S. Fish and Wildlife Service. 2019b. Apportionments/funding index. https://www.fws.gov/wsfrprograms/subpages/grantprograms/FundingIndex.htm. Accessed 27 January 2020.

U.S. Fish and Wildlife Service. 2019c. Wildlife & Sport Fish Restoration Program, State Wildlife Grant Program: overview. https://wsfrprograms.fws.gov/Subpages/GrantPrograms/SWG/SWG.htm. Accessed 20 February 2020.

U.S. Fish and Wildlife Service. 2020a. CITES, species information, animals, furbearers. https://www.fws.gov/international/animals/fur-bearers.html. Accessed 27 January 2020.

U.S. Fish and Wildlife Service. 2020b. Endangered species, mammals, *Canada lynx*. https://www.fws.gov/mountain-prairie/es/canadaLynx.php. Accessed 4 April 2020.

U.S. Fish and Wildlife Service. 2020c. Tribal wildlife grants. https://www.fws.gov/nativeamerican/grants.html. Accessed 30 March 2020.

U.S. Fish and Wildlife Service. 2020d. Working with tribes. https://www.fws.gov/endangered/what-we-do/tribal-partnership-stories.html. Accessed 31 March 2020.

U.S. Fish and Wildlife Service, and U.S. Census Bureau. 1991. National survey of fishing, hunting, and wildlife-associated recreation. U.S. Department of the Interior, and U.S. Department of Commerce, Washington, D.C., USA.

U.S. Fish and Wildlife Service, and U.S. Census Bureau. 2006. National survey of fishing, hunting, and wildlife-associated recreation. U.S. Department of the Interior, and U.S. Department of Commerce, Washington, D.C., USA.

U.S. Fish and Wildlife Service, and U.S. Census Bureau. 2016. National survey of fishing, hunting, and wildlife-associated recreation. U.S. Department of the Interior, and U.S. Department of Commerce, Washington, D.C., USA.

Urban, M. C. 2007. The Growth-predation risk trade-off under a growing gape-limited predation threat. Ecology 88:2587–2597.

Uusi-Heikkila, S., A. R. Whiteley, A. Kuparinen, S. Matsumura, P. A. Venturelli, C. Wolter, J. Slate, C. R. Primmer, T. Meinelt, S. S. Killen, D. Bierbach, G. Polverino, A. Ludwig, and R. Arlinghaus. 2015. The evolutionary legacy of size-selective harvesting extends from genes to populations. Evolutionary Applications 8:597–620.

Uusi-Heikkilä, S., C. Wolter, T. Klefoth, and R. Arlinghaus. 2008. A behavioral perspective on fishing-induced evolution. Trends in Ecology & Evolution 23:419–421.

Van de Walle, J., G. Pigeon, A. Zedrosser, J. E. Swenson, and F. Pelletier. 2018. Hunting regulation favors slow life histories in a large carnivore. Nature Communications 9:1–10.

Van Deelen, T. R., B. J. Dhuey, C. N. Jacques, K. R. McCaffery, R. E. Rolley, and K. Warnke. 2010. Effects of earn-a-buck and special antlerless-only seasons on Wisconsin's deer harvests. The Journal of Wildlife Management 74:1693–1700.

Vantassel, S. M., T. L. Hiller, K. D. Powell, and S. E. Hygnstrom. 2010. Using advancements in cable-trapping to overcome barriers to furbearer management in the United States. The Journal of Wildlife Management 74:934–939.

Vaske, J. J. 2008. Survey research and analysis: applications in parks, recreation and human dimensions. Venture Publishing Inc., State College, Pennsylvania. USA.

Vaske, J. J., and M. J. Manfredo. 2012. Social psychological considerations in wildlife management. Pages 43–57 *in* D. J. Decker, S. J. Riley, and W. F. Siemer, editors. Human dimensions of wildlife management, 2nd edition. John Hopkins University Press, Baltimore, Maryland, USA.

Veech, J. A. 2006. Increasing and declining populations of northern bobwhites inhabit different types of landscapes. The Journal of Wildlife Management 70:922–930.

Ver Steeg, J. M., and R. C. Elden, compilers. 2002. Harvest information program: evaluation and recommendations. International Association of Fish and Wildlife Agencies, Migratory Shore and Upland Game Bird Working Group, Ad Hoc Committee on HIP, Washington, D.C., USA.

VerCauteren, K. C., C. W. Anderson, T. R. Van Deelen, D. Drake, W. D. Walter, S. M. Vantassel, and S. E. Hygnstrom. 2011. Regulated commercial harvest to manage overabundant white-tailed deer: an idea to consider? Wildlife Society Bulletin 35:185–194.

Verhulst, P.-F. 1845. Recherches mathématiques sur la loi d'accroissement de la population [Mathematical researches into the law of population growth]. Nouveaux Mémoires de l'Adadémie Royale de Sciences et Belle-Letters de Bruxelles 18:8.

Villegas-Rios, D., E. Moland, and E. M. Olsen. 2017. Potential of contemporary evolution to erode fishery benefits from marine reserves. Fish and Fisheries 18:571–577.

Virginia Department of Game and Inland Fisheries. 2014. A study report on the effects of removing the prohibition against hunting over bait in Virginia. Virginia Department of Game and Inland Fisheries, Richmond, Virginia, USA.

Vitali, T. R. 1990. Sport hunting: moral or immoral? Environmental Ethics 12:69–82.

von Winterfeldt, D., and W. Edwards. 1986. Decision analysis and behavioral research. Cambridge University Press, Cambridge, England, United Kingdom.

Voyles, L., and L. Chase. 2017. The state conservation machine. The Association of Fish & Wildlife Agencies, and Arizona Game and Fish Department, Washington, D.C., USA.

Vrtiska, M. P. 2021. Harvest management of migratory game birds. *in* K. L. Pope and L. A. Powell, editors. Harvest of fish and wildlife: new paradigms for sustainable management. CRC Press, Boca Raton, Florida, USA.

Vrtiska, M. P., J. H. Gammonley, L. W. Naylor, and A. H. Raedeke. 2013. Economic and conservation ramifications from the decline of waterfowl hunters. Wildlife Society Bulletin 37:380–388.

Vrtiska, M. P., and S. L. Oldenburger. State management of migratory game. 2018. Pages 135–150 *in* T. Ryder, editor. State wildlife management and conservation. John Hopkins University Press, Baltimore, Maryland, USA.

Waggerman, G. L., G. E. Homerstad, R. R. George, and W. A. Shifflett. 1994. White-tipped dove. Pages 53–59 *in* T. C. Tacha and C. E. Braun, editors. Migratory shore and upland game bird management in North America. International Association of Fish and Wildlife Agencies, Washington, D.C., USA.

Walker, B., C. S. Holling, S. Carpenter, and A. Kinzig. 2004. Resilience, adaptability and transformability in social-ecological systems. Ecology and Society 9:2.

Wallace, M. S., H. L. Stribling, and H. A. Clonts. 1989. Factors influencing land access selection by hunters in Alabama. Transactions of the North American Wildlife and Natural Resources Conference 54:183–189.

Wallingford, B. D., D. R. Diefenbach, E. S. Long, C. S. Rosenberry, and G. L. Alt. 2017. Biological and social outcomes of antler point restriction harvest regulations for white-tailed deer. Wildlife Monographs 196:1–26.

Wallis, O. L. 1971. Management of aquatic resources and sport fishing in national parks by special regulations. Paper presented at the 51st Annual Meeting of the Western Division of the American Fisheries Society, Colorado, USA.

Walsh, M. R., S. B. Munch, S. Chiba, and D. O. Conover. 2006. Maladaptive changes in multiple traits caused by fishing: impediments to population recovery. Ecology Letters 9:142–148.

Walters, C. 1986. Adaptive management of renewable resources. Macmillian, New York, New York, USA.

Walters, C. J., and C. S. Holling. 1990. Large-scale management experiments and learning by doing. Ecology 71:2060–2968.

Warchol, G. L., and B. R. Johnson. 2011. Securing national resources from theft: an exploratory theoretical analysis. Journal of Applied Security Research 6:273–300.

Ward, H. G. M., M. S. Allen, E. V. Camp, N. Cole, L. M. Hunt, B. Matthias, J. R. Post, K. Wilson, and R.

Arlinghaus. 2016. Understanding and managing social-ecological feedbacks in spatially structured recreational fisheries: the overlooked behavioral dimension. Fisheries 41:524–535.

Ward, K. J., R. C. Stedman, A. Luloff, J. S. Shortle, and J. C. Finley. 2008. Categorizing deer hunters by typologies useful to game managers: a latent-class model. Society & Natural Resources 21:215–229.

Ward, M. 2015. Fisheries management plan for Leech Lake 2016-2020. Minnesota Department of Natural Resources, Section of Fisheries. St Paul, Minnesota, USA.

Washington Department of Fish and Wildlife. 2020a. Puget Sound/Strait of Juan de Fuca halibut seasons and regulations. https://wdfw.wa.gov/fishing/regulations/halibut/puget-sound. Accessed 30 July 2020.

Washington Department of Fish and Wildlife. 2020b. Washington game bird and small game hunting regulations: effective July 1, 2020 - June 30, 2021. Washington Department of Fish and Wildlife, Olympia, Washington, USA.

Waters, D. S., R. L. Noble, and J. E. Hightower. 2005. Fishing and natural mortality of adult largemouth bass in a tropical reservoir. Transactions of the American Fisheries Society 134:563–571.

Watkins, C., N. C. Poudyal, C. Caplenor, D. Buehler, and R. Applegate. 2018. Motivations and support for regulations: a typology of eastern wild turkey hunters. Human Dimensions of Wildlife 23:433–445.

Weaver, J. K., and H. S. Mosby. 1979. Influence of hunting regulations on Virginia wild turkey populations. The Journal of Wildlife Management 43:128–135.

Webb, M. A., and R. A. Ott, Jr. 1991. Effects of length and bag limits on population structure and harvest of white crappies in three Texas reservoirs. North American Journal of Fisheries Management 11:614–622.

Webb, S., and R. Anderson. 2016. Trappers as citizen scientists: collaborative efforts are promoting wolverine conservation. The Wildlife Professional 10:30–33.

Webb, S. L., S. Demarais, B. K. Strickland, R. W. DeYoung, B. P. Kinghorn, and K. L. Gee. 2012. Effects of selective harvest on antler size in white-tailed deer: a modeling approach. The Journal of Wildlife Management 76:48–56.

Weber, M. J., M. L. Brown, D. H. Wahl, and D. E. Shoup. 2015. Metabolic theory explains latitudinal variation in common carp populations and predicts responses to climate change. Ecosphere 6:1–16.

Weber, M. J., M. J. Hennen, M. L. Brown, D. O. Lucchesi, and T. R. St. Sauver. 2016. Compensatory response of invasive common carp *Cyprinus carpio* to harvest. Fisheries Research 179:168–178.

Weckel, M., R. F. Rockwell, A. Wincorn. 2011. The sustainability of controlled archery programs: the motivation and satisfaction of suburban hunters. Wildlife Society Bulletin 35:330–337.

Weidel, B. C., D. C. Josephson, and C. E. Kraft. 2007. Littoral fish community response to smallmouth bass removal from an Adirondack lake. Transactions of the American Fisheries Society 136:778–789.

Wellman, J. D. 1979. Recreational response to privacy stress: a validation study. Journal of Leisure Research 11:61–73.

Wellsmith, M. 2011. Wildlife crime: the problems of enforcement. European Journal on Criminal Policy and Research 17:125–148.

Wenger, S. J., D. J. Isaak, J. B. Dunham, K. D. Fausch, C. H. Luce, H. M. Neville, B. E. Rieman, M. K. Young, D. E. Nagel, D. L. Horan, and G. L. Chandler. 2011. Role of climate and invasive species in structuring trout distributions in the interior Colombian River basin, USA. Canadian Journal of Fisheries and Aquatic Science 68:988–1008.

White, G. C., and K. P. Burnham. 1999. Program MARK: survival estimation from populations of marked animals. Bird Study 46:S120–S139.

White, G. C., L. Cordes, and T. W. Arnold. 2013. Band reporting rates of waterfowl: does individual heterogeneity bias estimated survival rates? Ecology and Evolution 3:4215–4220.

White, G. C., D. J. Freddy, R. B. Gill, and J. H. Ellenberger. 2001. Effect of adult sex ratio on mule deer and elk productivity in Colorado. The Journal of Wildlife Management 65:543–551.

White, H. B., G. R. Batcheller, E. K. Boggess, C. L. Brown, J. W. Butfiloski, T. A. Decker, J. D. Erb, M. W. Fall, D. A. Hamilton, T. L. Hiller, G. F. Hubert, Jr., M. J. Lovallo, J. F. Olson, and N. M. Roberts. 2021. Best management practices for trapping furbearers in the United States. Wildlife Monographs 207:3–59.

Whitehead, J. C., W. B. Clifford, and T. J. Hoban. 2001. Willingness to pay for a saltwater recreational fishing license: a comparison of angler groups. Marine Resource Economics 16:177–194.

Whiteside, R. W., and F. S. Guthery. 1983. Ring-necked pheasant movements, home ranges, and habitat use in west Texas. The Journal of Wildlife Management 47:1097–1104.

Whittaker, D. G., and S. Torres. 1998. Ballot initiatives and natural resource management: some opinions on processes, impacts, and experience. Human Dimensions of Wildlife 3:1–7.

Wilberg, M. J., J. R. Bence, B. T. Eggold, D. Makauskas, and D. F. Clapp. 2005. Yellow Perch dynamics in Southwestern Lake Michigan during 1986-2002. North American Journal of Fisheries Management 25:1130–1152.

Wilberg, M. J., J. T. Thorson, B. C. Linton, and J. Berkson. 2009. Incorporating time-varying catchability into population dynamic stock assessment models. Reviews in Fisheries Science 18:7–24.

Wilcox, B. A., and D. D. Murphy. 1985. Conservation strategy: the effects of fragmentation on extinction. The American Naturalist 125:879–887.

Wilde, G. R. 1997. Largemouth bass fishery responses to length limits. Fisheries 22(6):14–23.

Wilde, G. R. 1998. Tournament-associated mortality in black bass. Fisheries 23(10):12–22.

Wilde, G. R., and K. L. Pope. 2008. A simple model for predicting survival of angler-caught and released largemouth bass. Transactions of the American Fisheries Society 137:834–840.

Wilde, G. R., K. L. Pope, and B. W. Durham. 2003. Lure-size restrictions in recreational fisheries. Fisheries 28(6):18–26.

Wildlife Management Institute. 2010. The deer management program of the Pennsylvania Game Commission: a comprehensive review and evaluation. The Wildlife Management Institute, Washington, D.C., USA.

Wildlife Society, The. 2020. Recovering America's Wildlife Act. https://wildlife.org/policy/recovering-americas-wildlife-act/. Accessed 27 October 2020.

Wilkins, E. J., and H. M. Miller. 2018, Public views of wetlands and waterfowl conservation in the United States: results of a survey to inform the 2018 update of the North American Waterfowl Management Plan: U.S. Geological Survey Open-File Report 2017–1148. https://doi.org/10.3133/ofr20171148. Accessed 17 April 2020.

Wilkins, K. A. 2007. Analyses of western Gulf Coast mottled duck harvest data. U.S. Fish and Wildlife Service, Division of Migratory Bird Management, Laurel, Maryland, USA.

Williams, B. K. 1996. Adaptive optimization and the harvest of biological populations. Math Bioscience 136:1–20.

Williams, B. K. 1997. Approaches to the management of waterfowl under uncertainty. Wildlife Society Bulletin 25:714–720.

Williams, B. K. 2011. Adaptive management of natural resources–framework and issues. Journal of Environmental Management 92:1346–1353.

Williams, B. K., and F. A. Johnson. 1995. Adaptive management and the regulation of waterfowl harvests. Wildlife Society Bulletin 23:430–436.

Williams, B. K., F. A. Johnson, and K. Wilkins. 1996. Uncertainty and the adaptive management of waterfowl harvests. The Journal of Wildlife Management 60:223–232.

Williams, B. K., M. D. Koneff, and D. A. Smith. 1999. Evaluation of waterfowl conservation under the North American waterfowl management plan. The Journal of Wildlife Management 64:417–440.

Williams, B. K., J. D. Nichols, and M. J. Conroy. 2002. Analysis and management of animal populations. Academic Press, San Diego, California, USA.

Williams, B. K., R. C. Szaro, and C. D. Shapiro. 2007. Adaptive management: the U.S. Department of the Interior technical guide. Adaptive Management Working Group, U.S. Department of the Interior, Washington, D.C., USA.

Williams, C. K., R. S. Lutz, and R. D. Applegate. 2004. Winter survival and additive harvest in northern bobwhite coveys in Kansas. The Journal of Wildlife Management 68:94–100.

Williams, C. K., R. S. Lutz, R. D. Applegate, and D. H. Rusch. 2000. Habitat use and survival of northern bobwhite (*Colinus virginianus*) in cropland and rangeland ecosystems during the hunting season. Canadian Journal of Zoology 78:1562–1566.

Williams, G. L. 1991. An example of simulation models as decision tools in wildlife management. Wildlife Society Bulletin 9:101–107.

Williams, J. E., R. N. Williams, R. F. Thurow, L. Elwell, D. P. Philipp, F. A. Harris, J. L. Kershner, P. J. Martinez, D. Miller, G. H. Reeves, C. A. Frissell, and J. R. Sedell. 2011. Native fish conservation areas: a vision for large-scale conservation of native fish communities. Fisheries 36:266–288.

Williams, S. C., A. J. DeNicola, T. Almendinger, and J. Maddock. 2013. Evaluation of organized hunting as a management technique for overabundant white-tailed deer in suburban landscapes. Wildlife Society Bulletin 37:137–145.

Williamson, S. J. 1998. Origins, history, and current use of ballot initiatives in wildlife management. Human Dimensions of Wildlife 3:51–59.

Williamson, S. J., S. Adair, K. L. Brown, and J. Turner. 2001. Contributions made by hunters toward conservation of the North American landscape. Transactions of the North American Wildlife and Natural Resources Conference 66:255–269.

Willis, D. 1989. Understanding length limit regulations. The In-Fisherman: The Journal of Freshwater Fishing 87:30–41.

Willisch, C. S., I. Biebach, U. Koller, T. Bucher, N. Marreros, M.-P. Ryser-Degiorgis, L. F. Keller, and P.

Neuhaus. 2012. Male reproductive pattern in a polygynous ungulate with a slow life-history: the role of age, social status and alternative mating tactics. Evolutionary Ecology 26:187–206.

Wilson, K. L., A. Foos, O. E. Barker, A. Farineau, J. De Gisi, and J. R. Post. 2020. Social-ecological feedbacks drive spatial exploitation in a northern freshwater fishery: a halo of depletion. Journal of Applied Ecology 57:206–218.

Wilson, R. S., and J. L. Arvai. 2006. Evaluating the quality of structured environmental management decisions. Environmental Science and Technology 40:4831–4837.

Winder, V. L., K. M. Carrlson, A. J. Gregory, C. A. Hagen, D. A. Haukos, D. C. Kesler, L. C. Larsson, T. W. Matthews, L. B. McNew, M. A. Patten, J. C. Pitman, L. A. Powell, J. A. Smith, T. Thompson, D. H. Wolfe, and B. K. Sandercock. 2015. Factors affecting female space use in ten populations of prairie chickens. Ecosphere 6:166.

Winemiller, K. O., and K. A. Rose. 1992. Patterns of life-history diversification in North American fishes: implications for population regulation. Canadian Journal of Fisheries and Aquatic Science 49:2196–2218.

Winkler, R., and K. Warnke. 2013. The future of hunting: an age-period-cohort analysis of deer hunter decline. Population and Environment 34:460–480.

Winston, W. L., and J. B. Goldberg. 2004. Operations research: applications and algorithms, volume 3. Thomson/Brooks/Cole, Belmont, California, USA.

Winter, J. D. 1996. Advances in underwater biotelemetry. Pages 555–590 *in* B. R. Murphy and D. W. Willis, editors. Fisheries techniques, 2nd edition. American Fisheries Society, Bethesda, Maryland, USA.

Winther, E., C. M. Barr, C. Miller, and C. Wheaton. 2020. Report on the predation index, predator control fisheries, and program evaluation for the Columbia River Basin northern pikeminnow sport reward program. U.S. Department of Energy, Bonneville Power Administration, 2019 Annual Report, Project Number 1990-077-00, Portland, Oregon, USA.

Wirsing, A. J., M. R. Heithaus, A. Frid, and L. M. Dill. 2008. Seascapes of fear: evaluating sublethal predator effects experienced and generated by marine mammals. Marine Mammal Science 24:1–15.

Wisconsin Department of Natural Resources. 2020. Guide to Wisconsin hook and line fishing regulations 2020-2021. https://dnr.wisconsin.gov/topic/Fishing/regulations. Accessed 4 August 2020.

Witmer, G. W., and D. S. DeCalesta. 1991. The need and difficulty of bringing the Pennsylvania deer herd under control. Page 130–137 *in* P. D. Curtis, M. J. Fargione, and J. E. Caslick, editors. Fifth eastern wildlife damage control conference. Cornell Cooperative Extension, Ithaca, New York, USA.

Witter, D. J., D. J. Case, J. H. Gammonley, and D. Childress. 2006. Social factors in waterfowl management: conservation goals, public perceptions and hunter satisfaction. Transactions of the North American Wildlife and Natural Resources Conference 71:233–248.

Wolfe, R. J. 2006. Playing with fish and other lessons from the North. The University of Arizona Press, Tucson, Arizona, USA.

Wood, S., and F. Scheipl. 2017. gamm4: Generalized additive mixed models using "mgcv" and "lme4." https://CRAN.R-project.org/package=gamm4.

Woods, G. R., D. C. Guynn, W. E. Hammitt, and M. E. Patterson. 1996. Determinants of participant satisfaction with quality deer management. Wildlife Society Bulletin 24:318–324.

Woolf, A., and J. L. Roseberry. 1998. Deer management: our profession's symbol of success or failure? Wildlife Society Bulletin 26:515–521.

Woolsey, N. G. 1985. Coyote field manual. Arizona Game and Fish Department, Special Report 15, Phoenix, Arizona, USA.

Wright, B. A., R. A. Kaiser, and J. E. Fletcher. 1988. Hunter access decisions by rural landowners: an east Texas example. Wildlife Society Bulletin 16:152–158.

Wright, B. A., E. B. D. Rodgers, and K. F. Backman. 2001. Assessing the temporal stability of hunting participation and the structure and intensity of constraints: a panel study. Journal of Leisure Research 33:450–469.

Wright, R. A., and C. E. Kraft. 2012. Stocking strategies for recreational small impoundments. Pages 155–180 *in* J. W. Neal and D. W. Willis, editors. Small impoundment management in North America. American Fisheries Society, Bethesda, Maryland, USA.

Wszola, L. S., and J. J. Fontaine. 2021. Harvest regulations in evolving fisheries. *in* K. L. Pope and L. A. Powell, editors. Harvest of fish and wildlife: new paradigms for sustainable management. CRC Press, Boca Raton, Florida, USA

Wszola, L. S., L. F. Gruber, E. F. Stuber, L. N. Messinger, C. J. Chizinski, and J. J. Fontaine. 2020a. Use and expenditures on public access hunting lands. Journal of Outdoor Recreation and Tourism 29:100256.

Wszola, L. S., A. E. Madsen, E. F. Stuber, C. J. Chizinski, J. J. Lusk, J. S. Taylor, K. L. Pope, and J. J.

Fontaine. 2020b. Public access for pheasant hunters: understanding an emerging need. The Journal of Wildlife Management 84:45–55.

Wszola, L. S., E. F. Stuber, C. J. Chizinski, J. J. Lusk, and J. J. Fontaine. 2019. Prey availability and accessibility drive hunter movement. Wildlife Biology 2019:wlb.00526.

Wu, L. 2017. Lionfish derby: dynamic optimization model of a fishing tournament to control local invasive species. Thesis, Cornell University, Ithaca, New York, USA.

Yeiser, J. M., J. J. Morgan, D. L. Baxley, R. B. Chandler, and J. A. Martin. 2018. Private land conservation has landscape-scale benefits for wildlife in agroecosystems. Journal of Applied Ecology 55:1930–1939.

Yi, Y. 2012. A critical review of customer satisfaction. Pages 68–123 *in* V. A. Zeithaml, editor. Review of marketing 1990. Marketing Classics Press, Chicago, Illinois, USA.

Ylikarjula, J., M. Heino, and U. Dieckmann. 1999. Ecology and adaptation of stunted growth in fish. Evolutionary Ecology 13:433–453.

Young, D. D., and R. D. Boertje. 2008. Recovery of low bull:cow ratios of moose in interior Alaska. Alces 44:65–71.

Young, J. C., A. Jordan, K. R. Searle, A. Butler, D. S. Chapman, P. Simmons, and A. D. Watt. 2013. Does stakeholder involvement really benefit biodiversity conservation? Biological Conservation 158:359–370.

Young, S. P., and E. A. Goldman. 1944. The wolves of North America. American Wildlife Institute, Washington, D.C. USA.

Zablan, M. A., C. E. Braun, and G. C. White. 2003. Estimation of greater sage-grouse survival in North Park, Colorado. The Journal of Wildlife Management 67:144–154.

Zedrosser, A., S. M. J. G. Steyaert, H. Gossow, and J. E. Swenson. 2011. Brown bear conservation and the ghost of persecution past. Biological Conservation 144:2163–2170.

Zhou, S., A. D. M. Smith, A. E. Punt, A. J. Richardson, M. Gibbs, E. A. Fulton, S. Pascoe, C. Bulman, P. Bayliss, and K. Sainsbury. 2010. Ecosystem-based fisheries management requires a change to the selective fishing philosophy. Proceedings of the National Academy of Sciences 107:9485–9489.

Zimmerman, G. S., T. J. Moser, W. L. Kendall, P. F. Doherty, Jr., G. C. White, and D. F. Caswell. 2009. Factors influencing reporting and harvest probabilities in North American geese. The Journal of Wildlife Management 73:710–719.

Zimmerman, G. S., J. R. Sauer, G. S. Boomer, P. K. Devers, and P. R. Garrettson. 2017. Integrating breeding bird survey and demographic data to estimate wood duck population size in the Atlantic Flyway. Condor 119:616–628.

Zimmermann, F., and C. Jørgensen. 2015. Bioeconomic consequences of fishing-induced evolution: a model predicts limited impact on net present value. Canadian Journal of Fisheries and Aquatic Sciences 72:612–624.

Zimmermann, F., and C. Jørgensen. 2017. Taking animal breeding into the wild: regulation of fishing gear can male fish stocks evolve higher productivity. Marine Ecology Progress Series 563:185–195.

Zimmer-Shaffer, S. A., J. T. Briggler, and J. J. Millspaugh. 2014. Modeling the effects of commercial harvest on population growth of river turtles. Chelonian Conservation and Biology 13:227–236.

Zinn, H. C. 2003. Hunting and sociodemographic trends: older hunters from Pennsylvania and Colorado. Wildlife Society Bulletin 31: 1004–1014.

Zinn, H. C., M. J. Manfredo, and S. C. Barro. 2002. Patterns of wildlife value orientations in hunters' families. Human Dimensions of Wildlife 7:147–162.

Zipkin, E. F., C. E. Kraft, E. G. Cooch, and P. J. Sullivan. 2009. When can efforts to control nuisance and invasive species backfire? Ecological Applications 19:1585–1595.

Zoë, A., J. C. R. Smart, S. Yearley, R. J. Irvine, and P. C. L. White. 2010. Identifying conflicts and opportunities for collaboration in the management of a wildlife resource: a mixed-methods approach. Wildlife Research 37:647–657.

Zornes, M., J. Short, D. Lutz, W. Rudd, and S. DeCecco. 2012. A critical review of mule deer antler point regulations, application, and effectiveness. Wyoming Game and Fish Department Report, Cheyenne, Wyoming, USA.

Zwick, R. R., B. Muth, and D. Solan. 2006. A longitudinal comparison of activities and motives of Vermont trappers: 1994, 2000, and 2005. Pages 24–32 *in* R. Burns and K. Robinson, compilers. Proceedings of the 2006 Northeastern Recreation Research Symposium. U.S. Department of Agriculture, Forest Service, General Technical Report NRS-P-14, Newtown Square, Pennsylvania, USA.

Index

Note: bold page numbers refer to figures and tables